Organic Reactions

Organic Reactions

VOLUME 11

EDITORIAL BOARD

Arthur C. Cope, *Editor-in-Chief*

Roger Adams
A. H. Blatt
Virgil Boekelheide

T. L. Cairns
David Y. Curtin
Carl Niemann

ADVISORY BOARD

Louis F. Fieser
John R. Johnson

Frank C. McGrew
Harold R. Snyder

ASSOCIATE EDITORS

Donald R. Baer
L. Guy Donaruma
Walter Z. Heldt

Andrew S. Kende
Christian S. Rondestvedt, Jr.
Peter A. S. Smith

Elmer R. Trumbull

FORMER MEMBERS OF THE BOARD, NOW DECEASED

Homer Adkins

Werner E. Bachmann

NEW YORK
JOHN WILEY & SONS, INC.
LONDON

COPYRIGHT © 1960
BY
ROGER ADAMS

All Rights Reserved

This book or any part thereof must not be reproduced in any form without the written permission of the publisher.

Library of Congress Catalog Card Number: 42-20265

PRINTED IN THE UNITED STATES OF AMERICA

PREFACE TO THE SERIES

In the course of nearly every program of research in organic chemistry the investigator finds it necessary to use several of the better-known synthetic reactions. To discover the optimum conditions for the application of even the most familiar one to a compound not previously subjected to the reaction often requires an extensive search of the literature; even then a series of experiments may be necessary. When the results of the investigation are published, the synthesis, which may have required months of work, is usually described without comment. The background of knowledge and experience gained in the literature search and experimentation is thus lost to those who subsequently have occasion to apply the general method. The student of preparative organic chemistry faces similar difficulties. The textbooks and laboratory manuals furnish numerous examples of the application of various syntheses, but only rarely do they convey an accurate conception of the scope and usefulness of the processes.

For many years American organic chemists have discussed these problems. The plan of compiling critical discussions of the more important reactions thus was evolved. The volumes of *Organic Reactions* are collections of chapters each devoted to a single reaction, or a definite phase of a reaction, of wide applicability. The authors have had experience with the processes surveyed. The subjects are presented from the preparative viewpoint, and particular attention is given to limitations, interfering influences, effects of structure, and the selection of experimental techniques. Each chapter includes several detailed procedures illustrating the significant modifications of the method. Most of these procedures have been found satisfactory by the author or one of the editors, but unlike those in *Organic Syntheses* they have not been subjected to careful testing in two or more laboratories. When all known examples of the reaction are not mentioned in the text, tables are given to list compounds which have been prepared by or subjected to the reaction. Every effort has been made to include in the tables all such compounds and references; however, because of the very nature of the reactions discussed and their frequent use as one of the several steps of syntheses in which not all of the intermediates have been isolated, some instances may well have been missed. Nevertheless, the investigator will be able to use the tables and their accompanying bibliographies in place of most or all of the literature search so often required.

Because of the systematic arrangement of the material in the chapters and the entries in the tables, users of the books will be able to find information desired by reference to the table of contents of the appropriate chapter. In the interest of economy the entries in the indices have been kept to a minimum, and, in particular, the compounds listed in the tables are not repeated in the indices.

The success of this publication, which will appear periodically, depends upon the cooperation of organic chemists and their willingness to devote time and effort to the preparation of the chapters. They have manifested their interest already by the almost unanimous acceptance of invitations to contribute to the work. The editors will welcome their continued interest and their suggestions for improvements in *Organic Reactions*.

CONTENTS

CHAPTER	PAGE
1. The Beckmann Rearrangement—*L. Guy Donaruma and Walter Z. Heldt*	1
2. The Demjanov and Tiffeneau-Demjanov Ring Expansions—*Peter A. S. Smith and Donald R. Baer*	157
3. Arylation of Unsaturated Compounds by Diazonium Salts—*Christian S. Rondestvedt, Jr.*	189
4. The Favorskiĭ Rearrangement of Haloketones—*Andrew S. Kende*	261
5. Olefins from Amines: The Hofmann Elimination Reaction and Amine Oxide Pyrolysis—*Arthur C. Cope and Elmer R. Trumbull*	317
Author Index, Volumes 1–11	495
Chapter Index, Volumes 1–11	497
Subject Index, Volume 11	499

CHAPTER 1

THE BECKMANN REARRANGEMENT

L. Guy Donaruma and Walter Z. Heldt*

Explosives Department, E. I. du Pont de Nemours and Company, Inc.

CONTENTS

	PAGE
Introduction	2
Stereochemistry of the Rearrangement	4
Mechanism	5
Scope and Limitations	14
Aliphatic Ketoximes	14
Aliphatic Aromatic Ketoximes	17
Diaryl Ketoximes	21
Alicyclic Ketoximes	26
Heterocyclic Ketoximes	32
Oximes of Polyfunctional Ketones	35
Cleavage of Oximes and Related Compounds Derived from Benzoins and α-Diketones	38
Aldoximes	41
Carbon-Nitrogen Rearrangements of Oxime Derivatives and Related Compounds	45
Oxime Esters	45
Imines and N-Halo Imines	46
Nitrones	47
Nitroles	48
Derivatives of Hydroxamic Acids	48
Hydrazones and Semicarbazones	48
Related Carbon-Nitrogen Rearrangements	49
Stereochemistry of Oximes	51
Preparation of Oximes	55

* The authors wish to express their thanks to Miss Diamond C. Ascani, Dr. George R. Coraor, and Dr. Thomas H. Regan for their help in proofreading the manuscript. Miss Adele MacIntyre also deserves acknowledgement for aiding in the preparation of the final manuscript.

	PAGE
EXPERIMENTAL CONDITIONS	55
Catalyst and Solvent	55
Temperature	56
Rearrangement of Oximes by Phosphorus Pentachloride	56
Rearrangement of Oximes by Concentrated Sulfuric Acid	57
EXPERIMENTAL PROCEDURES	57
Homodihydrocarbostyril	57
Phenanthridone	57
δ-Valerolactam	57
ε-Caprolactam	58
Acetanilide	58
Pivalanilide	58
Heptanamide	58
TABULAR SURVEY	59
Table I. Aliphatic Ketoximes	60
Table II. Aliphatic Aromatic Ketoximes	67
Table III. Diaryl Ketoximes	83
Talbe IV. Alicyclic Ketoximes	96
Table V. Steroid Oximes	114
Table VI. Heterocyclic Ketoximes	119
Table VII. Monoximes of Diketones	126
Table VIII. Dioximes of Diketones	127
Table IX. Quinone Oximes	131
Table X. Cleavage of Oximes and Oxime Derivatives	135
Table XI. Aldoximes	145
Table XII. Nitrones	149

INTRODUCTION

The rearrangement of a ketoxime to the corresponding amide was discovered in 1886 by E. Beckmann[1] and is known as the Beckmann rearrangement. The rearrangement is brought about by acids including

$$\begin{array}{c}R\\ \diagdown\\ C{=}NOH \xrightarrow{\text{Acid}} R'CONHR \text{ and/or } RCONHR'\\ \diagup\\ R'\end{array}$$

R and R' = H, aryl, alkyl.

Lewis acids. The more common rearranging agents are concentrated sulfuric acid, phosphorus pentachloride in ether, and Beckmann's mixture, hydrogen chloride in a mixture of acetic acid and acetic anhydride.

[1] Beckmann, Ber., **19**, 988 (1886); **20**, 1507 (1887).

Since the discovery of the reaction, numerous publications have appeared which deal with the mechanism of the reaction, the determination of the stereochemical configurations of the oximes employed, and the synthetic applications of the reaction. The Beckmann rearrangement is used frequently to determine the structure of ketones, by identification of the acid and amine obtained by hydrolysis of the amide formed by the rearrangement.

Blatt,[2] Jones,[3] and, more recently, Knunyants[4] have summarized the published literature concerning the Beckmann rearrangement up to 1948.

There is no uniform convention for the designation of the stereochemistry of oximes in the literature. In this review the following conventions are used:

(a) The configuration of a ketoxime is referred to as *syn* or *anti* when the hydroxyl group is *cis* or *trans*, respectively, to the first group named following the prefix *syn* or *anti* in the name of the compound.

$$CH_3CCH_2CH_3$$
$$\parallel$$
$$HON$$

syn-Methyl ethyl ketoxime

anti-2-Methylcyclohexanone oxime

(b) The configuration of aldoximes is referred to as *syn* or *anti* to the hydrogen of the aldoxime. In the older literature aldoxime configurations are often referred to as α (*syn*) or β (*anti*).

$$C_6H_5CH$$
$$\parallel$$
$$NOH$$

syn-Benzaldoxime

$$C_6H_5CH$$
$$\parallel$$
$$HON$$

anti-Benzaldoxime

(c) The nomenclature used in the literature for designating the configurations of benzoin oximes, benzil oximes, and benzil dioximes has been retained.

α-Benzoin oxime   β-Benzoin oxime   α-Benzil monoxime

[2] Blatt, *Chem. Revs.*, **12**, 215 (1933).
[3] Jones, *Chem. Revs.*, **35**, 335 (1944).
[4] Knunyants and Fabrichnyi, *Uspekhi Khim.*, **18**, 633 (1949) [*C.A.*, **45**, 6572 (1951)].

$$\underset{\text{β-Benzil monoxime}}{\underset{\text{HON}}{\text{C}_6\text{H}_5\text{COCC}_6\text{H}_5}} \qquad \underset{\text{α-Benzil dioxime}}{\underset{\text{HON} \quad \text{NOH}}{\text{C}_6\text{H}_5\text{C}-\text{CC}_6\text{H}_5}}$$

$$\underset{\text{β-Benzil dioxime}}{\underset{\text{NOH} \quad \text{HON}}{\text{C}_6\text{H}_5\text{C}\text{——}\text{CC}_6\text{H}_5}} \qquad \underset{\text{γ-Benzil dioxime}}{\underset{\text{HON} \quad \text{HON}}{\text{C}_6\text{H}_5\text{C}\text{——}\text{CC}_6\text{H}_5}}$$

STEREOCHEMISTRY OF THE REARRANGEMENT

Two stereoisomeric forms of an aldoxime or an unsymmetrical ketoxime are possible. Therefore, theoretically, the Beckmann rearrangement may occur with either a *syn* or an *anti* migration:

$$\underset{\text{RNH}}{\overset{\text{O}=\text{CR}'}{|}} \xleftarrow{\underset{\text{migration}}{Anti}} \underset{\text{NOH}}{\overset{\text{RCR}'}{\|}} \xrightarrow{\underset{\text{migration}}{Syn}} \underset{\text{HNR}'}{\overset{\text{RC}=\text{O}}{|}}$$

Beckmann assumed that the rearrangement occurs stereospecifically with *syn* migration, and the configurations assigned to the parent oximes up to about 1923 are based upon this assumption. In 1921 Meisenheimer carefully determined the configuration of β-benzil monoxime and rearranged the oxime with phosphorous pentachloride in ether.[5] No

$$\underset{\text{O}}{\underset{\text{N}\diagdown\diagup}{\overset{\|}{\text{H}_5\text{C}_6\text{C}\text{——}\text{CC}_6\text{H}_5}}} \xrightarrow{\text{O}_3} \underset{\text{O}\quad\text{O}}{\underset{\text{N}\quad\text{CC}_6\text{H}_5}{\overset{\|\quad\quad\|}{\text{H}_5\text{C}_6\text{C}\text{——}\text{CC}_6\text{H}_5}}}$$

↓ aq. NaOH

$$\underset{\text{C}_6\text{H}_5\text{NH}\quad\text{O}}{\underset{|}{\overset{\text{O}}{\overset{\|}{\text{C}\text{——}\text{CC}_6\text{H}_5}}}} \xleftarrow{\text{PCl}_5,\text{ ether}} \underset{\text{N}\quad\text{O}\diagdown\text{OH}}{\overset{\|\quad\|}{\text{C}_6\text{H}_5\text{C}\text{——}\text{CC}_6\text{H}_5}}$$

isomerization of the carbon-nitrogen bond occurred during the ozonolysis of 3,4,5-triphenylisoxazole. The product obtained from the ozonolysis, upon mild hydrolysis, yielded β-benzil monoxime. The rearrangement of the oxime gave only benzoylformanilide. Therefore Meisenheimer concluded that rearrangement must proceed with *anti* migration.

[5] Meisenheimer, *Ber.*, **54**, 3206 (1921).

THE BECKMANN REARRANGEMENT

When other acids such as Beckmann's mixture,[6,7] sulfuric acid, or its salts[8,9] are used as rearranging agents, products stemming from a possible *syn* and/or *anti* migration are isolated. The *syn* migration may be explained by assuming that isomerization of the oxime occurs *prior* to rearrangement.

MECHANISM

The mechanism of the Beckmann rearrangement consists essentially of the formation of an electron-deficient nitrogen atom by the partial ionization of the oxygen-nitrogen bond of the oxime with a simultaneous intramolecular migration of the group *anti* to the departing hydroxyl

$$\underset{\text{I}}{\underset{\text{NOH}}{\overset{\text{R'CR}}{\|}}} \xrightleftharpoons{H^{\oplus}} \underset{\text{II}}{\underset{\overset{\oplus}{NOH_2}}{\overset{\text{R'CR}}{\|}}} \longrightarrow \underset{\text{III}}{\underset{N----OH_2}{\overset{CR}{\underset{R'^{\oplus}}{\|}}}}$$

$$\underset{\text{III}}{} \longrightarrow \underset{\text{IV}}{\underset{R'N}{\overset{RC^{\oplus}OH_2}{\|}}} \xrightarrow{-H^{\oplus}} \underset{\text{V}}{\underset{R'N}{\overset{RCOH}{\|}}} + H^{\oplus}$$

$$\underset{\text{V}}{} \xrightleftharpoons{} \underset{\text{VI}}{\underset{R'NH}{\overset{RC=O}{|}}}$$

group. Rearrangement of II and III proceeds essentially as an intramolecular displacement, whereby R', if optically active, retains its optical activity.[10,11] Thus the oxime of (+)-3-ethylheptan-2-one (VII) has been rearranged to furnish the levorotatory amide (VIII). The amide (VIII) also was obtained from (+)-2-ethylhexanoic acid (IX) via the Hofmann degradation which is known to proceed with retention of configuration. (See equation on p. 6.)

The first product of the rearrangement is always an imine derivative (IV or V), which usually rearranges rapidly to the corresponding amide.

[6] Brown, van Gulick, and Schmidt, *J. Am. Chem. Soc.*, **77**, 1094 (1955).
[7] Smith, *Ber.*, **24**, 4025 (1891).
[8] von Auwers and Jordan, *Ber.*, **58**, 26 (1925).
[9] Kauffmann, *Ann.*, **344**, 30 (1906).
[10] Kenyon and Campbell, *J. Chem. Soc.*, **1946**, 25.
[11] Kenyon and Young, *J. Chem. Soc.*, **1941**, 263.

$$CH_3(CH_2)_3CH(C_2H_5)CO_2H \xrightarrow[\text{2. Hofmann rearrangement}]{\text{1. Amide}} CH_3(CH_2)_3CH(C_2H_5)NH_2$$
$$(+) \qquad\qquad\qquad\qquad\qquad\qquad (+)$$
$$\text{IX}$$

1. Acid bromide ↓ 2. Cd(CH$_3$)$_2$ $\qquad\qquad$ Hydrolysis ↑↓ Acetylation

$$CH_3(CH_2)_3CH(C_2H_5)COCH_3 \xrightarrow[\text{2. Beckmann rearrangement}]{\text{1. Oxime}} CH_3(CH_2)_3CH(C_2H_5)NHCOCH_3$$
$$(+) \qquad\qquad\qquad\qquad\qquad\qquad (-)$$
$$\text{VII} \qquad\qquad\qquad\qquad\qquad\qquad \text{VIII}$$

The presence of an imine intermediate in the rearrangement was demonstrated by Kuhara, who showed that diphenyl ketoxime benzenesulfonate (X) rearranged initially to N-phenylbenzimidobenzenesulfonate (XI), which in turn rearranged to N-benzenesulfonyl benzanilide (XII).[12]

$$\underset{\underset{\text{X}}{NOSO_2C_6H_5}}{\overset{C_6H_5CC_6H_5}{\|}} \xrightarrow{\text{Room temp.}} \underset{\underset{\text{XI}}{C_6H_5N}}{\overset{C_6H_5COSO_2C_6H_5}{\|}} \rightarrow \underset{\underset{\text{XII}}{C_6H_5NSO_2C_6H_5}}{\overset{C_6H_5C=O}{|}}$$

The existence of an imine intermediate was further indicated by the isolation of imine derivatives (XIII) formed by displacement of the sulfonyl ester by strong nucleophilic agents,[13] and by the formation of

$$\text{XI} \xrightarrow{HOCH_3} \underset{\underset{\text{XIII}}{C_6H_5N}}{\overset{C_6H_5COCH_3}{\|}} + HOSO_2C_6H_5$$

[Structures XIV, XV, XVI: cyclohexanone N-sulfonate derivative → cyclic imine with OSO$_2$C$_6$H$_5$ → tetrazole fused ring via HN$_3$]

XIV $\qquad\qquad$ XV $\qquad\qquad$ XVI

$$+ HOSO_2C_6H_5$$

tetrazoles in the presence of hydrazoic acid.[14,15] Tetrazoles (XVI) are not formed from oximes or amides except under the conditions of the

---

[12] Kuhara, Matsuimya, and Matsunami, *Mem. Coll. Sci. Kyoto Imp. Univ.*, **1**, 105 (1914) [*C.A.*, **9**, 1613 (1915)].

[13] Oxley and Short, *J. Chem. Soc.*, **1948**, 1514.

[14] Csuros, Zech, and Zech, *Acta Chim. Acad. Sci. Hung.*, **1**, 83 (1951) [*C.A.*, **46**, 5003 (1952)].

[15] Burke and Herbst, *J. Org. Chem.*, **20**, 726 (1955).

Beckmann rearrangement.[14] Other nucleophiles which have been employed are phenol, primary and secondary amines, and phenyl sulfamide.[13]

Chapman contributed greatly to the elucidation of electronic effects involved in the rearrangement of substituted benzophenone oxime ethers (XVII).[16] No acid catalyst was required to bring about the rearrangement of XVII to XVIII. The rate of rearrangement increased with

$$p\text{-XC}_6\text{H}_4\text{CC}_6\text{H}_4\text{Y-}p \qquad p\text{-YC}_6\text{H}_4\text{C}{=}\text{O}$$
$$\underset{\text{XVII}}{\overset{\|}{\text{NOC}_6\text{H}_2(\text{NO}_2)_3}} \rightarrow \underset{\text{XVIII}}{\overset{|}{p\text{-XC}_6\text{H}_4\text{NC}_6\text{H}_2(\text{NO}_2)_3}}$$

increasing electron-supplying power of X and was slightly increased by increased electron-supplying power of Y. An increase in the dielectric constant of the medium appeared to augment the rate of rearrangement. Therefore Chapman concluded that the rate-determining step in the rearrangement must be the partial ionization of the nitrogen-oxygen bond of the oxime ether with simultaneous migration of the aryl group *anti* to the picryl group.[16] Furthermore, Kuhara had demonstrated earlier that the rates of rearrangement of a series of esters of benzophenone oxime in chloroform were proportional to the acid strength of the esterifying acid.[17,18] The ease of rearrangement therefore increases with the dissociation constant of the esterifying acid.

$$\text{C}_6\text{H}_5\text{SO}_3\text{H} > \text{ClCH}_2\text{CO}_2\text{H} > \text{C}_6\text{H}_5\text{CO}_2\text{H} > \text{CH}_3\text{CO}_2\text{H}$$

Because of the multitude of possible intermediates involved in the Beckmann rearrangement the rate-determining step of the rearrangement (I to VI) depends upon the reaction temperature, the solvent, and the catalyst employed. In fact, two intermediates in the reaction sequence (I to VI) may rearrange with approximately equal rates and the determination of the rate-determining step may become quite difficult. The rate-determining step may precede the rearrangement (I to II), may proceed simultaneously with the migration of R' (II to III), or may follow the rearrangement (III to VI) depending upon the oxime, acid, and other reaction conditions employed.

The rate-determining process precedes the rearrangement when an oxonium salt (XX) is formed from a nitronium salt (XIX).[19] The salt

---

[16] Chapman and Fidler, *J. Chem. Soc.*, **1936**, 448.

[17] Kuhara and Todo, *Mem. Coll. Sci., Kyoto Imp. Univ.*, **2**, 387 (1910) [*C.A.*, **5**, 1278 (1911)].

[18] Kuhara and Watanabe, *Mem. Coll. Sci., Kyoto Imp. Univ.*, **9**, 349 (1913) [*C.A.*, **11**, 579 (1917)].

[19] Hauser and Hoffenberg, *J. Org. Chem.*, **20**, 1482, 1491 (1955); Hoffenberg and Hauser, *ibid.*, **20**, 1496 (1955).

XIX must first rearrange to XX before undergoing the Beckmann rearrangement. Similarly, two types of antimony pentachloride adducts

$$\underset{\text{XIX}}{\overset{R'}{\underset{R}{>}}C=\overset{\oplus}{N}\underset{\ominus BF_3}{\overset{OH}{<}}} \rightarrow \underset{\text{XX}}{\overset{R'}{\underset{R}{>}}C=N\overset{OH}{\underset{}{<}}\overset{\oplus}{\underset{\ominus}{\overset{}{BF_3}}}}$$

(XXI and XXII) are formed with benzophenone oxime methyl ether.[20] The adduct XXII is formed in concentrated solution from antimony pentachloride and benzophenone oxime methyl ether. The adduct XXI

$$\underset{\text{XXI}}{\left[\overset{C_6H_5}{\underset{C_6H_5}{>}}C=\overset{\oplus}{N}\overset{OCH_3}{\underset{H}{<}}\right]SbCl_6^-} \qquad \underset{\text{XXII}}{\underset{C_6H_5NHSbCl_6}{\overset{C_6H_5COCH_3}{\|}}}$$

is formed in dilute solution under otherwise identical conditions and cannot be rearranged to benzanilide. These results appear to indicate that, while in dilute solution the stable nitronium adduct XXI is formed, in concentrated solution the corresponding oxonium salt is formed and rearranges rapidly to XXII. Other examples are the addition products (XXIII and XXIV) formed by the reaction of antimony pentachloride with chlorimines.[21]

$$\underset{\text{XXIII}}{\overset{R'}{\underset{R}{>}}C=\overset{\oplus}{N}\overset{Cl}{\underset{SbCl_5^\ominus}{<}}} \qquad \underset{\text{XXIV}}{\overset{R'}{\underset{R}{>}}C=N\overset{\overset{\oplus}{Cl}}{\underset{}{<}}SbCl_5^\ominus}$$

The rate-determining step of the rearrangement (I to VI) may be the formation of oxime imino ethers (XXVI),[22] oxime anhydrides (XXVII),[23]

[20] Theilacker, Gerstenkorn, and Gruner, *Ann.*, **563**, 109 (1949).
[21] Theilacker, *Angew Chem.*, **51**, 834 (1938); Theilacker and Mohl, *Ann.*, **563**, 99 (1949).
[22] Chapman, *J. Chem. Soc.*, **1935**, 1223.
[23] Stephen and Staskun, *J. Chem. Soc.*, **1956**, 980.

or oxime sulfonates (XXVIII),[24-27] which rearrange rapidly after the oxime derivative is formed.

The occurrence of intermediates such as XXVI and XXVII was suggested by the strong catalytic effect of N-phenylbenzimidoyl chloride

$$I + ClCR'' \underset{\underset{XXV}{NR'''}}{\overset{\|}{\longrightarrow}} \underset{XXVI}{\overset{R'}{\underset{R}{\diagdown}} C=N-OCR''} + HCl$$

$$2\ I \xrightarrow{\text{Dehydrating agent}} \left(\underset{R}{\overset{R'}{\diagdown}} C=N \right)_2 O + H_2O$$
XXVII

$$I \xrightarrow{H_2SO_4} \underset{R}{\overset{R'}{\diagdown}} C=N\underset{OH_2}{\overset{\oplus}{\diagdown}} \xrightarrow{\text{Low }[H^+]} \underset{R}{\overset{R'}{\diagdown}} C\overset{\oplus}{=}N \cdots OH_2$$
III

$$\xrightarrow[\text{or } SO_3]{\text{High }[H^+]} \underset{R}{\overset{R'}{\diagdown}} C=N \diagdown OSO_3H$$
XXVIII

upon the rearrangement of benzophenone oxime in ether and by the fact that one mole of a Lewis acid, such as phosphorus pentachloride, rearranges two moles of ketoxime to a mixture containing the corresponding amide and oxime imino ether in approximately the same amounts.[22, 23]

Ogata and others found that the rate of rearrangement of ketoximes in sulfuric acid is first order and follows the Hammett acidity function (H_0) up to 65% of sulfuric acid.[24, 27-29] They suggested that at low acid concentrations the concentration of XXVIII is low and that the

[24] Pearson and Ball, *J. Org. Chem.*, **14**, 118 (1949).
[25] Wichterle and Rocek, *Chem. Listy*, **45**, 257, 379 (1951) [*C.A.*, **46**, 10809 (1952)].
[26] Rocek and Bergl, *Chem. Listy*, **47**, 472 (1953) [*C.A.*, **48**, 3279 (1954)].
[27] Ogato, Okano, and Matsumoto, *J. Am. Chem. Soc.*, **77**, 4643 (1955).
[28] Sluiter, *Rec. trav. chim.*, **24**, 372 (1905).
[29] Hammett and Deyrup, *J. Am. Chem. Soc.*, **54**, 2721 (1932).

rate-determining step may be the dissociation of III.[27] At higher acid concentrations, the rearrangement is no longer dependent upon H_0 and the rate-determining step appears to be exclusively the formation of XXVIII.[30]

If the formation of II is simple and without complication, II → III can be identified as the rate-determining step. Rearrangement of oxime picrates[16, 30–35] and oxime tosylates[36, 37] in nonpolar solvents proceeds without the formation of III as the slow step. The reaction products isolated are N-substituted amides, and the rearrangement of the imine intermediate IV (H of $\overset{\oplus}{O}H_2$ is replaced by either 2,4,6-$C_6H_2(NO_2)_3$ or p-$CH_3C_6H_4SO_2$) to VI is rapid compared to the transition II to III.[37, 38] Recently the transition state III for the Beckmann rearrangement was suggested.[30, 34–37, 39–41]

Such a transition state (or transitory intermediate) is similar to the phenonium ion occurring in anchimerically assisted rearrangements[42] or the azacyclopropene ring system isolated in the Neber rearrangement.[41] The following evidence argues for the formation of III as a transition state in the rate-determining step: the rate of rearrangement of a series of substituted *anti* acetophenone oxime picrates in 1,4-dichlorobutane depends strongly upon the nature of the p-substituent.[43] The reaction constant, ρ, calculated from the Hammett plot[44] was found to be -4.1, which is comparable to the ρ values found for typical electrophilic aromatic substitution reactions;[45, 46] and the rate-determining step under these conditions appears to be the electrophilic attack of nitrogen on the benzene ring as described by III.

Ortho substituents greatly increase the rate of rearrangement of substituted acetophenone oximes (or picryl ethers) in relation to the corresponding *meta* or *para* substituents.[40, 43, 47] This effect is attributed to the

[30] Huisgen, *Angew. Chem.*, **69**, 341 (1957).
[31] Chapman and Howis, *J. Chem. Soc.*, **1933**, 806.
[32] Chapman, *J. Chem. Soc.*, **1934**, 1550.
[33] Chapman, *Chem. & Ind.*, (*London*), **1935**, 463.
[34] Huisgen, Ugi, Assemi, and Witte, *Ann.*, **602**, 127 (1957).
[35] Huisgen, *Chimia (Switz.)*, **10**, 266 (1956).
[36] W. Z. Heldt, unpublished results.
[37] Heldt, *J. Am. Chem. Soc.*, **80**, 5880, 5972 (1958).
[38] Chapman, *J. Chem. Soc.*, **1927**, 1743.
[39] Pearson, Baxter, and Martin, *J. Org. Chem.*, **17**, 1511 (1952).
[40] Pearson and Cole, *J. Org. Chem.*, **20**, 488 (1955).
[41] Cram, *J. Am. Chem. Soc.*, **74**, 2137 (1952); Cram and Hatch, *ibid.*, **75**, 33 (1953).
[42] Winstein, Morse, Grunwald, Schreiber, Corse, Marshall, James, Trifan, Brown, Schlesinger, and Ingraham, *J. Am. Chem. Soc.*, **74**, 1113–1164 (1952).
[43] Huisgen, Witte, Walz, and Jira, *Ann.* **604**, 191 (1957).
[44] Hammett, *Physical Organic Chemistry*, p. 184, McGraw-Hill, New York, 1940.
[45] Roberts, Sanford, Sixma, Cerfontain, and Zagt, *J. Am. Chem. Soc.*, **76**, 4525 (1954).
[46] Kuivila and Benjamin, *J. Am. Chem. Soc.*, **77**, 4834 (1955).
[47] Pearson and Watts, *J. Org. Chem.*, **20**, 494 (1955).

steric interaction between the *ortho*-substituted ring and the oxime group, resulting in the loss of coplanarity of the latter with the benzene ring.

The *ortho* substituent increases the potential energy of the oxime because of the partial loss of resonance stabilization; the oxime resembles the transition state where the azacyclopropene ring is perpendicular to the benzene ring system. The electronic effect of the *ortho* substituent appears to contribute only slightly to this increase of the rate of rearrangement.[48]

The *ortho* effect accounts for the spontaneous rearrangement of di-ortho-substituted acetophenone oximes when treated with hydroxylamine hydrochloride.[49–52]

The steric requirements for this transition state III were nicely demonstrated in the benzcycloalkanone oxime system.[53] The stereochemistry of XXX requires that the methylene group attached to the

phenyl group and the one attached to the azacyclopropene ring be in the planes of the respective rings, which in turn are perpendicular to each other. This requirement is fulfilled without straining the molecule only if n is eight or more in XXX.

The sequence of rate constants for the *anti* form of XXIX represented in the table on p. 12 indicates that the formation of an azacyclopropene ring system in the transition state (or transitory intermediate) appears to be correct.

The table also indicates the relative rates of aryl versus alkyl migration.

[48] Huisgen, Witte, and Jira, *Chem. Ber.* **90,** 1850 (1957).
[49] Kadesch, *J. Am. Chem. Soc.*, **66,** 1207 (1944).
[50] Feith and Davies, *Ber.*, **24,** 3546 (1891).
[51] Chichibabin, *Bull. soc. chim. France*, [4] **51,** 1436 (1932).
[52] Pearson and Greer, *J. Am. Chem. Soc.*, **77,** 6649 (1955).
[53] Huisgen, Witte, and Ugi, *Chem. Ber.*, **90,** 1844 (1957).

Rates of Rearrangement of Benzcyloalkanone Oxime Picryl
Ethers XXX in 1,4-Dichlorobutane[53]

Configuration	n	$k_1 \times 10^6 \text{ sec}^{-1}$ (at 70°)
Anti	5	Too slow to be measured
Anti	6	< 0.02
Anti	7	1,865
Anti	8	429,000
Syn	7	6.43
Syn	8	2.96

The *anti* form of XXIX, with $n = 8$, rearranges 140,000 times faster than the corresponding *syn* form. Contrariwise, the rate of rearrangement of acetophenone oxime picryl ether (aryl migration) is only 3.4 times faster than the rate of rearrangement of cyclopentadecanone oxime picryl ether (alkyl migration). Whereas, in acetophenone oxime, the oxime double bond is conjugated with the benzene ring, such an effect is much diminished in XXIX where $n = 8$. The system present in XXIX therefore appears to give a better picture of alkyl versus aryl migration than the acetophenone oxime system.[53] In almost all investigations reported in the literature, the rate-determining step is either I → II or II → III. Only in one case, the acetolysis of cyclopentanone oxime *p*-toluenesulfonate, did the rate-determining step appear to follow III. The slow step in this reaction appears to be the solvolysis of the ion pair III ($OH_2 = OTs$).[37]

The reaction medium profoundly influences the products and the rate of rearrangement. A recent study of the products formed from a number of cyclohexanone oxime esters in aqueous solution shows that three classes of oxime esters yielding different products may be distinguished:[54]

(a) Oxime esters which hydrolyze in dilute acids or bases to regenerate the oxime and the acid. Esters of cyclohexanone oxime derived from acetic, butyric, oxalic, sulfuric, dithionic, and *o*-toluenesulfonic acids fall in this group.

(b) Oxime esters which in dilute acidic or basic solution generate undetermined peroxy compounds or perhaps nitrogen oxides. Cyclohexanone oxime benzoate and anhydride belong in this group.

(c) Oxime esters which undergo the Beckmann rearrangement. Cyclohexanone oxime benzenesulfonate, β-naphthalenesulfonate, *p*-toluenesulfonate, and picryl ether are in this group. The rate of rearrangement in this group decreases in the following sequence:

$$C_6H_5SO_2 > \beta\text{-}C_{10}H_7SO_2 > p\text{-}CH_3C_6H_4SO_2 > 2,4,6\text{-}(O_2N)_3C_6H_2$$

[54] Csuros, Zech, Dely, and Zalay, *Acta Chim. Acad. Sci. Hung.*, **1**, 66 (1951) [*C.A.*, **46** 5003 (1952)].

The yield of ε-caprolactam produced from this group of esters was independent of the esterifying group. The same results were obtained in 10% aqueous sulfuric acid solution and 10% sodium hydroxide solution. The yields were in the range 75–80%.

The rate of rearrangement of picryl ethers of benzophenone oxime in various solvents decreases in the following order:[31, 32]

$$CH_3CN > CH_3NO_2 > (CH_3)_2CO > C_6H_5Cl > \text{nonpolar solvents}$$

Therefore the rate of rearrangement is roughly proportional to the dielectric constant of the solvent. Since the rate-determining step in the rearrangement of an oxime picrate involves the partial ionization of the nitrogen-oxygen bond of the oxime,[16] it is probably the ionizing power of the solvent rather than the dielectric constant which determines the rate of rearrangement. Similarly, the rate of rearrangement of cyclohexanone oxime with sulfur trioxide is faster in sulfuric acid[55] than in nonpolar solvents such as carbon disulfide or chlorinated hydrocarbons.[56, 57]

Solvents of high nucleophilic power, such as water, amines, or alcohols, both increase the rate of rearrangement and compete for the imine intermediate.[13, 36] The second effect arrests the reaction at the imine stage as indicated by the following equations.[58]

$$\underset{R}{\overset{R'}{\diagdown}}C=N\underset{OSO_2C_6H_5}{} \longrightarrow \underset{R}{\overset{R'}{\diagup}}C=N\underset{OSO_2C_6H_5}{} \xrightarrow{C_6H_5NH_2}$$

$$\underset{R}{\overset{C_6H_5NH}{|}}C=N\overset{R'}{\diagup} + C_6H_5SO_3H$$

The ability of the solvent to interact with the intermediate probably increases with the nucleophilic power of the solvent.[36, 58] Solvolysis of ketoxime sulfonates is used extensively as a preparative method for imines.[13] Furthermore, several other reactions may be promoted selectively by different solvents. Cyclohexanone oxime sulfonate is probably an intermediate formed in the rearrangement of cyclohexanone

---

[55] Giltges and Welz (to Farbenfabriken Baeyer), Ger. pat. appl. F 11,979 (1954).
[56] Wichterle (Chemicke Zavody), U.S. pat. 2,573,374 (1951) [*C.A.*, **46**, 7585 (1952)].
[57] Blaser and Tischberek (to Henkel and Cie G.m.b.H.), Ger. pat. appl. H 9,265 and H 8,640 (1951).
[58] Atherton, Morrison, Cremyln, Kenner, Todd, and Webb, *Chem. & Ind.* (*London*), **1955**, 1183.

oxime in sulfuric acid.[24] When cyclohexanone oxime was rearranged in sulfuric acid, a trace (1×10^{-4} mole) of octahydrophenazine was formed.[59] Rearrangement of cyclohexanone oxime sulfonate in aqueous dioxane increased the yield of octahydrophenazine to 7%.[60] Perhaps the formation of octahydrophenazine proceeds in a manner analogous to the Neber rearrangement.[61] Similarly, a trace of aniline, 0.1 mole per cent, was

$$R'CH_2CR \atop \| \atop NOSO_3H \xrightarrow{\text{Base}} R'CH{-}CR \atop \diagdown N \diagup \longrightarrow R'CH{-}CR \atop | \quad \| \atop NH_2 \quad O$$

isolated from the Beckmann rearrangement of the same oxime in concentrated sulfuric acid,[59] the source possibly being a little-understood aromatization reaction of cyclic ketoximes.[62, 63]

SCOPE AND LIMITATIONS

Under the proper conditions, most oximes will undergo the normal Beckmann rearrangement to yield an amide or a mixture of amides. The generality of the reaction makes it difficult to consider the scope and limitations other than by noting specific instances where the normal products were not obtained or where oximes were rearranged under unusual conditions.

Aliphatic Ketoximes

The Beckmann rearrangement has been applied to a wide variety of aliphatic ketoximes employing many different acidic materials as catalysts.

$$RCR' \atop \| \atop NOH \xrightarrow{\text{Catalyst}} RCONHR' \text{ and/or } R'CONHR$$

where catalyst = PCl_5;[64] $R = CH_3$, $R' = n\text{-}C_3H_7$, $n\text{-}C_4H_9$, $n\text{-}C_5H_{11}$, $n\text{-}C_6H_{13}$; $R = n\text{-}C_4H_9$, $R' = n\text{-}C_4H_9$; $R = C_2H_5$, $R' = n\text{-}C_3H_7$. Yields range from 70 to 84%.

where catalyst = H_2SO_4;[64,65] $R = CH_3$, $R' = CH_3$, $n\text{-}C_3H_7$, $n\text{-}C_9H_{19}$; $R = C_2H_5$, $R' = n\text{-}C_3H_7$. Yields range from 85 to 100%.

where catalyst = BF_3;[19] $R = CH_3$, $R' = C_6H_5CH_2$. Yield is $\approx 50\%$.

---

[59] Schaffler and Ziegenbein, *Chem. Ber.*, **88**, 767 (1955).
[60] Smith, *J. Am. Chem. Soc.*, **70**, 323 (1948).
[61] Hatch and Cram, *J. Am. Chem. Soc.*, **75**, 38 (1953).
[62] Beringer and Ugelow, *J. Am. Chem. Soc.*, **75**, 2635 (1953).
[63] Horning, *Chem. Revs.*, **33**, 89 (1943).

One of the more unusual catalysts is metallic copper. Products that result from the rearrangement of dibenzyl ketoxime (XXXI) followed by reduction, dehydration, and/or hydrolysis of the rearrangement products were formed when the gaseous oxime was passed over copper at 200° in the presence of hydrogen.[66,67] When acetoxime was subjected to the

$$(C_6H_5CH_2)_2C=NOH$$
XXXI $\xrightarrow[200°]{Cu, H_2}$ $C_6H_5CH_2CO_2H + C_6H_5CH_2CONH_2 + C_6H_5CH_2CN$

same conditions, only reduction and hydrolysis of the oxime occurred.[66] An attempted rearrangement of the cuprous chloride complex of acetoxime gave inconclusive results.[68]

Catalysis of the rearrangement is often quite specific. Phosphorus pentachloride rearranges dibenzalacetone oxime (XXXII) to N-styrylcinnamamide, but concentrated sulfuric acid causes cyclization to the

$(C_6H_5CH=CH)_2C=NOH$
XXXII

$\xrightarrow{PCl_5 / Ether}$ $C_6H_5CH=CHCONHCH=CHC_6H_5$

$\xrightarrow{H_2SO_4}$

[isoxazoline ring with H_5C_6, O, N, $CH=CHC_6H_5$]
XXXIII

$C_6H_5CH=CHCCH_3$ with HON‖
XXXIV

$\xrightarrow{PCl_5}$ $C_6H_5CH=CHCONHCH_3$

$\xrightarrow{H_2SO_4}$ [isoxazoline ring with H_5C_6, O, N, CH_3]

isoxazoline XXXIII.[69] *syn*-Benzalacetone oxime (XXXIV) behaves similarly under identical conditions. This behavior is fairly general for oximes of α,β-unsaturated ketones.[69, 70]

Many abnormal products of the Beckmann rearrangement arise from dehydration or analogous reactions. Ethyl α,α-dibenzylacetoacetate oxime loses a molecule of ethanol to yield the isoxazolone (XXXV).[71]

[64] McLaren and Schachat, *J. Org. Chem.*, **14**, 254 (1949).
[65] Wallach, *Ann.* **312**, 171 (1900).
[66] Yamaguchi, *Bull. Chem. Soc. Japan*, **1**, 35 (1926) [*C.A.*, **21**, 75 (1927)].
[67] Yamaguchi, *Bull. Chem. Soc. Japan*, **1**, 54 (1926) [*C.A.*, **21**, 75 (1927)].
[68] Comstock, *Am. Chem. J.*, **19**, 484 (1897).
[69] von Auwers and Brink, *J. prakt. Chem.*, [2] **133**, 154 (1932).
[70] Blatt and Stone, *J. Am. Chem. Soc.*, **53**, 1133, 4134 (1931).
[71] Felkin, *Compt. rend.*, **227**, 510 (1948).

$$\underset{\underset{\text{HON}}{\|}}{\text{CH}_3\text{CC}(\text{CH}_2\text{C}_6\text{H}_5)_2\text{CO}_2\text{C}_2\text{H}_5} \xrightarrow[\text{H}_2\text{SO}_4]{85\%}$$

XXXV (isoxazoline structure with $(C_6H_5CH_2)_2$, CH_3, O, N)

The oxime of N-p-tolylmesoxalamide (XXXVI) gives N-p-tolylcyanoformamide when treated with phosphorus pentachloride.[72]

$$\underset{\underset{\text{NOH}}{\|}}{\text{H}_2\text{NCOCCONHC}_6\text{H}_4\text{CH}_3\text{-}p} \xrightarrow[\text{Ether}]{\text{PCl}_5} \text{NCCONHC}_6\text{H}_4\text{CH}_3\text{-}p + \text{NH}_3 + \text{CO}_2$$

XXXVI

Oximes of α-keto acids decarboxylate and dehydrate successively to form nitriles[73, 74] as shown in the following equation:

$$\underset{\underset{\text{NOH}}{\|}}{\text{RCCO}_2\text{H}} \xrightarrow{\text{Catalyst}} \text{RCN} + \text{CO}_2 + \text{H}_2\text{O}$$

R = CH_3, C_2H_5, i-C_3H_7, n-C_4H_9, n-C_6H_{13}, $HO_2C(CH_2)_3$, $HO_2C(CH_2)_4$.
Catalyst = CH_3COCl; $(CH_3CO)_2O$; H_2SO_4.

6-Methyl-5-hepten-2-one oxime yields the dihydropyridine XXXVII when treated with phosphorus pentoxide.[75] Similarly, oximes (XXXVIII,

$$\underset{\underset{\text{NOH}}{\|}}{(\text{CH}_3)_2\text{C}=\text{CHCH}_2\text{CH}_2\text{CCH}_3} \xrightarrow{\text{P}_2\text{O}_5} \text{CH}_3\text{CONHCH}_2\text{CH}_2\text{CH}=\text{C}(\text{CH}_3)_2 \xrightarrow{-\text{H}_2\text{O}}$$

XXXVII (dihydropyridine with three CH_3 groups)

XXXIX, XLI) containing an aryl group on the carbon atom β to the oximino group yield isoquinoline derivatives when treated with phosphorus pentoxide or phosphorus pentachloride.[76–78]

[72] Plowman and Whitley, *J. Chem. Soc.*, **125**, 587 (1924).
[73] Dieckmann, *Ber.*, **33**, 579 (1900).
[74] Locquin, *Bull. soc. chim. France*, [3] **31**, 1068 (1904).
[75] Wallach, *Ann.*, **319**, 77 (1901).
[76] Goldschmidt, *Ber.*, **28**, 818 (1895).
[77] Kaufmann and Rodsevic, *Ber.*, **49**, 675 (1916).
[78] Whaley and Govindachari, in Adams, *Organic Reactions*, Vol. VI, p. 77, John Wiley & Sons, New York, 1951.

C$_6$H$_5$CH=CHCCH$_3$ (HON=) XXXVIII →[P$_2$O$_5$ / Infusorial earth] [C$_6$H$_5$CH=CHCONHCH$_3$] → (isoquinoline)

XXXIX: 3,4-methylenedioxyphenyl-CH$_2$CH$_2$C(=NOH)CH$_3$

XXXIX →[PCl$_5$ / Benzene] 3,4-methylenedioxyphenyl-CH$_2$CH$_2$NHCOCH$_3$ +

3,4-methylenedioxyphenyl-CH$_2$CH$_2$CONHCH$_3$ + XL (6,7-methylenedioxy-1-methyl-3,4-dihydroisoquinoline)

XXXIX →[P$_2$O$_5$ / Heat] XL

XLI: 3,4-dimethoxyphenyl-CH$_2$CH$_2$C(=NOH)CH$_3$ →[P$_2$O$_5$ / Toluene] 6,7-dimethoxy-1-methyl-3,4-dihydroisoquinoline

Aliphatic Aromatic Ketoximes

The Beckmann rearrangement of acetophenone and related oximes has been studied extensively. The rearrangement products formed from this type of oxime are anilides, benzamides, or mixtures of the two. The anilide is the product isolated in most of the recorded reactions. The

CH$_3$CAr(=NOH) →[Catalyst] CH$_3$CONHAr and/or ArCONHCH$_3$ 75–100%

Catalyst = HF, BF$_3$, CH$_3$COCl, SOCl$_2$, CF$_3$CO$_2$H, PCl$_5$, C$_6$H$_5$SO$_2$Cl, H$_2$SO$_4$.
Ar = C$_6$H$_5$, p-CH$_3$OC$_6$H$_4$, o-CH$_3$C$_6$H$_4$, p-CH$_3$C$_6$H$_4$, mesityl, 2-naphthyl, p-xenyl.

rearrangement has been effected with a large number of catalysts.[18, 19, 79–84] Even catalysts like copper[85] or Japanese acid earth[86] will rearrange acetophenone oxime.

$$\text{CH}_3\text{CC}_6\text{H}_5 \atop \| \atop \text{HON}$$
$\xrightarrow{\text{Cu, H}_2, 200°}$ $\text{C}_6\text{H}_5\text{CO}_2\text{H} + \text{C}_6\text{H}_5\text{CN}$

$\xrightarrow[\text{earth, 180°}]{\text{Japanese acid}}$ $\text{C}_6\text{H}_5\text{CO}_2\text{H} + \text{CH}_3\text{CO}_2\text{H}, \text{C}_6\text{H}_5\text{NH}_2, \text{C}_6\text{H}_5\text{CN},$
$\text{C}_6\text{H}_5\text{COCH}_3 + \text{CH}_3\text{CONHC}_6\text{H}_5$

Sulfuric acid is not a good catalyst if the aryl group is substituted with an alkoxyl group.[87]

$$\text{ArCH}=\text{CHAr}' \atop \| \atop \text{HON}$$
$\xrightarrow{\text{PCl}_5}$ ArCH=CHCONHAr′

$\xrightarrow{\text{H}_2\text{SO}_4}$ Sulfonation products

Ar = C_6H_5, o-$\text{CH}_3\text{OC}_6\text{H}_4$, m-$\text{CH}_3\text{OC}_6\text{H}_4$; Ar′ = C_6H_5, o-$\text{CH}_3\text{OC}_6\text{H}_4$, m-$\text{CH}_3\text{OC}_6\text{H}_4$.

Products which appear to have been formed as a result of the Beckmann rearrangement have been obtained by refluxing ether solutions of lithium aluminum hydride and certain substituted acetophenone oximes.[88, 89]

$$\text{ArCCH}_3 \atop \| \atop \text{NOH}$$
$\xrightarrow[\text{Ether}]{\text{LiAlH}_4}$ [ArNHCOCH$_3$] → ArNHC$_2$H$_5$ + ArCH(NH$_2$)CH$_3$
$\qquad\qquad\qquad\qquad\qquad\qquad$ 15–59% $\qquad$ 4–50%

Ar = C_6H_5, p-XC$_6$H$_4$(X = F, Cl, Br, I), p-$\text{CH}_3\text{OC}_6\text{H}_4$, p-$\text{CH}_3\text{C}_6\text{H}_4$.

A number of investigators have observed the spontaneous rearrangement of di-o-methyl-substituted acetophenone oximes when the parent ketones were treated with hydroxylamine salts.[49–52] As discussed earlier on p. 11, an explanation of these observations may be that the ortho-substituent decreases coplanarity of the oximino side chain with the

---

[79] Bachmann and Barton, *J. Org. Chem.*, **3**, 300 (1938).
[80] Stephen and Bleloch, *J. Chem. Soc.*, **1931**, 886.
[81] Beckmann and Wegerhoff, *Ann.*, **252**, 1, 11 (1889).
[82] Huber (to du Pont), U.S. pat. 2,721,199 (1955) [*C.A.*, **50**, 10762 (1956)].
[83] Hudlicky, *Collection Czechoslov. Chem. Communs.*, **16–17**, 611 (1951–1952) [*C.A.*, **47**, 8012 (1953)].
[84] Swaminathan, *Science and Culture (Calcutta)*, **12**, 199 (1946) [*C.A.*, **41**, 2402 (1947)].
[85] Yamaguchi, *Mem. Coll. Sci., Kyoto Imp. Univ.*, **7A**, 281 (1924) [*C.A.*, **18**, 2880 (1924)].
[86] Inoue, *Bull. soc. chim. Japan*, **1**, 177 (1926) [*C.A.*, **21**, 892 (1927)].
[87] von Auwers and Brink, *Ann.*, **493**, 218 (1932).
[88] Larsson, *Svensk. Kem. Tidskr.*, **61**, 242 (1949) [*C.A.*, **44**, 1898 (1950).]
[89] Lyle and Troscianiec, *J. Org. Chem.*, **20**, 1757 (1955).

THE BECKMANN REARRANGEMENT

$$R\text{-}C_6H_2(CH_3)_2\text{-}COCH_3 \xrightarrow{(NH_3OH)^{\oplus}} R\text{-}C_6H_2(CH_3)_2\text{-}NHCOCH_3$$

R = H or CH$_3$.

aromatic ring.[52] Therefore resonance stabilization of the oxime is impeded and the rearrangement proceeds at an abnormally high rate.

α,β-Unsaturated ketoximes yield isoxazolines with sulfuric acid[70] as do similar compounds discussed in the aliphatic series.[69] However, ring formation did not occur under similar conditions with the oxime of α-bromobenzal-*p*-bromoacetophenone.[70]

$$(C_6H_5)_2C{=}CHC(\text{NOH})C_6H_5 \xrightarrow[\text{Ether}]{PCl_5} (C_6H_5)_2C{=}CHCONHC_6H_5$$

$$(C_6H_5)_2C{=}CHC(\text{NOH})C_6H_5 \xrightarrow{H_2SO_4} \text{isoxazoline with }(C_6H_5)_2\text{ and }C_6H_5$$

$$C_6H_5CH{=}CHC(\text{NOH})C_6H_4Br\text{-}p \xrightarrow{H_2SO_4} \text{isoxazoline product}$$

The formation of amidines was observed when aliphatic aromatic ketoximes were rearranged by treatment with thionyl chloride in ether.[80]

$$ArC(\text{NOH})R \xrightarrow[\text{Ether}]{SOCl_2} RCONHAr + RC(=NAr)(NHAr)$$

R = CH$_3$, C$_2$H$_5$, *n*-C$_3$H$_7$, C$_6$H$_5$CH$_2$; Ar = C$_6$H$_5$, *p*-CH$_3$C$_6$H$_4$.

Certain acetophenone oximes containing a tertiary α-carbon atom form olefins and benzonitrile on treatment with thionyl chloride.[90]

$$(CH_3)_2C(C_6H_5)C(\text{NOH})C_6H_5 \xrightarrow[\text{Benzene}]{SOCl_2} CH_2{=}C(CH_3)C_6H_5 + C_6H_5CN$$

$$C_6H_5C(\text{NOH})\text{-cyclohexyl-}C_6H_5 \xrightarrow[\text{Benzene}]{SOCl_2} \text{cyclohexenyl-}C_6H_5 + C_6H_5CN$$

[90] Lyle and Lyle, *J. Org. Chem.*, **18**, 1058 (1953).

When dilute hydrochloric acid is used as a catalyst for rearrangement, hydrolysis to the parent ketone is the principal reaction.[91]

$$\text{ArCR}(\text{NOH}) \xrightarrow[\text{HCl}]{18\%} \text{ArCOR} + \text{ArNH}_2 + \text{RCO}_2\text{H} + \text{NH}_2\text{OH}$$

Another hydrolysis reaction which has been observed is the formation of N-phenyloxalamide (XLII) by treatment of benzoyl cyanide oxime with phosphorus pentachloride.[92] Other catalysts gave no reaction.

$$\text{C}_6\text{H}_5\text{CCN}(\text{NOH}) \xrightarrow[\text{Ether}]{\text{PCl}_5} [\text{C}_6\text{H}_5\text{NHCOCN}] \xrightarrow{\text{H}_2\text{O}} \text{C}_6\text{H}_5\text{NHCOCONH}_2$$
$$\text{XLII}$$

The o- and p-chlorobenzoyl cyanide oximes failed to rearrange.

Oximes of o-hydroxyacetophenones (XLIII) yield benzoxazoles (XLIV) when subjected to the conditions of the Beckmann rearrangement.[8, 91]

[structures XLIII → intermediate → XLIV]

Catalyst = Beckmann's mixture, PCl_5, $KHSO_4$. R = CH_3 or H.

The hydrochlorides of the same oximes rearrange to benzoxazoles on heating. Another unusual reaction was disclosed by Busch and his co-workers who tentatively formulated the structure of the uncharacterized product XLV as an "anhydroöxime."[93, 94]

$$p\text{-RC}_6\text{H}_4\text{NHCH}_2\text{CC}_6\text{H}_5(\text{HON}) \xrightarrow[\text{Ether}]{\text{PCl}_5} p\text{-RC}_6\text{H}_4\text{N}\overset{\text{CH}_2\text{—CC}_6\text{H}_5}{\text{——N} \to \text{O}}$$

R = CH_3, CH_3O XLV

Both the syn- and anti-oximes of benzoylformic acid undergo successive decarboxylation and dehydration to yield nitriles when treated with benzenesulfonyl chloride in sodium hydroxide.[95]

$$\underset{syn\ or\ anti}{\text{C}_6\text{H}_5\text{C}(\text{NOH})\text{CO}_2\text{H}} \xrightarrow[\text{NaOH}]{\text{C}_6\text{H}_5\text{SO}_2\text{Cl}} [\text{C}_6\text{H}_5\text{CH}(\text{NOH})] + \text{CO}_2 \xrightarrow{-\text{H}_2\text{O}} \text{C}_6\text{H}_5\text{CN}$$

[91] von Auwers, Lechner, and Bundesman, Ber., **58**, 36 (1925).
[92] Zimmermann, J. prakt. Chem., [2] **66**, 353 (1902).
[93] Busch, Stratz, Unger, Reichald, and Eckhardt, J. prakt. Chem., [2] **150**, 1 (1937).
[94] Busch and Kammerer, Ber., **63**, 649 (1930).
[95] Werner and Piguet, Ber., **37**, 4295 (1904).

Diaryl Ketoximes

In general, diaryl ketoximes can be rearranged easily with the common catalysts to yield an amide or mixture of amides.[8, 79, 96–105]

$$(C_6H_5)_2C\!=\!NOH \xrightarrow{\text{Catalyst}} C_6H_5CONHC_6H_5$$
$$70\text{--}100\%$$
Catalyst = HF, HCl, HBr, H_3PO_4-P_2O_5, PCl_5, CH_3COCl.

$$\underset{\underset{NOH}{\|}}{ArCAr'} \xrightarrow{PCl_5} ArCONHAr' \text{ and/or } Ar'CONHAr$$

Ar = C_6H_5.
Ar' = p-ClC$_6$H$_4$, o-BrC$_6$H$_4$, p-NO$_2$C$_6$H$_4$, o-HOC$_6$H$_4$, p-CH$_3$OC$_6$H$_4$, o-H$_2$NC$_6$H$_4$, p-CH$_3$C$_6$H$_4$, p-C$_6$H$_5$C$_6$H$_4$, 1-phenanthryl.

A number of unusual catalysts have been employed in the rearrangement of diaryl ketoximes; for example, benzophenone oxime was converted to benzanilide by the chlorides of K, Mg, Li, Hg, Fe(III), and Al, though their sulfates, hydroxides, and oxides were ineffective.[99] Chloral will rearrange benzophenone oxime hydrochloride to benzanilide.[106]

Thiobenzanilide was obtained from benzophenone oxime, phosphorus pentasulfide being used as a rearrangement catalyst.[107,108] When a mixture of phosphorus pentasulfide and phosphorus pentoxide was employed, the intermediate XLVI was isolated.[107, 108]

$$[(C_6H_5)_2C\!=\!N\!-\!S]_2PO_2H \xleftarrow[P_2S_5]{P_2O_5} (C_6H_5)_2C\!=\!NOH \xrightarrow{P_2S_5} C_6H_5CSNHC_6H_5$$
XLVI

(with Heat arrows via intermediate $[(C_6H_5)_2C\!=\!NSH]$, P_2S_5)

---

[96] Bachmann and Boatner, *J. Am. Chem. Soc.*, **58**, 2097 (1936).
[97] Hantzch, *Ber.*, **24**, 13 (1891).
[98] Meisenheimer and Kappler, *Ann.*, **539**, 99 (1939).
[99] Beckmann and Bark, *J. prakt. Chem.*, [2] **105**, 327 (1923).
[100] Beckmann, *Ber.*, **20**, 2580 (1887).
[101] Meisenheimer and Meis, *Ber.*, **57**, 289 (1924).
[102] Lehmann, *Angew. Chem.*, **36**, 360 (1923).
[103] Kardos, *Ber.*, **46**, 2086 (1913).
[104] Simons, Archer, and Randall, *J. Am. Chem. Soc.*, **62**, 485 (1940).
[105] Kuhara and Kainosho, *Mem. Coll. Sci., Kyoto Imp. Univ.*, **1906–1907**, 254 [*C.A.*, **1**, 2882 (1907)].
[106] Kuhara, Agatsuma, and Araki, *Mem. Coll. Sci., Kyoto Imp. Univ.*, **3**, No. 1, 1 (1917) [*C.A.*, **13**, 119 (1919)].
[107] Dodge, *Ann.*, **264**, 184 (1891); Ciusa, *Atti reale accad. Lincei*, [5] **15, II**, 379 (1906) (*Chem. Zentr.*, **1907, I**, 28).
[108] Kuhara and Kashima, *Mem. Coll. Sci., Kyoto Imp. Univ.*, **4**, 69 (1919) [*C.A.*, **15**, 69 (1921)].

Spontaneous formation of the amides obtainable by rearrangement of the oximes of 2,2′,4′-trimethylbenzophenone oxime and 2,4,6-trimethylbenzophenone oxime was observed when the parent ketones were heated with an aqueous solution of hydroxylamine hydrochloride.[7] The previously cited explanations (p. 11) for similar phenomena also may

$$\underset{O}{ArCAr'} \xrightarrow[120°]{NH_2OH \cdot HCl,\ H_2O} ArCONHAr' + Ar'CONHAr$$

Ar = C_6H_5, Ar′ = mesityl; Ar = o-tolyl, Ar′ = 2,4-$(CH_3)_2C_6H_3$.

apply here.[52] 4,4′-Bis(dimethylamino)benzophenone (Michler's ketone) also undergoes spontaneous rearrangement when treated with hydroxylamine hydrochloride.[109]

The aromatic ketoximes sometimes yield products resulting from the reaction of the catalyst with the oxime or amide. For example, acetanilide was isolated from the rearrangement of benzophenone oxime with acetic anhydride.[100] The chlorine-containing products XLVII and, perhaps, XLVIII have been isolated from the rearrangement of 2-nitrofluorenone oxime with phosphorus pentachloride.[110] On further reaction both XLVII and XLVIII gave only the phenanthridone XLIX. More recent work has indicated that both XLVIII and its isomer L can be isolated

[109] Morin, Warner, and Poirier, *J. Org. Chem.*, **21**, 616 (1956).
[110] Moore and Huntress, *J. Am. Chem. Soc.*, **49**, 2618 (1927).

from the reaction of 2-nitrofluorenone oxime with phosphorus pentachloride and phosphorus oxychloride.[111]

L

Phosphorus pentachloride was the only catalyst with which intermediate products could be isolated from p-chlorobenzophenone oxime.[81] Concentrated sulfuric acid and Beckmann's mixture both yielded only p-chlorobenzanilide.

$$p\text{-ClC}_6\text{H}_4\overset{\text{NOH}}{\underset{\|}{\text{C}}}\text{C}_6\text{H}_5 \begin{cases} \xrightarrow[\text{Ether}]{\text{PCl}_5} \text{C}_6\text{H}_5\text{C(Cl)}\!=\!\text{NC}_6\text{H}_4\text{Cl-}p \\ \xrightarrow[\substack{\text{or} \\ \text{Beckmann's} \\ \text{mixture}}]{\text{H}_2\text{SO}_4} \text{C}_6\text{H}_5\text{CONHC}_6\text{H}_4\text{Cl-}p \end{cases}$$

The formation of these chlorine-containing products might be rationalized in the following manner.

$$\underset{\text{NOH}}{\overset{\text{ArCAr}}{\|}} \xrightarrow{\text{Phosphorus halide(s)}} \begin{bmatrix} \text{Ar} \cdots \overset{\text{CAr}}{\underset{\overset{\|}{\text{N}}}{}} \\ \oplus \end{bmatrix} \longrightarrow \begin{bmatrix} \text{ArC}\!=\!\text{NAr} \\ \oplus \end{bmatrix}$$
LI   LII

$$\text{LI} \xrightarrow{\text{Cl}^-} \underset{\text{NCl}}{\overset{\text{ArCAr}}{\|}}$$

$$\text{LII} \xrightarrow{\text{Cl}^-} \underset{\overset{|}{\text{Cl}}}{\text{ArC}\!=\!\text{NAr}} \xrightarrow{\text{H}_2\text{O}} \begin{bmatrix} \text{ArC}\!=\!\text{NAr} \\ | \\ \text{OH} \\ + \\ \text{HCl} \end{bmatrix} \longrightarrow \text{ArCONHAr}$$

[111] Nunn, Schofield, and Theobald, *J. Chem. Soc.*, **1952**, 2797.

Some of the products obtained from the reaction of Grignard reagents with oximes may have been formed as the result of a Beckmann rearrangement.[112, 113]

$$(C_6H_5)_2C=NOH \xrightarrow{CH_3MgI \text{ or } C_2H_5MgI} [C_6H_5CONHC_6H_5] \rightarrow C_6H_5COR + C_6H_5NH_2$$
$$R = CH_3 \text{ or } C_2H_5$$

Amidines occur as by-products of the rearrangement of diaryl ketoximes.[80] Benzophenone oxime and p-ethoxybenzophenone oxime both yielded amidines as well as amides when treated with thionyl chloride.

$$(C_6H_5)_2C=NOH \xrightarrow[\text{Benzene}]{SOCl_2} C_6H_5CONHC_6H_5 + C_6H_5C\begin{smallmatrix}NC_6H_5\\NHC_6H_5\end{smallmatrix}$$

$$p\text{-}C_2H_5OC_6H_4\overset{NOH}{\underset{\|}{C}}C_6H_5 \xrightarrow[\text{Ether}]{SOCl_2} p\text{-}C_2H_5OC_6H_4CONHC_6H_5 + C_6H_5CONHC_6H_4OC_2H_5\text{-}p$$

$$p\text{-}C_2H_5OC_6H_4C\begin{smallmatrix}NC_6H_5\\NHC_6H_5\end{smallmatrix} + C_6H_5C\begin{smallmatrix}NC_6H_4OC_2H_5\text{-}p\\NHC_6H_4\text{—}OC_2H_5\text{-}p\end{smallmatrix}$$

anti-2-Hydroxybenzophenone oxime (LIII) yielded 2-phenylbenzoxazole, possibly due to dehydration of the amide formed by the rearrangement.[114] The *syn*-oxime (LIV) yielded the anilide of salicylic acid. In

[structures showing LIII (anti) reacting with PCl₅/Ether to intermediate then $-H^{\oplus}$ giving 2-phenylbenzoxazole + aminophenol]

[structures showing LIV (syn) reacting with PCl₅/Ether to give the salicylanilide]

[112] Grammaticakis, *Compt. rend.*, **210**, 716 (1940).
[113] Hoch, *Compt. rend.*, **203**, 799 (1936).
[114] Kohler and Bruce, *J. Am. Chem. Soc.*, **53**, 1569 (1931).

an analogous reaction, 2-phenylbenzimidazole (LV) was obtained from 2-aminobenzophenone oxime.[115] The formation of benzoxazoles or benzimidazoles from *anti*-2-hydroxy or 2-amino aryl ketoximes, respectively, is a general reaction;[116] a rationalization of the reaction has been suggested. The *syn*-oximes give the normal rearrangement products.[114]

Phthalanilide (LVI) can be prepared from 2-carboxybenzophenone oxime.[101]

Under the conditions of the Beckmann rearrangement, oximes of 1-aroylanthraquinones (LVII) yield *peri*-benzoylene-9-morphanthridones.[117–119]

Ar = C_6H_5, p-$CH_3C_6H_4$, 2,4- and 2,5-$(CH_3)_2C_6H_3$, 2,4,6-$(CH_3)_3C_6H_2$.
Catalyst = H_2SO_4, CH_3CO_2H, KI, HCl.

[115] von Auwers and Jordan, *Ber.*, **57**, 800 (1924).
[116] Blatt, *J. Org. Chem.*, **20**, 591 (1955).
[117] Scholl, Semp, and Stix, *Ber.*, **64**, 71 (1931).
[118] Scholl, Stephani, and Stix, *Ber.*, **64**, 315 (1931).
[119] Scholl, Mueller, and Donat, *Ber.*, **64**, 639 (1931).

The Beckmann rearrangement of certain 2-methyl-1-aroylanthraquinones (LVIII) yields 1-carboxy-2-methylanthraquinone carboxylic acids rather than *peri*-benzoylene-9-morphanthridones.

Ar = C_6H_5, 2,4- and 2,5-$(CH_3)_2C_6H_3$.

Alicyclic Ketoximes

Alicyclic ketoximes rearrange to yield lactams.

The reaction is very general for rings of all sizes.[57, 82, 83, 120–129]

Where $n = 3$, catalyst = HF, H_2SO_4, H_3PO_4-P_2O_5.
Where $n = 4$, catalyst = HF, H_2SO_4, $NaHSO_4$, CF_3CO_2H, SO_3, $SOCl_2$.
Where $n = 5$, catalyst = HF, H_3PO_4, SO_3.
Where $n = 6$, catalyst = H_2SO_4.
Where $n = 13$, catalyst = H_2SO_4.

[120] (To I. G. Farben), Ger. pat. appl., I 63,377 (1938).
[121] Novotny, U.S. pat. 2,579,851 (1951).
[122] Ruzicka, Goldberg, Hurbin, and Boeckenoogen, *Helv. Chim. Acta*, **16**, 1323 (1933).
[123–129] (See p. 27.)

The rearrangement of cyclohexanone oxime to ε-caprolactam, which is typical of the entire alicyclic series, has been studied in great detail and thus serves as a very broad standard of comparison for the other alicyclic ketoximes.

Cyclohexanone oxime rearranges to ε-caprolactam under almost any conditions known to effect the Beckmann transformation. The most common catalyst is sulfuric acid, but the use of this reagent is subject to certain difficulties. The yield of ε-caprolactam at a given temperature is dependent upon the strength of the acid employed.[130] At 100°, 97.5% acid gave an 83.4% yield of the lactam. The yield of the lactam gradually diminished to 64.5% as the acid strength was lowered to 85%. The loss of product was accounted for by hydrolysis of the oxime to cyclohexanone. Silicon dioxide was present in the reaction mixture as an accelerator and to absorb water.

The temperature at which the rearrangement is carried out is also important. With 80–85% sulfuric acid as a catalyst the yield of ε-caprolactam was 75% at 120°, 95% at 140°, and 85% at 160°.[131] The temperature of the usually highly exothermic reaction can be easily controlled by using the proper solvent,[56, 57, 120, 132–140] additives,[139–141] or equipment.[141–145]

[123] Horning and Stromberg, *J. Am. Chem. Soc.*, **74**, 2680 (1952).
[124] (To Maatschappij voor Kolenbewerking), Brit. pat. 719,109 (1954) [*C.A.*, **49**, 5043 (1955)].
[125] Stickdorn (to Deutsche Hydrierwerke G.m.b.H.), Ger. pat. 920,072 (1954).
[126] Hudlicky, *Chem. Listy*, **46**, 92 (1946) [*C.A.*, **47**, 8013 (1953)].
[127] (To Deutsche Hydrierwerke Aktiengesellschaft), Fr. pat. 892,603 (1944).
[128] Runge and Maas, *Chem. Tech. (Berlin)*, **5**, 421 (1953) [*C.A.*, **49**, 3845 (1955)].
[129] Kipping, *J. Chem. Soc.*, **65**, 490 (1894).
[130] Hajime, Tatsuo, and Nakamura (to Dai-Nippon Celluloide), Jap. pat. 157,331 (1943).
[131] (To Zellwolle and Kunstseide-Ring G.m.b.H.), Ger. pat. appl. Z 1,391 (1942).
[132] (To Société des Usines Chimiques Rhône-Poulenc), Brit. pat. 594,263 (1947) [*C.A.*, **42**, 2268 (1948)].
[133] (To Deutsche Hydrierwerke A. G.), Fr. pat. 894,102 (1944).
[134] (To Phrix-Werke A. G.), Fr. pat. 903,790 (1945).
[135] (To Deutsche Hydrierwerke A. G.), Ger. pat. 875,811 (1953).
[136] Welz (to Farbenfabriken Baeyer), Ger. pat. appl. F 7,449 (1951).
[137] (To Deutsche Hydrierwerke), Ger. pat. appl. D 4,334 (1952).
[138] Moncrieff and Young (to Brit. Celanese Ltd.), U.S. pat. 2,423,200 (1947) [*C.A.*, **41**, 6577 (1947)].
[139] Lincoln and Cohn (to Brit. Celanese Ltd.), U.S. pat. 2,723,266 (1955) [*C.A.*, **50**, 15580 (1956)].
[140] (To Deutsche Hydrierwerke A. G.), Ger. pat. 859,167 (1952).
[141] Johnson and MacCormack (to du Pont), U.S. pat. 2,487,246 (1949) [*C.A.*, **44**, 2016 (1950)].
[142] (To Bata A. G.), Fr. pat. 896,244 (1945).
[143] (To Bata A. G.), Fr. pat. 900,577 (1945).
[144] Klar and Hilgetag (to I. G. Farbenind.), Ger. pat. 735,727 (1943) [*C.A.*, **38**, 2663 (1944)].
[145] (To Thuringische Zellwolle), Ger. pat. appl. T 4,820 (1941).

Under certain conditions, cyclohexanone oxime yields the cleavage product 5-cyano-1-pentene (LIX).[146-148] Five- and seven-membered ring ketoximes also yield related nitriles (LX, LXI) under similar conditions.[146-148]

Certain spirane oximes (LXII, LXIII) yield unusual products when treated with polyphosphoric acid or thionyl chloride.[149] Similarly, camphor oxime (LXIV) and β-pericyclocamphenone oxime form nitriles when treated with catalysts known to cause the Beckmann rearrangement.[150, 151] These reactions are analogous to those described earlier on p. 19.[90] The formation of ω-olefinic nitriles and other cleavage products from alicyclic ketoximes is known.[147, 140-155] Under the conditions used to prepare the ω-olefinic nitriles (LIX-LXI), aromatic compounds

[146] Lazier and Rigby (to du Pont), U.S. pat. 2,234,566 (1941) [*C.A.*, **35**, 3650 (1941)].
[147] Wallach, *Ann.*, **309**, 1 (1889).
[148] Davydoff, *Chem. Tech. (Berlin)*, **7**, 647 (1955) [*C.A.*, **50**, 10678 (1956)].
[149] Hill and Conley, *Chem. & Ind. (London)*, **1956**, 1314.
[150] Borsche and Sander, *Ber.*, **48**, 117 (1915).
[151] Bredt and Holz, *J. prakt. Chem.*, [2] **95**, 133 (1917).
[152] Lyle, Fielding, Cauquil, and Rouzand, *J. Org. Chem.*, **20**, 623 (1955).
[153] Wallach and Kempe, *Ann.*, **329**, 82 (1903).
[154] Meisenheimer and Theilacker, *Ann.*, **493**, 33 (1932).
[155] Rupe and Splittgerber, *Ber.*, **40**, 4313 (1907).

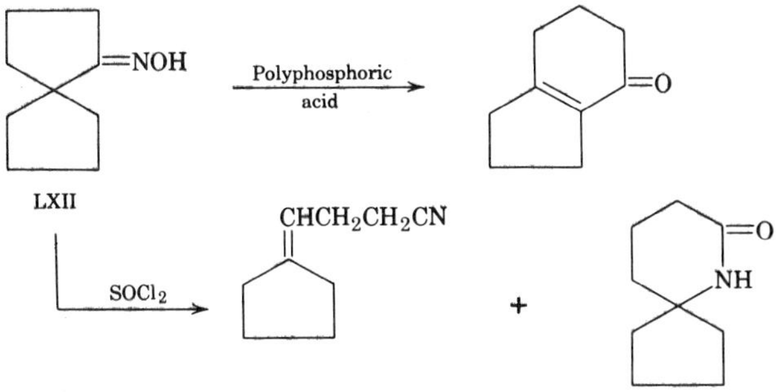

THE BECKMANN REARRANGEMENT

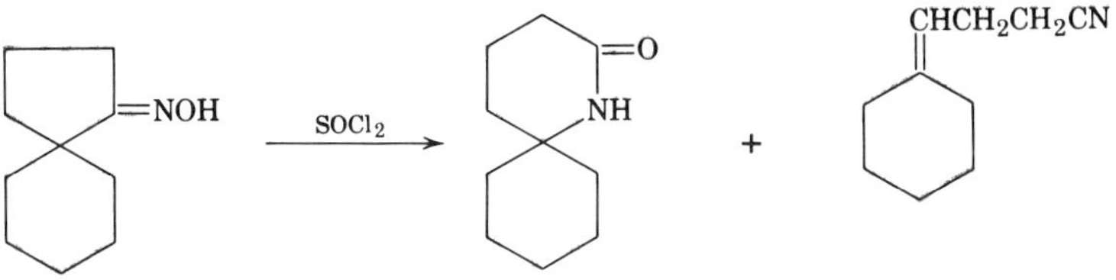

[Scheme showing LXII spiro[3.3] ketoxime → polyphosphoric acid → bicyclic enone]

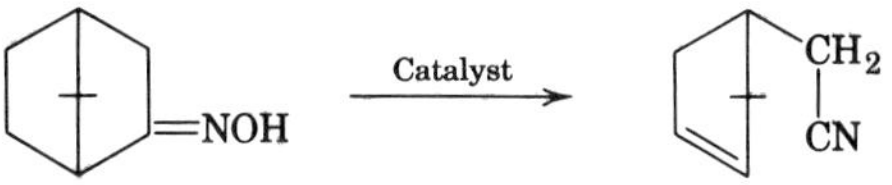

[Scheme: LXII with SOCl₂ → cyclopentylidene-CHCH₂CH₂CN + spiro lactam]

[Scheme: LXIII spiro[4.5] ketoxime with SOCl₂ → spiro lactam + cyclohexylidene-CHCH₂CH₂CN]

[Scheme: LXIV with Catalyst → aromatic CH₂CN product]

Catalyst = PCl₅, C₆H₅SO₂Cl, H₂SO₄.

(LXV–LXVII) are also formed.[65, 147, 156, 157] Other examples of aromatization are known.[59, 65, 147, 156, 157] They are illustrated by the following equations.

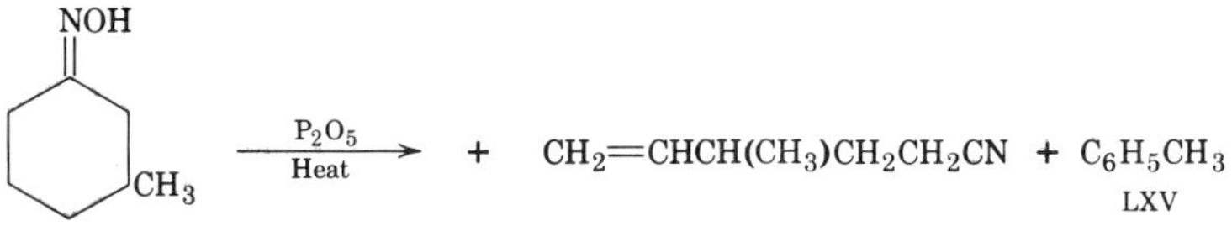

[Scheme: 2-methylcyclohexanone oxime with P₂O₅/Heat → + CH₂=CHCH(CH₃)CH₂CH₂CN + C₆H₅CH₃ (LXV)]

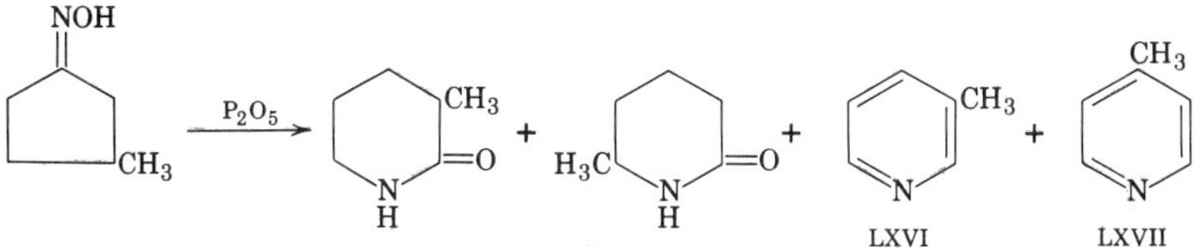

[Scheme: 2-methylcyclopentanone oxime with P₂O₅ → methylpiperidone + methylpiperidone isomer + methylpyridine (LXVI) + methylpyridine (LXVII)]

[156] Wolff, *Ann.*, **322**, 351 (1902).
[157] Wallach, *Ann.*, **346**, 266 (1906).

The aromatization of cyclohexenone oximes (LXVIII, LXIX) is a general reaction.[156–161]

Cyclohexanone oxime forms octahydrophenazine and aniline in small amounts under the conditions of the Beckmann transformation.[59]

The two hydrindone oximes, LXX, and LXXI, yield unusual products when treated with acetyl chloride.[162]

---

[158] Schroeter, Gluschke, Gotsky, Huang, Irmisch, Laves, Schrader, and Stier, *Ber.*, **63**, 1308 (1930).
[159] Hardy, Ward, and Day, *J. Chem. Soc.*, **1956**, 1979.
[160] Bhatt, *Experientia*, **13**, 70 (1957) [*C.A.*, **51**, 17857 (1957)].
[161] Vanags and Vitols, *J. Gen. Chem. U.S.S.R.*, **25**, 1953 (1955) [*C.A.*, **50**, 8644 (1956)].
[162] Leuchs and Rauch, *Ber.*, **48** 1531 (1915).

Cyclohexanone oxime can be rearranged to ε-caprolactam in the vapor phase in the presence of dehydration catalysts.[163,164] Cyclohexanone oxime can also be converted to hexamethylene diamine in the vapor phase.[165]

$$\text{cyclohexanone oxime} \xrightarrow[\text{Vapor phase}]{\text{CuCO}_3, \text{H}_2, \text{NH}_3} \left[\text{caprolactam} \right] \longrightarrow \text{H}_2\text{N(CH}_2)_6\text{NH}_2$$

In a somewhat similar fashion, 1-menthone oxime yields small amounts of the azacycloheptene LXXII.[166]

$$\text{menthone oxime} \xrightarrow[200°]{\text{Cu, H}_2} \underset{\text{LXXII}}{\text{azacycloheptene}} + \text{other products}$$

ε-Aminocaproic acid (LXXIII) can be prepared directly from cyclohexanone oxime by refluxing with 70% sulfuric acid.[120]

$$\text{cyclohexanone oxime} \xrightarrow{70\% \text{ H}_2\text{SO}_4} \left[\text{caprolactam} \right] \xrightarrow{\text{H}_2\text{O}} \underset{\text{LXXIII}}{\text{H}_2\text{N(CH}_2)_5\text{CO}_2\text{H}}$$

Simultaneous oximation of cyclohexanone and rearrangement of the oxime formed *in situ* has been accomplished with the use of hydroxylamine and sulfuric acid,[121, 167, 168] and by employing primary nitroparaffin as a source of hydroxylamine.[169] δ-Valerolactam can be prepared from cyclopentanone under the same conditions.[168]

[163] (To I. G. Farbenind.), Fr. pat. 895,509 (1945).
[164] Hopff and Drossbach (to I. G. Farbenind.), Ger. pat. 752,574 (1944).
[165] (To I. G. Farbenind. A. G.), Fr. pat. 896,330 (1945).
[166] Komatsu and Kurata, *Mem. Coll. Sci., Kyoto Imp. Univ.*, **7**, 151 (1924) [*C.A.*, **18**, 2149 (1924)].
[167] Novotny, U.S. pat. 2,569,114 (1951) [*C.A.*, **46**, 5078 (1952)].
[168] (To Bata), Brit. pat. appl. 33,342 (1948).
[169] Hass and Riley, *Chem. Revs.*, **32**, 373 (1943).

Nitrocyclohexane can be converted to ε-caprolactam by passing the vaporized nitroparaffin over a dehydration catalyst.[170] Sodium *aci*-nitrocyclohexane gives ε-caprolactam when added to hot oleum containing sulfur.[171] In this case, the intermediate oxime is probably formed by the self-reduction of the *aci*-salt.[172,173]

Steroid oximes rearrange to lactams.[174–178]

Heterocyclic Ketoximes

The classification of heterocyclic ketoximes here is purely arbitrary. Included are ketoximes which contain a hetero atom within a ring system in any portion of the molecule.

In general, ketoximes containing a variety of hetero atoms and ring

---

[170] England (to du Pont), U.S. pat. 2,634,269 (1953) [*C.A.*, **48**, 2767 (1954)].
[171] (To I. G. Farbenind. A. G.), Fr. pat. 977,095 (1951) [*C.A.*, **47**, 9998 (1953)].
[172] Schickh (Badische Anilin und Soda Fabrik), U.S. pat. 2,712,032 (1955).
[173] Donaruma and Huber, *J. Org. Chem.*, **21**, 965 (1956).
[174] Regan and Hayes, *J. Am. Chem. Soc.*, **78**, 639 (1956).
[175] Kaufmann, *J. Am. Chem. Soc.*, **73**, 1779 (1951).
[176] Anliker, Muller, Wohlfahrt, and Heusser, *Helv. Chim. Acta*, **38**, 1399, 1404 (1955).
[177] Schmidt-Thomé, *Ber.*, **88**, 895 (1955).
[178] Julian, Cole, Meyer, and Magnani, U.S. pat. 2,531,441 (1950) [*C.A.*, **45**, 2988 (1951)].

members undergo the Beckmann rearrangement in the normal manner to yield amides or mixtures of isomeric amides. The usual catalysts and solvents employed in the rearrangement of other types of oximes may be used to rearrange heterocyclic ketoximes.

In certain cases, abnormal products may be formed by interaction of the oxime or product with the catalyst or because of elimination, cleavage, polymerization, or hydrolysis reactions of the oxime or amides in the reaction mixture.

The oxime of N-phenacylisoquinolinium chloride (LXXIV), when rearranged with phosphorus pentachloride, yields a chlorination product

of the expected amide.[179] The oxime of 5-benzoyl-8-hydroxyquinoline (LXXV) yields a ring-sulfonated anilide upon rearrangement with sulfuric acid.[180]

N-Methyl-4-phenyl-4-benzoylpiperidine oxime (LXXVI) undergoes an elimination reaction of the type previously described on p. 19 to yield an

olefin and a nitrile.[90] Another example of nitrile formation is shown by formulas LXXVII and LXXVIII.[181]

---

[179] Ihlder, *Arch. Pharm.*, **240**, 691 (1902) (*Chem. Zentr.*, **1903**, I, 402).
[180] Matsumura and Sone, *J. Am. Chem. Soc.*, **52**, 4433 (1930); **53**, 1493 (1931).
[181] Rabe and Ritter, *Ann.*, **350**, 180 (1906).

N-Methyl-4-piperidone oxime yields a polymer when treated with polyphosphoric acid.[182]

The dioxime of 2-methyl-3-acetyl-6-methylbenzopyran-4-one forms an oxadiazine (LXXIX) when treated with sulfuric acid, acetyl chloride, or phosphorus pentachloride.[183]

Cuskohygrine oxime (LXXX) yields only cuskohygrine when treated with phosphorus pentachloride, hydrogen chloride, or sulfuric acid.[184]

[182] Barkenbus, Diehl, and Vogel, *J. Org. Chem.*, **20**, 871 (1955).
[183] Wittig and Bangert, *Ber.*, **58**, 2627 (1925).
[184] Hess and Fink, *Ber.*, **53**, 781 (1920).

Oximes of Polyfunctional Ketones

Oximes of ketones containing two or more carbonyl groups will rearrange to yield amides. The notable exceptions to this statement occur, for the most part, with oximes derived from α-diketones.

It has been demonstrated that the monoxime of an α-diketone may rearrange to yield one of two possible amides, depending on the configuration of the oxime.[95, 185–187]

$$\underset{\substack{\| \\ \text{NOH} \\ \alpha\text{-oxime}}}{\text{RCOCR}'} \xrightarrow{\text{Catalyst}} \text{RCONHCOR}'$$

$$\underset{\substack{\| \\ \text{HON} \\ \beta\text{-oxime}}}{\text{RCOCR}'} \xrightarrow{\text{Catalyst}} \text{RCOCONHR}'$$

However, in many cases cleavage to a nitrile and an acid accompanies rearrangement or is the main reaction.[95, 99, 188–193]

$$\underset{\substack{\| \\ \text{NOH}}}{\text{RCOCR}'} \rightarrow \text{RCO}_2\text{H} + \text{R}'\text{CN}$$

These cleavage reactions are sometimes referred to as "second-order" Beckmann rearrangements.[95] This phenomenon is not confined to monoximes of α-diketones and, therefore, is discussed in more detail later (p. 38).

The Beckmann rearrangement of monoximes of diketones in which the two carbonyl groups are not adjacent to each other proceeds in the conventional manner.[95, 194–196]

$$\underset{\substack{\| \\ \text{HON}}}{\text{RC—(CR}'_2)_n\text{COR}} \rightarrow \text{RNHCO(CR}'_2)_n\text{COR} \quad \text{and/or} \quad \text{RCONH(CR}'_2)_n\text{COR}$$

R' = alkyl, aryl, or H.

---

[185] Meisenheimer and Lange, *Ber.*, **57**, 282 (1924).
[186] Rule and Thompson, *J. Chem. Soc.*, **1937**, 1761.
[187] Francesconi and Pirrazoli, *Gazz. chim. ital.*, **33**, 36 (1903).
[188] Borsche and Sander, *Ber.*, **47**, 2815 (1914).
[189] Bulow and Grotrosky, *Ber.*, **34**, 1479 (1901).
[190] Brady and Bishop, *J. Chem. Soc.*, **1926**, 810.
[191] Meisenheimer, Beisswenger, Kauffmann, Kummer, and Link, *Ann.*, **468**, 202 (1929).
[192] Bishop and Brady, *J. Chem. Soc.*, **121**, 2364 (1922).
[193] Taylor, *J. Chem. Soc.*, **1931**, 2018.
[194] Finzi, *Gazz. chim. ital.*, **42**, 356 (1912).
[195] Beckmann and Liesche, *Ber.*, **56**, 1 (1923).
[196] Raphael and Vogel, *J. Chem. Soc.*, **1952**, 1958.

Monoximes of diketones appear to react abnormally chiefly by cleavage reactions. However, a few unusual products arising by reaction of the oxime or the rearrangement product with the catalyst have been recorded.

The α-diketone monoxime LXXXa, in which the locations of the methoxy and methylenedioxy groups have not been established, yielded the acyl derivative LXXXI upon refluxing with acetic anhydride.[197]

$$(CH_3O)(CH_2O_2)C_6H_2CCOCH_3 \xrightarrow[\text{At reflux}]{(CH_3CO)_2O} (CH_3O)(CH_2O_2)C_6H_2CON(COCH_3)_2$$
$$\underset{\text{HON}}{\|}$$
LXXXa LXXXI

5-Phenyl-5-oximinopentan-2-one and α-benzil monoxime have been reported to yield imido esters (LXXXII, LXXXIII) when rearranged with benzenesulfonyl chloride in the presence of base.[95, 194] Similar products

$$C_6H_5CCH_2CH_2COCH_3 \xrightarrow[\text{Base}]{C_6H_5SO_2Cl} C_6H_5SO_2OCCH_2CH_2COCH_3$$
$$\underset{\text{NOH}}{\|} \qquad \underset{C_6H_5N}{\|}$$
LXXXII

$$C_6H_5CCOC_6H_5 \xrightarrow[\text{Base}]{C_6H_5SO_2Cl} C_6H_5SO_2OCCOC_6H_5$$
$$\underset{\text{HON}}{\|} \qquad \underset{C_6H_5N}{\|}$$
α-oxime LXXXIII

have been obtained with phosphorus pentachloride as a catalyst. N-Benzoylbenzimido chloride (LXXXIV) has been obtained from benzil monoxime in this manner.[198]

$$C_6H_5CCOC_6H_5 \xrightarrow[\text{Ether}]{PCl_5} C_6H_5C{=}NCOC_6H_5$$
$$\underset{\text{HON}}{\|} \qquad \underset{\text{Cl}}{|}$$
α-oxime LXXXIV

Similarly the preparation of LXXXV from the monoxime of 2,4-dinitrobenzil and phosphorus pentachloride has been reported.[192]

$$2,4\text{-}(O_2N)_2C_6H_3COCC_6H_5 \xrightarrow[\text{Ether}]{PCl_5} 2,4\text{-}(O_2N)_2C_6H_3CON\overset{ClCC_6H_5}{\overset{\|}{}}$$
$$\underset{\text{NOH}}{\|}$$
LXXXV

[197] Rimini, *Gazz. chim. ital.*, **35,** 406 (1905).
[198] Beckmann and Sandel, *Ann.*, **296,** 279 (1897).

THE BECKMANN REARRANGEMENT 37

The behavior of dioximes of diketones is similar to that of the corresponding monoximes. Dioximes of α-diketones usually do not yield amides under the conditions of the Beckmann rearrangement.

1,2,4-Oxadiazoles (LXXXVI) apparently are formed when α-diketone dioximes are treated with reagents known to cause rearrangement of oximes.[187,199-202] The reaction probably involves a Beckmann rearrangement followed by dehydration. Under similar and sometimes identical conditions furazans (LXXXVII) may be formed by elimination of water

from the oximino groups.[199-203] The configuration of the dioxime may determine whether a furazan or an oxadiazine will be formed. However, there is not sufficient information concerning the stereochemistry of dioximes to enable one to make valid statements on this subject.

α-Benzil dioxime (LXXXVIII) has been reported to yield three different products under closely related conditions.[200,203,204]

[199] Ponzio, *Gazz. chim. ital.*, **62**, 854 (1932).
[200] Ponzio, *Gazz. chim. ital.*, **62**, 1025 (1932).
[201] Gastaldi, Langiane, and Sircona, *Gazz. chim. ital.*, **56**, 550 (1926).
[202] Brady and Muers, *J. Chem. Soc.*, **1930**, 216.
[203] Gunter, *Ber.*, **21**, 516 (1888).
[204] Gunter, *Ann.*, **252**, 44 (1889).

Dioximes of diketones usually rearrange in the normal manner when other groups are interposed between the oximino functions.[122, 205-207] However, abnormal reactions other than cleavage can occur.[208]

$$\underset{\underset{\text{HON}}{\|}\underset{\text{NOH}}{\|}}{\text{RC}(\text{CR}'_2)_n\text{CR}} \rightarrow \begin{array}{c} \text{RNHCO}(\text{CR}'_2)_n\text{CONHR} \\ \text{and/or} \\ \text{RNHCO}(\text{CR}'_2)_n\text{NHCOR} \\ \text{and/or} \\ \text{RCONH}(\text{CR}'_2)_n\text{NHCOR} \end{array}$$

Attempts to rearrange trioximes or derivatives of trioximes have been reported.[206, 209] Investigation of higher homologs has not been reported.

Cleavage of Oximes and Related Compounds Derived from Benzoins and α-Diketones

In previous portions of the text, the cleavage of oximes to yield nitriles has been discussed.[65, 90, 146-151, 181] These cleavages may be related to the more generally known cleavage of benzil- and benzoin-type oximes which has been termed a "second-order" Beckmann rearrangement.

In 1904 and 1905 Werner, Piguet, and Deutscheff found that, when the monoximes of benzil (LXXXIX, XC) were treated with benzenesulfonyl chloride, the normal rearrangement products (N-benzoylbenzamide and benzoylformanilide) were not obtained.[95, 210] Instead, a mixture of benzonitrile and benzoic acid was isolated from the rearrangement of α-benzil monoxime (LXXXIX), and phenyl isocyanide and benzoic acid were obtained from β-benzil monoxime (XC).[95] The oximes of benzoin

$$\underset{\underset{\text{HON}}{\|}}{\text{C}_6\text{H}_5\text{CCOC}_6\text{H}_5} \xrightarrow[\text{Pyridine}]{\text{C}_6\text{H}_5\text{SO}_2\text{Cl}} \text{C}_6\text{H}_5\text{CN} + \text{C}_6\text{H}_5\text{CO}_2\text{H}$$
α-oxime
LXXXIX

$$\underset{\underset{\text{NOH}}{\|}}{\text{C}_6\text{H}_5\text{CCOC}_6\text{H}_5} \xrightarrow[\text{Pyridine}]{\text{C}_6\text{H}_5\text{SO}_2\text{Cl}} \text{C}_6\text{H}_5\text{NC} + \text{C}_6\text{H}_5\text{CO}_2\text{H}$$
β-oxime
XC

---

[205] Knunyants and Fabrichnyi, *Doklady Akad. Nauk S.S.S.R.*, **68**, 701 (1949) [*C.A.*, **44**, 1918 (1950)].
[206] Milane and Venturello, *Gazz. chim. ital.*, **66**, 808 (1936).
[207] Anderson, Fritz, and Scotoni, *J. Am. Chem. Soc.*, **79**, 6511 (1957).
[208] Mamlok, *Bull. soc. chim. France*, **1956**, 1182.
[209] Schenck, *Z. physiol. Chem.*, **89**, 360 (1914).
[210] Werner and Deutscheff, *Ber.*, **38**, 69 (1905).

(XCI, XCII) were cleaved to benzaldehyde and benzonitrile or phenyl isocyanide depending upon the configuration of the oxime.[210] α-Benzfuroin oxime under similar conditions yielded benzaldehyde and 2-cyanofuran, while β-benzfuroin oxime yielded benzaldehyde but no carbylamine.[210]

$$C_6H_5CH(OH)CC_6H_5 \xrightarrow[\text{Base}]{C_6H_5SO_2Cl} C_6H_5CN + C_6H_5CHO$$
$$\underset{\substack{\text{α-oxime}\\\text{XCI}}}{\overset{\|}{\text{NOH}}}$$

$$C_6H_5CH(OH)CC_6H_5 \xrightarrow[\text{Base}]{C_6H_5SO_2Cl} C_6H_5NC + C_6H_5CHO$$
$$\underset{\substack{\text{β-oxime}\\\text{XCII}}}{\overset{\|}{\text{HON}}}$$

The cleavage of oximes and their parent ketones was later studied in considerable detail.[211] The accompanying formulations illustrate the behavior of several oximes toward benzenesulfonyl chloride.

$$(C_6H_5)_2C(OH)CC_6H_5 \xrightarrow{C_6H_5SO_2Cl} (C_6H_5)_2CO + C_6H_5CN + H_2O$$
$$\underset{\text{α-oxime}}{\overset{\|}{\text{NOH}}}$$

$$C_6H_5C(CH_3)(OH)CC_6H_5 \xrightarrow{C_6H_5SO_2Cl} C_6H_5COCH_3 + C_6H_5CN + H_2O$$
$$\underset{\text{α-oxime}}{\overset{\|}{\text{NOH}}}$$

$$C_6H_5CH=CHC-C_6H_4Br\text{-}p \xrightarrow{C_6H_5SO_2Cl} C_6H_5CH=CHCONHC_6H_4Br\text{-}p$$
$$\overset{\text{HON}}{\overset{\|}{}}$$

Benzil can be cleaved with potassium cyanide to benzaldehyde and benzoic acid.[212] Benzoin yielded small amounts of benzaldehyde under similar conditions.[213, 214] Phenylbenzoin (XCIII) and methylbenzoin (XCIV) also can be cleaved with potassium cyanide.[211]

[211] Blatt and Barnes, *J. Am. Chem. Soc.*, **56**, 1148 (1934).
[212] Jourdan, *Ber.*, **16**, 659 (1883).
[213] Buck and Ide, *J. Am. Chem. Soc.*, **53**, 2350 (1931).
[214] Buck and Ide, *J. Am. Chem. Soc.*, **53**, 2784 (1931).

$$2(C_6H_5)_2C(OH)COC_6H_5 \xrightarrow{KCN} 2(C_6H_5)_2CO + C_6H_5CH(OH)COC_6H_5$$
XCIII

$$2C_6H_5C(CH_3)(OH)COC_6H_5 \xrightarrow{KCN} 2C_6H_5COCH_3 + C_6H_5CH(OH)COC_6H_5$$
XCIV

α-Benzil monoxime and α-benzoin oxime also undergo cleavage when treated with potassium cyanide.[211] However, no isonitrile could be

$$\underset{\underset{NOH}{\|}}{C_6H_5COCC_6H_5} \xrightarrow{KCN} C_6H_5CHO + C_6H_5CN$$

$$\underset{\underset{NOH}{\|}}{C_6H_5CH(OH)CC_6H_5} \xrightarrow{KCN} C_6H_5CN + C_6H_5CHO$$

detected from the reaction of the β-form of either oxime with potassium cyanide. Benzonitrile was isolated from β-benzil monoxime. A mechanism has been proposed to account for the formation of benzonitrile from β-benzil monoxime.[215]

Although a large number of benzoin and benzil oximes and their esters are known to undergo cleavage,[95, 210, 211, 216–219] not enough is yet known about the structural factors in the oxime to specify the scope of the process in a satisfactory manner.

α-Nitroso-β-naphthol (XCV) yields o-cyanocinnamoyl chloride when treated with benzenesulfonyl chloride in pyridine.[95, 195, 219] 2,3-Dimethoxy-6-carboxyphenylacetonitrile is obtained from the indandione monoxime (XCVI) on treatment with p-toluenesulfonyl chloride in aqueous sodium hydroxide.[220] Furoin oxime[210] appears to yield 2-furyl isocyanide

---

[215] Tessieri and Oakwood, "The Cleavage of β-Benzil Monoximes," presented at the 112th A.C.S. Meeting, New York, 1947.
[216] Buck and Ide, *J. Am. Chem. Soc.*, **53**, 1912 (1931).
[217] Meisenheimer and Lamparter, *Ber.*, **57**, 276 (1924).
[218] Gheorghiu and Cozubschi-Scuirevici, *Bull. soc. sci. Cluj.*, *Rumanie*, **24**, 15 (1942) [*C.A.*, **38**, 3276 (1944)].
[219] Borsche and Sander, *Ber.*, **47**, 2815 (1914).
[220] Chakravarti and Swaminathan, *J. Indian Chem. Soc.*, **11**, 101 (1934).

under similar conditions, and phenanthraquinone monoxime yields 2-cyano-2'-carboxybiphenyl.[95]

XCVI → (with p-$CH_3C_6H_4SO_2Cl$) → product (95%)

3-Oximinoisatin (XCVII) yields o-isocyanatobenzonitrile when treated with phosphorus pentachloride.[99, 188] Under similar conditions,

XCVII → (PCl_5) → o-CN-C_6H_4-NCO → (H_2O, $-CO_2$) → $CO(NHC_6H_4CN$-$o)_2$

N-methyl-3-oximinoisatin (XCVIII) yields o-cyano-N-methylphenylcarbamyl chloride.[188] 2,3-Dihydro-2-oxo-3-oximinobenzothiophene (XCIX) yields o-cyanophenylsulfenyl chloride under the same conditions.[188]

XCIX → (PCl_5) → o-CN-C_6H_4-SCl

Aldoximes

Under the proper conditions aldoximes will undergo the Beckmann rearrangement to yield amides.

$$RCH\!=\!NOH \xrightarrow{\text{Catalyst}} RCONH_2 \text{ and/or } HCONHR$$

R = CH_3, n-C_3H_7, $C_6H_5CH\!=\!CH$, C_6H_5, p-ClC_6H_4, m-$O_2NC_6H_4$, o-HOC_6H_4, p-$CH_3OC_6H_4$, p-$(CH_3)_2NC_6H_4$.

Catalysts include: Ni, Cu, BF_3, CF_3CO_2H, PCl_5, H_2SO_4.

Usually, only the unsubstituted amide is formed. Only rarely has the isolation of a substituted formamide been recorded.[221, 222]

Benzamide was obtained as one of the products formed by passing benzaldoxime and hydrogen over copper at 200°.[223, 224]

$$C_6H_5CH=NOH \xrightarrow[200°]{Cu,\ H_2} C_6H_5CONH_2 + C_6H_5CN + C_6H_5CO_2H$$

Similarly, pyrolysis of the sodium salt of benzaldoxime yielded benzamide along with benzoic acid, benzonitrile, ammonia, and benzamidine.[225]

Aldoximes can be rearranged to amides with Raney nickel catalysts.[226, 227] The intermediate complex C was described as a red oil. Traces of iron

$$RCH=NOH \xrightarrow{Raney\ Ni} \underset{C}{[complex]} \to RCONH_2$$

$R = C_6H_5,\ n\text{-}C_6H_{13},\ C_6H_5CH_2CH_2,\ C_6H_5CH=CH,\ 2\text{-furyl}.$

and aluminum in the Raney nickel may actually catalyze the transformation of the nickel complex to the amide. Tetrakis(furfuraldoxime)

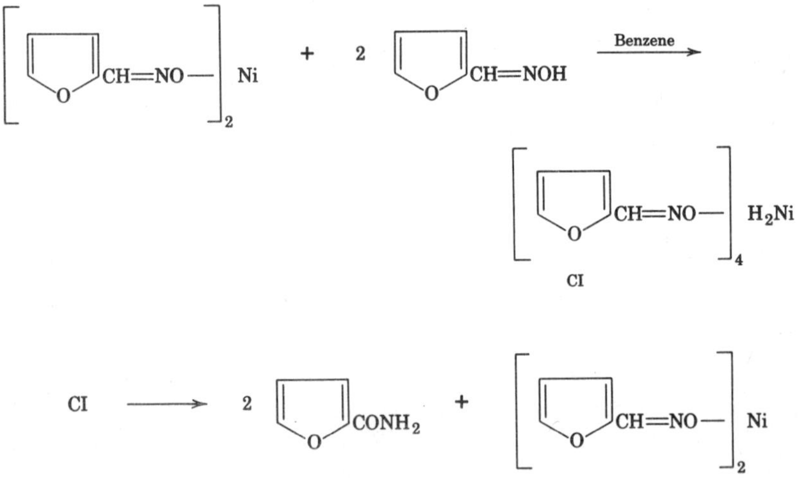

[furan-CH=NO—]₂Ni + 2 furan-CH=NOH $\xrightarrow{Benzene}$

[furan-CH=NO—]₄H₂Ni

CI

CI $\longrightarrow$ 2 furan-CONH₂ + [furan-CH=NO—]₂Ni

---

[221] Hantzsch and Lucas, *Ber.*, **28**, 744 (1895).
[222] Horning and Stromberg, *J. Am. Chem. Soc.*, **74**, 5151 (1952).
[223] Yamaguchi, *Bull. Chem. Soc. Japan*, **1**, 35 (1926) [*C.A.*, **21**, 75 (1927)].
[224] Yamaguchi, *Mem. Coll. Sci., Kyoto Imp. Univ.*, **9A**, 33 (1925) [*C.A.*, **19**, 3261 (1925)].
[225] Komatsu and Hiraidzumi, *Mem. Coll. Sci., Kyoto Imp. Univ.*, **8A**, 273 (1925) [*C.A.*, **19**, 2475 (1925)].
[226] Paul, *Compt. rend.*, **204**, 363 (1937).
[227] Paul, *Bull. soc. chim. France*, [5] **4**, 1115 (1937).

nickel (CI) can be decomposed to yield pyromucamide and bis(furfuraldoxime) nickel.[228] This evidence suggests that a nickel complex may be present as a reaction intermediate as postulated by Paul.[227]

Some other unusual catalysts which are known to rearrange aldoximes to amides are cuprous chloride and cuprous bromide,[68] both of which rearrange benzaldoxime to benzamide. Cinnamaldoxime is known to form a complex (CII) with cuprous bromide that can be converted to cinnamamide by heating in toluene.[68]

$$[C_6H_5CH{=}CHCH{=}NOH]CuBr \xrightarrow[\text{Toluene}]{\text{Heat}} C_6H_5CH{=}CHCONH_2$$
$$\text{CII}$$

Phenylglyoxaldoxime (CIII) can be converted to benzoylformamide with sodium bisulfite.[229]

$$C_6H_5COCH{=}NOH \xrightarrow{NaHSO_3} C_6H_5C(OH)(SO_3Na)CH(SO_3Na)(NHSO_3Na)$$
$$\text{CIII} \qquad\qquad\qquad \downarrow 20\% \; H_2SO_4$$
$$C_6H_5COCONH_2$$

Aldoximes can be dehydrated readily by acidic reagents to form nitriles.

$$RCH{=}NOH \xrightarrow{\text{Acid}} RCN + H_2O$$

Therefore nitriles are often formed from aldoximes under the conditions of the Beckmann rearrangement.[221, 223, 230–236]

Isoquinoline (CIV) is formed when cinnamaldoxime is treated with certain catalysts known to cause the Beckmann rearrangement.[237, 238]

$$C_6H_5CH{=}CHCH{=}NOH \xrightarrow[\text{acids}]{P_2O_5 \text{ or}} \left[\begin{array}{c} \text{(intermediate)} \end{array} \right] \xrightarrow{-H_2O} \text{CIV}$$

---

[228] Bryson and Dwyer, *J. Proc. Roy. Soc. N.S. Wales*, **74**, 471 (1941) [*C.A.*, **35**, 4768 (1941)].
[229] Kodama, *J. Chem. Soc. Japan*, **44**, 339 (1923) [*C.A.*, **17**, 3023 (1923)].
[230] Meisenheimer, Zimmermann, and von Kummer, *Ann.*, **446**, 205 (1926).
[231] Pawlewski, *Anz. Akad. Wiss. Krakau*, **1903**, 8 (*Chem. Zentr.*, **1903, I**, 837).
[232] von Auwers and Hugel, *J. prakt. Chem.*, [2] **143**, 179 (1935).
[233] von Auwers and Wolter, *Ann.*, **492**, 283 (1932).
[234] Steinkopf and Bohrmann, *Ber.*, **41**, 1044 (1908).
[235] Meisenheimer, Theilacker, and Beisswenger, *Ann.*, **495**, 249 (1932).
[236] Wohl and Losanitsch, *Ber.*, **40**, 4723 (1907).
[237] Bamberger and Goldschmidt, *Ber.*, **27**, 1954 (1894).
[238] Komatsu, *Mem. Coll. Sci., Kyoto Imp. Univ.*, **7**, 147 (1924) [*C.A.*, **18**, 2126 (1924)].

This is analogous to the formation of isoquinolines from β-phenyl α,β-unsaturated ketoximes.[76-78] This is an example of a reaction in which the formamide rather than the unsubstituted amide may be formed, *in situ*, by the rearrangement.[221, 224, 225]

o-Azidobenzaldoxime (CV) can be rearranged thermally to *o*-azidobenzamide and other products.[239]

$$\text{CV} \xrightarrow{\text{Heat}} o\text{-}N_3C_6H_4CO_2H + o\text{-}N_3C_6H_4CONH_2 +$$

$$o\text{-}H_2NC_6H_4CO_2H + o\text{-}H_2NC_6H_4CH{=}NOH + \text{(indazole N-oxide)}$$

o-Aminobenzaldoxime (CVI) does not rearrange with Beckmann's mixture; instead it yields the oxadiazacycloheptatriene CVII.[240]

$$\text{CVI} \xrightarrow[\text{(CH}_3\text{CO)}_2\text{O}]{\text{HCl, CH}_3\text{CO}_2\text{H}} \text{CVII}$$

6-(N-Oximinoglyoxal)aminotetralin (CVIII) undergoes a normal Beckmann rearrangement followed by cyclization when treated with 90% sulfuric acid.[241]

$$\text{CVIII} \xrightarrow{\underset{\text{H}_2\text{SO}_4}{90\%}} [\text{—NHCOCONH}_2]$$

[239] Bamberger and Demuth, *Ber.*, **35**, 1885 (1902).
[240] Meisenheimer and Diedrich, *Ber.*, **57**, 1715 (1924).
[241] Von Braun, Rohmer, Jungmann, Zobel, Brauns, Bayer, Stuckenschmidt, and Reutter, *Ann.*, **451**, 1 (1926).

Carbon-Nitrogen Rearrangements of Oxime Derivatives and Related Compounds

Oxime Esters. Oxime esters are converted, under the proper conditions, to amides.[12-14, 19, 43, 60, 158, 242-245]

$$\underset{RCR}{\overset{NOX}{\|}} \rightarrow \left[\underset{RC=NR}{\overset{OX}{|}} \right] \xrightarrow{H_2O} RCONHR + XOH$$

X = Acyl, benzenesulfonyl, p-toluenesulfonyl, picryl, etc.

Acids,[19, 23, 36, 54, 60, 245] bases,[13, 54] and materials of high solvolytic power such as water or alcohols[54, 158, 242] will facilitate the transformation. The behavior of the oxime esters in the rearrangement is analogous to that of oximes. Abnormal products formed under rearranging conditions are, in general, similar to those formed from oximes: amidines,[13] phenazines,[60] isoxazoles,[246] nitriles,[216] imino ethers,[13, 58] or lactams and other solvolysis products.[37, 158] Oxime sulfonates or arylsulfonates can be rearranged merely by heating the ester in solution.[247, 248]

In the presence of strong bases, oxime arylsulfonates are converted to α-aminoketones.[249-259] This reaction has become known as the Neber

$$(RCH_2)_2C{=}NOSO_2Ar \xrightarrow{KOR} RH_2C\underset{N}{\overset{}{\diagdown\diagup}}CHR \xrightarrow[H\oplus]{H_2O} RCH_2COCH(NH_2)R$$

rearrangement. The reaction is general for most oxime arylsulfonates having hydrogen atoms on the carbon atom adjacent to the one bearing

---

[242] Knunyants and Fabrichnyi, *Doklady Akad. Nauk S.S.S.R.*, **68**, 528 (1949) [*C.A.*, **44**, 1469 (1950)].
[243] Huntress and Walker, *J. Am. Chem. Soc.*, **70**, 3702 (1948).
[244] Wege, *Ber.*, **24**, 3537 (1891).
[245] Lindemann and Romanoff, *J. prakt. Chem.*, [2] **122**, 214 (1929).
[246] Hill and Hale, *Am. Chem. J.*, **29**, 253 (1903).
[247] Scheuing and Walach, Ger. pat. 579,227 [*C.A.*, **27**, 4630 (1933)].
[248] Knoll, Ger. pat. 574,943 (1933) (*Chem. Zentr.*, **1933, I**, 4040).
[249] Neber, U.S. pat. 2,055,583 (1936) [*C.A.*, **30**, 7583 (1936)].
[250] Neber and von Friedolsheim, *Ann.*, **449**, 109 (1926).
[251] Neber and Uber, *Ann.*, **467**, 52 (1928).
[252] Neber and Burgard, *Ann.*, **493**, 281 (1932).
[253] Neber and Huh, *Ann.*, **515**, 283 (1935).
[254] Neber, Hartung, and Ruopp, *Ber.*, **58**, 1234 (1925).
[255] Geissman and Armen, *J. Am. Chem. Soc.*, **77**, 1623 (1955).
[256] Neber, Burgard, and Thier, *Ann.*, **526**, 277 (1936).
[257] Neber (to Zellwolle and Kunstseide-Ring G.m.b.H.), Ger. pat. 870,415 (1953) (*Chem. Zentr.*, **1954**, 1598).
[258] Baumgarten and Bower, *J. Am. Chem. Soc.*, **76**, 4561 (1954).
[259] Cram and Hatch, *J. Am. Chem. Soc.*, **75**, 33 (1953).

the oximino group. Recently Baumgarten and Bower[258] have found that under similar conditions certain N,N-dichloroamines will form products characteristic of the Neber rearrangement.

$$\text{C}_6\text{H}_{11}\text{NH}_2 \xrightarrow[\text{H}_2\text{O}]{\text{Cl}_2} \text{C}_6\text{H}_{11}\text{NCl}_2 \xrightarrow[\text{H}_2\text{O}]{\text{NaOCH}_3} \left[\text{cyclohexanone NCl}\right] \longrightarrow \left[\text{azirine}\right] \xrightarrow[\text{H}^{\oplus}]{\text{H}_2\text{O}} \text{2-aminocyclohexanone}$$

Acidic catalysts that rearrange oximes will also convert oxime ethers to amides.[260-263]

$$\underset{\text{R}}{\overset{\text{NOR}'}{\|}}\text{C}\text{R} + \text{H}_2\text{O} \xrightarrow{\text{Acid}} \text{RCONHR} + \text{R}'\text{OH}$$

Imines and N-Halo Imines. The reaction of N-chlorobenzophenone imine (CIX) with potassium hydroxide to yield aniline and with antimony pentachloride to yield benzanilide or p-chlorobenzanilide has been reported.[21, 90]

$$(\text{C}_6\text{H}_5)_2\text{C}=\text{NCl} \quad \text{CIX}$$

- $\xrightarrow[\text{Fuse}]{\text{KOH}}$ $\text{C}_6\text{H}_5\text{NH}_2$
- $\xrightarrow[\text{CCl}_4,\ 25°]{\text{SbCl}_5}$ $p\text{-ClC}_6\text{H}_4\text{NHCOC}_6\text{H}_5$
- $\xrightarrow[\substack{\text{CCl}_2=\text{CCl}_2 \\ 40-50°}]{\text{SbCl}_5}$ $\text{C}_6\text{H}_5\text{CONHC}_6\text{H}_5$

Dimesityl ketimine was converted to the amide (CX) with hydrogen peroxide in glacial acetic acid.[264]

[260] Theilacker, Gerstenkorn, and Gruner, *Ann.*, **563**, 104 (1949).
[261] Hudlicky and Hokr, *Collection Czechoslov. Chem. Communs.*, **14**, 561 (1949) [*C.A.*, **44**, 5826 (1950)].
[262] Perold and von Reiche, *J. Am. Chem. Soc.*, **79**, 465 (1957).
[263] Donaruma, *J. Org. Chem.*, **22**, 1024 (1957).
[264] Hauser and Hoffenberg, *J. Am. Chem. Soc.*, **77**, 4885 (1955).

$$\left[\underset{\underset{CH_3}{|}}{\overset{\overset{CH_3}{|}}{H_3C-C_6H_2}} -C=NH \right]_2 \xrightarrow[CH_3CO_2H]{H_2O_2} \underset{\underset{CH_3}{|}}{\overset{\overset{CH_3}{|}}{H_3C-C_6H_2}} -CONH- \underset{\underset{CH_3}{|}}{\overset{\overset{CH_3}{|}}{C_6H_2}} -CH_3$$

CX

Nitrones. Nitrones are converted to amides when treated with catalysts which are acidic, or basic, or are esterifying agents.[265–276] In fact, some nitrones will yield amides when heated in solution.[268] Monosubstituted nitrones (CXI) apparently undergo rearrangement,[265–271, 274, 275]

$$\underset{CXI}{RCH=\overset{\overset{O}{\uparrow}}{N}R'} \rightarrow RCONHR'$$

$$\underset{CXII}{R_2C=\overset{\overset{O}{\uparrow}}{N}R'} \rightarrow RCONHR + R'NH_2$$

while disubstituted nitrones (CXII) are known to disproportionate to yield an amide and an amine[272] and to rearrange to oxime ethers.[273]

Intermediate solvolysis products of monosubstituted nitrones, e.g., CXIII, have been isolated.[269] The group on the nitrogen does not appear to migrate during the rearrangement of a monosubstituted nitrone.[265–271, 274, 275]

$$ArCH=\overset{\overset{O}{\uparrow}}{N}C_6H_5 \xrightarrow[CH_3OH]{KCN} \underset{CXIII}{Ar\overset{\overset{OCH_3}{|}}{C}=NC_6H_5} \xrightarrow[H_2O]{H^\oplus} ArCONHC_6H_5$$

Ar = o-, m-, or p-$O_2NC_6H_4$.

[265] Alessandrini, *Gazz. chim. ital.*, **51**, 75 (1921).
[266] Barrow, Griffiths, and Bloom, *J. Chem. Soc.*, **121**, 1713 (1922).
[267] Tonasescu and Nanu, *Ber.*, **72**, 1083 (1939).
[268] Tonasescu and Nanu, *Ber.*, **75**, 650 (1942).
[269] Bellavita, *Gazz. chim. ital.*, **65**, 755, 889, 897 (1935); *Atti congr. nazl. chim. pura ed appl.*, 5th Congr., Rome, **1935**, Part I, 285 (1936) [*C.A.*, **30**, 2935, 3419–3420 (1936)].
[270] Brady, Dunn, and Goldstein, *J. Chem. Soc.*, **1926**, 2411.
[271] Krohnke, *Chem. Ber.*, **80**, 298 (1947).
[272] Exner, *Collection Czechoslov. Chem. Communs.*, **16**, 258 (1951) [*C.A.*, **47**, 5884 (1953)].
[273] Cope and Haven, *J. Am. Chem. Soc.*, **72**, 4897 (1950).
[274] Beckmann, *Ber.*, **37**, 4136 (1904).
[275] Scheiber and Brandt, *J. prakt. Chem.*, [2] **78**, 80 (1908).
[276] Splitter and Calvin, *J. Org. Chem.*, **23**, 651 (1958).

These observations suggest that the reaction is not similar mechanistically to the Beckmann rearrangement and that it may be the oxygen that migrates or is exchanged by solvolysis. Perhaps oxaziranes are intermediates in this transformation.[276]

Nitroles. Products which may be the result of a Beckmann rearrangement are formed by the thermal decomposition of nitroles.[277, 278]

$$HCNO_2 \text{ (=NOH)} \xrightarrow{\text{Heat}} HN{=}C{=}O + HNO_2$$

$$CH_3CNO_2 \text{ (=NOK)} \xrightarrow{\text{Heat}} CH_3N{=}C{=}O + KNO_2$$

Derivatives of Hydroxamic Acids. 1,2,4-Oxadiazoles (CXIV) have been prepared from α-oximino hydroxamic acids, acid chlorides, amides, and anilides.[199, 200]

$$ArC(=NOH)-CX(=NOH) \xrightarrow{POCl_3 \text{ or } PCl_5} \left[\begin{array}{c} ArC=N \\ | \quad | \\ N \quad OH \\ \diagdown C \diagup \\ | \\ X \end{array} \right] \xrightarrow{-H_2O} \begin{array}{c} ArC=N \\ | \quad | \\ N \quad O \\ \diagdown C \diagup \\ | \\ X \end{array}$$

CXIV

Ar = C$_6$H$_5$, p-CH$_3$C$_6$H$_4$, C$_6$H$_5$CO.
X = NH$_2$, NHC$_6$H$_5$, Cl.

Hydroxamic acid amides also undergo the Beckmann rearrangement to yield unsymmetrical ureas; the reaction is known as the Tiemann reaction.[279]

$$RCNH_2 \text{ (=NOH)} \xrightarrow{C_6H_5SO_2Cl} RCNH_2 \text{ (=NOSO}_2\text{C}_6\text{H}_5\text{)} \xrightarrow{CHCl_3} C_6H_5SO_2OH + RNHCONH_2$$

CXV

Hydrazones and Semicarbazones. When hydrazones and semicarbazones are treated with nitrous acid[280-282] or heated with strong

---

[277] Wieland, *Ber.*, **42**, 803 (1909).
[278] Hantzch and Kanasirski, *Ber.*, **42**, 889 (1909).
[279] Partridge and Turner, *J. Pharm. Pharmacol.*, **5**, 103 (1953) [*C.A.*, **47**, 12278 (1953)].
[280] Pearson, Carter, and Greer, *J. Am. Chem. Soc.*, **75**, 5905 (1953).
[281] Pearson and Greer, *J. Am. Chem. Soc.*, **71**, 1895 (1949).
[282] Carter, *J. Org. Chem.*, **23**, 1409 (1958).

acids,[283-285] products characteristic of the Beckmann rearrangement are sometimes formed.

$$R_2C{=}NNH_2 \xrightarrow{\underset{-N_2}{HONO}}$$

$$R_2C{=}NNHCONH_2 \xrightarrow{\underset{-N_2,\ -CO_2}{HONO,\ H_2O}} \quad [R_2C{=}\overset{\oplus}{N}] \xrightarrow{H_2O} RCONHR$$

$$R_2C{=}NNH_2 \xrightarrow{\underset{Heat}{Acid}}$$

The reactions employing nitrous acid have been used to prepare benzanilides and perhaps are involved in the mechanism of certain reactions which yield ε-caprolactam.[286-288]

X = H or CONH$_2$

Acids and anilines can be obtained by heating p-chlorobenzophenone hydrazones to 450° in the presence of zinc chloride.[285]

These reactions may be related to the Beckmann rearrangement because rearrangement of an alkyl group to an electron-deficient nitrogen atom occurs.

Related Carbon-Nitrogen Rearrangements

The Lossen (CXV),[289] Curtius (CXVI),[290] and Hofmann (CXVII)[291] reactions are mechanistically related to the Beckmann rearrangement in that the three reactions all proceed via the migration of a group from a carbon atom to an electron-deficient nitrogen atom. Since there is only

---

[283] Steiglitz and Senior, *J. Am. Chem. Soc.*, **38**, 2727 (1916).
[284] Smith and Most, *J. Org. Chem.*, **22**, 358 (1957).
[285] Xanthopaulos, *Abstr. of Theses, University of Chicago, Science Series*, **4**, 195 (1925) [*C.A.*, **22**, 3639 (1928)].
[286] Ohashi (to East Asia Synthetic Chem. Ind.), Jap. pat. 125(1952) [*C.A.*, **48**, 1430 (1954)].
[287] Donaruma (to Du Pont), U.S. pat. 2,777,841 (1956) [*C.A.*, **51**, 10565 (1957)].
[288] Donaruma (to Du Pont), U.S. pat. 2,763,644 (1956) [*C.A.*, **51**, 5822 (1957)].
[289] Yale, *Chem. Revs.*, **33**, 243 (1943).
[290] Smith, in Adams, *Organic Reactions*, Vol. III, p. 337, John Wiley & Sons, New York, 1946.
[291] Wallis and Lane, in Adams, *Organic Reactions*, Vol. III, p. 267, John Wiley & Sons, New York, 1946.

$$\begin{array}{l}\text{RCONHOH} \xrightarrow{\text{Heat}}_{-H_2O}\\ \text{CXV}\\ \text{RCON}_3 \xrightarrow{\text{Heat}}_{-N_2}\\ \text{CXVI}\\ \text{RCONH}_2 \xrightarrow{\text{NaOBr}}\\ \text{CXVII}\end{array} \longrightarrow [\text{RCON}^\oplus] \longrightarrow \text{RNCO} \xrightarrow{H_2O} \text{RNH}_2 + CO_2$$

one group which can migrate in these three reactions, there are no stereochemical factors present as in the Beckmann rearrangement and only a single product can be formed. This statement also holds true for one phase of the Schmidt reaction, the reaction of hydrazoic acid with carboxylic acids (CXVIII).[292]

$$\text{RCO}_2\text{H} \xrightarrow[\text{HN}_3]{H_2SO_4} [\text{RCONN}_2]^\oplus \xrightarrow{-N_2} \overset{\oplus}{\text{RCONH}} \rightarrow \overset{\oplus}{\text{CONHR}} \xrightarrow{-H^\oplus} \text{RN}{=}\text{C}{=}\text{O}$$
CXVIII   H

$$\text{RN}{=}\text{C}{=}\text{O} \xrightarrow{H_2O} \text{RNH}_2 + CO_2$$

However, when ketones are treated with hydrazoic acid, the possibility of migration of one of two groups arises.

$$\text{RCOR}' + \text{HN}_3 \xrightarrow{H^\oplus} \begin{bmatrix} \text{R} & \text{OH} \\ & \diagdown \mid \diagup \\ & \text{CNHN}_2 \\ & \diagup \\ \text{R}' & \end{bmatrix}^\oplus \xrightarrow{-N_2} \begin{array}{c} \text{R} \quad \text{OH} \\ \diagdown \mid \diagup \\ \text{CNH} \\ \diagup \oplus \\ \text{R}' \end{array} \rightarrow$$

$$\begin{array}{cc} \text{OH} & \\ \mid & \\ \text{RCNHR}' & \text{and/or} \quad \text{R}'\overset{\oplus}{\text{CNHR}} \\ \oplus & \mid \\ & \text{OH} \\ \text{CXIX} & \text{CXX} \end{array}$$

$$\text{CXIX and/or CXX} \xrightarrow{-H^\oplus} \text{RCONHR}' \text{ and/or } \text{R}'\text{CONHR}$$

Aldehydes usually form nitriles when treated with hydrazoic acid.[292]

When hydrazoic acid or one of its salts is added to a system in which the Beckmann rearrangement is being carried out, tetrazoles (CXXI) are

$$R_2C{=}\text{NOH} \xrightarrow{\text{Catalyst}} [\text{R}\overset{\oplus}{\text{C}}{=}\text{NR}] \xrightarrow{\text{HN}_3} \begin{array}{c} \text{RC}{=}\text{N} \\ \mid \quad \diagdown \\ \quad \quad \text{NH} \\ \text{RN}{=}\text{N} \diagup \\ \text{CXXI} \end{array}$$

[292] Wolff, in Adams, *Organic Reactions*, Vol. III, p. 307, John Wiley & Sons, New York, 1946.

formed.[14, 248, 293-298] The reaction is applicable to a large number of oximes and oxime derivatives, particularly alicyclic ketoximes.

STEREOCHEMISTRY OF OXIMES

The Beckmann rearrangement has important synthetic uses. Since the rearrangement is stereospecific, a brief review of the stereochemistry of oximes is in order.

The oximation of ketones and aldehydes when measured in buffered systems appears to be an equilibrium reaction at low pH values and may become irreversible at pH 7.[299, 300] Optimum yields of oximes in such buffered systems are obtained at about pH 4.5.[299] The rates for oxime formation and oxime hydrolysis appear to be quite rapid.[299, 301, 302]

$$RCOR' + NH_2OH \rightleftharpoons RR'C(OH)(NHOH) \rightleftharpoons \underset{\underset{NOH}{\|}}{RCR'} \text{ and/or } \underset{\underset{HON}{\|}}{RCR'} + H_2O$$

Few investigators have attempted to determine the ratio of *syn* to *anti* isomers formed on oximation. This may be due to the fact that adequate methods for the analysis of such systems were not available until recently. Often only one stereoisomeric form is isolated The composition of the equilibrium mixture of oximes of unsymmetrical ketones frequently appears to be determined by stereochemical considerations.[79, 96, 303, 304, 304a]

[293] Harrill, Herbst, and Roberts, *J. Org. Chem.*, **15**, 58 (1950).
[294] Boehringer, Brit. pat. 309,949 (1929) (*Chem. Zentr.*, **1930, I,** 287).
[295] Knoll, Ger. pat. 538,981 (1931) (*Chem. Zentr.*, **1932, I,** 1297).
[296] Boehringer, Fr. pat. 645,265 (1928) (*Chem. Zentr.*, **1929, I,** 2586).
[297] Boehringer, Ger. pat. 543,026 (1928) [*C.A.*, **26**, 3263 (1932)].
[298] Boehringer, Brit. pat. 285,080 (1927) [*C.A.*, **22**, 4538 (1928)].
[299] Oländer, *Z. physik. Chem.*, **129**, 1 (1927).
[300] Fitzpatrick and Gettler, *J. Am. Chem. Soc.*, **78**, 530 (1956).
[301] Craft, Landrum, Suratt, and Lester, *J. Am. Chem. Soc.*, **73**, 4462 (1951).
[302] Vavon and Montheard, *Compt. rend.*, **207**, 926 (1938).
[303] Ungnade and McLaren, *J. Org. Chem.*, **10**, 29 (1945).
[304] Decombe, Jacquemain, and Rabinovitch, *Bull. soc. chim. France*, **1948,** 447.
[304a] Hantsch, *Ber.*, **24**, 4018 (1891).

However, resonance and inductive effects often influence the configuration of the oxime formed as the result of the stabilization of one stereoisomer by hydrogen bonding.[305, 306]

X—C$_6$H$_4$—CH=CHC—CH$_2$
                    ‖ |
                    NOH   H
                    Syn

X = OCH$_3$; syn 65%, anti 35%.
X = NO$_2$; syn only.

The configuration of an oxime may be determined by chemical or physical methods or both. Ring cleavage of the corresponding isoxazole[5, 307, 308] has frequently been employed for this purpose.

H$_5$C$_6$C———CC$_6$H$_5$ H$_5$C$_6$C———CC$_6$H$_5$
   ‖   ‖   →O$_3$ ‖   ‖
   N   CC$_6$H$_5$ N   CC$_6$H$_5$
    \\   /   ‖
     O   O   O

Other chemical methods employed are ring closure to the corresponding isoxazole,[116, 230, 309] or formation of coordination compounds with metal ions.[310, 311]

O$_2$N—C$_6$H$_3$(Br)—CC$_6$H$_5$ →NaOH→   O$_2$N—C$_6$H$_3$—C(C$_6$H$_5$)=N—O (isoxazole ring)
              ‖
              HON

Some of the physical methods used for the determination of the configuration of an oxime are dipole measurements[312, 313] and infrared,[314, 315] ultraviolet,[316] and nuclear magnetic resonance spectroscopy.[317]

[305] Corbett and Davy, J. Chem. Soc., **1955**, 296.
[306] Brady and Benger, J. Chem. Soc., **1953**, 3612.
[307] Kohler, J. Am. Chem. Soc., **46**, 1733 (1924).
[308] Kohler and Richtmyer, J. Am. Chem. Soc., **50**, 3092 (1928).
[309] Brady and Bishop, J. Chem. Soc., **127**, 1357 (1925).
[310] Brady and Muers, J. Chem. Soc., **1930**, 1599.
[311] Chugaev, Ber., **41**, 1678 (1923).
[312] Sutton and Taylor, J. Chem. Soc., **1931**, 2190.
[313] Sutton and Taylor, J. Chem. Soc., **1933**, 63.
[314] Palm and Werbin, Can. J. Chem., **31**, 1004 (1953).
[315] Palm and Werbin, Can. J. Chem., **32**, 858 (1954).
[316] Brady and Grayson, J. Chem. Soc., **1933**, 1037.
[317] Phillips, Ann. N.Y. Acad. Sci., **70**, 817 (1958).

Much experimental work has been reported in the older literature on the isomerization of oximes. Unfortunately, because many of the authors were not able to employ pure reagents, the conclusions drawn from their work frequently are questionable.

The equilibrium distribution of the two isomeric oximes appears to depend to a high degree upon the structure of the oxime, the acid employed in the reaction, and the reaction medium. Isomerization of one oxime form to the other may be effected by acids in nonpolar solvents[97, 221] or bases in ionizing solvents.[318-321] The stability of the *syn* oxime relative to the *anti* oxime depends upon steric and electrostatic effects. *syn-t*-Butyl phenyl ketoxime appears to isomerize prior to rearrangement when Beckmann's mixture is used as the reagent. Under similar conditions *syn*-isopropyl phenyl ketoxime yields only the normal products expected from *trans* migration.[6] The relative stabilities of monosubstituted benzophenone oximes also have been investigated.[7]

$$R\text{-}C_6H_4\text{-}C(=NOH)C_6H_5 \xrightarrow{H^\oplus} R\text{-}C_6H_4\text{-}C(=NOH)C_6H_5$$

Syn   Anti

$R = CH_3, C_2H_5, n\text{-}C_3H_7$

The *anti* oximes were more stable and their stability increased with the electron-releasing effect of the substituent ($CH_3 > C_2H_5 > n\text{-}C_3H_7$).

The importance of reaction medium upon the relative stability of two isomeric oximes is exemplified by the isomerization of mesitylaldoxime.[221]

$$2,4,6\text{-}(CH_3)_3C_6H_2CH(=NOH)\cdot HCl \underset{\text{Dry ether, HCl}}{\overset{\text{Wet ether, HCl}}{\rightleftarrows}} 2,4,6\text{-}(CH_3)_3C_6H_2CH(=NOH)\cdot HCl$$

In wet ethereal solution, the *syn*-aldoxime appears to be the more stable; in dry ethereal solution the *anti* oxime is the more stable form.

Recently it has been shown that the more stable *syn*-2-chlorobenzaldoxime was converted to the *anti*-oxime by equimolar amounts of hydrogen chloride or boron trifluoride in ether[19] (see equations on p. 54). A salt was formed which precipitated and displaced the equilibrium in favor of the *anti* oxime salt. The less stable *anti* form was isomerized to the *syn* form in ethanol or water by catalytic amounts of hydrochloric

---

[318] Patterson and Montgomery, *J. Chem. Soc.*, **101**, 2100 (1912).
[319] Hauser and Jordan, *J. Am. Chem. Soc.*, **58**, 1304, (1936).
[320] Brady and Thomas, *J. Chem. Soc.*, **1922**, 2098.
[321] Gilman, *Organic Chemistry*, John Wiley and Sons, New York, 1943, Vol. I, p. 472.

acid or by traces of boron trifluoride in ether. The equilibrium appears to be displaced in favor of the *syn* oxime because the acid catalyst is removed continuously from the *syn* oxime by the nucleophilic solvent. This example may explain the larger number of similar isomerizations effected by acids in different media.

$$p\text{-ClC}_6\text{H}_4\text{CH} \atop \| \atop \underset{\oplus}{\text{Cl}^{\ominus}\text{HNOH}} \rightleftarrows \left[{p\text{-ClC}_6\text{H}_4\overset{\oplus}{\text{CH}} \atop | \atop \text{Cl}^{\ominus}\text{HNOH}} \underset{\underset{\text{Ethanol or water}}{\text{HCl (catalytic)}}}{\overset{\text{HCl (equimolar)}}{\underset{\text{Ether}}{\longrightarrow}}} {p\text{-ClC}_6\text{H}_4\overset{\oplus}{\text{CH}} \atop | \atop \text{HONHCl}^{\ominus}} \right] \rightleftarrows {p\text{-ClC}_6\text{H}_4\text{CH} \atop \| \atop \underset{\oplus}{\text{HONHCl}^{\ominus}}}$$

$$\searrow \text{Base} \qquad\qquad\qquad\qquad \text{Base} \swarrow$$

$$\underset{Syn}{p\text{-ClC}_6\text{H}_4\text{CH} \atop \| \atop \text{NOH}} \qquad\qquad\qquad \underset{Anti}{p\text{-ClC}_6\text{H}_4\text{CH} \atop \| \atop \text{HON}}$$

Isomerization in alkaline media has been observed quite frequently. Electrostatic repulsion appears to play an important role in these isomerizations.[7,116,322] Such effects may be prevented by conversion to the corresponding oxime ether.

$$\underset{\text{NOH}}{\text{C}_6\text{H}_5\text{CCO}_2\text{H} \atop \|} \underset{\text{HCl}}{\overset{\text{NaOH}}{\rightleftarrows}} \underset{\text{Na}^{\oplus}\overset{\ominus}{\text{ON}}}{\text{C}_6\text{H}_5\text{CCO}_2^{\ominus}\text{Na}^{\oplus} \atop \|}$$

Little is known about the function of temperature and catalyst upon isomerization of oximes.[116]

The effect of the reaction medium on the distribution of products from the Beckmann rearrangement is very important. Rearrangements by phosphorus pentachloride in benzene and in ether proceed without isomerization provided the reaction is carried out at or below room temperature.[8,79,323] A solvent of high dielectric constant or a solvent of high nucleophilic power and/or solvolytic power may favor the isomerization considerably. Whereas *syn-t*-butyl phenyl ketoxime is rearranged by phosphorus pentachloride in ether without isomerization, hydrogen chloride in acetic acid isomerizes the oxime *before* rearrangement.[6] An increase in the acid concentration of the rearranging agent increases the amount of isomerization preceding the rearrangement. Eighty-five per cent sulfuric acid rearranges methyl *n*-propyl ketoxime to

---

[322] Hantsch, *Ber.*, **25**, 2164 (1892).
[323] Blakey, Jones, and Scarborough, *J. Chem. Soc.*, **1927**, 2865.

N-n-propylacetamide.[64] Rearrangement with 93% sulfuric acid yields both isomeric amides.[64] In view of these observations, oxime configurations determined on the basis of *anti* rearrangement should be considered highly suspect unless it has been shown previously that the rearrangement conditions will not isomerize the oxime in question. Phosphorus pentachloride in ether at or below room temperature appears to be a system wherein no isomerization occurs.[5–7, 69, 70,79, 323] However, possible exceptions to this statement are known.[7,323a] Hydrogen chloride, in acetic acid or ethanol,[6, 115] and sulfuric acid[64, 243] isomerize oximes prior to rearrangement. Before 1921 some oxime configurations were determined on the assumption that *cis* migration occurs during rearrangement.[2] Therefore oxime configurations determined up to 1924 may not be correct.

PREPARATION OF OXIMES

Oximes can be prepared conveniently from the reaction of aldehydes or ketones with hydroxylamine salts in the presence of a base (i.e., pyridine or sodium hydroxide).[309, 324] Oximes can also be prepared by the reduction of nitroparaffins[325–332] or the reaction of nitroparaffin *aci* salts with acid solutions of hydroxylamine salts,[333] and by nitrosation of carbon atoms.[334]

EXPERIMENTAL CONDITIONS

Catalyst and Solvent. The basis for the choice of catalyst and solvent can best be illustrated by describing the results which might be expected from certain catalysts and solvents.

Phosphorus pentachloride in ether appears to favor a stereospecific rearrangement.[70,79] Therefore, for determining the configuration of an

[323a] Terent'ev and Makarova, *Zhur. Obshchei Kkim.*, **21**, 270 (1951) [*C.A.*, **45**, 7105 (1951)].

[324] Shriner, Fuson, and Curtin, *The Systematic Identification of Organic Compounds*, p. 254, John Wiley & Sons, New York, 1956.

[325] Hopff, Reidel, and v. Schickh (to Badische Anilin und Soda Fabrik), Ger. pat. 922,709 1955) (*Chem. Zentr.*, **1955**, 5183).

[326] Weise (to Farbenfabriken Bayer), Ger. pat. 917,426 (1954) (*Chem. Zentr.*, **1954**, 10816).

[327] Weise (to Farbenfabriken Bayer), Ger. pat. 916,948 (1954) (*Chem. Zentr.*, **1954**, 10816).

[328] Welz (to Farbenfabriken Bayer), Ger. pat. 910,647 (1954) (*Chem. Zentr.*, **1954**, 6344).

[329] Welz and Giltges (to Farbenfabriken Bayer), Ger. pat. 877,304 (1953) (*Chem. Zentr.*, **1953**, 6567).

[330] Ufer (to Badische Anilin und Soda Fabrik), Ger. pat. 877,303 (1953) (*Chem. Zentr.*, **1953**, 8208).

[331] Weist (to Badische Anilin und Soda Fabrik), Ger. pat. 855,555 (1952) (*Chem. Zentr.*, **1954**, 1591).

[332] Welz (to Farbenfabriken Bayer), Ger. pat. 855,253 (1952) (*Chem. Zentr.*, **1954**, 1351).

[333] Hopff and Schickh (to Badische Anilin und Soda Fabrik), Ger. pat. 900,094 (1953) (*Chem. Zentr.*, **1954**, 9393).

[334] Touster, in Adams, *Organic Reactions*, Vol. VII, p. 346, John Wiley & Sons, New York, 1953.

oxime on the basis of *anti* migration, this system would seem to be preferred.

If a high yield of amide is desired, polyphosphoric acid and fuming sulfuric acid are recommended as catalysts.[222, 335] With these catalysts, hydrolysis of the oxime to the ketone and of the amide to the acid and amine is negligible.

Hydrolysis of the amide formed *in situ* to the acid and amine can be achieved by employing 70% sulfuric acid as a catalyst.[120] Likewise, solvolysis of oxime sulfonates to obtain imino ethers,[13,37,158] and amidines,[13] can be achieved by employing solvents such as alcohols, phenols, or amines, respectively, in the presence of a suitable catalyst.

Steroids rearrange best if the acid chloride of a weak sulfonic acid, such as *p*-acetamidobenzenesulfonyl chloride, is used as a catalyst.[174–178]

Temperature. The optimum temperature for a given rearrangement is important for a high yield of product. The optimum temperature at which a Beckmann rearrangement must be carried out depends on the nature of the oxime, the product, the catalyst, and the solvent and often cannot be predicted accurately. However, when sulfuric acid is used as a catalyst, the rearrangement usually proceeds best between 100° and 140°.

Catalysts like phosphorus pentachloride,[70] hydrogen fluoride,[83,104,126] and sulfur trioxide[55,57,336] enable one to carry out the reaction near or below room temperature.

Temperature can also be controlled by employing the proper reactor,[142–145] by using solvents,[56,57,132–136] and by adding inorganic salts,[139,140] or other additives[141] to the rearrangement system.

Rearrangement of Oximes by Phosphorus Pentachloride

A large number of oximes have been rearranged to amides with phosphorus pentachloride as a catalyst.[70]

The usual procedure is to dissolve the oxime in absolute ether and cool the solution in an ice bath. Excess phosphorus pentachloride is added to the cold solution, which is then allowed to warm to room temperature. If the reaction is vigorous, further cooling may be necessary. The mixture is allowed to stand at room temperature for several hours and is then poured over crushed ice. The ether can be evaporated by directing an air stream over it. If the product is a solid, it can. be removed by filtration and recrystallized. A liquid product can be isolated by solvent extraction. The extract should be dried and, after the solvent has been removed, the residue can be purified by distillation.

[335] Horning, Stromberg, and Lloyd, *J. Am. Chem. Soc.*, **74**, 5153 (1952).
[336] Potts (Henkel and Cie. G.m.b.H.), Brit. pat. 732,899 (1955) [*C.A.*, **50**, 5738 (1956)].

Rearrangement of Oximes by Concentrated Sulfuric Acid

Fifty grams of the oxime is added in small portions to 50 g. of well-stirred concentrated sulfuric acid, the temperature of the solution being held below 25° by external cooling. When all the oxime has dissolved, the solution is added dropwise to 25 g. of concentrated sulfuric acid at 120–130°. The temperature of the reaction mixture is held at 120–130° for an additional five to ten minutes and then brought down to below 36°. At this temperature or below, the pH of the reaction mixture is adjusted to 6 with 28% aqueous ammonia. The mixture is extracted several times with chloroform or another suitable solvent; the combined extracts are dried, and the solvent removed by distillation. The residue can be recrystallized or distilled.

This procedure is a slight modification of that described by Wiest[337] and is applicable to most oximes. The yields range from 50 to 90%.

EXPERIMENTAL PROCEDURES

Homodihydrocarbostyril (Rearrangement of 1-Tetralone Oxime by Polyphosphoric Acid).[335] Four grams of 1-tetralone oxime was heated with 120 g. of polyphosphoric acid for ten minutes at 120–130°. The solution was cooled, treated with 350 ml. of water, and extracted with chloroform. After the chloroform solution was washed, dried, and evaporated, there remained 3.64 g. (91%) of slightly discolored crystalline material, m.p. 135.5–138°. Recrystallization from ethanol provided colorless homodihydrocarbostyril, m.p. 142.5–143°. The aqueous solution remaining after the chloroform extraction was made alkaline with 25% aqueous potassium hydroxide and subjected to continuous ether extraction. The ether furnished 0.19 g. of a red oil, which was not characterized but which may have contained β-naphthylamine.

Phenanthridone (Rearrangement of Fluorenone Oxime by Polyphosphoric Acid).[335] A mixture of 2.00 g. of fluorenone oxime and 60 g. of polyphosphoric acid was heated with manual stirring to 175–180° and maintained at this temperature for a few minutes. The resulting solution was cooled and treated with 300 ml. of water. The product separated in crystalline form and was removed by filtration. After washing and drying, there was obtained 1.85 g. (93%) of phenanthridone, m.p. 286–289°.

δ-Valerolactam (Rearrangement of Cyclopentanone Oxime with Benzenesulfonyl Chloride and Sodium Hydroxide).[257] To a cold solution containing 26 g. of sodium hydroxide, 200 ml. of water, and 49 g. of

---

[337] Wiest (to Alien Property Custodian), U.S. pat. 2,351,381 (1944) [*C.A.*, **38,** 5225 (1944)].

cyclopentanone oxime was added 115 g. of benzenesulfonyl chloride. The mixture was allowed to stand for twelve hours in an ice bath and was then neutralized and extracted with chloroform. The solvent was removed by distillation, and the residue distilled to yield 47.6 g. (95%) of δ-valerolactam, b.p. 95°/10 mm.

ε-Caprolactam (Direct Preparation from Cyclohexanone Using Nitromethane as a Source of Hydroxylamine).[167] To 500 g. of well-stirred concentrated sulfuric acid heated to 125°, 305 g. of nitromethane was added dropwise with external cooling when necessary to hold the temperature of the acid at 125–130°. After an additional five minutes at 125–130°, 440 g. of cyclohexanone was added slowly to the mixture, which was again heated when necessary to hold the temperature at 120–125°. When the addition of the ketone was complete, the temperature of the mixture was held at 120–125° for five minutes. The reaction mixture was then cooled to below 36° and held at that temperature or below while it was neutralized with 28% aqueous ammonia. The mixture was filtered and the filtrate extracted several times with chloroform. The chloroform extract was dried and the solvent removed by distillation. The residue was distilled to yield 360 g. (79%) of ε-caprolactam, b.p. 138°/10 mm.

Acetanilide (Rearrangement of Acetophenone Oxime by Trifluoroacetic Acid).[82] A solution of 25 g. of acetophenone oxime in 60 g. of trifluoroacetic acid was slowly added to 38 g. of boiling trifluoroacetic acid. The reaction temperature increased from 72° to 108°. After digestion at this temperature for one-half hour, the excess acid was removed by distillation under reduced pressure, and the residue recrystallized from a methanol-water mixture to yield 22.8 g. (91%) of acetanilide.

Pivalanilide (Rearrangement of Pivalophenone Oxime by Hydrogen Chloride in Acetic Acid).[6] Into a solution of 1.0 g. of pivalophenone oxime in 15 ml. of acetic acid, hydrogen chloride was bubbled for fifteen minutes. The mixture was allowed to stand overnight. It was then heated to boiling for five minutes and poured over ice. The mixture was neutralized with dilute aqueous sodium hydroxide and extracted with ether. The extract was dried and the solvent removed to yield 0.94 g. (94%) of pivalanilide, m.p. 118–141°. After one recrystallization from heptane the pivalanilide melted at 117–124°.

Heptanamide (Rearrangement of Heptanaldoxime by Raney Nickel).[226, 227] The solid mass obtained by heating 5.0 g. of heptanaldoxime with 1 g. of Raney nickel at 100° for ninety minutes was triturated with ether to separate the catalyst from the product. The ether was evaporated to yield 5 g. (100%) of crystals melting at 93°. By treatment with activated charcoal and then by recrystallization from benzene, heptanamide was obtained as silky white platelets, m.p. 95°.

TABULAR SURVEY OF THE BECKMANN REARRANGEMENT

The data listed in the twelve tables that follow represent a compilation of most of the available publications concerning the Beckmann rearrangement from 1887 to 1957. The authors feel that the data are reasonably complete, but some publications were undoubtedly missed.

The tables are arranged in the order in which different classes of oximes were discussed in the text. Oxime ethers and esters are listed with the ketoximes from which they are derived. The compounds within a class are listed in order of increasing number of ketone carbon atoms. To find a compound in the tables all that is required is to know the number of carbon atoms in the parent ketone and to look up this number in the proper table. For instance, cyclohexanone oxime, cyclohexanone oxime methyl ether, and cyclohexanone oxime p-toluenesulfonate are all in Table IV in the six-carbon-atom group. The tables include the name of the oxime or starting material, the product(s) formed by rearrangement, the conditions and reagents employed (catalyst(s), solvent(s)), the percentage yield of product, and the pertinent reference(s) when this information was available.

TABLE I

ALIPHATIC KETOXIMES

No. of C Atoms	Starting Material	Products (% Yield)	Catalysts and Experimental Conditions	References
C_2	Potassium methyl nitrole	CH_3NCO, KNO_2		278
C_3	Acetoxime	Acetone and isopropylamine	Cu, H_2 (carrier gas)	66
		N-Methylacetamide	H_2SO_4, CH_3CO_2H	65
		Diphenyl N-methylacetamidoyl phosphate	Diphenylphosphochloridate, pyridine	338
	Acetoxime benzenesulfonate	N-Methylacetimino phenyl ether (100)	C_6H_5OH, $C_6H_5CH_3$	13
		N-Methylacetimino p-tolyl ether (90)	p-$CH_3C_6H_4OH$, $C_6H_5CH_3$	13
		N-Benzenesulfonyl-N'-methylacetamidine (35)	$C_6H_5SO_2NH_2$, pyridine	13
		N-Benzenesulfonyl-N,N'-dimethylacetamidine (43)	$C_6H_5SO_2NH_2$, methylamine	13
		N-2-Pyridyl-N'-methylacetamidine (84)	2-Aminopyridine	13
		N-2-Furfuryl-N'-methylacetamidine (69)	Furfurylamine	13
		N,N-3-Oxapentamethylene-N'-methylacetamidine (40) and 4-(1'-methyliminoethyl)morpholine	Morpholine	13
		N,N-Diphenyl-N'-methylacetamidine (82)	$(C_6H_5)_2NH$	13
		N-Methylacetamidine (21)	Aq. ammonia	13
		N-Cyclohexyl-N'-methylacetamidine (75)	Cyclohexylamine	13

THE BECKMANN REARRANGEMENT

	Oxime	Product (%)	Reagent	Reference
		N-Phenyl-N'-methyl-acetamidine (85)	Aniline	13
	bis-Acetoxime copper(I) chloride	1,5-Dimethyl-1,2,3,4-tetrazole	NaN_3, C_2H_5OH	296
		Methylamine	$CHCl_3$	106
		Unidentified product	$C_6H_5CH_3$	68
C_4	Methyl ethyl ketoxime	N-Ethylacetamide (81)	PCl_5, $(C_2H_5)_2O$	64
		Ethylamine (66) and methylamine (33)	PCl_5	340
	Methyl ethyl ketoxime benzenesulfonate	N-Cyclohexyl-N'-ethyl-acetamidine	Cyclohexylamine	13
		Tetrabenzylpyrophosphate* (38)	CH_3CN, $(C_2H_5)_3N$	338
	Propionylformic acid oxime	C_2H_5CN, CO_2, and H_2O	H_2SO_4	74
C_5	Methyl n-propyl ketoxime	N-n-Propylacetamide (84)	PCl_5, $(C_2H_5)_2O$	64
		N-n-Propylacetamide (88)	93% H_2SO_4	64
		N-n-Propylacetamide	HCl, $(CH_3CO)_2O$, CH_3CO_2H	100
		Methylamine and ethylamine	PCl_5	340
	Methyl isopropyl ketoxime	N-Isopropylacetamide (83)	PCl_5, $(C_2H_5)_2O$	64, 340
		N-Isopropylacetamide (88)	85% H_2SO_4	64
	Methyl cyclopropyl ketoxime	N-Cyclopropylacetamide (80)	PCl_5, $(C_2H_5)_2O$	64
		N-Methylcyclopropanecarboxamide (80)	$C_6H_5SO_2Cl$, $(C_2H_5)_2O$	341
	Diethyl ketoxime	N-Cyclopropylacetamide (35)	PCl_5, $(C_2H_5)_2O$	341
		N-Ethylpropionamide	H_2SO_4, CH_3CO_2H	65
		N-Ethylpropionamide (—, 97)	SO_3, SO_2; $ClSO_3H$, SO_3	342, 456
	Diethyl ketoxime benzenesulfonate	N,N-Diphenyl-N'-ethyl-propionamidine (80)	$(C_6H_5)_2NH$	13
		N-Cyclohexyl-N'-ethyl-propionamidine (78)	Cyclohexylamine	13

Note: References 338 to 593 are on pp. 152–156.

* The isolation of the amide was not reported.

TABLE I—Continued
ALIPHATIC KETOXIMES

No. of C Atoms	Starting Material	Products (% Yield)	Catalysts and Experimental Conditions	References
C_5 (continued)	α-Oximinovaleric acid	n-C_3H_7CN, CO_2, and H_2O	H_2SO_4	74
	β-Methyl-α-oximinobutyric acid	i-C_3H_7CN, CO_2, and H_2O	H_2SO_4	74
C_6	Levulinic acid oxime	N-Methylsuccinamic acid (50)	H_2SO_4	339
	Methyl n-butyl ketoxime	N-n-Butylacetamide (74)	PCl_5, $(C_2H_5)_2O$	64
	Pinacolone oxime	N-t-Butylacetamide	PCl_5, $(C_2H_5)_2O$	343
	Ethyl n-propyl ketoxime	N-n-Propylpropionamide (74)	PCl_5, $(C_2H_5)_2O$	64
		N-n-Propylpropionamide (92)	93% H_2SO_4	64
	Ethyl acetoacetate oxime sulfonate	Unidentified product	HCl (4N), CH_3CO_2H	60
	α-Oximinocaproic acid	n-C_4H_9CN, CO_2, and H_2O	H_2SO_4	74
	γ-Methyl-α-oximinovaleric acid	β-Methylbutyronitrile, CO_2, and H_2O	H_2SO_4	74
	α-Oximinoadipic acid	γ-Cyanobutyric acid	$(CH_3CO)_2O$	73
	Acetonyltrimethylammonium chloride oxime	$[(CH_3)_3NCH_2CONHCH_3]Cl^{\ominus}$	PCl_5; CH_3COCl, $(CH_3CO)_2O$; H_2SO_4; C_6H_5COCl	344
	Acetonyltrimethylammonium bromide oxime	$[(CH_3)_3NCH_2CONHCH_3]Br^{\ominus}$	PCl_5; H_2SO_4; CH_3COCl,† $(CH_3CO)_2O$	344
C_7	Methyl n-amyl ketoxime	N-n-Amylacetamide (76)	PCl_5, $(C_2H_5)_2O$	64
	Di-n-propyl ketoxime	N-n-Propyl-n-butyramide	H_2SO_4, CH_3CO_2H	65
	Diisopropyl ketoxime	Isobutyric acid and isopropylamine	CH_3COCl	346
	Dicyclopropyl ketoxime	N-Cyclopropylcyclopropanecarboxamide (65)	$C_6H_5SO_2Cl$, 62% dioxane	347

THE BECKMANN REARRANGEMENT

	Starting Material	Product	Reagent	Ref.
	δ-Methyl-α-oximinocaproic acid	γ-Methylvaleronitrile, CO_2, and H_2O	H_2SO_4	74
	α-Oximinopimelic acid	δ-Cyanovaleric acid	$(CH_3CO)_2O$	73
C_8	Methyl n-hexyl ketoxime	N-n-Hexylacetamide (73)	PCl_5, $(C_2H_5)_2O$	64, 340
	2-Oximino-3,4,4-trimethylpentane	N-(1,2,2-Trimethylpropyl)acetamide (36)	PCl_5, $(C_2H_5)_2O$	348
	2-Oximino-4,4-dimethylhexane	N-(2,2-Dimethylbutyl)acetamide (20)	PCl_5, $(C_2H_5)_2O$	348
		N-(2,3-Dimethylbutyl)acetamide (20)		348
	2-Methyl-2-hepten-6-one oxime	Dihydrocollidone	$C_6H_5SO_2Cl$	75
	α-Oximinocaprylic acid	n-$C_6H_{13}CN$, CO_2, and H_2O	$(CH_3CO)_2O$	73
	Di-n-butyl ketoxime	N-n-Butylvaleramide (40)	PCl_5, $(C_2H_5)_2O$	64
C_9	Ethyl cyclohexyl ketoxime	N-Cyclohexylpropionamide	PCl_5, $(C_2H_5)_2O$	349
	(+)-2-Oximino-3-ethylheptane	(−)-N-Acetyl-3-aminoheptane	PCl_5, $(C_2H_5)_2O$	11
	dl-2-Oximino-3-ethylheptane	dl-N-Acetyl-3-aminoheptane	PCl_5, $(C_2H_5)_2O$	11
	d + dl-2-Oximino-3-ethylheptane	d + dl-N-Acetyl-3-aminoheptane	PCl_5, $(C_2H_5)_2O$	11
	Phenylacetone oxime	N-Benzylacetamide (40)	BF_3, CH_3CO_2H	19
	Phenylacetone oxime sulfonate	2,5-Diphenyl-3,6-dimethylpiperazine	HCl (4N), CH_3CO_2H	60
C_{10}	(+)-Methyl α-phenylethyl ketoxime	(−)-N-α-Phenylethylacetamide	H_2SO_4, $(C_2H_5)_2O$	10
	α-Oximino-β-methylpelargonic acid	α-Methylcaprylonitrile, CO_2, and H_2O	H_2SO_4	74
	Ethyl α,α-diethyl-β-oximinobutyrate	$CH_3CONHC(C_2H_5)_2CO_2C_2H_5$‡	85% H_2SO_4	71

Note: References 338 to 593 are on pp. 152–156.

† Benzoyl chloride did not bring about rearrangement.
‡ Phosphorus pentachloride, sulfuric acid, and hydrochloric acid were not satisfactory catalysts.

TABLE I—Continued
ALIPHATIC KETOXIMES

No. of C Atoms	Starting Material	Products (% Yield)	Catalysts and Experimental Conditions	References
C_{10} (continued)	Benzalacetone oxime	Quinoline	P_2O_5, infusorial earth	76
	syn-Methyl styryl ketoxime	N-Styrylacetamide	PCl_5, $(C_2H_5)_2O$	69
	anti-Methyl styryl ketoxime	N-Methylcinnamamide	PCl_5, $(C_2H_5)_2O$	69
		3-Methyl-4-phenylisoxazole	H_2SO_4	69
	syn-Methyl 4-nitrostyryl ketoxime	N-4-Nitrostyrylacetamide	PCl_5, $(C_2H_5)_2O$	305
	α-Chlorobenzalacetone oxime	N-Phenylacetylacetamide	PCl_5, $(C_2H_5)_2O$	69
	α-Bromobenzalacetone oxime	N-Phenylacetylacetamide	PCl_5, $(C_2H_5)_2O$	69
	5-Keto-3,4,6-trimethyl-heptanoic acid oxime	Isobutyric acid § and isopropylamine	$p\text{-}CH_3C_6H_4SO_2Cl$, pyridine	350
C_{11}	Methyl n-nonyl ketoxime	N-Nonylacetamide	80% H_2SO_4	351
		$n\text{-}C_9H_{19}CONHCH_3$, $CH_3CONHC_9H_{19}\text{-}n$	H_2SO_4	352
	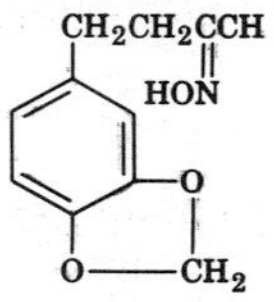 (structure shown)	β-(3-Piperonyl)propionic acid N-methyl amide (20), N-β-(3-piperonyl)acetamide (45), 1-methyl-6,7-methylene-dioxyisoquinoline (15)	PCl_5, C_6H_6	77
		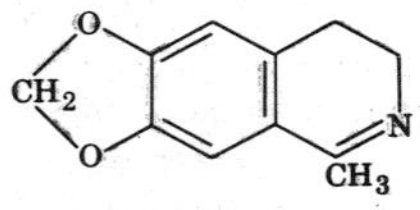 (4)	P_2O_5, $C_6H_5CH_3$	77

	Oxime	Reagent	Product(s)	Ref.
	syn-Methyl 4-methoxystyryl ketoxime	PCl_5, $(C_2H_5)_2O$	N-4-Methoxystyrylacetamide	305
	anti-Methyl 4-methoxystyryl ketoxime	PCl_5, $(C_2H_5)_2O$	N-Methyl-(4-methoxy)-cinnamamide	305
C_{12}	6-Cyclohexyl-6-oxocaproic acid oxime	PCl_5, $(C_2H_5)_2O$	N-Cyclohexyladipic acid monoamide	354
	Methyl β-(3,4-dimethoxyphenyl)ethyl ketoxime	P_2O_5, $C_6H_5CH_3$	1-Methyl-3,4-dihydro-6,7-dimethoxyisoquinoline	77
	p-Dimethylaminobenzalacetone oxime		Failed to react	355
C_{13}	Methyl undecyl ketoxime	H_2SO_4, CH_3CO_2H	Undecylamine and lauric acid	356
		PCl_5, $(C_2H_5)_2O$	N-Undecylacetamide	357
C_{14}	Ethyl α,α-di-n-butyl-β-oximinobutyrate	85% H_2SO_4	$CH_3CONH(C_4H_9-n)_2CO_2C_2H_5$	71
	α-Kessylketoxime ($C_{14}H_{23}NO_2$)	H_2SO_4	Isoxime (m.p. 160°), nitrile (m.p. 155°)	358
C_{15}	Dibenzyl ketoxime	Cu, H_2 (carrier gas), 200°	Phenylacetamide (11) Phenylacetic acid (12) Phenylacetonitrile (13) Dibenzyl ketone (64)	67
	Dibenzyl ketoxime benzenesulfonate	H_2O	N-Benzylphenylacetamide	106
	1,1-Diphenylacetone oxime p-toluenesulfonate	Pyridine	N-Acetylbenzhydrylamine (35)	61
C_{17}	Methyl n-pentadecyl ketoxime	H_2SO_4, CH_3CO_2H	n-Pentadecylamine and palmitic acid*	356
	2-Oximino-3,3-dibenzyl propane	$SOCl_2$	No reaction	231

Note: References 338 to 593 are on pp. 152–156.

* The isolation of the amide was not reported.

§ The amide was hydrolyzed to yield the product(s).

TABLE I—Continued
ALIPHATIC KETOXIMES

No. of C Atoms	Starting Material	Products (% Yield)	Catalysts and Experimental Conditions	References
C_{17} (continued)	Dibenzalacetone oxime	Unidentified product	H_2SO_4	360
		3-Phenyl-5-styrylisoxazoline	H_2SO_4	69
		N-Styrylcinnamamide	PCl_5, $(C_2H_5)_2O$	69
C_{18}	Ethyl pentadecyl ketoxime	Pentadecylamine and palmitic acid§	H_2SO_4, CH_3CO_2H	356
	3-Oximinostearic acid	N-Tetradecylsuccinamide	H_2SO_4	359
	10-Oximinostearic acid	n-Octylamine, 9-aminononanoic acid, sebacic acid, pelargonic acid§	H_2SO_4	359
C_{19}	n-Propyl pentadecyl ketoxime	Pentadecylamine and palmitic acid§	H_2SO_4, CH_3CO_2H	356
C_{20}	Ethyl n-heptadecyl ketoxime	n-Heptadecylamine and stearic acid§	H_2SO_4, CH_3CO_2H	356
	Ethyl α,α-dibenzyl-β-oximinobutyrate	3-Methyl-4,4-dibenzyl-5-isoxazolone (25–40)	85% H_2SO_4	71
C_{29}	10-Nonacosanone oxime	N-Nonyleicosanamide	H_2SO_4, CH_3CO_2H	361
		Mixture of amides	H_2SO_4, CH_3CO_2H	356
	β,β,β′,β′-Tetraphenyldiethyl ketoxime	N-β,β-Diphenylethyl-β′,β′-diphenylpropionamide	PCl_5, $(C_2H_5)_2O$	362
C_{31}	Palmitone oxime	N-n-Pentadecylpalmitamide	H_2SO_4, CH_3CO_2H	356
C_{50}	$CH_3C(=NOH)C_{47}H_{93}$⟨CO–O⟩	$HO_2CC_{47}H_{93}$⟨CO§–O⟩	PCl_5, $(C_2H_5)_2O$	363

Note: References 3338 to 593 are on pp. 152–156.

§ The amide was hydrolyzed to yield the product(s).

TABLE II
Aliphatic Aromatic Ketoximes

No. of C Atoms	Starting Material	Products (% Yield)	Catalysts and Experimental Conditions	References
C_8	Acetophenone oxime	Acetanilide (39)	CH_3COCl	18,100
		Acetanilide (41)	C_6H_5COCl	18
		Acetanilide (40)	$ClCH_2COCl$	18
		Acetanilide (98)	$C_6H_5SO_2Cl$	18
		Acetanilide (70–80)	$C_6H_5SO_2Cl, (C_2H_5)_2O;$ $PCl_5, (C_2H_5)_2O$	6
		Acetanilide	$PCl_5, (C_2H_5)_2O$	81
		Acetanilide (80), diphenylacetamidine (15–20)	$SOCl_2, (C_2H_5)_2O$	80
		Acetanilide	HCl; HBr; HI	102
		Acetanilide	$H_2SO_4(5M)$	1
		Acetanilide (65)	Anhydrous HF	83
		Acetanilide (87–98)	BF_3, CH_3CO_2H	19
		Acetanilide	$HCl, (CH_3CO)_2O,$ CH_3CO_2H	100
		Acetanilide (91, 53)	CF_3CO_2H	82, 409
		Acetanilide (33) and 4-chloroacetanilide (22)	$Cl_2 \cdot Br_2, SO_2$	364
		N-Ethylaniline (9) and α-phenylethylamine (30)	$LiAlH_4, (C_2H_5)_2O$	89
		N-Ethylaniline (10–15) and α-phenylethylamine	$LiAlH_4, (C_2H_5)_2O$	88
		C_6H_5CN and $C_6H_5CO_2H$	Cu, H_2	85
		$CH_3CO_2H, C_6H_5CN, NH_3, C_6H_5CO_2H,$ $C_6H_5COCH_3, C_6H_5NH_2,$ and $CH_3CONHC_6H_5$	Japanese acid earth (Al_2O_3)	86

Note: References 338 to 593 are on pp. 152–156.

TABLE II—Continued
Aliphatic Aromatic Ketoximes

No. of C Atoms	Starting Material	Products (% Yield)	Catalysts and Experimental Conditions	References
C$_8$ (continued)	Acetophenone oxime (continued)	Acetanilide	H$_3$BO$_3$-Al$_2$O$_3$	148
		2-Methyl-3-phenyl-5-ethyl-6-methyl-4-pyrimidone (65)*	PCl$_5$, (C$_2$H$_5$)$_2$O	365
		No reaction	(CH$_3$CO)$_2$O	80, 100
	Acetophenone oxime hydrochloride	Acetanilide and diphenylacetamidine		102
	Acetophenone oxime hydrobromide	Acetanilide		102
	Acetophenone oxime cuprous chloride complex	Unidentified product	C$_6$H$_5$CH$_3$	68
	Acetophenone oxime sulfonate	Acetanilide (77)	HCl, dioxane	60
	Potassium acetophenone oxime sulfonate	1-Phenyl-5-methyltetrazole (72)	Alkali or acid and NaN$_3$	14
	Acetophenone oxime methanesulfonate	N,N′-Diphenylacetamidine (24)	C$_6$H$_5$NH$_2$	13
	Acetophenone oxime benzenesulfonate	Acetanilide (27)		13
		Tetrabenzyl pyrophosphate (16)†	Dibenzyl hydrogen phosphate, CH$_3$CN, (C$_2$H$_5$)$_3$N	338
	Acetophenone oxime p-toluenesulfonate	N,N′-Diphenylacetamidine (95)	C$_6$H$_5$NH$_2$	13
	Acetophenone oxime picryl ether	N-Picrylacetanilide‡		43
	Methyl 4-fluorophenyl ketoxime	N-Ethyl-p-fluoroaniline (8) and α-4-fluorophenylethylamine (35)	LiAlH$_4$, (C$_2$H$_5$)$_2$O	89
	Methyl 4-fluorophenyl ketoxime picryl ether	N-Picryl-4-fluoroacetanilide‡		43
	Methyl 2-chlorophenyl ketoxime	N-Acetyl-2-chloroaniline (90)	H$_2$SO$_4$	40
		Methyl 2-chlorophenyl ketone	18% HCl	91

THE BECKMANN REARRANGEMENT

Oxime	Product	Reagent	Ref.
Methyl 2-chlorophenyl ketoxime picryl ether	N-Picryl-2-chloroacetanilide‡		43
Methyl 3-chlorophenyl ketoxime picryl ether	N-Picryl-3-chloroacetanilide‡		43
Methyl 4-chlorophenyl ketoxime	N-Ethyl-p-chloroaniline (7) and α-4-chlorophenylethylamine (50)	$LiAlH_4$, $(C_2H_5)_2O$	89
Methyl 4-chlorophenyl ketoxime picryl ether	N-Picryl-4-chloroacetanilide‡		43
Methyl 4-bromophenyl ketoxime	N-Ethyl-p-bromoaniline (5) and α-4-bromophenylethylamine (35)	$LiAlH_4$, $(C_2H_5)_2O$	89
Methyl 4-bromophenyl ketoxime benzenesulfonate	N,N'-bis(4-Bromophenyl)acetamidine (96)	$p\text{-}BrC_6H_4NH_2$	13
Methyl 2-iodophenyl ketoxime	Methyl 2-iodophenyl ketone	18% HCl	91
Methyl 4-iodophenyl ketoxime	N-Ethyl-p-iodoaniline (7) and α-4-iodophenylethylamine (14)	$LiAlH_4$, $(C_2H_5)_2O$	89
Methyl 4-iodophenyl ketoxime picryl ether	N-Picryl-4-iodoacetanilide‡		43
Methyl 2-nitrophenyl ketoxime	N-Acetyl-2-nitroaniline (86) Methyl 2-nitrophenyl ketone	H_2SO_4 18% HCl	40 91
Methyl 2-nitrophenyl ketoxime picryl ether	N-Picryl-2-nitroacetanilide‡		43
Methyl 4-nitrophenyl ketoxime picryl ether	N-Picryl-4-nitroacetanilide‡		43
Methyl 2-hydroxyphenyl ketoxime	Methyl 2-hydroxyphenyl ketone	18% HCl	91
Methyl 2-hydroxy-5-nitrophenyl ketoxime acetate	3-Methyl-5-nitrobenzisoxazole and 2-hydroxy-5-nitroacetanilide	Not specified	245

Note: References 338 to 593 are on pp. 152–156.

* The amide was not isolated. The intermediate chlorimide was treated with an α-alkyl-β-aminocrotonate ester to yield the 4-pyrimidone.

† Isolation of the amide was not reported.

‡ The picryl ethers were rearranged in 85–90% yield by heating in ethylene dichloride or another chlorinated hydrocarbon.

TABLE II—Continued
Aliphatic Aromatic Ketoximes

No. of C Atoms	Starting Material	Products (% Yield)	Catalysts and Experimental Conditions	References
C_8 (continued)	Methyl 2-aminophenyl ketoxime	Unidentified product	18% HCl	91
		N-Acetyl-o-phenylenediamine and $C_{10}H_{10}NO_2$	P_2O_5; $ZnCl_2$; HCl, $(CH_3CO)_2O$, CH_3CO_2H	367
	Methyl 2-bromo-5-nitrophenyl ketoxime (syn-anti mixture)	2-Bromo-5-nitroaniline (50)	H_2SO_4	368
	syn-Methyl 2-bromo-5-nitrophenyl ketoxime	2-Bromo-5-nitroacetanilide (77)	PCl_5, $(C_2H_5)_2O$	368
		2-Bromo-5-nitroacetanilide (58)	PCl_5, $(C_2H_5)_2O$	368
		2-Bromo-5-nitroaniline	H_2SO_4	368
	syn-Chloromethyl phenyl ketoxime	Chloroacetanilide	PCl_5, $(C_2H_5)_2O$	369
	Chloromethyl phenyl ketoxime picryl ether	N-Picrylchloroacetanilide (100)‡		48
	syn-Bromomethyl phenyl ketoxime	Bromoacetanilide	PCl_5, $(C_2H_5)_2O$	369
	Chloromethyl 4-chlorophenyl ketoxime	N-Chloroacetyl-4-chloroaniline	H_2SO_4	370
	Chloromethyl 4-bromophenyl ketoxime	N-Chloroacetyl-4-bromoaniline	H_2SO_4	370
	anti-Bromomethyl 3-nitrophenyl ketoxime	N-Bromoacetyl-3-nitroaniline	PCl_5, $(C_2H_5)_2O$	369
	Bromomethyl 4-chlorophenyl ketoxime	N-Bromoacetyl-4-chloroaniline	H_2SO_4	370
	Dibromomethyl 4-bromophenyl ketoxime	N-Dibromoacetyl-4-bromoaniline	H_2SO_4	370
	Benzoylformic acid oxime (syn or anti)	Benzonitrile	$C_6H_5SO_2Cl$, pyridine	95
	Benzoyl cyanide oxime	N-Phenyloxalamide	PCl_5, $(C_2H_5)_2O$	92

THE BECKMANN REARRANGEMENT

Ketoxime	Product (% yield)	Reagent	Ref.
o-Chlorobenzoyl cyanide oxime	No reaction	PCl_5, $(C_2H_5)_2O$	92
p-Chlorobenzoyl cyanide oxime	No reaction	PCl_5, $(C_2H_5)_2O$	92
C₉			
Ethyl phenyl ketoxime	Propionanilide (65–80)	PCl_5, $(C_2H_5)_2O$; $C_6H_5SO_2Cl$	6
	Propionanilide (85) and N,N'-diphenylpropionamidine (15)	$SOCl_2$, $(C_2H_5)_2O$	80
Ethyl phenyl ketoxime picryl ether	N-Picryl-n-propionanilide (96)‡		48
Methyl o-anisyl ketoxime	Sulfonation products	H_2SO_4	40
	Methyl o-anisyl ketone and anisidine	18% HCl	91
Methyl o-anisyl ketoxime picryl ether	N-Picryl-2-methoxyacetanilide‡		43
Methyl m-anisyl ketoxime	Methyl m-anisyl ketone	18% HCl	91
Methyl m-anisyl ketoxime picryl ether	N-Picryl-3-methoxyacetanilide‡		43
Methyl p-anisyl ketoxime	p-Anisidine (75–85)	PCl_5, $(C_2H_5)_2O$	84
	N-Ethyl-p-anisidine (59) and α-anisylethylamine (4)	$LiAlH_4$, $(C_2H_5)_2O$	89
	Acet-p-aniside (99)	Polyphosphoric acid	123
Methyl p-anisyl ketoxime picryl ether	N-Picryl-4-methoxyacetanilide‡		43
Methyl o-tolyl ketoxime	Methyl o-tolyl ketone and o-toluidine	18% HCl	91
	N-Acetyl-o-toluidine (100)	H_2SO_4	40
Methyl o-tolyl ketoxime picryl ether	N-Picryl-2-methylacetanilide‡		43
Methyl m-tolyl ketoxime	Methyl m-tolyl ketone	18% HCl	91
Methyl m-tolyl ketoxime picryl ether	N-Picryl-3-methylacetanilide‡		43
Methyl p-tolyl ketoxime	N-Ethyltoluidine (30) and α-tolylethylamine (17)	$LiAlH_4$, $(C_2H_5)_2O$	89
	N-Acetyl-p-toluidine (80) and N,N'-di-p-tolylacetamidine (20)	$SOCl_2$, $(C_2H_5)_2O$	80

Note: References 338 to 593 are on pp. 152–156.

‡ The picryl ethers were rearranged in 85–90% yield by heating in ethylene dichloride or another chlorinated hydrocarbon.

TABLE II—Continued
ALIPHATIC AROMATIC KETOXIMES

No. of C Atoms	Starting Material	Products (% Yield)	Catalysts and Experimental Conditions	References
C₉ (continued)	Methyl p-tolyl ketoxime (continued)	2-Methyl-3-p-tolyl-5-ethyl-6-methyl-4-pyrimidone (65)*§	PCl₅, (C₂H₅)₂O	81, 365
	Methyl p-tolyl ketoxime picryl ether	N-Picryl-4-methylacetanilide‡		43
	Methyl 2-methyl-4-hydroxyphenyl ketoxime	Methyl 2-methyl-4-hydroxyphenyl ketone and 4-hydroxy-6-methylaniline	18% HCl	91
	Methyl 2-hydroxy-3-methylphenyl ketoxime	Methyl 2-hydroxy-3-methylphenyl ketone	18% HCl	91
	Methyl 3-methyl-4-hydroxyphenyl ketoxime	Methyl 3-methyl-4-hydroxyphenyl ketone	18% HCl	91
	Methyl 2-hydroxy-4-methylphenyl ketoxime	Methyl 2-hydroxy-4-methylphenyl ketone	18% HCl	91
	Methyl 2-hydroxy-5-methylphenyl ketoxime	Methyl 2-hydroxy-5-methylphenyl ketone 2,5-Dimethylbenzoxazole	18% HCl	91
			P₂O₅; P₂O₅, (C₂H₅)₂O; KHSO₄; PCl₅, (C₂H₅)₂O	8
	Methyl 2-hydroxy-5-methylphenyl ketoxime hydrochloride	Methyl 2-hydroxy-5-methylphenyl ketone, 2-hydroxy-5-methylbenzoic acid, 2-hydroxy-5-methylbenzanilide, 2-hydroxy-5-methylaniline, and, 2,5-dimethylbenzoxazole	H₂O	8
C₁₀	n-Propyl phenyl ketoxime	N,N′-Diphenylbutyramidine (80) and butyranilide (20)	SOCl₂, (C₂H₅)₂O	80, 372

THE BECKMANN REARRANGEMENT

Oxime	Product	Reagent	Ref.
n-Propyl phenyl ketoxime picryl ether	2-n-Propyl-3-phenyl-5-ethyl-6-methyl-4-pyrimidone (72)§	PCl_5, $(C_2H_5)_2O$	365
	N-Picryl-n-butyranilide (88)‡		48
syn-Isopropyl phenyl ketoxime	Isobutyranilide	$C_6H_5SO_2Cl$, pyridine	373
syn-Isopropyl phenyl ketoxime picryl ether	N-Picrylisobutyranilide (81)‡		48
anti-Isopropyl phenyl ketoxime	N-Isopropylbenzamide (31)	$C_6H_5SO_2Cl$, pyridine	373
anti-Isopropyl phenyl ketoxime picryl ether	N-Picryl-N-isopropyl benzamide (84)‡		48
Ethyl 2-fluoro-5-methylphenyl ketoxime	2-Fluoro-5-methylaniline	PCl_5, $(C_2H_5)_2O$	374
Ethyl 4-fluoro-6-methylphenyl ketoxime	4-Fluoro-6-methylaniline	PCl_5, $(C_2H_5)_2O$	374
Methyl p-phenetyl ketoxime	p-Phenetidine (80)	PCl_5, $(C_2H_5)_2O$	84
Methyl 2,3-dimethylphenyl ketoxime	Methyl 2,3-dimethylphenyl ketone	18% HCl	91
Methyl 2,4-dimethylphenyl ketoxime	2,4-Dimethylacetanilide	PCl_5	375
	Methyl 2,4-dimethylphenylketone and 2,4-dimethylaniline	18% HCl	91
Methyl 2,6-dimethylphenyl ketoxime	Methyl 2,6-dimethylphenyl ketone	18% HCl	91
Methyl 2-methoxy-3-methylphenyl ketoxime	Methyl 2-methoxy-3-methylphenyl ketone and 2-methoxy-3-methylaniline	18% HCl	91
Methyl 2-methoxy-4-methylphenyl ketoxime	2-Methoxy-4-methylaniline and methyl 2-methoxy-4-methylphenyl ketone	18% HCl	91

Note: References 338 to 593 are on pp. 152–156.

* The amide was not isolated. The intermediate chlorimide was treated with an α-alkyl-β-aminocrotonate ester to yield the 4-pyrimidone.
‡ The picryl ethers were rearranged in 85–90% yield by heating in ethylene dichloride or another chlorinated hydrocarbon.
§ The 5-methyl pyrimidone can be made in the same fashion using the proper crotonate ester.

TABLE II—Continued

ALIPHATIC AROMATIC KETOXIMES

No. of C Atoms	Starting Material	Products (% Yield)	Catalysts and Experimental Conditions	References
C_{10} (continued)	Methyl 2-methoxy-5-methylphenyl ketoxime	Methyl 2-methoxy-5-methylphenyl ketone and 2-methoxy-5-methylaniline	18% HCl	91
	Methyl 2-methyl-4-methoxyphenyl ketoxime	Methyl 2-methyl-4-methoxyphenyl ketone and 2-methyl-4-methoxyaniline	18% HCl	91
	Methyl 2-hydroxy-3,5-dimethylphenyl ketoxime	Methyl 2-hydroxy-3,5-dimethylphenyl ketone	18% HCl	91
	syn-Methyl 2-hydroxy-4,6-dimethylphenyl ketoxime	2,4,6-Trimethylbenzoxazole (100)	HCl, CH_3CO_2H; HCO_2H, H_2O	8
		2,4,6-Trimethylbenzoxazole	PCl_5, $(C_2H_5)_2O$; heat; $KHSO_4$	8
	syn-Methyl 2-hydroxy-4,6-dimethylphenyl ketoxime hydrochloride	No reaction	HCl, $(CH_3CO)_2O$, CH_3CO_2H; HCO_2H, H_2O	8
	anti-Methyl 2-hydroxy-4,6-dimethylphenyl ketoxime	2,4,6-Trimethylbenzoxazole		
	anti-Methyl 2-hydroxy-4,6-dimethylphenyl ketoxime hydrochloride	2,4,6-Trimethylbenzoxazole	PCl_5, $(C_2H_5)_2O$; $KHSO_4$	8
C_{11}	syn-Isobutyl phenyl ketoxime	N-Methylisovaleranilide (65–72)	PCl_5, $(C_2H_5)_2O$; $C_6H_5SO_2Cl$, $(C_2H_5)_2O$	6
		N-Isobutylbenzamide (70)	HCl, CH_3CO_2H	6
	anti-Isobutyl phenyl ketoxime	N-Isobutylbenzamide (28–72)	PCl_5, $(C_2H_5)_2O$; $C_6H_5SO_2Cl$, $(C_2H_5)_2O$	6
		N-Isobutylbenzamide (80)	HCl, CH_3CO_2H	6

n-Propyl 2-hydroxy-5-methylphenyl ketoxime	n-Propyl 2-hydroxy-5-methylphenyl ketone	18% HCl	91
Methyl 2-methoxy-3,5-dimethylphenyl ketoxime	Methyl 2-methoxy-3,5-dimethylphenyl ketone and 2-methoxy-3,5-dimethylaniline	18% HCl	91
Methyl 2-methoxy-4,6-dimethylphenyl ketoxime	Methyl 2-methoxy-4,6-dimethylphenyl ketone and 2-methoxy-4,6-dimethylaniline	18% HCl	91
Ethyl 2,4-dimethylphenyl ketoxime	Ethyl 2,4-dimethylphenylketone and 2,4-dimethylaniline	18% HCl	91
Methyl mesityl ketone	Mesidine and acetic acid	$NH_2OH \cdot HCl$	50
Methyl mesityl ketoxime	Acetomesidide (94)	H_2SO_4	52
	Acetomesidide (86)	BF_3	264
5-Acetylindane oxime	N-Acetyl-5-aminoindane	PCl_5, $(C_2H_5)_2O$	376
	5-Aminoindane‖	PCl_5, $(C_2H_5)_2O$	377
4-Nitro-5-acetylindane oxime	4-Nitro-5-(N-acetylamino)indane	PCl_5, $(C_2H_5)_2O$	378
Methyl 4-carbethoxyphenyl ketoxime picryl ether	N-Picryl-4-carbethoxyacetanilide‡		43
C₁₂			
Methyl 2,5-diethylphenyl ketoxime	Methyl 2,5-diethylphenyl ketone and 2,5-diethylaniline	18% HCl	91
n-Propyl 2-methoxy-5-methylphenyl ketoxime	n-Propyl 2-methoxy-5-methylphenyl ketone and 2-methoxy-5-methyl aniline	18% HCl	91
Methyl 2-ethoxy-3,4-dimethylphenyl ketoxime	2-Ethoxy-4,5-dimethylaniline	$SOCl_2$, $CHCl_3$	377a
Methyl 1-naphthyl ketoxime	Acetic acid (99) and 1-naphthoic acid (1)‖	PCl_5, C_6H_6	79
Methyl 2-naphthyl ketoxime	Acetic acid (99) and 2-naphthoic acid (1)‖	PCl_5, C_6C_6	79

Note: References 338 to 593 are on pp. 152–156.

‡ The picryl ethers were rearranged in 85–90% yield by heating in ethylene dichloride or another chlorinated hydrocarbon.
‖ The amide was hydrolyzed to the product(s) without prior isolation.

TABLE II—Continued

ALIPHATIC AROMATIC KETOXIMES

No. of C Atoms	Starting Material	Products (% Yield)	Catalysts and Experimental Conditions	References
C_{12} (continued)	Methyl 2-naphthyl ketoxime sulfonate	N-Acetyl-2-naphthylamine (87)	HCl (4N), dioxane	60
	β-Naphthacyl bromide oxime	β-Naphthylamine (61)	None given	590
	β-Naphthacyl iodide oxime	β-Naphthylamine (71)	None given	590
	2-Acetyl-5,6,7,8-tetrahydronaphthalene oxime	N-Acetyl-2-amino-5,6,7,8-tetrahydronaphthalene		381
	1-Acetylazulene oxime	1-Acetamidoazulene (16)	PCl_5, $(C_2H_5)_2O$	382
	6-p-Anisyl-5-ketovaleric acid oxime	N-(4-Methoxyphenyl)glutaramic acid	BF_3, $(C_2H_5)_2O$	384
C_{13}	6-p-Phenetyl-5-ketovaleric acid oxime	N-(4-Ethoxyphenyl)glutaramic acid	BF_3, $(C_2H_5)_2O$	384
	Cyclohexyl phenyl ketoxime	N-Cyclohexylbenzamide	PCl_5, $(C_2H_5)_2O$	349, 383
	sym-Ethyl 3,5-dimethoxy-4-ethyl-phenyl ketoxime	3,5-Dimethoxy-4-ethylaniline (68)	PCl_5, $(C_2H_5)_2O$	379
	1-Methoxy-4-acetylnaphthalene oxime	1-(N-Acetylamino)-4-methoxy-naphthalene (55–60)	PCl_5, $(C_2H_5)_2O$	84
	1-Methoxy-2-acetylnaphthalene oxime picryl ether	N-Picryl-N-(1-methoxy-2-naphthyl)-acetamide‡		48
	3-Methoxy-2-acetylnaphthalene oxime picryl ether	N-Picryl-N-(3-methoxy-2-naphthyl)-acetamide‡		48
C_{14}	Benzyl phenyl ketoxime	Phenylacetanilide	$C_6H_5SO_2Cl$, pyridine	95
		Phenylacetanilide (60)	PCl_5, $(C_2H_5)_2O$	385, 203
		Phenylacetanilide (80–85) and N,N'-diphenylphenylacetamidine (15–20)	$SOCl_2$, $(C_2H_5)_2O$	80
	Benzyl phenyl ketoxime picryl ether	N-Picryl-N-phenylacetanilide (88)‡		48
	sym-Benzyl 2-chlorophenyl ketoxime	2'-Chloro-2-phenylacetanilide (65)	PCl_5, $(C_2H_5)_2O$	371
	Benzyl 4-chlorophenyl ketoxime	$C_6H_5CH_2CONHC_6H_4Cl$-4	PCl_5, $(C_2H_5)_2O$	371

THE BECKMANN REARRANGEMENT

Oxime	Product	Reagent	Ref.
syn-2-Chlorobenzyl phenyl ketoxime	2-ClC₆H₄CH₂CONHC₆H₅	PCl₅, (C₂H₅)₂O	371
syn-4-Chlorobenzyl phenyl ketoxime	4-ClC₆H₄CH₂CONHC₆H₅	PCl₅, (C₂H₅)₂O	371
2-Chlorobenzyl 2-chlorophenyl ketoxime	2-ClC₆H₄CH₂CONHC₆H₄Cl-2 and 2-ClC₆H₄CONHCH₂C₆H₄Cl-2	PCl₅, (C₂H₅)₂O	371
syn-4-Chlorobenzyl 4-chlorophenyl ketoxime	4-ClC₆H₄CH₂CONHC₆H₄Cl-4 (55–80)	PCl₅, (C₂H₅)₂O	371
syn-2-Chlorobenzyl 4-chlorophenyl ketoxime	2-ClC₆H₄CH₂CONHC₆H₄Cl-4 (55–80)	PCl₅, (C₂H₅)₂O	371
syn-4-Chlorobenzyl 2-chlorophenyl ketoxime	4-ClC₆H₄CH₂CONHC₆H₄Cl-2 (65)	PCl₅, (C₂H₅)₂O	371
Methyl 2,5-di-n-propylphenyl ketoxime	Methyl 2,5-di-n-propylphenyl ketone and 2,5-di-n-propylaniline	18% HCl	91
Phenyl anilinomethyl ketoxime	Anilinoacetanilide	PCl₅, (C₂H₅)₂O	93
Methyl p-xenyl ketoxime	Acetic acid (99) and 4-carboxybiphenyl (1)‖	PCl₅, C₆H₆	79
Methyl p-xenyl ketoxime picryl ether	N-Picryl-4-phenylacetanilide‡		43
Cyclopropyl β-naphthyl ketoxime	β-Naphthoic acid‖	Polyphosphoric acid	387
5-Acetyl-6-nitroacenaphthene oxime	5-Acetamido-6-nitroacenaphthene (95)	HCl, CH₃CO₂H, (CH₃CO)₂O	366
4-Acetyl-s-hydrindacene oxime	4-Acetamido-s-hydrindacene (81)	HCl, (CH₃CO)₂O, CH₃CO₂H	388
Benzoylformanilide oxime methyl ether	Oxalic acid dianilide	PCl₅	202
C₁₅ syn-Benzyl 4-methoxyphenyl ketoxime	C₆H₅CH₂CONHC₆H₄OCH₃-4 (60)	PCl₅, (C₂H₅)₂O	385
	N-Benzyl-p-methoxybenzamide	C₆H₅SO₂Cl, aq. NaOH	390
syn-4-Methoxybenzyl phenyl ketoxime	2-(p-Methoxyphenyl)acetanilide	PCl₅, (C₂H₅)₂O	390
anti-4-Methoxybenzyl phenyl ketoxime	N-(4-Methoxybenzyl)benzamide	C₆H₅SO₂Cl, aq. NaOH	390

Note: References 338 to 593 are on pp. 152–156.

‡ The picryl ethers were rearranged in 85–90% yield by heating in ethylene dichloride or another chlorinated hydrocarbon.
‖ The amide was hydrolyzed to the product(s) without prior isolation.

TABLE II—Continued

ALIPHATIC AROMATIC KETOXIMES

No. of C Atoms	Starting Material	Products (% Yield)	Catalysts and Experimental Conditions	References
C_{15} (continued)	syn-2-Chlorobenzyl 4-methoxyphenyl ketoxime	2-$ClC_6H_4CH_2CONHC_6H_4OCH_3$-4 (60)	PCl_5, $(C_2H_5)_2O$	385
	syn-4-Chlorobenzyl 4-methoxyphenyl ketoxime	2-$ClC_6H_4CH_2CONHC_6H_4OCH_3$-4 4-$ClC_6H_4CH_2CONHC_6H_4OCH_3$-4 (60)	$C_6H_5SO_2Cl$, aq. NaOH PCl_5, $(C_2H_5)_2O$	390 385
	syn-Benzyl 3,4-methylenedioxyphenyl ketoxime	N-Piperonylphenylacetamide	$C_6H_5SO_2Cl$, aq. NaOH	390
	syn-2-Chlorobenzyl 3,4-methylenedioxyphenyl ketoxime	2-$ClC_6H_4CH_2CONHC_6H_3(O_2CH_2)$-3,4	PCl_5, $(C_2H_5)_2O$; $C_6H_5SO_2Cl$, aq. NaOH	390
	Phenyl p-anisidinomethyl ketoxime	p-Anisidinoacetanilide	PCl_5, $(C_2H_5)_2O$	93, 94
	Phenyl p-toluidinomethyl ketoxime	C_6H_5C—CH_2 || $O \leftarrow N$—$NC_6H_5CH_3$-p	PCl_5, $(C_2H_5)_2O$	93, 94
	syn-Methyl 1-(2-hydroxy-3-carbethoxy)naphthyl ketoxime	N-Acetyl-2-hydroxy-3-carbethoxy-1-naphthylamine (30)	HCO_2H	235
	syn-Methyl 1-(2-hydroxy-3-carbethoxy)naphthyl ketoxime	N-Acetyl-2-hydroxy-3-carbethoxy naphthylamine (100)	PCl_5, dioxane	235
	anti-Methyl 1-(2-hydroxy-3-carbethoxy)naphthyl ketoxime	N-Methyl-2-hydroxy-3-carbethoxy-1-naphthamide (40)	PCl_5, dioxane	235
	Phenyl styryl ketoxime	Cinnamanilide	PCl_5, $(C_2H_5)_2O$	391
	anti-Phenyl styryl ketoxime	Cinnamanilide	PCl_5, $(C_2H_5)_2O$	392
	syn-Styryl 2-chlorophenyl ketoxime	N-2-Chlorophenylcinnamamide	PCl_5, $(C_2H_5)_2O$	70
	syn-Styryl 2-bromophenyl ketoxime	N-2-Bromophenylcinnamamide	PCl_5, $(C_2H_5)_2O$	70
	Styryl 4-bromophenyl ketoxime	N-4-Bromophenylcinnamamide and N-styryl-4-bromobenzamide	PCl_5, $(C_2H_5)_2O$	70

Ketoxime	Product	Reagent	Ref.
syn-Styryl 4-bromophenyl ketoxime	N-4-Bromophenylcinnamamide (100)	PCl$_5$, (C$_2$H$_5$)$_2$O	70
anti-Styryl 4-bromophenyl ketoxime	3-p-Bromophenyl-5-phenylisoxazoline	H$_2$SO$_4$	70
α-Bromostyryl phenyl ketoxime	3-p-Bromophenyl-5-phenylisoxazoline	H$_2$SO$_4$	70
α-Bromostyryl 4-bromophenyl ketoxime	Benzoic acid	PCl$_5$, (C$_2$H$_5$)$_2$O	70, 391
	p-Bromobenzoic acid ‖	PCl$_5$, (C$_2$H$_5$)$_2$O	70
α,β-Dibromo-β-phenylethyl phenyl ketoxime	No reaction	H$_2$SO$_4$	70
	α,β-Dibromo-β-phenylpropionanilide	PCl$_5$, (C$_2$H$_5$)$_2$O	87, 391
syn-α,β-Dibromo-β-phenylethyl-4-bromophenyl ketoxime	N-4-Bromophenyl-α,β-dibromo-β-phenylpropionamide	PCl$_5$, (C$_2$H$_5$)$_2$O	70
anti-α,β-Dibromo-β-phenylethyl-4-bromophenyl ketoxime	N-Styryl-p-bromobenzamide	PCl$_5$, (C$_2$H$_5$)$_2$O	70
3,4-(CH$_3$O)$_2$C$_6$H$_3$CC$_6$H$_5$ ‖ NOH	Two amides		379
C$_{16}$ 3,5-Diphenylisoxazoline	4-Phenyl-3,4-dihydrocarbostyril	HI	393
β-Phenylbutyrophenone oxime	β-Phenyl-n-butyranilide (43)	SOCl$_2$, (C$_2$H$_5$)$_2$O	394
2-Chlorobenzyl 3,4-dimethoxyphenyl ketoxime	2-ClC$_6$H$_4$CH$_2$CONHC$_6$H$_3$(OCH$_3$)$_2$-3,4	PCl$_5$, (C$_2$H$_5$)$_2$O	390
Benzyl 4-dimethylaminophenyl ketoxime	C$_6$H$_5$CH$_2$CONHC$_6$H$_4$N(CH$_3$)$_2$-4	C$_6$H$_5$SO$_2$Cl, aq. NaOH	390
2-Chlorobenzyl 4-dimethylaminophenyl ketoxime	2-ClC$_6$H$_4$CH$_2$CONHC$_6$H$_4$N(CH$_3$)$_2$-4	PCl$_5$, (C$_2$H$_5$)$_2$O	390
C$_6$H$_5$SCH$_2$CC$_6$H$_3$(OCH$_3$)$_2$-3,4 ‖ NOH	Two amides		395

Note: References 338 to 393 are on pp. 152–156.

‖ The amide was hydrolyzed to the product(s) without prior isolation.

TABLE II—Continued
Aliphatic Aromatic Ketoximes

No. of C Atoms	Starting Material	Products (% Yield)	Catalysts and Experimental Conditions	References
C_{16} (continued)	Methyl 4′-ethyl-p-xenyl ketoxime	4′-Ethyl-p-xenylacetamide	PCl_5, $(C_2H_5)_2O$	396
	Methyl 1-anthryl ketoxime	1-Aminoanthracene (20) and 1-carboxyanthracene‖	PCl_5, C_6H_6	79
	Methyl 1-phenanthryl ketoxime	N-1-Phenanthrylacetamide (71) and N-methyl-1-phenanthramide	PCl_5, C_6H_6	96
	Methyl 2-phenanthryl ketoxime	N-2-Phenanthrylacetamide (81) and N-methyl-2-phenanthramide (1)	PCl_5, C_6H_6	96
	Methyl 3-phenanthryl ketoxime	N-3-Phenanthrylacetamide (87) and N-methyl-3-phenanthramide (2)	PCl_5, C_6H_6	96
	Methyl 9-phenanthryl ketoxime	N-9-Phenanthrylacetamide (50) and N-methyl-9-phenanthramide (6)	PCl_5, C_6H_6	96
	Styryl o-anisyl ketoxime	N-o-Anisylcinnamide	PCl_5, $(C_2H_5)_2O$	87
		Sulfonation products	H_2SO_4	87
	Styryl m-anisyl ketoxime	N-m-Anisylcinnamide	PCl_5, $(C_2H_5)_2O$	87
	Styryl p-anisyl ketoxime	N-p-Anisylcinnamide	PCl_5, $(C_2H_5)_2O$	87, 391
	o-Methoxystyryl phenyl ketoxime	o-Methoxycinnamanilide	PCl_5, $(C_2H_5)_2O$	87
		Sulfonation Products	H_2SO_4	87
	m-Methoxystyryl phenyl ketoxime	m-Methoxycinnamanilide	PCl_5, $(C_2H_5)_2O$	87
		Sulfonation products	H_2SO_4	87
	β-Methylstyryl phenyl ketoxime	β-Methylcinnamanilide	PCl_5, $(C_2H_5)_2O$	397
	Phenyl α-phenyl-α-methylethyl ketoxime	2-Phenylpropylene and benzonitrile	$SOCl_2$, C_6H_6	90
		$C_6H_5C(CH_3)_2CONHC_6H_5$ (80)	HCl, CH_3CO_2H	90
	α-Bromostyryl p-anisyl ketoxime	β-Bromocinnamic acid aniside	PCl_5, $(C_2H_5)_2O$	391
	β-Bromostyryl p-anisyl ketoxime	p-Anisic acid‖	PCl_5, $(C_2H_5)_2O$	391

	Oxime	Product	Reagent	Ref.
C_{17}	4-Methoxybenzyl 3,4-dimethoxyphenyl ketoxime	No reaction	PCl_5, $(C_2H_5)_2O$, C_6H_6	398
	4-Methyl-9-acetyl-1,2,3,4-tetrahydrophenanthrene oxime	4-Methyl-9-(N-acetylamino)-1,2,3,4-tetrahydrophenanthrene (63)		378
	7-Acetyl-9-methyl-1,2,3,4-tetrahydrophenanthrene oxime	7-(N-Acetylamino)-9-methyl-1,2,3,4-tetrahydrophenanthrene		378
	syn-Styryl p-phenetyl ketoxime	N-p-Phenetylcinnamamide	H_2SO_4	399
	![structure: OH / CH₂C(=NOH)C₆H₅ on cyclohexane]	2-Hydroxyapocamphane-1-acetanilide (28), camphenecarboxanilide (10), 2-hydroxyapocamphone-1 acetic acid	PCl_5, $(C_2H_5)_2O$	400
C_{18}	3,4-$(CH_3O)_2C_6H_3SCH_2$-$CC_6H_3(OCH_3)_2$-2,4 ‖ NOH	Two unidentified products		395
	9,14-Benz-12-acetylacenaphthene oxime	9,14-Benz-12-acetamidoacenaphthene	PCl_5, $(C_2H_5)_2O$	401
	7-Ethyl-9-acetyl-1,2,3,4-tetrahydrophenanthrene oxime	Unidentified product		378
	2-Chloro-3-acetyl-9,10-dimethylanthracene oxime	2-Chloro-3-amino-9,10-dimethylanthracene	H_2SO_4	402
	1-Chloro-4-acetyl-9,10-dimethylanthracene oxime	1-Chloro-4-amino-9,10-dimethylanthracene	H_2SO_4	402

Note: References 338 to 593 are on pp. 152–156.

‖ The amide was hydrolyzed to the product(s) without prior isolation.

TABLE II—Continued
Aliphatic Aromatic Ketoximes

No. of C Atoms	Starting Material	Products (% Yield)	Catalysts and Experimental Conditions	References
C_{19}	2,9,10-Trimethyl-3-acetylanthracene oxime	2-Methyl-3-amino-9,10-dimethylanthracene	H_2SO_4	402
	p-$CH_3C_6H_4C(CH_3)_2CH_2(=NOH)$-$C_6H_4CH_3$-$p$	p-$CH_3C_6H_4C(CH_3)_2CONHC_6H_4CH_3$-$p$ (87)	PCl_5, $(C_2H_5)_2O$	588
C_{20}	6-Acetylchrysene oxime	6-(N-Acetylamino)chrysene (96)	PCl_5, $(C_2H_5)_2O$	403
	1-Methyl-2-acetyl-7-isopropylphenanthrene oxime	1-Methyl-2-acetamido-7-isopropylphenanthrene (95)	PCl_5, $(C_2H_5)_2O$	587
	6-Propionylchrysene oxime	6-(N-Propionylamino)chrysene	PCl_5, $(C_2H_5)_2O$	403
	β,β-Diphenylethyl phenyl ketoxime	β,β-Diphenylpropionanilide	PCl_5, $(C_2H_5)_2O$	404
	Benzaldesoxybenzoin oxime	Benzoic acid and benzyl phenyl ketone‖	PCl_5, $(C_2H_5)_2O$	70
		Unidentified product	H_2SO_4	70
	β-Phenylbenzalacetophenone oxime	β-Phenylcinnamanilide (100)	PCl_5, $(C_2H_5)_2O$	70
		3-Phenyl-5,5-diphenylisoxazoline	H_2SO_4	70
C_{22}	3-Bromoacetylhexosterol dimethyl ether oxime	3-Bromoacetamidohexosterol dimethyl ether	PCl_5, $(C_2H_5)_2O$	405
	3-Acetylhexosterol dimethyl ether oxime	3-Acetamidohexosterol dimethyl ether (80)	PCl_5, $(C_2H_5)_2O$	405
C_{23}	3-n-Propionylhexosterol dimethyl ether oxime	3-Propionamidohexosterol dimethyl ether	PCl_5, $(C_2H_5)_2O$	405
C_{24}	3-n-Butyrylhexosterol dimethyl ether oxime	3-Butyramidohexosterol dimethyl ether	PCl_5, $(C_2H_5)_2O$	405
C_{29}	3-n-Pelargonylhexosterol dimethyl ether oxime	3-n-Pelargonamidohexosterol dimethyl ether	PCl_5, $(C_2H_5)_2O$	405

Note: References 338 to 593 are on pp. 152–156.

‖ The amide was hydrolyzed to the product(s) without prior isolation.

TABLE III

DIARYL KETOXIMES

No. of C Atoms	Starting Material	Products (% Yield)	Catalysts and Experimental Conditions	References
C_{13}	Benzophenone oxime	Benzanilide (100,84)	PCl_5, $(C_2H_5)_2O$; PCl_5, then H_2O	104, 411
		Benzanilide	PCl_5; $POCl_3$	406
		Benzanilide (quant.)	Polyphosphoric acid	123
		Benzanilide (72)	HF, CH_3CO_2H	104
		Benzanilide (quant.)	HCl; HBr	102
		Benzanilide	HI	102
		Benzanilide (quant.)	HCl, xylene	102
		Benzanilide (quant.)	HCl, CH_3CO_2H, $(CH_3CO)_2O$	100
		Benzanilide	H_2SO_4; CH_3COCl; $C_6H_5SO_2Cl$ or p-$CH_3C_6H_4SO_2Cl$, aq. NaOH	1, 95, 100, 105, 244
		Benzanilide (50–90)	CH_3COCl or $ClCH_2COCl$, $CHCl_3$	17
		Benzanilide (88)	CF_3CO_2H	409
		Benzanilide (70–96)	BF_3, CH_3CO_2H	19
		Benzanilide	BF_3, $(C_2H_5)_2O$	384
		Benzanilide (20)	Benzophenone oxime hydrochloride	408
		Benzanilide (48)*	Picric acid, CH_3NO_2	22
		Benzanilide	Various metal halides†	68, 99

Note: References 338 to 593 are on pp. 152–156.

* The phenylbenzimino picrate formed was hydrolyzed to the product.
† The oxides and halides used, the conditions and yields when given follow: KCl at 150–160°, 33%; $MgCl_2$ at 170°, 33%; $ZnCl_2$ at 120–130°, 86%; $AlCl_3$ at 100–110°, 86%; $FeCl_2$ at 165–170°, 66%; $FeCl_3$, 60%; HgCl, 70%; $HgCl_2$, 86%; $SbCl_3$, 80%.

TABLE III—Continued

DIARYL KETOXIMES

No. of C Atoms	Starting Material	Products (% Yield)	Catalysts and Experimental Conditions	References
C_{13} (continued)	Benzophenone oxime (continued)	Benzanilide	Al_2O_3	410
		Benzanilide and N,N'-diphenylbenzamidine	$SOCl_2$, $(C_2H_5)_2O$	80
		$C_6H_5CH_2CH_2C_6H_5$, $C_6H_5C_6H_5$, $(C_6H_5)_2CO$, $(C_6H_5)_2C=NH$	Cu in H_2 atm. at 200°	66
		Acetophenone and aniline	CH_3MgI, $(C_2H_5)_2O$	113
		$CH_3CH_2COC_6H_5$ and aniline	CH_3CH_2MgI, $(C_2H_3)_2O$	113
		Diphenyltetrazole	SO_2Cl_2, then NaN_3 in $CHCl_3$	296
		2,3-Diphenyl-5-ethyl-6-methyl-4-pyrimidone (55)	PCl_5, $(C_2H_5)_2O$	365
		Thiobenzanilide (92, —)	P_2S_5, CS_2; P_2S_5, C_6H_6	107
		$(C_6H_5)_2C=NS_2P(O)OH$	P_2S_5, $(C_2H_5)_2O$	108
		$C_6H_5CONHC_6H_4Cl$-2 (28) and $C_6H_2CONHC_6H_4Cl$-4 (52)	$Cl_2 \cdot Br_2$, SO_2	364
		$C_6H_5CONHC_6H_4Br$-4 (52)	Br_2, SO_2	364
		Benzanilide (48)	H_2O, (boil)	408
	Benzophenone oxime hydrochloride	Benzanilide (100)	$ZnCl_2$, $CHCl_3$	106
		Benzanilide (91)	Chloral (10 min. at 90°)	106
		Benzanilide	$CHCl_3$	106
		Trifluoroboron benzanilide complex (85)	BF_3	412
	Benzophenone oxime methyl ether	Benzanilide (53)‡	BF_3, CH_3CO_2H	19
		N-Phenylbenzimido methyl ether (63)‡	$SbCl_5$, C_6H_5Cl (110°)	260
		Benzophenone oxime methyl ether hexachloroantimonate acid salt (52)	$SbCl_5$, $CHCl_3$ (wet)	260

Benzophenone oxime methyl ether hexachlorostibine salt	N-Phenylbenzimino methyl ether hexachloroantimonate acid salt (72)	$SbCl_5$, C_6H_5Cl	20, 260
N-Chlorobenzohydrylidenimine	Aniline; methyl benzoate (72)	Tartaric acid, H_2O	20, 260
	p-Chlorobenzanilide (5)‡	$SbCl_5$, CCl_4	21
	Benzanilide (75)‡	$SbCl_5$, $CHCl_2CHCl_2$ (40–45°)	21
	Aniline	KOH (fuse)	106
Benzophenone oxime acetate	Benzanilide‡	HCl (gas), $CHCl_3$	105
	Benzylamine (79%) and N-phenylbenzylamine	$LiAlH_4$, tetrahydrofuran	413
	Benzanilide (70)‡	BF_3, CH_3CO_2H	19
	Benzanilide	BF_3	19
	Benzanilide	$C_6H_5SO_3H$	106
	Benzanilide (96)	HCl (gas), $CHCl_3$	106
Benzophenone oxime benzenesulfonate	Benzanilide	Aq. NaOH	244
	Benzanilide and benzenesulfonic acid*	$CHCl_3$	12
	Benzanilide‡	$CHCl_3$	106
	N-Phenylbenzimino phenyl ether	C_6H_5OH, C_6H_6	13
	N,N′-Diphenylbenzamidine (92)	$C_6H_5NH_2$, C_6H_6	13
	N-Benzoyl-N,N′-diphenylbenzamidine	$(C_6H_5)_2NH$	13
	N-Phenylbenzimino ethyl ether	Pyridine, C_2H_5OH	13
	N-Phenylbenzamidine (18)	NH_3, C_6H_6	13
	N-Phenyl-N,N′-diethylbenzamidine (90)	$(C_2H_5)_2NH$, C_6H_6	13

Note: References 338 to 593 are on pp. 152–156.

* The phenylbenzimino-benzenesulfonate formed was hydrolyzed to the product.
‡ The products were obtained by treating the reaction mixture with water.

TABLE III—Continued

DIARYL KETOXIMES

No. of C Atoms	Starting Material	Products (% Yield)	Catalysts and Experimental Conditions	References
C_{13} (continued)	Benzophenone oxime benzenesulfonate (continued)	N'-Phenyl-N,N-pentamethylenebenzamidine (89)	Piperidine, C_6H_6	13
		N-Phenyl-N'-benzylbenzamidine (93)	$C_6H_5CH_2NH_2$, C_6H_6	13
		N-Phenyl-N'-p-tolylbenzamidine (100)	p-$H_2NC_6H_4CH_3$, C_6H_6	13
		N-Phenyl-N'-o-chlorophenylbenzamidine (94)	o-$H_2NC_6H_4Cl$, C_6H_6	13
		N-Phenyl-N'-p-chlorophenylbenzamidine (96)	p-$H_2NC_6H_4Cl$, C_6H_6	13
		N,N,N'-Triphenylbenzamidine (83)	Aniline, C_6H_6	13
		N-Phenyl-N'-2-pyridylbenzamidine (20)	Pyridine, C_6H_6	13
	Benzophenone oxime p-toluenesulfonate	Benzanilide	Aq. NaOH	244
	Benzophenone oxime picryl ether	N-(2,4,6-Trinitrophenyl)benzanilide	Acetone	414
		Benzanilide (50)	Aq. acetone	13
	Benzophenone oxime β-naphthalenesulfonate	Benzanilide	Aq. NaOH	244
	Benzophenone oxime α-phenylimidobenzyl ether	Benzanilide (100)	Conc. HCl	22
		Benzanilide (45)	HCl, $(C_2H_5)_2O$	22
		Benzoyl-s-diphenylbenzylamidine	H_2SO_4, $(C_2H_5)_2O$	22
	Benzophenone oxime diphenylphosphorochloridate	Benzanilide	Al_2O_3	410
	Benzophenone	Benzanilide (91)	Polyphosphoric acid, CH_3NO_2	415

2-Chlorobenzophenone oxime	2-Chlorobenzanilide and aniline	PCl_5, $(C_2H_5)_2O$	101
	2-Chlorobenzophenone	18% HCl	91
4-Chlorobenzophenone oxime	Benzoic acid (44%) and 4-chlorobenzoic acid (56%)‡ p-$ClC_6H_4C(Cl)=NC_6H_5$	PCl_5, C_6H_6	79
		PCl_5, $(C_2H_5)_2O$	81
	4-Chlorobenzanilide	HCl (gas), $(CH_3CO)_2O$, CH_3CO_2H; H_2SO_4	81
4,4'-Dichlorobenzophenone oxime	4,4'-Dichlorobenzanilide	PCl_5, $(C_2H_5)_2O$	416
2-Bromobenzophenone oxime	2-Bromobenzanilide (100)	18% HCl	101
	2-Bromobenzophenone	18% HCl	91
2-Nitrobenzophenone oxime	2-Nitrobenzophenone	18% HCl	91
syn-4-Nitrobenzophenone oxime	4-Nitrobenzanilide	PCl_5, $(C_2H_5)_2O$	417
	4-Nitrobenzanilide (94)	$POCl_3$, $(C_2H_5)_2O$	418
$anti$-4-Nitrobenzophenone oxime	4'-Nitrobenzanilide (90)	$POCl_3$	418
	4'-Nitrobenzanilide	PCl_5, $(C_2H_5)_2O$	417, 419
2-Hydroxybenzophenone oxime	Salicylanilide (62)	PCl_5, $(C_2H_5)_2O$	101
syn-2-Hydroxybenzophenone oxime	Salicylanilide (45, –)	PCl_5, $(C_2H_5)_2O$	114
$anti$-2-Hydroxybenzophenone oxime	2'-Hydroxybenzanilide	PCl_5, $(C_2H_5)_2O$	114
	2-Phenylbenzoxazole (42) and o-aminophenol	PCl_5, $(C_2H_5)_2O$	114
syn-4-Hydroxybenzophenone oxime	4-Hydroxybenzanilide	PCl_5, $(C_2H_5)_2O$	7
$anti$-4-Hydroxybenzophenone oxime	4'-Hydroxybenzanilide	PCl_5, $(C_2H_5)_2O$	7
2-Aminobenzophenone oxime	2-Aminobenzanilide	PCl_5, $(C_2H_5)_2O$	101
syn-2-Aminobenzophenone oxime	2-Phenyl-4,5-benzimidazole	HCl, C_2H_5OH	115
$anti$-2-Aminobenzophenone oxime	2-Phenyl-4,5-benzimidazole	HCl, C_2H_5OH	115
2-Chloro-5-nitrobenzophenone oxime	2-Chloro-5-nitrobenzanilide (63) and $C_{13}H_{10}O_6N_2PCl \cdot H_2O$	PCl_5, $(C_2H_5)_2O$	230
2-Bromo-5-nitrobenzophenone oxime	2-Bromo-5-nitrobenzanilide (77) and $C_{13}H_{10}O_6N_2PBr$ (14)	PCl_5, $(C_2H_5)_2O$	230

Note: References 338 to 593 are on pp. 152–156.

‡ The products were obtained by treating the reaction mixture with water.

TABLE III—Continued

DIARYL KETOXIMES

No. of C Atoms	Starting Material	Products (% Yield)	Catalysts and Experimental Conditions	References
C_{13} (continued)	Fluorenone oxime	Phenanthridone (84)	PCl_5, $POCl_3$	110
	Fluorenone	Phenanthridone (67)	Polyphosphoric acid, CH_3NO_2	415
	2-Nitrofluorenone oxime	9-Aza-10-chloro-2-nitrophenanthrene (16) and 2-nitrofluorenone-9-imino chloride (84)	PCl_5, $POCl_3$	110
		9-Aza-10-chloro-2-nitrophenanthrene (58) and 10-aza-9-chloro-2-nitrophenanthrene (29)	PCl_5, $POCl_3$	111
	3-Nitrofluorenone oxime	10-Aza-9-oxo-3-nitro-9,10-dihydrophenanthrene (87)	PCl_5, $POCl_3$	111
C_{14}	2-Methylbenzophenone oxime	o-Toluic acid (77) and benzoic acid (23)‡	PCl_5, C_6H_6	7, 79
	3-Methylbenzophenone oxime	m-Toluic acid (50) and benzoic acid (50)‡	PCl_5, C_6H_6	79
	4-Methylbenzophenone oxime	4-$CH_3C_6H_4CONHC_6H_5$	PCl_5, $(C_2H_5)_2O$; HCl CH_3CO_2H, $(CH_3CO)_2O$	81
		p-Toluic acid (52) and benzoic acid (48)‡	PCl_5, C_6H_6	79
		$C_6H_5CONHC_6H_4CH_3$-4 (100)	PCl_5, C_6H_6	97
		$C_6H_5CONHC_6H_4CH_3$-4	C_6H_5COCl, C_6H_6; CH_3COCl; $(CH_3CO)_2O$; $POCl_3$	97
		4-$CH_3C_6H_4CONHC_6H_5$ and $C_6H_5CONHC_6H_4CH_3$-4	PCl_5, $(C_2H_5)_2O$	97

Oxime	Product	Reagent	Reference
4-Methoxybenzophenone oxime	Benzoic acid (51) and 4-methoxybenzoic acid (49).‡	PCl$_5$, C$_6$H$_6$	97
2-Carboxybenzophenone oxime	Phthalanilide	H$_2$SO$_4$	101
2′-Carboxy-4′-hydroxybenzophenone oxime	4-Hydroxyphthalanilide	None given	583
anti-Phenyl 2-hydroxy-5-methylphenyl ketoxime	2-Hydroxy-5-methylbenzanilide and/or 5-methyl-2-phenylbenzoxazole	PCl$_5$, (C$_2$H$_5$)$_2$O	8, 584
syn-3-Bromo-4-methoxybenzophenone oxime	3-Bromo-4-methoxybenzanilide	PCl$_5$, (C$_2$H$_5$)$_2$O	323
anti-3-Bromo-4-methoxybenzophenone oxime	3′-Bromo-4′-methoxybenzanilide	PCl$_5$, (C$_2$H$_5$)$_2$O	323
syn-3-Iodo-4-methoxybenzophenone oxime	3-Iodo-4-methoxybenzanilide	PCl$_5$, (C$_2$H$_5$)$_2$O	323
anti-3-Iodo-4-methoxybenzophenone oxime	3′-Iodo-4′-methoxybenzanilide	PCl$_5$, (C$_2$H$_5$)$_2$O	323
syn-3-Nitro-4-methoxybenzophenone oxime	3-Nitro-4-methoxybenzanilide	PCl$_5$, (C$_2$H$_5$)$_2$O	323
2-Methoxy-5-nitrobenzophenone oxime	2-Methoxy-5-nitrobenzanilide	PCl$_5$, (C$_2$H$_5$)$_2$O	230
2-Bromo-2′-hydroxy-5′-methyl-5-nitrobenzophenone oxime	Unidentified product	PCl$_5$, (C$_2$H$_5$)$_2$O	420
syn-3,5-Dichloro-4-methoxybenzophenone oxime	3,5-Dichloro-4-methoxybenzanilide	PCl$_5$, (C$_2$H$_5$)$_2$O	323

C$_{15}$

Oxime	Product	Reagent	Reference
syn-4-Ethylbenzophenone oxime	4-Ethylbenzanilide	PCl$_5$, (C$_2$H$_5$)$_2$O	7
anti-4-Ethylbenzophenone oxime	4′-Ethylbenzanilide (100)	PCl$_5$, (C$_2$H$_5$)$_2$O	7
4-Ethoxybenzophenone oxime	4- and 4′-Ethoxybenzanilide	SOCl$_2$	80
syn-4-Ethoxybenzophenone oxime	4-Ethoxybenzanilide (90)	SOCl$_2$, (C$_2$H$_5$)$_2$O	80
anti-4-Ethoxybenzophenone oxime	4′-Ethoxybenzanilide	SOCl$_2$	80

Note: References 338 to 593 are on pp. 152–156.

‡ The products were obtained by treating the reaction mixture with water.

TABLE III—Continued

DIARYL KETOXIMES

No. of C Atoms	Starting Material	Products (% Yield)	Catalysts and Experimental Conditions	References
C_{15} (continued)	4-Dimethylaminobenzophenone oxime	Benzanilide and p-dimethylaminoaniline (75)	PCl_5, C_2H_5OH	421
	syn-4-Dimethylaminobenzophenone oxime	4-(Dimethylamino)benzanilide (75)	PCl_5, $CHCl_3$	422
	anti-4-Dimethylaminobenzophenone oxime	4'-(Dimethylamino)benzanilide (80)	PCl_5, $CHCl_3$	422
	2,4-Dimethylbenzophenone oxime	2,4-Dimethylbenzanilide (34)	H_2SO_4, CH_3CO_2H	117
		2,4-Dimethylbenzophenone	18% HCl	91
	anti-2,4-Dimethylbenzophenone oxime	2,4-Dimethylbenzanilide and 2',4'-dimethylbenzanilide	PCl_5, $(C_2H_5)_2O$; CH_3COCl (room temp.)	7
		2',4'-Dimethylbenzanilide	PCl_5, $(C_2H_5)_2O$ ($-20°$)	7
	syn-2,4-Dimethylbenzophenone oxime	2,4-Dimethylbenzanilide	PCl_5, $(C_2H_5)_2O$	7
	2,4'-Dimethylbenzophenone oxime	2,4'-Dimethylbenzanilide		423
	2,5-Dimethylbenzophenone oxime	2,5-Dimethylbenzophenone and methylaniline	18% HCl	91
	4,4'-Dimethylbenzophenone oxime	2,3-Di-p-tolyl-5-ethyl-6-methyl-4-pyrimidone (60)§ ‖	PCl_5, $(C_2H_5)_2O$	365
	syn-2,4-Dimethoxybenzophenone oxime	2,4-Dimethoxybenzoic acid	H_2SO_4	9
	anti-2,4-Dimethoxybenzophenone oxime	Benzoic acid	H_2SO_4	9
	4,4'-Dimethoxybenzophenone oxime	4,4'-Dimethoxybenzanilide	Polyphosphoric acid; PCl_5, $(C_2H_5)_2O$	123, 424

Oxime	Product	Reagent	Reference
2-Hydroxy-3,5-dimethylbenzo- phenone oxime	2-Phenyl-5,7-dimethylbenzoxazole (34)	PCl_5, $(C_2H_5)_2O$	420
3-Hydroxy-4,6-dimethylbenzo- phenone oxime	3'-Hydroxy-2',4'-dimethylbenzanilide and trace of 3-hydroxy-2,4-di- methylbenzanilide	CH_3COCl, $(CH_3CO)_2O$, CH_3CO_2H	420
2-Bromo-2'-methoxy-5'-methyl- benzophenone oxime	2-Bromo-2'-methoxy-5'-methyl benzanilide	PCl_5, $CHCl_3$	420
C_{16} *syn*-4-*n*-Propylbenzophenone oxime	4-*n*-Propylbenzanilide	PCl_5, $(C_2H_5)_2O$	7
anti-4-*n*-Propylbenzophenone oxime	4'-*n*-Propylbenzanilide	PCl_5, $(C_2H_5)_2O$	7
syn-4-Isopropylbenzophenone oxime	4-Isopropylbenzanilide (100)	PCl_5, $(C_2H_5)_2O$	7
anti-4-Isopropylbenzophenone oxime	4- and 4'-Isopropylbenzanilide	PCl_5, $(C_2H_5)_2O$	7
syn-3-Methoxy-4,6-dimethylbenzo- phenone oxime	3-Methoxy-4,6-dimethylbenzanilide (100)	PCl_5, $(C_2H_5)_2O$	420
2-Carboxy-2',4'-dimethylbenzo- phenone oxime	Phthalic acid and 2,4-xylidine	H_2SO_4	101
2,2',4'-Trimethylbenzophenone	2,2',4'-Trimethylbenzanilide and 2,4,2'-trimethylbenzanilide	Aq. $NH_2OH \cdot HCl$	7
2,4,6-Trimethylbenzophenone	2,4,6- and 2',4',6'-Trimethylbenz- anilide	Aq. $H_2NOH \cdot HCl$	7
2,4,6-Trimethylbenzophenone oxime	2',4',6'-Trimethylbenzanilide (94)	BF_3, CH_3CO_2H; PCl_5	264
5-Hydrindenyl phenyl ketoxime	5-Hydrindanilide	HCl, $(CH_3CO_2)_2O$, CH_3CO_2H	386
2,2',4,4'-Tetramethoxybenzophenone oxime	2,2',4,4'-Tetramethoxybenzanilide	PCl_5, $(C_2H_5)_2O$	9
C_{17} *syn*-3-Ethoxy-4,6-dimethylbenzo- phenone oxime	3-Hydroxy-4,6-dimethylbenzoic acid (100) and aniline (100)	CH_3COCl, $(CH_3CO)_2O$, CH_3CO_2H	420
syn-Phenyl 1-naphthyl ketoxime	N-1'-Naphthylbenzamide	P_2O_5, $(C_2H_5)_2O$	425

Note: References 338 to 593 are on pp. 152–156.

§ The amide was not isolated. The intermediate chlorimide was treated with an α-alkyl-β-aminocrotonate ester to yield the 4-pyrimidone.

‖ The 5-methyl derivative can be prepared by analogous reaction.

92   ORGANIC REACTIONS

TABLE III—*Continued*

DIARYL KETOXIMES

No. of C Atoms	Starting Material	Products (% Yield)	Catalysts and Experimental conditions	References
C_{17} *(continued)*	*anti*-Phenyl 1-naphthyl ketoxime	1-Naphthanilide	P_2O_5, $(C_2H_5)_2O$	425
	syn-Phenyl 2-naphthyl ketoxime	N-2′-Naphthylbenzamide	PCl_5, $(C_2H_5)_2O$	426
	anti-Phenyl 2-naphthyl ketoxime	2-Naphthanilide	PCl_5, $(C_2H_5)_2O$	426
	syn-2-(5,6,7,8-Tetrahydronaphthyl) phenyl ketoxime	2(5,6,7,8-Tetrahydronaphth)anilide		381
	anti-2-(5,6,7,8-Tetrahydronaphthyl) phenyl ketoxime	N-2-(5,6,7,8-Tetrahydronaphthyl)-benzamide		381
	meso-Benzanthrone	8-(*o*-Carboxyphenyl)-1-naphthylamine	PCl_5, $POCl_3$	427
	4,4′-Bis(dimethylamino)benzophenone	4,4′-Bis(dimethylamino)benzanilide	$NH_2OH \cdot HCl$, C_2H_5OH	109, 428
	4,4′-Bis(dimethylamino)thiobenzophenone	4,4′-Bis(dimethylamino)benzanilide	$NH_2OH \cdot HCl$, C_2H_5OH	428, 429
	4,4′-Bis(dimethylamino)benzophenone oxime	4,4′-Bis(dimethylamino)benzanilide (85, 61)	$SOCl_2$, CCl_4	109, 428
C_{18}	4-*t*-Butyl-4′-methylbenzophenone oxime	4′-*t*-Butyl-4-methylbenzanilide	PCl_5, C_6H_6	430
C_{19}	Dimesityl ketimine	2,2′,4,4′,6,6′-Hexamethylbenzanilide	H_2O_2, CH_3CO_2H	264
	5,5′-Diindanyl ketoxime	N′-5′-Indanyl-5-indancarboxylic acid amide	PCl_5, $(C_2H_5)_2O$	376
	p-Xenyl phenyl ketoxime	Biphenyl-4-carboxylic acid (51) and benzoic acid (49)‡	PCl_5, C_6H_6	79
	syn-p-Xenyl phenyl ketoxime	4-Phenylbenzanilide (100)	PCl_5, C_6H_6	79
	anti-p-Xenyl phenyl ketoxime	N-*p*-Xenylbenzamide (100)	PCl_5, C_6H_6	79
C_{20}	*p*-Xenyl *o*-tolyl ketoxime	Biphenyl-4-carboxylic acid (34) and *o*-toluic acid (64)‡	PCl_5, C_6H_6	79

THE BECKMANN REARRANGEMENT

Oxime	Products	Reagent	Reference
p-Xenyl m-tolyl ketoxime	m-Toluic acid (47) and p-phenylbenzoic acid (53)‡	PCl$_5$, C$_6$H$_6$	79
p-Xenyl p-tolyl ketoxime	p-Toluic acid (34) and p-phenylbenzoic acid (53)‡	PCl$_5$, C$_6$H$_6$	79
2-(p-Nitrobenzoyl)benzophenone oxime	2-(p-Nitrobenzoyl)benzanilide (50)		431
C$_{21}$			
4,4'-Di-t-butylbenzophenone oxime	4,4'-Di-t-butylbenzanilide	PCl$_5$, C$_6$H$_6$	430
4,4'-Bis(diethylamino)benzophenone	4,4'-Bis(diethylamino)benzanilide	NH$_2$OH·HCl, C$_2$H$_5$OH	432
Di-1-naphthyl ketoxime	N-1'-Naphthyl-1-naphthamide	PCl$_5$, (C$_2$H$_5$)$_2$O	433
Di-2-naphthyl ketoxime	N-2'-Naphthyl-2-naphthamide	PCl$_5$, (C$_2$H$_5$)$_2$O	433
2-Naphthyl 1-naphthyl ketoxime	N-2'-Naphthyl-1-naphthamide	PCl$_5$, (C$_2$H$_5$)$_2$O	433
1-Phenanthryl phenyl ketoxime	N-1-Phenanthrylbenzamide (18) and N-phenyl-1-phenanthramide (82)	PCl$_5$, C$_6$H$_6$	96
2-Phenanthryl phenyl ketoxime	N-2-Phenanthrylbenzamide (44) and N-phenyl-2-phenanthramide (56)	PCl$_5$, C$_6$H$_6$	96
3-Phenanthryl phenyl ketoxime	N-3-Phenanthrylbenzamide (37) and N-phenyl-3-phenanthramide (63)	PCl$_5$, C$_6$H$_6$	96
9-Phenanthryl phenyl ketoxime	N-9-Phenanthrylbenzamide (4) and N-phenyl-9-phenanthramide (96)	PCl$_5$, C$_6$H$_6$	96
1-Benzoylanthraquinone oxime	(46)	H$_2$SO$_4$, CH$_3$CO$_2$H	117, 118

Note: References 338 to 593 are on pages 152–156.

‡ The products were obtained by treating the reaction mixture with water.

TABLE III—Continued

DIARYL KETOXIMES

No. of C Atoms	Starting Material	Products (% Yield)	Catalysts and Experimental Conditions	References
C_{21} (continued)	1-Benzoylanthraquinone oxime	Anthraquinone-1-carboxylic acid (55), anthraquinone-1-carboxanilide (35), and trace of $C_{21}H_{11}ON$	Aq. HCl	117–119
C_{22}	2-Methyl-1-benzoylanthraquinone oxime	2-Methylanthraquinone-1-carboxylic acid	H_2SO_4, CH_3CO_2H	117–119
	p-Toluylanthraquinone oxime	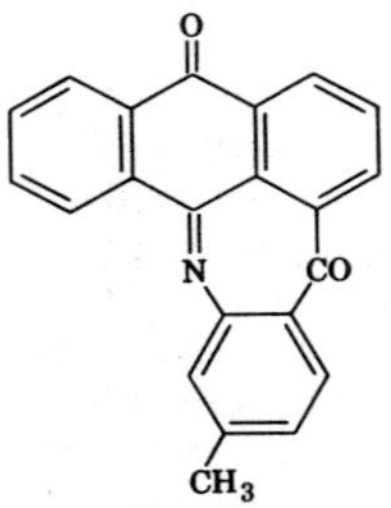 (50)	H_2SO_4, CH_3CO_2H	117–119
C_{23}	1-(2,4-Dimethylbenzoyl)-anthraquinone oxime	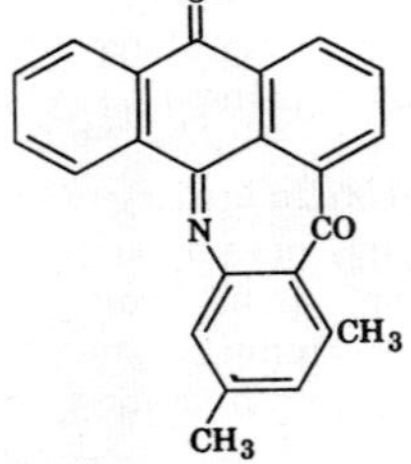 (50)	H_2SO_4, CH_3CO_2H	117–119

1-(2,5-Dimethylbenzoyl)-anthraquinone oxime	3,4-Dimethylanthraquinone-1-carboxanilide (small)	HCl, C$_2$H$_5$OH	117–119
	(50) [structure]	H$_2$SO$_4$, CH$_3$CO$_2$H	117–119
C$_{24}$ Mesitoylanthraquinone oxime	Starting material (50)	H$_2$SO$_4$, CH$_3$CO$_2$H	117–119
2-Methyl-1-(2,4-dimethylbenzoyl)-anthraquinone oxime	2-Methylanthraquinone-1-carboxylic acid (trace)	H$_2$SO$_4$, CH$_3$CO$_2$H	117–119
	2,2′,4′-Trimethylanthraquinone-1-carboxanilide	HCl, C$_2$H$_5$OH	117–119
2-Methyl-1-(2,5-dimethylbenzoyl)-anthraquinone oxime	2-Methylanthraquinone-1-carboxylic acid (trace)	H$_2$SO$_4$, CH$_3$CO$_2$H	117–119
C$_{25}$ *m*-Terphenylyl phenyl ketoxime	N-*m*-Terphenylylbenzamide (50)	PCl$_5$, (C$_2$H$_5$)$_2$O	434

Note: References 338 to 593 are on pp. 152–156.

TABLE IV
Alicyclic Ketoximes

No. of C Atoms	Starting Material	Products (% Yield)	Catalysts and Experimental Conditions	References
C₅	Cyclopentanone oxime	δ-Valerolactam (60, 98, 92)	H₂SO₄	65, 124, 438
		δ-Valerolactam (94)	92% H₂SO₄*	144
		δ-Valerolactam (53)	90% H₂SO₄	435
		δ-Valerolactam	80–90% H₂SO₄; 80% H₂SO₄	436, 437, 462
		δ-Valerolactam	H₂SO₄*	441
		δ-Valerolactam	Na₂SO₄·3H₂SO₄	140
		δ-Valerolactam	Aq. H₂SO₄; CH₃CO₂H	138
		δ-Valerolactam (82)	H₂SO₄, CH₃CH₂CO₂H	137
		δ-Valerolactam	H₂SO₄, fatty acids	439
		δ-Valerolactam (82)	Metaphosphoric acid; 270°§	125
		δ-Valerolactam (74)	Polyphosphoric acid	123
		δ-Valerolactam (47)	SOCl₂, CHCl₃	440
		δ-Valerolactam (37)	Br₂, SO₂	364
		δ-Valerolactam (74)	HF	83, 126
		5-Benzamidovaleric acid (71)	H₂SO₄, then NaOH and C₆H₅COCl	585
		⌬=NH₂ (47)	NH₃	13
		⌬=NHC₆H₅ (57)	C₆H₅NH₂	13

		Tetramethylenetetrazole	
Cyclopentanone oxime sulfonate			H_2SO_4, NaN_3; $CHCl_3$, $ClSO_3H$, NaN_3	248
Cyclopentanone oxime benzenesulfonate	Tar	Aq. HCl, dioxane	60
	δ-Valerolactam (93)	Dibenzyl hydrogen phosphate; $(C_2H_5)_3N$, CH_3NO_2	338
Cyclopentanone oxime p-nitrobenzenesulfonate	(structure: $(CH_2)_5$ ring with C—OPCH$_2$C$_6$H$_5$ / OCH$_2$C$_6$H$_5$ and N)	$(C_6H_5CH_2O)_2PO_2NH_4$, $C_2H_5NO_2$ or CH_3CN	58
Cyclopentanone	δ-Valerolactam (99)	CH_3CN, $CHCl_3$†	442
	δ-Valerolactam (100)	H_2SO_4, $(NH_2OH)_2 \cdot H_2SO_4$	121
	δ-Valerolactam (81)	Oleum, $(NH_2OH)_2 \cdot H_2SO_4$	168
	δ-Valerolactam	H_2SO_4, CH_3NO_2	443
Nitrocyclopentane	5-Methyl-5-valerolactam (61–76%)‡	75% H_2SO_4	444
2-Methylcyclopentanone oxime	β- and γ-Picoline, pentenonitrile	P_2O_5	147
3-Methylcyclopentanone oxime	3-Methyl-5-valerolactam and 4-methyl-5-valerolactam	80% H_2SO_4	65, 445

C$_6$: (structural formula shown: H$_2$C$_5$—NH—CO, H$_2$C$_4$—CHCH$_3$, CH$_2$ — cyclohexanone with NH, numbered positions 1–6)

Note: References 338 to 593 are on pp. 152–156.

* Special equipment or procedure was employed.
† Dibenzyl hydrogen phosphate and selected amines and solvents gave similar results.
‡ Substituted lactams are named according to the following system: 2-methyl-5-valerolactam is

§ · This reaction was run in the vapor phase under reduced pressure.

TABLE IV—Continued
ALICYCLIC KETOXIMES

No. of C Atoms	Starting Material	Products (% Yield)	Catalysts and Experimental Conditions	References
C_6 (continued)	Cyclohexanone oxime	ε-Caprolactam (41)	$K_2S_2O_7$, pumice, H_2§	164
		ε-Caprolactam (60)	$KHSO_4$, pumice, vacuum§	453
		ε-Caprolactam (95)	$ClSO_3H$, alone or with SO_3	456
		ε-Caprolactam (45)	SO_3 or $SOCl_2$; SO_2	128, 342
		ε-Caprolactam (50–60)	$SOCl_2$, SO_2	463
			$SOCl_2$ alone or with $CHCl_3$	54
		ε-Caprolactam (93)	$C_6H_5SO_2Cl$, aq. NaOH	257, 446
				457
		ε-Caprolactam	CH_3SO_2Cl or $p\text{-}CH_3C_6H_4SO_2Cl$, aq. NaOH	457
		ε-Caprolactam (67, 46)	HF	83, 126
		ε-Caprolactam (92)	Anhydrous HF	458
		ε-Caprolactam (82–87)	CF_3CO_2H	82
		ε-Caprolactam (85)	70% $HClO_4$, CH_3CO_2H	135
		ε-Caprolactam	89% H_3PO_4, C_6H_6, or $CHCl_3$	54
		ε-Caprolactam (85)	Orthophosphoric acid§	125, 133
		ε-Caprolactam (89)	Polyphosphoric acid	123
		ε-Caprolactam (85)	$NaHSO_4$, H_3PO_4, polyphosphoric acid, $H_4P_2O_7$§	127

ε-Caprolactam (50–60)	P_2O_5, $POCl_3$, PCl_3, PBr_3, $SOCl_2$	463
ε-Caprolactam (56–79)	B_2O_3 (21.5–36.5% on Al_2O_3)‖	148
ε-Caprolactam (41)	BPO_4, H_2, NH_3‖	164
ε-Caprolactam	NH_3, SiO_2, 200–500°‖	146
ε-Aminocaproic acid (88)	70% H_2SO_4	120
ε-Caprolactam (70–98)	H_2SO_4*	15, 455, 143, 124, 464, 465, 441, 142, 445
ε-Caprolactam (59–99)	H_2SO_4	65, 447, 439, 142, 460, 445, 438, 449
ε-Caprolactam (92, 87–90)	H_2SO_4, cyclohexane, $C_2H_4Cl_2$; H_2SO_4, CH_3CO_2H	132, 137, 467
ε-Caprolactam	25–28% H_2SO_4 CH_3CO_2H	138
ε-Caprolactam (60)	60% oleum	452
ε-Caprolactam (91)	60–90% H_2SO_4*	144
ε-Caprolactam (80–97)	75–96.4% H_2SO_4	444
ε-Caprolactam (80)	80% H_2SO_4*	145, 462
ε-Caprolactam (75–95)	80–85% H_2SO_4	131
ε-Caprolactam (89)	80–94% H_2SO_4	436

Note: References 338 to 593 are on pp. 152–156.

* Special equipment or procedure was employed.

§ This reaction was run in the vapor phase under reduced pressure.

‖ Vapor phase reaction.

TABLE IV—Continued
ALICYCLIC KETOXIMES

No. of C Atoms	Starting Material	Products (% Yield)	Catalysts and Experimental Conditions	References
C_6 (continued)	Cyclohexanone oxime (continued)	ε-Caprolactam (90)	86% H_2SO_4	437
		ε-Caprolactam (96, 66, —)	90% H_2SO_4	144,* 450, 454
		ε-Caprolactam (87)	90–96% H_2SO_4	435
		ε-Caprolactam (78)	95% H_2SO_4	451
		ε-Caprolactam (96)	98% H_2SO_4; 100% H_2SO_4	141, 337
		ε-Caprolactam (96)	Oleum, SO_3	55
		ε-Caprolactam	Oleum, CCl_4, C_6H_6, or other hydrocarbons	56
		ε-Caprolactam (90–98)	1–60% Oleum	459
		ε-Caprolactam (90–94)	6–60% Oleum, $C_6H_5NO_2$, or 1-nitro-1-methylcyclopentane	136
		ε-Caprolactam (87–97)	15% Oleum	461
		ε-Caprolactam (98)	65% Oleum	466
		ε-Caprolactam (good, —)	SO_3, CS_2	57, 468
		ε-Caprolactam (100, —, —)	SO_3, CH_2ClCH_2Cl; SO_3, $CCl_2=CCl_2$; SO_3, chlorinated hydrocarbon	57, 336
		ε-Caprolactam (93)	SO_3, SO_2	463
		ε-Caprolactam	SO_3, SO_2, fluorinated or chlorinated hydrocarbons	342, 336

Substrate	Reagents	Reference
ε-Caprolactam	NH_4HSO_4, H_2SO_4, SO_2, CS_2, or chlorinated hydrocarbon	139
ε-Caprolactam (74)	$NH_4HSO_4 \cdot H_2SO_4$	469
ε-Caprolactam	$Na_2SO_4 \cdot 3H_2SO_4$	140
ε-Caprolactam (83–85)	85–97.5% H_2SO_4, SiO_2; H_2SO_4, SiO_2	130, 451
ε-Caprolactam	$KHSO_4$, pumice, H_2 or NH_3	163
ε-Caprolactam (30–70)	$Cl_2 \cdot Br_2$ or $Cl_2 \cdot I_2$ or $Br_2 \cdot I_2$ or $Br_2 \cdot SOCl_2$, with SO_2	364
δ-Caprolactam	None given	592
ε-Aminocaproic acid (good)	Oleum, then water	134
1,6-Hexamethylenediamine	$CuCO_3$ on SiO_2, H_2 or NH_3*§¶	165
Pentamethylenetetrazole (95)	H_2SO_4, NaN_3, $CHCl_3$; $ClSO_3$, NaN_3	248
Pentamethylenetetrazole	$POCl_3$ or $SOCl_2$ with NaN_3 and $CHCl_3$	248
Cyclohexanone oxime hydrochloride	H_2SO_4; H_2SO_4, HCl	470, 591
Cyclohexanone oxime methyl ether	10% Oleum	261
	H_2SO_4	263
Cyclohexanone oxime allyl ether	10% Oleum	261
Cyclohexanone oxime picryl ether (77–79)	Aq. acid or base	54, 128

Note: References 338 to 593 are on pp. 152–156.

* Special equipment or procedure was employed.
§ This reaction was run in the vapor phase under reduced pressure.
¶ Other catalysts, such as H_3PO_4-SiO_2, H_3BO_3-SiO_2, and H_4TiO_4-TiO_2-U_2O_5 were also used.

TABLE IV—Continued
ALICYCLIC KETOXIMES

No. of C Atoms	Starting Material	Products (% Yield)	Catalysts and Experimental Conditions	References
C_6 (continued)	Cyclohexanone oxime sulfonate	Octahydrophenazine (7)	HCl (4N), dioxane	60
		Tetrahydrophenazine (5)	Pyridine, CH_3OH	60
		ε-Caprolactam	Aq. HCl	14
	Cyclohexanone oxime potassium sulfonate	Pentamethylenetetrazole (70)	NaN_3; H_2O	248
		2-Iminohexamethyleneimine (50)	NH_3	13
	Cyclohexanone oxime benzenesulfonate	2-Anilinohexamethyleneimine (72)	$C_6H_5NH_2$	13
		ε-Caprolactam (77)	Aq. acid or base	54
		Cyclohexanone oxime	Aq. base or acid	54
	Cyclohexanone oxime o-toluenesulfonate	ε-Caprolactam (79)	Aq. acid or base	54
	Cyclohexanone oxime p-toluenesulfonate	Pentamethylenetetrazole	$NaNO_2$ and N_2H_4, CH_3CO_2H, $CHCl_3$	294
	Cyclohexanone oxime 2-naphthylsulfonate	ε-Caprolactam (78)	Aq. acid or base	54
	Cyclohexanone	ε-Caprolactam (87)	Oleum, $(NH_2OH)_2 \cdot H_2SO_4$	168
		ε-Caprolactam (90)	H_2SO_4, $(NH_2OH)_2 \cdot H_2SO_4$	121, 471
		ε-Caprolactam (79)	H_2SO_4, primary nitroparaffin	167
		ε-Caprolactam	$(NH_2OH)_2 \cdot H_2SO_4$, $(CH_3CO)_2O$, CH_3CO_2H**	472

Oxime	Product	Reagent	Ref.
Nitrocyclohexane	ε-Caprolactam (35)	SiO$_2$, N$_2$; BPO$_4$, N$_2$; phosphomolybdic acid, N$_2$; silicotungstic acid, N$_2$‖	170
	ε-Caprolactam (30)	20% Oleum, S	171
	ε-Caprolactam	H$_2$SO$_4$, C$_2$H$_5$NO$_2$	443
	ε-Caprolactam	Salt*	473
	ε-Caprolactam (74)	(NH$_2$OH)$_2$·H$_2$SO$_4$; 20% oleum	443
	ε-Caprolactam (60)	H$_2$SO$_4$, CH$_3$NO$_2$	443
syn-Cyclohexenone oxime	Δ^2-6-Caprolactam (25)	Polyphosphoric acid	475
anti-Cyclohexenone oxime	Unidentified product	Polyphosphoric acid	475
2-Chlorocyclohexanone oxime	Octahydrophenazine	Aq. HCl, dioxane	60
2-Ethylcyclopentanone oxime	5-Ethyl-5-valerolactam (61)	80% H$_2$SO$_4$	444
C$_7$ 2-Methylcyclohexanone oxime	6-Methyl-6-caprolactam (88–97)	85–96.4% H$_2$SO$_4$	444
	2-Methyl-6-caprolactam (70%) and 6-methyl-6-caprolactam (30%)	SO$_3$–H$_2$SO$_4$	476
	6-Methyl-6-caprolactam (67)	H$_2$SO$_4$	303
	2-Methyl-6-caprolactam and 6-methyl-6-caprolactam (50–80)	PCl$_5$, C$_6$H$_6$; H$_2$SO$_4$	477
	2-Methyl-6-caprolactam	H$_2$SO$_4$	477
	10-Methylpentamethylenetetrazole (61)	ClSO$_3$H, NaN$_3$, CH$_2$ClCH$_2$Cl	293
3-Methylcyclohexanone oxime	3-Methyl-6-caprolactam and 5-methyl-6-caprolactam	80% H$_2$SO$_4$	65, 445, 478
	5-Methyl-6-caprolactam	C$_6$H$_5$SO$_2$Cl, aq. NaOH	446

Note: References 338 to 593 are on pp. 152–156.

* Special equipment or procedure was employed.

‖ Vapor phase reaction.

** Monochloroacetic acid may be used in place of acetic acid and CH$_3$CONHOCOCH$_3$ may be used in place of hydroxylamine sulfate.

TABLE IV—Continued
ALICYCLIC KETOXIMES

No. of C Atoms	Starting Material	Products (% Yield)	Catalysts and Experimental Conditions	References
C_7 (continued)	3-Methylcyclohexanone oxime (continued)	7-Methylpentamethylenetetrazole (63)	$ClSO_3H$, NaN_3, CH_2ClCH_2Cl	293
		Toluene and hexenonitrile, lutidine and mixed lactams	P_2O_5	147
	4-Methylcyclohexanone oxime	4-Methyl-6-caprolactam (62)	H_2SO_4	303
		4-Methyl-6-caprolactam (good)	$C_6H_5SO_2Cl$, aq. KOH	446, 457
		4-Methyl-6-caprolactam (89)	90% H_2SO_4*	144
		8-Methylpentamethylenetetrazole (57)	$ClSO_3$, HN_3, aq. NaOH	293
	Cycloheptanone oxime	2-Oxoheptamethylenimine (92)	$SO_3 \cdot H_2SO_4$; 60% oleum	124, 459
		2-Oxoheptamethylenimine (50)	H_2SO_4	65, 147, 474, 445
		2-Oxoheptamethylenimine (80)	o-Phosphoric acid§	125
		2-Oxoheptamethylenimine (30)	HF	83, 126
		2-Oxoheptamethylenimine	H_2SO_4*	441
	Cycloheptanone	2-Oxoheptamethylenimine (93)	H_2SO_4, $(NH_2OH)_2 \cdot H_2SO_4$	121
C_8	2-n-Propylcyclopentanone oxime	5-n-Propyl-5-valerolactam (59)	80% H_2SO_4	444
	2-Ethylcyclohexanone oxime	6-Ethyl-6-caprolactam (99)	H_2SO_4, CCl_4	593
	3-Ethylcyclohexanone oxime	5-Ethyl-6-caprolactam (77)	H_2SO_4	303
	4-Ethylcyclohexanone oxime	4-Ethyl-6-caprolactam (90)	H_2SO_4	480
	trans-2,4-Dimethylcyclohexanone oxime	4,6-Dimethyl-6-caprolactam (53)	H_2SO_4	303

Oxime	Product	Reagent	Ref.
trans-2,5-Dimethylcyclohexanone oxime	3,5-Dimethyl-6-caprolactam (71)	H_2SO_4	303
3,4-Dimethylcyclohexanone oxime	Dimethyl-ε-caprolactam (67)	H_2SO_4	303
3,5-Dimethylcyclohexanone oxime	4,6-Dimethyl-6-caprolactam (60)	H_2SO_4	480
	7,9-Dimethylhexamethylenetetrazole (58)	$ClSO_3H$, NaN_3, CH_2ClCH_2Cl	293
cis-3,5-Dimethylcyclohexanone oxime	cis-4,6-Dimethyl-6-caprolactam (45)	H_2SO_4	303
Bicyclo[2.2.1]heptan-2-one oxime	2-Aza-3-oxobicyclo[2.2.1]octane (70–90)	85% H_2SO_4	481
Cycloöctanone oxime	2-Oxoöctamethylenimine (80, 85–90)	80% H_2SO_4	122
	2-Oxoöctamethylenimine (68)	90% H_2SO_4	474
Cycloöctanone oxime hydrochloride	2-Oxoöctamethylenimine (69)	H_2SO_4, HCl	591
Indanone oxime	4,5-Benzvalerolactam (10)	PCl_5, $(C_2H_5)_2O$	129
	Hydrocarbostyril	PCl_5	482
2-Oximinoindanone	o-Carboxyphenylacetamide	H_2SO_4	483
5-Oximinohydrindene	2-Oxo-5,6-cyclopentanohexamethylenimine (70)	$C_6H_5SO_2Cl$, aq. NaOH	484
1-Oximino-2-nitro-3-ketoindane	1-Chloro-3-nitro-4-hydroxyisoquinoline	HCl, $(CH_3CO)_2O$, CH_3CO_2H; $POCl_3$, CH_3CO_2H	161
C_9			
4-n-Propylcyclohexanone oxime	4-n-Propyl-6-caprolactam (76)	H_2SO_4	485
2-Isopropylcyclohexanone oxime	6-Isopropyl-6-caprolactam	H_2SO_4	480
3-Isopropylcyclohexanone oxime	8-Isopropylhexamethylenetetrazole (67)	$ClSO_3H$, HN_3, CH_2ClCH_2Cl	15, 293
4-Isopropylcyclohexanone oxime	4-Isopropyl-6-caprolactam (72)	H_2SO_4	480

Note: References 338 to 593 are on pp. 152–156.

* Special equipment or procedure was employed.
§ This reaction was run in the vapor phase under reduced pressure.

TABLE IV—Continued
ALICYCLIC KETOXIMES

No. of C Atoms	Starting Material	Products (% Yield)	Catalysts and Experimental Conditions	References
C_9 (continued)	3-Ethyl-5-methylcyclohexanone oxime	7-Methyl-9-ethylhexamethylenetetrazole (32)	$ClSO_3H$, NaN_3, CH_2ClCH_2Cl	15, 293
		7-Methyl-9-ethylhexamethylenetetrazole	$ClSO_3H$, NaN_3	293
	2,3,5-Trimethylcyclohexanone oxime	3,5,6-Trimethyl-6-caprolactam (73)	H_2SO_4	303
	2,4,6-Trimethylcyclohexanone oxime	2,4,6-Trimethyl-6-caprolactam (57)	H_2SO_4	303
	3,3,5-Trimethylcyclohexanone oxime	7,9,9-Trimethylhexamethylenetetrazole (72)	$ClSO_3H$, NaN_3, CH_2ClCH_2Cl	293
		3,3,5- and 3,3,5-Trimethyl-6-caprolactam	50% H_2SO_4	477
	3,5,5-Trimethyl-2-cyclohexen-1-one oxime (isophorone oxime)	3,3,5-Trimethyl-Δ^5- and 3,5,5-trimethyl-Δ^2-6-caprolactam	80–100% H_2SO_4	486
	syn-3,5,5-Trimethyl-2-cyclohexen-1-one oxime (syn-isophorone oxime)	3,5,5-Trimethyl-Δ^2-6-caprolactam (25)	PCl_5, $(C_2H_5)_2O$	487
	anti-3,5,5-Trimethyl-2-cyclohexen-1-one oxime (anti-isophorone oxime)	3,3,5-Trimethyl-Δ^5-6-caprolactam (20)	PCl_5, $(C_2H_5)_2O$	487
	4,4,6-Trimethylcyclohexanone oxime	3,3,5- and 3,3,5-Trimethyl-6-caprolactam	50% H_2SO_4	477
C_{10}	4-sec-Butylcyclohexanone oxime	8-sec-Butylhexamethylenetetrazole (50)	$ClSO_3H$, NaN_3, CH_2ClCH_2Cl	293
	t-Butylcyclohexanone oxime	t-Butyl-6-caprolactam (100)	H_2SO_4	337
	4-t-Butylcyclohexanone oxime	4-t-Butyl-6-caprolactam (82)	H_2SO_4	480
		8-t-Butylhexamethylenetetrazole (68)	$ClSO_3H$, NaN_3, CH_2ClCH_2Cl	293
	3-Methyl-3-n-propylcyclohexanone oxime	7-Methyl-7-isopropylhexamethylenetetrazole (37)	$ClSO_3H$, NaN_3, CH_2ClCH_2Cl	293

Oxime	Product	Reagents	Refs.
2-Isopropyl-5-methylcyclohexanone oxime	7-Methyl-10-isopropylhexamethylenetetrazole (27)	$ClSO_3H$, NaN_3, CH_2ClCH_2Cl	293
3-Isopropyl-5-methylcyclohexanone oxime	7-Methyl-9-isopropylhexamethylenetetrazole (50)	$ClSO_3H$, NaN_3, CH_2ClCH_2Cl	293
d-Carvone oxime	Unknown product $C_{10}H_6ClNO$	PCl_5	487
Tetrahydrocarvone oxime	Unidentified product	60% H_2SO_4, CH_3CO_2H; H_2SO_4	65, 445, 488
Pulenone oxime	3,6,6-Trimethyl-6-caprolactam	$POCl_3$, $CHCl_3$	153
	3,6-Dimethyl-5-heptenonitrile	$(CH_3CO)_2O$; PCl_5, $POCl_3$	153
Menthone oxime	3-Methyl-6-isopropyl-6-caprolactam and deeylenic acid, menthylamines and menthonitrile	H_2SO_4	445, 488
l-Menthone oxime	 structure: seven-membered ring with CH$_3$, C$_3$H$_7$-i, NH	Cu, H_2	166
syn-Isonitrosocamphor	1,2,2-Trimethylcyclopentane-1,3-dicarboximide	PCl_5, $(C_2H_5)_2O$	154
β-Thujone oxime	structures (two bicyclic lactams)	PCl_5, $CHCl_3$	489
2-Methyl-2-hydroxy-5-(2′-hydroxyisopropyl)cyclohexanone oxime	bicyclic lactone structure ††	HCl or HBr, CH_3CO_2H or C_6H_6	490

Note: References 338 to 593 are on pp. 152–156.
†† This product was obtained by hydrolysis of the lactam.

TABLE IV—Continued
ALICYCLIC KETOXIMES

No. of C Atoms	Starting Material	Products (% Yield)	Catalysts and Experimental Conditions	References
C_{10} (continued)	β-Dihydroumbellulone oxime	1-Isopropyl-5-methyl-2-azabicyclo-[4.1.0]-heptan-3-one	p-$CH_3C_6H_4SO_2Cl$ or p-$BrC_6H_4SO_2Cl$, pyridine	491
	β-Dihydroumbellulone oxime p-toluenesulfonate	1-Isopropyl-5-methyl-2-azabicyclo-[4.1.0]-heptan-3-one	HCl, pyridine	491
	anti-1,2-Benzcyclohexanone oxime picryl ether	N-Picryl-5,6-benz-6-caprolactam‡‡		53
	1-Tetralone oxime p-toluenesulfonate	6,7-Benz-6-caprolactam p-toluenesulfonic acid salt	C_6H_5OH	158
		6,7-Benz-6-caprolactim phenyl ether	C_6H_5OH	158
		⟨phenyl⟩-$NH_3^{\oplus}$ $(CH_2)_3CO_2CH_3$ $^{\ominus}O_3SC_6H_4CH_3$-p (100)	CH_3OH	158
		⟨phenyl⟩-$NH_3^{\oplus}$ $(CH_2)_3CO_2C_2H_5$ $^{\ominus}O_3SC_6H_4CH_3$-p (30)	C_2H_5OH	158
	7-Nitro-1-tetralone oxime	1-Amino-7-nitronaphthalene (10)	Polyphosphoric acid	492
	7-Nitro-1-tetralone oxime acetate	1-Amino-7-nitronaphthalene (45)	HCl	492
	7-Nitro-1-tetralone oxime phenylcarbamate	1-Amino-7-nitronaphthalene (22)	HCl, C_2H_5OH	492
	2-Tetralone oxime	o-(2-Aminoethyl)phenylacetolactam (78)	p-$CH_3C_6H_4SO_2Cl$, aq. NaOH	493

THE BECKMANN REARRANGEMENT

Oxime	Product	Conditions	Ref.
1-Tetralone oxime sulfonate	α-Naphthylamine (12)	Aq. HCl, dioxane	60
2-Tetralone oxime p-toluenesulfonate	3,4-Benz-6-caprolactam (78)	CH₃OH	242
2-Decalone oxime	(mixture, 90) — octahydro-benzazepinones	80% H₂SO₄	437
cis-2-Decalone oxime	(mixture, 93) — octahydro-benzazepinones	98–100% H₂SO₄	337
4-t-Amylcyclohexanone oxime	8-t-Amylhexamethylenetetrazole (57)	ClSO₃H, NaN₃, CH₂ClCH₂Cl	293
2-t-Butyl-4-methylcyclohexanone oxime	4-Methyl-6-t-butyl-6-caprolactam (69)	H₂SO₄	303
6-Methoxy-1-tetralone oxime p-toluenesulfonate	CH₃O–C₆H₄–(CH₂)₃CO₂CH₃ · NH₃⁺ ⁻O₃SC₆H₄CH₃-p	CH₃OH	158
8-Methyl-1-tetralone oxime	(benzazepinone with CH₃)	HCl, (CH₃CO)₂O, CH₃CO₂H	158

C₁₁

Note: References 338 to 593 are on pp. 152–156.
‡‡ The picryl ether was rearranged by heating in ethylene dichloride.

TABLE IV—Continued
ALICYCLIC KETOXIMES

No. of C Atoms	Starting Material	Products (% Yield)	Catalysts and Experimental Conditions	References
C_{11} (continued)	Thujamethone oxime	3-Isopropyl-4,5-dimethyl-5-valerolactam	H_2SO_4	489
	β-Thujamethone oxime	2,3-Dimethyl-4-isopropyl-5-valerolactam	66% H_2SO_4, CH_3CO_2H	494, 489
	syn-1,2-Benzcycloheptanone oxime picryl ether	N-Picryl-2,3-benz-7-enantholactam‡‡		53
	anti-1,2-Benzcycloheptanone oxime picryl ether	N-Picryl-6,7-benz-7-enantholactam‡‡		53
C_{12}	2-Cyclohexylcyclohexane oxime	6-Cyclohexyl-6-caprolactam (100)	H_2SO_4	354
	4-Cyclohexylcyclohexanone oxime	8-Cyclohexylhexamethylenetetrazole (51)	$ClSO_3H$, NaN_3, CH_2ClCH_2Cl	293
	5,8-Dimethyl-1-tetralone oxime acetate	![structure: bicyclic lactam with two CH3 groups and N-H]	HCl, $(CH_3CO)_2O$ CH_3CO_2H	158
	5,8-Dimethyl-1-tetralone oxime p-toluenesulfonate	![structure: open chain ammonium tosylate with CH3 groups and $(CH_2)_3CO_2CH_3$, $NH_3^{\oplus} \ominus O_3SC_6H_4CH_3$-p]	CH_3OH	158
	3-Carbomethoxy-1-tetralone oxime	3-Carbomethoxy-5,6-benz-6-caprolactam (50)	Polyphosphoric acid	495

THE BECKMANN REARRANGEMENT

	Oxime	Product	Reagent	Ref.
	syn-1,2-Benzcyclooctanone oxime picryl ether	N-Picryl-2,3-benz-8-caprylolactam ‡‡		53
	anti-1,2-Benzcyclooctanone oxime picryl ether	N-Picryl-7,8-benz-8-caprylolactam ‡‡		53
C_{13}	4-Cyclohexylmethylcyclohexanone oxime	4-Cyclohexylmethyl-6-caprolactam	H_2SO_4	480
	syn-3-Methyl-5-phenyl-2-cyclohexen-1-one oxime	3-Phenyl-5-methyl-Δ^5-6-caprolactam (25)	PCl_5, $(C_2H_5)_2O$, C_6H_6	487
	anti-3-Methyl-5-phenyl-2-cyclohexen-1-one oxime	3-Methyl-5-phenyl-Δ^2-6-caprolactam (15)	PCl_5, $(C_2H_5)_2O$, C_6H_6	487
	α-Ionone oxime	2,2,6-Trimethyl-4-cyclohexene-1-acetaldehyde (60)	PCl_5, $CHCl_3$	479
	3-Carbethoxy-1-tetralone oxime	3-Carbethoxy-5,6-benz-6-caprolactam (86)	Polyphosphoric acid	495
C_{14}	1,2,3,4,6,7,8,9-Octahydroanthracene-1-one oxime p-toluenesulfonate	$(CH_2)_3CO_2CH_3$ $NH_3^\oplus \ominus O_3SC_6H_4CH_3$-$p$	CH_3OH	158
		![structure with N=OC6H5]	C_6H_5OH	158

Note: References 338 to 593 are on pp. 152–156.

‡‡ The picryl ether was rearranged by heating in ethylene dichloride.

TABLE IV—Continued
ALICYCLIC KETOXIMES

No. of C Atoms	Starting Material	Products (% Yield)	Catalysts and Experimental Conditions	References
C_{14} (continued)	1,2,3,4,5,6,7,8-Octahydrophenanthracene-4-one oxime p-toluenesulfonate	[tetrahydronaphthalene with $NH_3^{\oplus} \;{}^{\ominus}O_3SC_6H_4CH_3$-$p$ and $(CH_2)_3CO_2CH_3$ substituents]	CH_3OH	158
C_{15}	cis-9-Keto-4a-methyl-1,2,3,4,4a,9,10,10a-octahydrophenanthrene oxime	[cis fused tricyclic lactam with H_3C, NH, C=O] (60)	Polyphosphoric acid	496
	trans-9-Keto-4a-methyl-1,2,3,4,4a,9,10,10a-octahydrophenanthrene oxime	[trans fused tricyclic lactam with H_3C, NH, C=O] (90)	Polyphosphoric acid	496
	Cyclopentadecanone oxime	2-Oxopentadecamethylenimine (90, 94)	H_2SO_4 conc., C_6H_6; H_2SO_4	122, 121

	Oxime	Reagent	Reference
	1,2,3,4,5,6,7,8-Octahydro-10-methoxyphenanthrene-4-one oxime acetate	HCl, (CH₃CO)₂O, CH₃CO₂H	158
C₁₆	8-Methylcyclopentadecanone oxime	H₂SO₄	497
	1,2,3,4,5,6,7,8-Octahydro-9-acetamidophenanthrene-1-one oxime p-toluenesulfonate	CH₃OH	158
C₁₇	2-(o-Carboxybenzyl)hydrindone oxime	CH₃COCl	162
	Dihydroisocumarin-1-hydrindone-3,2-spiran (35)		
C₂₂	3-Isopropyl-7-methyl-8-keto-8,9,10,11-tetrahydrobenz-anthracene oxime	PCl₅, C₆H₆; 50% H₂SO₄	499
	(40–50)		

Note: References 338 to 593 are on pp. 152–156.

TABLE V
STEROID OXIMES

No. of C Atoms	Starting Material	Products (% Yield)	Catalysts and Experimental Conditions	References
C_{18}	Estrone oxime	3-Hydroxy-13a-amino-13,17-seco-1,3,5(10) estratrien-17-oic acid 13,17-lactam (82.5)	$SOCl_2$, dioxane; 40°	174
	Estrone methyl ether oxime	O-Methyl estroic acid	PCl_5, $(C_2H_5)_2O$	500
		3-Methoxy-13a-amino-13,17-seco-1,3,5(10) estratrien-17-oic acid 13,17-lactam (80)	$SOCl_2$, dioxane	174
C_{19}	4-Androstene-13,17-dione 17-oxime	Δ^4-13a-Amino-13,17-seco-androsten-3-one-17-oic acid 13,17-lactam (50)	p-Acetamidobenzenesulfonyl chloride, pyridine	175
		Δ^4-13a-Amino-13,17-seco-androsten-3-one-17-oic acid (50)	p-Acetamidobenzenesulfonyl chloride, aq. NaOH	501
C_{21}	3-β-Hydroxypregnan-20-one oxime	3-β-Acetoxy-17-acetamidoetiocholanol	$POCl_3$, pyridine	502
	3-β-Hydroxy-5-pregnen-2-one oxime	17-Amino-5-androstene-3β-ol (60)		177
	3-β-Acetoxy-5-androsten-17-one oxime	3-β-Hydroxy-Δ^5-13a-amino-13,17-seco-androsten-17-oic acid lactam	$SOCl_2$, dioxane	174
		3-β-Hydroxy-Δ^5-13a-amino-13,17-seco-androsten-17-oic acid lactam (50, 73)	p-Acetamidobenzenesulfonyl chloride, pyridine	175, 176
		3-β-Hydroxy-Δ^5-13a-amino-13,17-seco-androsten-17-oic acid lactam (50)	p-Acetamidobenzenesulfonyl chloride, aq. NaOH	501
	3-β-Acetoxy-17-ketoandrostan oxime	3-β-Acetoxy-13a-amino-13,17-seco-androstane-17-oic acid 13,17-lactam	p-Acetamidobenzenesulfonyl chloride, pyridine	176

	Oxime	Product	Reagent	Ref.
	3-β-Acetoxy-5-androsten-16,17-dione 16-oxime	3-β-Acetoxy-16,17-seco-5-androsten-16,17-imide (68)	$SOCl_2$	174
C_{22}	i-Pregnenolone methyl ether oxime	6-Methoxy-i-androsten-17-amine	p-$CH_3C_6H_4SO_2Cl$, pyridine	178
	i-Pregnenolone methyl ether oxime p-toluenesulfonate	6-Methoxy-i-androsten-17-amine	p-$CH_3C_6H_4SO_2Cl$, pyridine	178
	3-Acetoxy-17-acetyl-1,3,5,16-estratetraene oxime	Esterone (40)	p-$NH_2C_6H_4SO_2Cl$, pyridine	503
	3-β-Acetoxy-5-pregnen-20-one oxime	17-Amino-Δ⁵-androstene-3-β-ol	$POCl_3$, pyridine	177
		17-Amino-Δ⁵-androstene-3-β-ol (87)	$C_6H_5SO_2Cl$ or p-$CH_3C_6H_4SO_2Cl$, pyridine	178
C_{23}		17-Amino-Δ⁵-androsten-3-β-ol (30–95)	$SOCl_2$, C_6H_6	177
		17-Amino-Δ⁵-androsten-3-β-ol (87)	$C_6H_5SO_2Cl$ or p-$CH_3C_6H_4SO_2Cl$, basic solvent*	178
	Acetylpregnenolone oxime	3-Oxy-17-aminoandrostene	$SOCl_2$, C_6H_6	504
	3-Acetoxyallopregnan-20-one oxime	3-Hydroxy-17-aminoandrostane	p-$CH_3C_6H_4SO_2Cl$, $HOCH_2CH_2NH_2$	178
		3-Hydroxy-17-aminoandrostane	p-$CH_3C_6H_4SO_2Cl$, pyridine	178
	3-β-Acetoxyallopregnan-20-one oxime	3-β-Acetoxy-17-acetaminoandrostane	p-$CH_3C_6H_4SO_2Cl$, pyridine	502
	3-β-Acetoxy-17-α-5-pregnen-20-one oxime	Dehydroepiandrosterone acetate	$POCl_3$, pyridine	505
	5,16-Pregnadien-3-β-ol-20-one 3-acetate oxime	Dehydroepiandrosterone	p-$H_2NC_6H_4SO_2Cl$, pyridine	503

Note: References 338 to 593 are on pp. 152–156.

* The solvents used were methanol, sodium ethoxide in ethanol, n-butylamine, cyclohexylamine, N-ethylcyclohexylamine, and sodium 1-hexoxide in 1-hexanol.

TABLE V—Continued

STEROID OXIMES

No. of C Atoms	Starting Material	Products (% Yield)	Catalysts and Experimental Conditions	References
C_{23} (continued)	7,16-Allopregnadien-3-β-ol-20-one acetate oxime	Δ^7-Androsten-3-β-ol-17-one (75)	p-$H_2NC_6H_4SO_2Cl$, pyridine	503
	16-Allopregnen-3-β,11α-20-one diacetate oxime	Androstan-3-β,11α-diol-17-one (50)	p-$H_2NC_6H_4SO_2Cl$, pyridine	503
	3-Acetoxy-5-ternorcholenyl methyl ketone oxime	3-Hydroxy-5-pregnen-20-amine	p-$CH_3C_6H_4SO_2Cl$, pyridine	178
	8,11-Diketolanostan-2-yl acetate 8-oxime	8,11-Diketo-8a-aza-β-homolanostan-2-yl actate (50)	PCl_5, C_6H_6 or petroleum ether	506
	8,11-Diketolanost-9-ene-2-yl acetate 8-oxime	8,11-Diketo-7a-aza-α-β-homolanost-9-en-2-yl acetate and 8,11-diketo-8a-aza-β-homolanost-9-en-2-yl acetate (55)	PCl_5, C_6H_6	506
	Desoxybilianic acid monoxime $C_{20}H_{33}\begin{cases}(CO_2H)_3\\(=NOH)\end{cases}$	Desoxybilianic acid isoxime $C_{20}H_{33}\begin{cases}(CO_2H)_3\\-CONH-\end{cases}$	90% H_2SO_4	507
	β-Cholantricarboxylic acid oxime $C_{20}H_{33}\begin{cases}(CO_2H)_3\\=NOH\end{cases}$	β-Cholantricarboxylic acid isoxime $C_{20}H_{33}\begin{cases}(CO_2H)_3\\-CONH_2\end{cases}$	90% H_2SO_4	508
	5-Pregnene-3-β,17α-diol-20-one-3-acetate oxime	Dehydroepiandrosterone acetate (98)	$POCl_3$, pyridine	509
	allo-Pregnan-3-β,17α-diol-20-one-3-acetate oxime	epi-Androsterone acetate (90)	$POCl_3$, pyridine	509

	Compound (oxime)	Structure	Product	Structure	Reagent	Refs.
C_{24}	Dehydrocholic acid dioxime	$C_{23}H_{35}\begin{Bmatrix}-CO_2H\\(=NOH)_2\end{Bmatrix}$	Dehydrocholic acid diisoxime	$C_{22}H_{35}\begin{Bmatrix}-CO_2H\\(-CONH-)_2\end{Bmatrix}$	90% H_2SO_4	510
	Dehydrocholic acid trioxime	$C_{23}H_{33}\begin{Bmatrix}-CO_2H\\(=NOH)_3\end{Bmatrix}$	Dehydrocholic acid isodioxime 12 (?)	$C_{23}H_{33}\begin{Bmatrix}-CO_2H\\-(CONH)_2-\\=NOH\end{Bmatrix}$	90% H_2SO_4	209, 511, 512
	Bilianic acid dioxime	$C_{21}H_{31}\begin{Bmatrix}(-CO_2H)_3\\(=NOH)_2\end{Bmatrix}$	Bilianic acid dioxime	$C_{19}H_{31}\begin{Bmatrix}(-CO_2H)_3\\(-CONH-)_2\end{Bmatrix}$	90% H_2SO_4	512
	Bilianic acid dioxime	$C_{20}H_{33}\begin{Bmatrix}(-CO_2H)_4\\=NOH\\-NH_2\end{Bmatrix}$	Bilianic acid isoxime amino carboxylic acid	$C_{19}H_{33}\begin{Bmatrix}(-CO_2H)_4\\-NH_2\\-CONH-\end{Bmatrix}$	H_2SO_4	513
	Isobilianic acid dioxime	$C_{21}H_{31}\begin{Bmatrix}(-CO_2H)_3\\(=NOH)_2\end{Bmatrix}$	Isobilianic acid isoxime	$C_{20}H_{31}\begin{Bmatrix}(-CO_2H)_3\\-CONH-\\=NOH\end{Bmatrix}$	90% H_2SO_4	514, 511
C_{25}	3-β-21-Diacetoxyallopregnan-20-one oxime		3-β-Acetoxy-17-aminoandrostane (96)		$POCl_3$, pyridine	177
	Estrone 3-benzoate oxime		3-Hydroxy-13α-amino-13,17-seco-1,3,5(10)-estratrien-17-oic acid 13,17-lactam (82.5)		p-$CH_3CONHC_6H_4SO_2Cl$ pyridine	174, 175
			3-Hydroxy-13α-amino 13,17-seco-1,3,5(10) estratrien-17-oic acid 13,17-lactam (50)		p-$CH_3CONHC_6H_4SO_2Cl$ aq. NaOH	501

Note: References 338 to 593 are on pp. 152–156.

TABLE V—*Continued*

STEROID OXIMES

No. of C Atoms	Starting Material	Products (% Yield)	Catalysts and Experimental Conditions	References
C_{29}	15-Keto-$\Delta^{8(14)}$-cholesten-3β-ol acetate oxime	15-Aza-16-keto-$\Delta^{8(14)}$-D-homocholesten-3β-ol acetate	p-CH$_3$C$_6$H$_4$SO$_2$Cl, pyridine	515
C_{31}	30-Nor-20-ketothurberogenin acetate oxime	Unidentified product (5)	POCl$_3$, pyridine	516
C_{34}	*syn*-16-Ketocholestan-3β-ol benzoate oxime	17-Aza-16-keto-D-homocholesten-3β-ol benzoate (55)	p-CH$_3$C$_6$H$_4$SO$_2$Cl, pyridine	515
	anti-16-Ketocholestan-3β-ol benzoate oxime	16-Aza-17-keto-D-homocholesten-3β-ol benzoate	p-CH$_3$C$_6$H$_4$SO$_2$Cl, pyridine	515
	16-Keto-Δ^{14}-cholestenyl benzoate oxime	17-Aza-16-keto-Δ^{14}-D-homocholesten-3β-ol benzoate (16)	p-CH$_3$C$_6$H$_4$SO$_2$Cl, pyridine	515

Note: References 338 to 593 are on pp. 152–156.

TABLE VI

HETEROCYCLIC KETOXIMES

No. of C Atoms	Starting Material	Products (% Yield)	Catalysts and Experimental Conditions	References
C_5	Tetrahydro-1,4-thiapyrone oxime	1-Aza-5-thiacycloheptan-2-one (85)	Polyphosphoric acid	182
	Tetrahydro-1,4-thiapyrone-1,1-dioxide oxime	Potassium 2-[2'-aminoethylsulfonyl]-propionate*	85% H_2SO_4	524
	Tetrahydro-1,4-pyrone oxime	1-Aza-5-oxacycloheptan-2-one	Polyphosphoric acid	182
	4-Piperidone oxime hydrochloride	1,5-Diazacycloheptan-2-one	$SOCl_2$	517
	Tetrahydro-2,6-dioxy-1,4-thiapyrone oxime	Unidentified product	Polyphosphoric acid	182
C_6	1-Methyl-4-piperidone oxime	Polymer	Polyphosphoric acid	182
	1-Methyl-4-piperidone oxime hydrochloride	1,5-Diaza-5-methylcycloheptan-2-one	$SOCl_2$	517
	2-Acetylthiophene oxime	2-Acetamidothiophene (55)	PCl_5, $(C_2H_5)_2O$	518
		2-Aminothiophene	PCl_5, $(C_2H_5)_2O$	518
	2-Hydroxyacetylfuran oxime p-toluenesulfonate	$CH_3COCOCH=CH(OCH_2CH_3)_2$	C_2H_5OH	407
	syn-Methyl 2-pyrryl ketoxime	2-C_4H_4NCONHCH_3	PCl_5, $(C_2H_5)_2O$	321, 586
	anti-Methyl 2-pyrryl ketoxime	2-C_4H_4NCONHCH_3	PCl_5, $(C_2H_5)_2O$	321, 586
	2-Oxo-3-acetyl-4-butyrolactone oxime	2-Acetoxy-3-acetamido-Δ^2-butyrolactone	CH_3COCl, H_2O	519
C_7	2,6-Dimethyl-4-piperidone oxime hydrochloride	1,4-Diaza-3,5-dimethylcycloheptan-2-one	$SOCl_2$	517
	Methyl-4-pyridyl ketoxime p-toluenesulfonate	4-Aminoacetylpyridine diethylketal	K, C_2H_5OH	520

Note: References 338 to 593 are on pp. 152–156.

* The amide was not isolated.

TABLE VI—Continued
HETEROCYCLIC KETOXIMES

No. of C Atoms	Starting Material	Products (% Yield)	Catalysts and Experimental Conditions	References
C_7 (continued)	2-Acetyl-5-methylfuran oxime p-toluenesulfonate	Starting material	C_2H_5OH	407
	2-Propionylfuran oxime p-toluene sulfonate	Ammonium p-toluenesulfonate and 2-propionylfuran	C_2H_5OH	407
	2,6-Dimethyl-1,4-pyrone oxime	1-Aza-4,6-dimethyl-5-oxacyclo-heptan-2-one (70)	Polyphosphoric acid	182
	2,3-Dimethyl-1,4-thiapyrone oxime	Unidentified product	H_2SO_4; $POCl_3$, HCl; PCl_5; CH_3COCl	99
	3-(Hydroxymethyl)-5,6-dihydro-1,4-pyrone-2-carboxylic acid lactone oxime	(structure with OCOCH3)	CH_3COCl	519
C_8	Acetonylpyridinium chloride oxime	Unidentified product	PCl_5, $POCl_3$	344
	2,2,5,5-Tetramethyl-3-oximinotetrahydrofuran	1-Aza-2,2,4,4-tetramethyl-3-oxacyclohexan-6-one (64)	77% H_2SO_4	523
	Acetonylpyridinium chloride oxime (structure with =NOH)	1-Oxa-3-aza-5,6-benzcyclohexane-2,4-dione (40)	PCl_5, petroleum ether	522
	2,2,5,5-Tetramethyl-4,5-dihydro-3(2h)-furanone oxime	Acetone (64), NH_3 (55), $(CH_3)_2C=CHCO_2H$	77% H_2SO_4	523

THE BECKMANN REARRANGEMENT

C₉	2,3,5,6-Tetramethyl-4-piperidone oxime hydrochloride	1,4-Diaza-2,3,5,6-tetramethylcyclo-heptan-2-one	SOCl₂	517
	4-methyl chromanone oxime (structure, CH₃ position) (45–50)	lactam (structure with NH, CH₃)	PCl₅, petroleum ether	522
	chromanone oxime (H₃C)	lactam (H₃C, NH)	PCl₅, petroleum ether	522
	chromanone oxime (H₃C) (50–52)	lactam (H₃C, NH)	PCl₅, petroleum ether	522
	4-Thiachromanone-1,1-dioxide oxime	Unidentified products	PCl₅; POCl₃; H₂SO₄	524
	4-Thiachromanone-1,1-dioxide oxime benzenesulfonate	Unidentified products	Polyphosphoric acid	524
	4-Thiachromanone-1,1-dioxide oxime 2-nitrobenzenesulfonate	2-[2′-Aminobenzenesulfonyl]-propionic acid lactam (43)	Aq. HCl	524
C₁₀	C₆H₅C(=NOH)—N=C(OH)—N=C(HN)—C(=O) (75–80)	C₆H₅CONH—C=N / HN—C=O / N—OH	PCl₅; CH₃COCl	525

Note: References 338 to 593 are on pp. 152–156.

TABLE VI—Continued
HETEROCYCLIC KETOXIMES

No. of C Atoms	Starting Material	Products (% Yield)	Catalysts and Experimental Conditions	References
C₁₁	2-Benzoylfuran oxime p-toluenesulfonate	Furanilide	C_2H_5OH	407
	2,3-Dimethylbenzopyrone oxime	Unidentified sulfonic acid	H_2SO_4†	99
	Methyl 5-(8-hydroxyquinolyl) ketoxime	5-Acetamido-8-hydroxyquinoline	$SOCl_2$, $(C_2H_5)_2O$; H_2SO_4; HCl, $(CH_3CO)_2O$, CH_3CO_2H	180
C₁₂	2-Benzoylthiophene oxime	Unidentified product	PCl_5	97
	syn-Phenyl 2-pyridyl ketoxime	2-Benzamidopyridine (68)	$SOCl_2$, $CHCl_3$	243
	syn-Phenyl 2-pyridyl ketoxime p-toluenesulfonate	Benzoic acid and 2-aminopyridine (90)	$CHCl_3$	243
	anti-Phenyl 2-pyridyl ketoxime	α-Picolinic acid anilide (86)	$SOCl_2$ or PCl_5, $CHCl_3$	243
	anti-Phenyl 2-pyridyl ketoxime p-toluenesulfonate	Benzoic acid and 2-aminopyridine (92)	$CHCl_3$	243
	Methyl 3-(2-methylquinolyl) ketoxime	2-Methyl-3-aminoquinoline	H_2SO_4	526
		2-Methyl-3-acetamidoquinoline	PCl_5, $POCl_3$	526
	6-Acetyl-4-chloroquinaldine oxime	6-Acetamido-4-chloroquinoline (79)	PCl_5, C_6H_6	589
	Ethyl 5-quinolyl ketoxime	N-Ethyl quinoline-5-carboxamide (80)	$SOCl_2$, $(C_2H_5)_2O$	528
	1-Hydroxy-5,6-benzisatin oxime	2,3-Naphthyleneurea		527
	2-p-Methoxybenzoylfuran oxime p-toluenesulfonate	![structure]$CONHC_6H_4OCH_3$-p	C_2H_5OH	407
C₁₃	Cuskohygrine oxime	Cuskohygrine	PCl_5	184
	2-Pyridylmethyl phenyl ketoxime	2-Pyridylacetanilide (90)	PCl_5, $(C_2H_5)_2O$	529
	2-Pyridyl 4-carboxyphenyl ketoxime	Terephthalic acid	PCl_5	530

THE BECKMANN REARRANGEMENT

Oxime	Product	Reagent	Ref.
4-Pyridyl 4-carboxyphenyl ketoxime	Terephthalic acid	PCl_5	530
![phenyl-NCH₂CC₆H₅=NOH, Cl⁻]	![phenyl-NCH₂CONHC₆H₅, Cl⁻]	PCl_5	531
![phenyl-NCH₂CC₆H₅=NOH, Br⁻]	Not isolated	PCl_5	521
	![phenyl-NCH₂CONHC₆H₅, HSO₄⁻]	H_2SO_4	521
2,6-Dimethyl-3-acetylchromone dioxime	2,6-Dimethyl-3,4-methyloxadiazino-chromone	H_2SO_4; PCl_5	183
Thiaxanthone-5,5-dioxide oxime	Thiaxanthone-5,5-dioxide and 2-(2′-aminobenzensulfonyl)benzoic acid	PCl_5; $POCl_3$	524
3-Acetyldibenzthiophene oxime	N-Acetyl-3-aminodibenzthiophene	HCl, $(CH_3CO)_2O$, CH_3CO_2H	532
C₁₄ 2-Acetylphenoxathiin oxime	2-Aminophenoxathiin (75)	PCl_5, C_6H_6	533
	2-Acetamidophenoxathiin (80)	PCl_5, $(C_2H_5)_2O$	534
C₁₅ 2-Benzoylbenzofuran oxime p-toluenesulfonate	2-Carboxanilidobenzofuran (84)	C_2H_5OH	407
![dibenzoselenophene with CCH₂CH₃=NOH]	![dibenzoselenophene with NH₂]	PCl_5, $(C_2H_5)_2O$	535

Note: References 338 to 593 are on pp. 152–156.

† There was no reaction with hydrogen chloride, acetyl chloride, or phosphorus pentachloride.

TABLE VI—Continued
Heterocyclic Ketoximes

No. of C Atoms	Starting Material	Products (% Yield)	Catalysts and Experimental Conditions	References
C_{16}	2-Acetyl-7-chloro-9-ethylcarbazole oxime	2-Acetamido-7-chloro-9-ethyl-carbazole	PCl_5, $(C_2H_5)_2O$	536
	Phenyl 5-(8-hydroxyquinolyl) ketoxime	5-Benzamido-8-hydroxyquinoline (100)	$SOCl_2$, $(C_2H_5)_2O$	180
		5-Benzamido-8-hydroxyquinoline	HCl, $(CH_3CO)_2O$, CH_3CO_2H	180
		Sulfonated benzamide	H_2SO_4	180
	4-Pyridyl α-naphthyl ketoxime	N-(4-Pyridyl)-α-naphthamide (90)	PCl_5	537
	3,6-Diacetyldibenzothiophene dioxime	N,N'-Diacetyl-3,6-diaminodibenzothiophene	HCl, $(CH_3CO)_2O$, CH_3CO_2H	532
	2,6-Diphenyltetrahydro-1,4-thiapyrone oxime	1-Aza-4,6-diphenyl-5-thiacyclo-heptan-2-one (75)	Polyphosphoric acid	182
	2,8-Diacetylphenoxathiin dioxime	2,8-Diaminophenoxathiin (75)	PCl_5, C_6H_6	533
C_{17}	6-Benzoylquinaldine	Quinaldine-6-carboxylic acid (50)	PCl_5, $(C_2H_5)_2O$	587
		Quinaldine-6-carboxylic acid and benzoic acid	PCl_5, $(C_2H_5)_2O$	589
	![structure with NCH₂CC₆H₅=NOH on naphthalene, Cl⁻]	![structure with NCH₂CCl₂NHC₆H₅, Cl⁻]	PCl_5, $(C_2H_5)_2O$	179
C_{18}	3-Benzoyl-6-phenylpyridine oxime	2-Phenylnicotinic acid anilide (100)	PCl_5	538

Dihydrocodeinone oxime	SOCl$_2$, POCl$_3$, PCl$_3$	539

(50) structure with NCH$_3$, CH$_2$, OHC, CN, CH$_3$O, OH

C$_{19}$	2-Phenyl-3-cyano-5,6-dihydro-6-benzoylpyran oxime	PCl$_5$, (C$_2$H$_5$)$_2$O	535
	2-Phenyl-3-cyano-5,6-dihydropyran-o-carboxylic acid anilide	SOCl$_2$, CHCl$_3$	
C$_{20}$	4-Quinolyl β-(1-methyl-3-vinyl-4-piperidinyl)ethyl ketoxime	PCl$_5$, CHCl$_3$	540
	4-Aminoquinoline (52) and 1-methyl-3-vinyl-4-(β-aminoethyl)piperidine (28)		541
	1-Aceto-6-acetylcodein oxime	PCl$_5$, CHCl$_3$	
	4-Carboxymethylquinoline (6) and 4-aminoquinoline (43) CH$_3$CONHC$_{18}$H$_{20}$O$_3$N	HCl, CH$_3$CO$_2$H	541 542
C$_{22}$	3,4-Diphenyl-5-benzoylisoxazole oxime	PCl$_5$	543

(95) structure with H$_2$N, Se, NH$_2$

H$_5$C$_6$—CONHC$_6$H$_5$ / N-O / H$_5$C$_6$ and   H$_5$C$_6$—C(=NCl)C$_6$H$_5$ / N-O / H$_5$C$_6$

Note: References 338 to 593 are on pp. 152–156.

TABLE VII
MONOXIMES OF DIKETONES

No. of C Atoms	Starting Material	Products (% Yield)	Catalysts and Experimental Conditions	References
C_9	$C_6H_5C(=NOH)COCH_3$	No reaction	10% H_2SO_4	544
C_{10}	$p\text{-}CH_3OC_6H_4C(=NOH)COCH_3$	$CH_3COCONHC_6H_4OCH_3\text{-}p$	10% H_2SO_4	544
C_{11}	1-Oximino-1-phenylpentan-4-one	$C_6H_5SO_2OC(CH_2)_2COCH_3$ $\Vert$ C_6H_5N and 4-ketovaleranilide	$C_6H_5SO_2Cl$, aq. NaOH	194
C_{14}	$CH_3O(CH_2O_2)C_6H_2C(=NOH)COCH_3$†	$CH_3O(CH_2O_2)C_6H_2CONHCOCH_3$	$(CH_3CO)_2O$	197
	α-Benzil monoxime	Dibenzamide	PCl_5, $(C_2H_5)_2O$	5, 544a
	β-Benzil monoxime	$C_6H_5C(Cl)=NCOC_6H_5$ Benzoylformanilide	PCl_5	198
	γ-Benzil monoxime	$C_6H_5SO_2OCCOC_6H_5$ $\Vert$ C_6H_5N	PCl_5, $(C_2H_5)_2O$ $C_6H_5SO_2Cl$, pyridine	5, 544a 95
C_{15}	$2,4\text{-}(O_2N)_2C_6H_3COC(=NOH)C_6H_5$	$2,4\text{-}(O_2N)_2C_6H_3COC(Cl)=NC_6H_5$	PCl_5, $(C_2H_5)_2O$	192
	$p\text{-}CH_3OC_6H_4C(=NOH)COC_6H_5$	p-Anisoylformanilide, p-anisic acid, and p-anisoylformic acid	PCl_5, $(C_2H_5)_2O$	185
	$p\text{-}CH_3OC_6H_4COC(=NOH)C_6H_5$	p-Anisic acid and benzoic acid	PCl_5, $(C_2H_5)_2O$	185
C_{21}	$C_6H_5COC(C_6H_5)=CHC(=NOH)C_6H_5$	$C_6H_5COC(C_6H_5)=CHNHCOC_6H_5$‡	$C_6H_5SO_2Cl$, pyridine	545

Note: References 338 to 593 are on pp. 152–156.

† The location of the methoxyl and methylenedioxy groups has not been established.
‡ The same reaction may be obtained using aqueous sodium hydroxide instead of pyridine.

TABLE VIII
Dioximes of Diketones

No. of C Atoms	Starting Material	Products (% Yield)	Catalysts and Experimental Conditions	References
C_6	1,4-Cyclohexanedione dioxime	Succinic acid, ethylene diamine, and alanine	$p\text{-}CH_3C_6H_4SOCl$, pyridine	204
	1,4-Cyclohexanedione dioxime dihydrochloride	1,4-Diamino-2-chlorobenzene	Polyphosphoric acid	208
C_8	Benzoylformohydroxamic acid oxime	4-Hydroxy-5-phenyl-1,2,3-oxadiazole	$POCl_3$	200
	α-Benzoylformohydroxamic acid oxime	3-Phenyl-5-hydroxy-1,2,4-oxadiazole and benzonitrile	PCl_5, $(C_2H_5)_2O$	200
		Monoanilide of oxalic acid monohydroxamic acid	PCl_5, $(C_2H_5)_2O$	201
	β-Benzoylformohydroxamic acid oxime	3-Phenyl-5-amino-1,2,4-oxadiazole	$POCl_3$	200
		Benzonitrile and 3-phenyl-5-hydroxy-1,2,4-oxadiazole	PCl_5, $(C_2H_5)_2O$	201
	α-Benzoylformohydroxamic acid oxime disodium salt	Isomeric β-oxime and 3-phenyl-5-hydroxy-1,2,4-oxadiazole	CH_3COCl, C_6H_6	201
	Benzoylformohydroxamic acid chloride oxime	Benzonitrile and 3-phenyl-5-chloro-1,2,4-oxadiazole	PCl_5, $(C_2H_5)_2O$	200
		Monoanilide of oxalic acid hydroxamic acid and 4-chloro-5-phenyl-1,2,3-oxadiazole	Steam distil	200
	α-Benzoylformohydroxamic acid amide oxime	4-Amino-5-phenyl-1,2,3-oxadiazole	$POCl_3$	199
	β-Benzoylformohydroxamic acid amide oxime	5-Amino-3-phenyl-1,2,4-oxadiazole	$POCl_3$	199

TABLE VIII—Continued
DIOXIMES OF DIKETONES

No. of C Atoms	Starting Material	Product (% Yield)	Catalysts and Experimental Conditions	References
C_9	α-Benzoylacetyl dioxime	3-Phenyl-5-methyl-1,2,4-oxadiazole	$POCl_3$	200
	β-Benzoylacetyl dioxime	3-Phenyl-5-methyl-1,2,4-oxadiazole	$POCl_3$	200
	α-p-Toluylformohydroxamic acid amide oxime	4-Amino-5-p-tolyl-1,2,3-oxadiazole	$POCl_3$	199
	β-p-Toluylformohydroxamic acid amide oxime	5-Amino-3-p-tolyl-1,2,4-oxadiazole	$POCl_3$	199
C_{10}	$C_6H_5COC(NOH)C(NOH)NH_2$	4-Amino-5-benzoyl-1,2,3-oxadiazole	$POCl_3$	199
	α-Phenyldiacetyl dioxime	3-Benzoyl-5-methyl-1,2,4-oxadiazole	$POCl_3$	200
	$H_3CC{-}C{-}CC_6H_5$ $\| \|$ $N NOH$ $\diagdown O \diagup$	$H_3CC{-}{-}CCONHC_6H_5$ $\| \|$ $N N$ $\diagdown O \diagup$	$POCl_3$	206
C_{14}	Benzil dioxime	Unidentified product	H_2SO_4	203, 204
		No reaction	$HCl, (CH_3CO)_2O, CH_3CO_2H$	203, 204
		3,5-Diphenyl-1,2,4-oxadiazole	PCl_5	204
		$C_6H_5C(Cl){=}NN{=}C(Cl)C_6H_5$	PCl_5, PBr_3 or $POCl_3, (C_2H_5)_2O$	203
	α-Benzil dioxime	4,5-Diphenyl-1,2,3-oxadiazole		
		3,5-Diphenyl-1,2,4-oxadiazole	$POCl_3; PCl_5, (C_2H_5)_2O$	200
		$C_6H_5C(Cl){=}N{-}N{=}C(Cl)C_6H_5$	$POCl_3; PCl_5, (C_2H_5)_2O$	200
		3,5-Diphenyl-1,2,4-oxadiazole	$POCl_3$	200
		N-Phenyl-N′-benzoylurea	$PCl_5, (C_2H_5)_2O$	200

β-Benzil dioxime	Aniline, carbon dioxide, sulfanilic acid, ammonia, and carbon monoxide	$POCl_3$; PCl_5, $(C_2H_5)_2O$ or C_6H_6; H_2SO_4; P_2O_5	203
	Dibenzamide	PCl_5	203
	Oxalic acid dianilide	$POCl_3$; PCl_5, $(C_2H_5)_2O$	200
α-Benzil dioxime monomethyl ether	3,5-Diphenyl-1,2,3-oxadiazole	PCl_5	202
β-Benzil dioxime monomethyl ether	$C_6H_5CCONHC_6H_5$ $\\|$ $NOCH_3$	PCl_5	202
γ-Benzil dioxime monomethyl ether	$C_6H_5CCONHC_6H_5$ $\\|$ CH_3ON	PCl_5	202
Benzoylformohydroxamic acid anilide oxime	4-Anilino-5-phenyl-1,2,3-oxadiazole	$POCl_3$	199
1,3-Diacetylazulene dioxime	1,3-Diacetamidoazulene (21) and 1-acetyl-3-acetamidoazulene (25)	PCl_5, $(C_2H_5)_2O$	207
1,3-Diacetylazulene dioxime diacetate	1,3-Diacetamidoazulene (6) and 1-acetyl-3-acetamidoazulene (16)	Aq. CH_3CO_2H	207
	1,3-Diacetamidoazulene (30), 1-acetyl-3-acetamidoazulene (30), and 1-acetyl-3-acetamidoazulene oxime (11)	Al_2O_3, aq. NaOH	207
	1,3-Diacetamidoazulene (0–50), 1-acetyl-3-acetamidoazulene (20–70), and 1-acetyl-3-acetamidoazulene oxime (2–30)	CH_3CO_2Na, $(CH_3CO)_2O$, C_2H_5OH; CH_3CO_2Na, CH_3CO_2H, C_2H_5OH; CH_3CO_2H, C_2H_5OH	207

TABLE VIII—Continued

DIOXIMES OF DIKETONES

No. of C Atoms	Starting Material	Products (% Yield)	Catalysts and Experimental Conditions	References
C_{16}	4,4'-Dimethoxybenzil α-dioxime	3,5-Di-p-anisyl-1,2,4-oxadiazole	$POCl_3$	200
		3,5-Di-p-anisyl-1,2,4-oxadiazole and oxalic acid di-p-aniside	$POCl_3$	200
C_{17}	1,5-Diphenylpentane-1,5-dione dioxime	Glutaric acid dianilide	$POCl_3$	206

C_{30}	Cyclotriconta-1,16-dione dioxime	CO—$(CH_2)_{14}$—NH
		  | |
		NH—$(CH_2)_{14}$—CO (85)
		or
		CO—$(CH_2)_{14}$—CO
		  | |
		NH—$(CH_2)_{14}$—NH
			H_2SO_4	122

TABLE IX

QUINONE OXIMES

No. of C Atoms	Starting Material	Products (% Yield)	Catalysts and Experimental Conditions	References
C_6	1,4-Benzoquinone monoxime	4,4'-Dihydroxyazoxybenzene (45)	$C_6H_5SO_2Cl$, pyridine	196
		1-Benzenesulfonoxy-4-nitrosobenzene; 1-aza-2,5-dioxo-3,6-cycloheptadiene	$C_6H_5SO_2Cl$, pyridine	195
	1,4-Benzoquinone dioxime	Unidentified product	HCl; CH_3CO_2H, $(CH_3CO)_2O$; PCl_5 $SOCl_2$; HCl $C_6H_5SO_2Cl$	195
		No reaction		
		1,4-Benzoquinone dioxime dibenzenesulfonate		
C_{10}	1,2-Naphthoquinone 2-oxime	Unidentified product, $C_{10}H_6ClNO$	PCl_5, petroleum ether	195
		![structure] (27-33)	$HCl, (CH_3CO)_2O$, CH_3CO_2H; C_6H_5COCl pyridine	195
	1,2-Naphthoquinone dioxime	![structure]	PCl_5, petroleum ether	195
	1,4-Naphthoquinone monoxime	1-Acetoxy-2,3-dichloro-2,3-dihydro-4-nitrosonaphthalene	$HCl, (CH_3CO)_2O$, CH_3CO_2H	195

TABLE IX—Continued
QUINONE OXIMES

No. of C Atoms	Starting Material	Products (% Yield)	Catalysts and Experimental Conditions	References
C_{10} *(continued)*	1,4-Naphthoquinone monoxime *(continued)*	1-Benzenesulfonoxy-4-nitroso-2,3-dihydronaphthalene	$C_6H_5SO_2Cl$, pyridine	195
		Unidentified product	CH_3COCl	195
	1,4-Naphthoquinone dioxime	2,3-Dihydro-1,4-naphthoquinone dioxime diacetate	HCl, $(CH_3CO)_2O$, CH_3CO_2H	195
C_{12}	Acenaphthenequinone monoxime		$C_6H_5SO_2Cl$, pyridine	95
			HCl, $(CH_3CO)_2O$, CH_3CO_2H	187
			Heat	186

C_{14}	Anthraquinone monoxime	(structure: CONH-xanthone-like with C=O)	Polyphosphoric acid	195
	Anthraquinone dioxime	Dianthranilide (85)	Polyphosphoric acid	546
	syn-1-Chloro-anti-5-chloroanthraquinone dioxime	(structure with Cl, N, C=O) or (58)	Polyphosphoric acid; $(CH_3CO)_2O$, CH_3CO_2H, HCl	195, 546
	anti-1-Chloro-anti-5-chloroanthraquinone dioxime	4,10-Dichloroanthranilide (72)	Polyphosphoric acid	546
	Phenanthraquinone monoxime	2,2′-Diphenic acid imide (80)	HCl, $(CH_3CO)_2O$, CH_3CO_2H	81
		2,2′-Diphenic acid imide (40–50) and 1-carboxyfluorenone amide (45)	H_2SO_4	81
	Phenanthraquinone dioxime	1-Carboxyfluorenenone (80)	HCl, $(CH_3CO)_2O$, CH_3CO_2H	81

Note: References 338 to 593 are on pp. 152–156.

TABLE IX—*Continued*

QUINONE OXIMES

No. of C Atoms	Starting Material	Products (% Yield)	Catalysts and Experimental Conditions	References
C_{16}	Aceanthrenequinone monoxime	1,9-Anthracenedicarboximide (100)	HCl, $(CH_3CO)_2O$, CH_3CO_2H; H_2SO_4	103
C_{18}	Chrysoquinone monoxime	2-(o-Benzamido)-1-naphthoic acid and 2-(o-benzoic acid)-1-naphthamide	H_2SO_4	547
		6-Benzo[3,4,b]fluorenonecarboxylic acid	H_2SO_4	548

Note: References 338 to 593 are on pp. 152–156.

TABLE X

Cleavage of Oximes and Oxime Derivatives
("Second Order" Beckmann Rearrangement)

No. of C Atoms	Starting Material	Products (% Yield)	Catalysts and Experimental Conditions	References
C_3	Isonitrosoacetone	Pyruvic acid	Isopropyl phosphonofluorodate, Na_2HPO_4	549
C_4	Diacetyl monoxime	Acetyl chloride and acetaldoxime	HCl	550
C_5	Cyclopentanone oxime	Pentenonitrile	P_2O_5	65
		δ-Valerolactam (27) and 4-pentenonitrile	B_2O_3 and Al_2O_3*	148
C_6	Cyclohexanone oxime	5-Hexenonitrile	P_2O_5; SiO_2, NH_3; B_2O_3-Al_2O_3*	146, 148
C_7	2-Methylcyclohexanone oxime	Heptenonitrile	P_2O_5	147
C_8	Isonitrosoacetophenone	Benzoic acid	Isopropyl phosphonofluorodate, Na_2HPO_4, $(CH_3)_2CHOH$	549
	$C_6H_5C(=NOH)C(=NOH)OH$	Benzonitrile and 3-phenyl-4-hydroxy-1,2,4-oxadiazole	PCl_5, $POCl_3$	200
	Isatin 3-monoxime	2-Cyanophenyl isocyanate	PCl_5; PCl_3	99, 188
	2-Keto-3-oximino-2,3-dihydrobenzothiophene	2-Cyanophenylsulfenyl chloride	PCl_5	188
C_9	Acetyl benzoyl dioxime NOH ‖ NOH	Benzonitrile and benzoyl chloride	PCl_5, $(C_2H_5)_2O$	200
	o-$NO_2C_6H_4CCOCO_2H$	2-Nitrobenzonitrile and oxalic acid		550a
	1-Methylisatin 3-oxime	2-Cyano-N-methylphenylcarbamyl chloride	PCl_5, $(C_2H_5)_2O$	188

Note: References 338 to 593 are on pp. 152–156.

* This reaction was run in the vapor phase under reduced pressure.

TABLE X—Continued
CLEAVAGE OF OXIMES AND OXIME DERIVATIVES
("Second Order" Bechmann Rearrangement)

No. of C Atoms	Starting Material	Products (% Yield)	Catalysts and Experimental Conditions	References
C_9 (continued)	Spiro-[4,4]-nonan-1-one oxime	$\Delta^{8,9}$-Hydrinden-4-one	Polyphosphoric acid	149
		6-Azaspiro-[4,5]-decan-7-one and 4-cyclopentylidenebutyronitrile	$SOCl_2$	149
C_{10}	Camphor oxime	Unidentified nitriles	$SOCl_2$; aq. HCl	231, 551
		2,3,3-Trimethylcyclopentane-4-acetic acid	Conc. HCl	552
		Unidentified nitrile and camphor oxime anhydride	Aq. HCl; H_2SO_4	553
		α-Campholenic amide, α-campholenic acid, campholenonitrile, and bornylamine	Cu, H_2 (200°)	556
	Isonitrosocamphor	1,2,2-Trimethyl-3-cyanocyclopentene-1-carboxylic acid (100)	PCl_5, ligroin; $(CH_3CO)_2O$	155
		2,3,3-Trimethyl-1-cyclopentene-4-carbonitrile (40)	PCl_5, $(C_2H_5)_2O$	150
	anti-α-Isonitrosocamphor	1,2,2-Trimethyl-3-cyanocyclopentane-1-carboxylic acid	PCl_5, $(C_2H_5)_2O$	154
		1,2,2-Trimethylcyclopentane-1,3-dicarboxylic acid and its anhydride	PCl_5, $(C_2H_5)_2O$	554
	syn-α-Isonitrosocamphor	1,2,2-Trimethyl-3-cyanocyclopentane-1-carboxylic acid and 1,2,2-trimethylcyclopentane-1,3-dicarboximide	PCl_5, $(C_2H_5)_2O$	154

Oxime	Products	Reagent	Ref.
anti-Isonitrosocamphor oxime	1,2,2-Trimethylcyclopentane-1,3-dicarboximide (50); 1,2,2-trimethyl-3-cyanocyclopentane-1-carboxylic acid (20); 1,2,2-trimethylcyclopentane-1,3-dicarboxylic acid (3)	Aq. H_2SO_4	554
	1,2,2-Trimethyl-3-cyanocyclopentane-1-carboxylic acid and 1,2,2-trimethylcyclopentane-1,3-dicarboxylic acid	Conc. H_2SO_4	554
	1,2,2-Trimethylcyclopentane-1,3-dicarboxylic acid	PCl_5, $(C_2H_5)_2O$	554
l-Menthone oxime	Menthononitrile and decylenic acid	Al_2O_3*	86
Camphenilone oxime	Camphocene nitrile (78–80) (structure not determined), and isocamphenyl oxime	CH_3COCl	555
Pinocamphone oxime	Pinocamphene nitrile	H_2SO_4; P_2O_5	557
β-peri-Camphanone oxime		Aq. H_2SO_4	151
1,2-Naphthoquinone 1-monoxime	2-Cyanocinnamoyl chloride	PCl_5, $(C_2H_5)_2O$	188
1,2-Naphthoquinone 2-monoxime	2-Chlorocarboxycinnamonitrile	PCl_5	188
α-Nitroso-β-naphthol	2-Cyanocinnamic acid	$C_6H_5SO_2Cl$, pyridine	95, 195, 210, 219
α-Furoin oxime	2-Isocyanofuran	$C_6H_5SO_2Cl$, aq. NaOH	210
β-Furoin oxime	2-Cyanofuran	$C_6H_5SO_2Cl$, aq. NaOH	210
1-Acetylisatin 3-oxime	2-Cyanophenyl isocyanate	PCl_5, $POCl_3$	99

Note: References 338 to 593 are on pp. 152–156.

* This reaction was run in the vapor phase under reduced pressure.

TABLE X—Continued
Cleavage of Oximes and Oxime Derivatives
("Second Order" Beckmann Rearrangement)

No. of C Atoms	Starting Material	Products (% Yield)	Catalysts and Experimental Conditions	References
C_{10} (continued)	Spiro-[4,5]-decan-1-one oxime,	3-Oxo-1-cyclodecene 7-Azaspiro-[5,5]-undecan-8-one and 4-cyclohexylidenebutyronitrile	Polyphosphoric acid $SOCl_2$	149 149
	Spiro-[4,5]-decan-6-one oxime	2-Cyclopentylidenecyclopentanone and δ-cyclopentylidenevaleramide	Polyphosphoric acid	149
C_{11}	$C_6H_5CH=CHCHCOCH_3$ ‖ NOH	Cinnamic acid (40)	$(COCl)_2$	558
	$p\text{-}CH_3C_6H_4C\underset{NOCOCH_3}{\overset{COCOCH_3}{=}}NOCOCH_3$	p-Toluinitrile and p-toluylformic acid oxime; 3-p-tolyl-5-hydroxy-1,2,4-oxadiazole	NaOH	568
	$C_6H_5COCCOCOCH_3$ ‖ NOH	Benzoyl cyanide, acetic acid, and carbon monoxide	C_6H_6	559
	Spiro-[5,5]-undecan-1-one oxime 3-Methylcamphor oxime	Bicyclo-[5.4.0]-10-undecene-4-one 2,3,3-Trimethyl-4-α-cyanoethyl-1-cyclopentene	Polyphosphoric acid Conc. HCl	149 560
C_{12}	$C_6H_5CHOHC\underset{HON}{=}\overset{O}{\underset{}{\bigcirc}}$	Benzaldehyde and 2-cyanofuran	$C_6H_5SO_2Cl$, aq. NaOH	210

THE BECKMANN REARRANGEMENT

C_6H_5CHOHC(=NOH)– (furan ring)	Benzaldehyde and 2-isocyanofuran	$C_6H_5SO_2Cl$, aq. NaOH	210
2a,3,4,5-Tetrahydro-4-oximino-5-acenaphthenone	7-Carboxy-1-indonacetonitrile (70)	$C_6H_5SO_2Cl$, pyridine	561
α-(N,N-Dimethylamino)ethyl piperonyl ketoxime	3,4-Methylenedioxybenzonitrile	$SOCl_2$, $CHCl_3$	380
C_{14}			
Benzoin oxime	Benzaldehyde and benzonitrile	$C_6H_5SO_2Cl$, aq. NaOH	95, 210
α-Benzoin oxime	Unidentified material	KCN	211
β-Benzoin oxime	Benzaldehyde and phenyl isocyanide	$C_6H_5SO_2Cl$, aq. NaOH	95, 210
α-Benzoin oxime acetate	Benzonitrile, benzaldehyde, and benzoin	Aq. NaOH	211
	Benzaldehyde and benzonitrile	Heat with water	211
	β-Benzoin oxime (100)	Aq. NaOH	211
	Benzaldehyde, benzonitrile, and phenyl isocyanide	KCN, aq. C_2H_5OH	211
α-Benzoin oxime mesitoate	Mesitoic acid, benzaldehyde, and benzonitrile	NaOH, CH_3OH; Na_2CO_3, CH_3OH	562
β-Benzoin oxime mesitoate	Mesitoic acid, benzaldehyde, and benzonitrile	NaOH, CH_3OH; Na_2CO_3, CH_3OH	562
Benzil monoxime	Benzonitrile and benzoic acid	$C_6H_5SO_2Cl$, pyridine	95
α-Benzil monoxime	Benzoic acid and benzonitrile	Aq. NaOH	211
β-Benzil monoxime	No reaction	Aq. NaOH	211
Benzil monoxime acetate	Benzonitrile and benzoic acid	10% NaOH	230
α-Benzil monoxime acetate	Benzoic acid and benzonitrile	Heat	211
β-Benzil monoxime acetate	No reaction	Heat	211
Benzil monoxime propionate	Benzoic acid and benzonitrile	Conc. NH_4OH	230
Benzil monoxime ethoxalate	Benzonitrile and benzoic acid	10% NaOH	230
Benzil monoxime benzoate	$C_6H_5C(Cl){=}NCOC_6H_5$	PCl_5, $(C_2H_5)_2O$	198

Note: References 338 to 593 are on pp. 152–156.

TABLE X—*Continued*

CLEAVAGE OF OXIMES AND OXIME DERIVATIVES

("Second Order" Beckmann Rearrangement)

No. of C Atoms	Starting Material	Products (% Yield)	Catalysts and Experimental Conditions	References
C_{14} (continued)	α-Benzil monoxime benzoate	Benzoic acid (90)	Aq. NaOH	5
	β-Benzil monoxime benzoate	Benzonitrile (80) and benzoic acid (94)	Aq. NaOH	5
	Benzil monoxime cinnamate	Benzonitrile, benzoic acid (100), and cinnamic acid (100)	10% NaOH	230
	γ-Benzil dioxime diacetate	3,4-Diphenylfurazan	Aq. NaOH	217
	α-Benzil dioxime dibenzoate	α-Benzil monoxime, benzoic acid, and aniline	25% NaOH	217
	β-Benzil dioxime dibenzoate	β-Benzil monoxime	15% NaOH	217
	γ-Benzil dioxime dibenzoate	3,4-Diphenylfurazan	Aq. NaOH	217
	syn-Phenyl 2,4-dinitrobenzoyl ketoxime	Benzonitrile and 2,4-dinitrobenzoic acid	PCl_5, $(C_2H_5)_2O$	190
		2-Hydroxy-4-nitrobenzonitrile	NaOH	192
		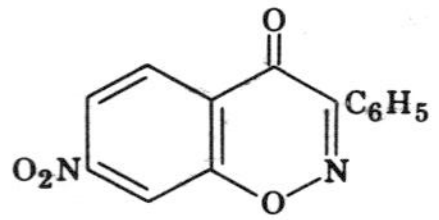 (structure shown)	C_2H_5OH	190
	anti-Phenyl 2,4-dinitrobenzoyl ketoxime	Benzoic acid and 2,4-dinitrobenzonitrile	PCl_5, $(C_2H_5)_2O$	190
		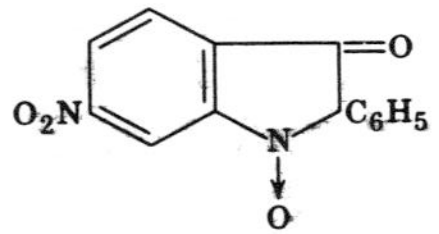 (structure shown)	$NaOC_2H_5$	190

Oxime	Products	Reagents	Ref.
C₆H₅CCl=NCOC₆H₃(NO₂)₂-2,4	Benzonitrile and 2,4-dinitrobenzoic acid	Aq. NaOH	192
C₆H₅C(Cl)=NOCOC₆H₅	Benzonitrile and benzoyl chloride	Heat	198
9,10-Phenanthraquinone monoxime	4-Cyanofluorenone	PCl₅, (C₂H₅)₂O	188
	2-Cyanobiphenyl-2′-carboxylic acid	C₆H₅SO₂Cl, pyridine	95
2-Nitro-9,10-phenanthraquinone 10-monoxime	2-Cyano-4-nitrobiphenyl-2′-carboxylic acid	C₆H₅SO₂Cl, pyridine	95
2,7-Dinitro-9,10-phenanthraquinone monoxime	2-Cyano-4,4′-dinitrobiphenyl-2′-carboxylic acid	C₆H₅SO₂Cl, pyridine	95
(structure: benzisoxazinone with C₆H₅, O₂N)	2-Hydroxy-4-nitrobenzoic acid and benzonitrile	NaOH	190
C₁₅			
Methylbenzoin oxime	Acetophenone, benzoin, and desylacetophenone	C₆H₅SO₂Cl, pyridine	211
C₆H₅CH—C=C₆H₄OCH₃-p OH HON	Benzaldehyde (94) and p-anisonitrile (98)	C₆H₅COCl, pyridine	216
syn-o-Tolyl benzoyl ketoxime	Benzoic acid and o-tolunitrile	NaOH	191
syn-Phenyl p-anisoyl ketoxime	Benzoic acid (99) and o-tolunitrile	NaOH, C₂H₅OH	191
anti-Phenyl p-anisoyl ketoxime benzoate	Benzonitrile, benzoic acid, and p-anisic acid	Aq. NaOH	185
	p-Anisoylformanilide (55), p-anisic acid, and p-anisoylformic acid	PCl₅, (C₂H₅)₂O	185
α-Piperidinylethyl piperonyl ketoxime	3,4-Methylenedioxybenzonitrile	SOCl₂, CHCl₃; p-CH₃C₆H₄SO₂Cl, aq. NaOH, acetone	380

Note: References 338 to 593 are on pp. 152–156.

TABLE X—Continued

Cleavage of Oximes and Oxime Derivatives
("Second Order" Beckmann Rearrangement)

No. of C Atoms	Starting Material	Products (% Yield)	Catalysts and Experimental Conditions	References
C_{16}	$C_6H_5CHOHCC_6H_4N(CH_3)_2$-$p$ ‖ NOH	Benzaldehyde (86) and p-(N,N-dimethylamino)phenyl isocyanide	$C_6H_5SO_2Cl$, pyridine	216
	$C_6H_5CHOHCC_6H_4N(CH_3)_2$-$p$ ‖ HON	p-(N,N-Dimethylamino)benzonitrile and benzoic acid	$C_6H_5SO_2Cl$, pyridine	563
	p-$(CH_3)_2NC_6H_4CHOHCC_6H_5$ ‖ HON	Benzoic acid and p-(N,N-dimethylamino)benzonitrile	$C_6H_5SO_2Cl$, pyridine	218
		Benzaldehyde and p-(N,N-dimethylamino)benzonitrile	$SOCl_2$, $CHCl_3$	564
	p-$(CH_3)_2NC_6H_4CHOHCC_6H_5$ ‖ NOH	Benzaldehyde and p-N,N-dimethylaminophenyl isocyanide	$C_6H_5SO_2Cl$, NaOH	216
	$C_6H_5CHOHCC_6H_3(CH_2OH)_2$-3,4 ‖ HON	Benzaldehyde (92) and 3,4-bis(hydroxymethyl)benzonitrile (98)	$C_6H_5SO_2Cl$, NaOH	216
	2-$ClC_6H_4CHOHCC_6H_3(OCH_3)_2$-3,4 ‖ HON	2-Chlorobenzaldehyde (78) and 3,4-dimethoxybenzonitrile (61)	$C_6H_5SO_2Cl$, NaOH	216
	2-$ClC_6H_4CHOHCC_6H_4N(CH_3)_2$-4 ‖ HON	2-Chlorobenzaldehyde (77) and 4-(N,N-dimethylamino)benzonitrile (83)	$C_6H_5SO_2Cl$, NaOH	216

THE BECKMANN REARRANGEMENT

Oxime	Product	Reagent	Ref.
2-ClC$_6$H$_4$CHOHCC$_6$H$_3$(CH$_2$OH)$_2$-3,4 ‖ HON	2-Chlorobenzaldehyde (65) and 3,4-bis(hydroxymethyl)benzonitrile (56)	C$_6$H$_5$SO$_2$Cl, NaOH	216
2-CH$_3$OC$_6$H$_4$CHOHCC$_6$H$_4$OCH$_3$-4 ‖ HON	2-Methoxybenzaldehyde (96) and anisole (98)	C$_6$H$_5$SO$_2$Cl, NaOH	216
Dioximinothebenone	Thebedinitrile (55)	p-CH$_3$C$_6$H$_4$SO$_2$Cl, pyridine	565
epi-Dioximinothebenone	epi-Thebedinitrile (28)	p-CH$_3$C$_6$H$_4$SO$_2$Cl, pyridine	565
C$_{17}$ C$_6$H$_5$CC(OCH$_3$)$_2$C$_6$H$_4$OCH$_3$-p ‖ NOH	Benzonitrile and p-anisic acid	HCl, (CH$_3$CO)$_2$O, CH$_3$CO$_2$H	185
C$_{18}$ 2-Methyl-7-isopropyl-9,10-phenanthraquinone 10-oxime	2-Cyano-3-methyl-4′-isopropyl-biphenyl-2′-carboxylic acid	C$_6$H$_5$SO$_2$Cl, pyridine	95
C$_{19}$ 2,2-Diphenylcycloheptanone oxime	7,7-Diphenylheptamide and unidentified product, C$_{19}$H$_{19}$NO	Polyphosphoric acid	149
	7,7-Diphenyl-6-heptenonitrile (50–97)†	SOCl$_2$, C$_6$H$_6$; HCl, CH$_3$CO$_2$H	152
1-Methyl-4-phenyl-4-benzoyl-piperidine oxime	C$_6$H$_5$CN and 1-methyl-4-phenyl-Δ^{3,4}-piperidine	SOCl$_2$, CCl$_4$	90
Isonitrosocinchotoxin	1-Methyl-3-vinyl-4-cyanomethyl-piperidine (29) and quinoline-4-carboxylic acid	PCl$_5$, CHCl$_3$	181
1-Phenyl-1-benzoylcyclohexane oxime	C$_6$H$_5$CN and 2-phenylcyclohexene	SOCl$_2$, C$_6$H$_6$	90

Note: References 338 to 593 are on pp. 152–156.
† No product was obtained using sulfuric acid.

TABLE X—Continued

CLEAVAGE OF OXIMES AND OXIME DERIVATIVES ("Second Order" Beckmann Rearrangement)

No. of C Atoms	Starting Material	Products (% Yield)	Catalysts and Experimental Conditions	References
C_{20}	Phenylbenzoin oxime	Benzophenone and benzoin	$C_6H_5SO_2Cl$, pyridine	211
C_{25}	Flavothebaone trimethyl ether desazo-4-methine oxime	Flavothebaone trimethyl ether desazaneomethine and acetonitrile	$SOCl_2$ ($-10°$)	566
		1,2,7,10-Tetramethoxy-11-vinyl chrysofluorene	$SOCl_2$ ($-5°$)	566
	Flavothebaone trimethyl ether hexahydrodesazomethine oxime	Flavothebaone trimethyl ether dihydrodesazaneomethine and acetonitrile	$SOCl_2$	566
	Flavothebaone trimethyl ether hexahydrodesazomethine oxime	Unidentified product, $C_{26}H_{31}NO_5$	$SOCl_2$	566
C_{27}	Flavothebaone trimethyl ether 4-methine oxime	Flavothebaone trimethyl ether neomethine and acetonitrile	$SOCl_2$	566

Note: References 338 to 593 are on pp. 152–156.

TABLE XI

ALDOXIMES

No. of C Atoms	Starting Material	Products (% Yield)	Catalysts and Experimental Conditions	References
C_1	Nitrole	Nitrous acid, isocyanic acid	Raney nickel	277
C_2	Acetaldoxime	Acetamide (88, 86)	Raney nickel	226, 227
		Acetaldehyde and unidentified amine	Cu, H_2 (carrier gas)	223
		Nitroacetonitrile	$SOCl_2, (C_2H_5)_2O$	234
	α-Nitroacetaldoxime		CF_3CO_2H	82
C_4	Butyraldoxime	Butyramide	Polyphosphoric acid	567
	Succinaldoxime	Succinimide (5)	Heat	225
C_5	γ-Methylbutyraldoxime sodium salt	γ-Methylbutyronitrile (97)	Raney nickel, alone or with ethanol	226, 227
	Furfuraldoxime	Pyromucamide (88, —)	Cu, H_2 (carrier gas)	66, 223
		Pyromucamide (45), furfural (55)	Heat, C_6H_6	228
	Nickel tetrakisfurfuraldoxime	Pyromucamide (50) and nickel bis-(furfuraldoxime)		
C_7	n-Heptanaldoxime	n-Heptanamide (90, 74)	Raney nickel; BF_3	227, 569
	Benzaldoxime	Benzamide (65)	Raney nickel, $(C_2H_5)_2O$	226
		Benzamide (75–76)	Raney nickel	226, 227
		Benzamide (98)	BF_3, CH_3CO_2H	569
			90% H_2SO_4; CuCl, CuBr, $SbCl_3$, $C_6H_5CH_3$	1, 68
		Benzoic acid, benzamide, phenylnitromethane, benzohydroxamic acid, and 3,5-diphenyl-1,2,4-oxadiazole	$K_2S_2O_5, H_2O, H_2SO_4$	570
		Benzamide (50) and benzoic acid (12)	H_2SO_4	571

Note: References 338 to 593 are on pp. 152–156.

TABLE XI—Continued
ALDOXIMES

No. of C Atoms	Starting Material	Products (% Yield)	Catalysts and Experimental Conditions	References
C₇ (continued)	Benzaldoxime (continued)	Benzamide (52), benzonitrile (58), and benzoic acid (21)	Cu, H₂ (carrier gas)	67
		Benzonitrile	H₃BO₃-Al₂O₃, vapor phase, 250°	148
	syn-Benzaldoxime	Benzamide, benzoic acid, and benzonitrile, and ammonia	Cu, H₂ (carrier gas)	224
	anti-Benzaldoxime	Benzamide, benzoic acid, and benzonitrile	Cu, H₂ (carrier gas)	224
		Benzonitrile, sulfur dioxide, and hydrogen chloride	SOCl₂	231
	Sodium benzaldoxime	Benzamide (5), benzonitrile (86), benzoic acid (7), and ammonia	Heat	225
	4-Chlorobenzaldoxime	4-Chlorobenzamide (95)	BF₃	569
	3-Nitrobenzaldoxime	3-Nitrobenzamide (98, —)	BF₃; H₂SO₄	569, 571
	Salicylaldoxime	Salicylamide (47)	BF₃	569
	2-Azidobenzaldoxime	2-Oxy-1,2-benzodiazole, 2-azidobenzamide, 2-aminobenzaldehyde, 2-azidobenzoic acid, and anthranilic acid	Heat; aq. NaOH	239
	2-Aminobenzaldoxime	1-Acetylbenzodiazole or

![structure](structure.png)

| HCl; (CH₃CO)₂O, CH₃CO₂H | 240 |

	Oxime	Product	Reagent	Ref.
	2-Chloro-5-nitrobenzaldoxime	2-Chloro-5-nitrobenzonitrile (95)	PCl_5, $(C_2H_5)_2O$	230
	2-Chloro-5-nitrobenzaldoxime acetate	2-Chloro-5-nitrobenzoic acid	HCl	309
	syn-2,6-Dichloro-3-nitrobenz-aldoxime	2,6-Dichloro-3-nitrobenzonitrile	PCl_5, $(C_2H_5)_2O$	572
	Benzohydroxamic acid amide	Benzamide, benzoic acid, and benzonitrile	Cu, H_2 (carrier gas)	66
C_8	Anisaldoxime	Anisamide (70)	BF_3	569
	3-Methoxybenzaldoxime hydrochloride	3-Methoxybenzamide		573
	Piperonaldoxime	Unidentified product	BF_3	569
	Phenylglyoxal dioxime	3-Phenylfurazan	C_6H_5COCl, pyridine	574
	Phenylglyoxal monoxime	Benzoylformamide	$NaHSO_3$, 20% H_2SO_4	229
	1-Ethyl-3,4-dehydropiperidine-3-carboxaldehyde oxime	1-Ethyl-3,4-dehydro-3-cyanopiperidine hydrochloride	$SOCl_2$	236
	N-Glyoxyliminoaniline	Isatin	H_2SO_4	575
	N-α-Bromoglyoxyl-o-toluidine oxime	5-Bromo-7-methylisatin	H_2SO_4	576
	4-Dimethylaminobenzaldoxime	4-Dimethylaminobenzamide (95)	BF_3	569
C_9	Cinnamaldoxime	Isoquinoline	P_2O_5	237, 238
		Cinnamaldehyde, cinnamonitrile	Cu, H_2 (carrier gas)	66
		Cinnamamide (30)	Raney nickel	226, 227
		Isoquinoline	H_2SO_4	233
	Bis(cinnamaldoxime)copper(I) bromide	Cinnamamide	Heat, $C_6H_5CH_3$	68
	β-Chlorocinnamaldoxime	trans-β-Chlorocinnamonitrile (48)	PCl_5, $(C_2H_5)_2O$	232
	cis-anti-β-Chlorocinnamaldoxime	trans-β-Chlorocinnamonitrile	PCl_5, $(C_2H_5)_2O$	232
	cis-syn-β-Chlorocinnamaldoxime	trans-β-Chlorocinnamonitrile	PCl_5, $(C_2H_5)_2O$	232
	N-Glyoxylimino-3-chloro-6-methoxyaniline	4-Chloro-7-methoxyisatin	H_2SO_4	575

Note: References 338 to 593 are on pp. 152–156.

TABLE XI—Continued

ALDOXIMES

No. of C Atoms	Starting Material	Products (% Yield)	Catalysts and Experimental Conditions	References
C_9 (continued)	cis-α-Bromocinnamaldoxime	trans-α-Bromocinnamonitrile	PCl_5, $(C_2H_5)_2O$	391

	$C_6H_5N=CHCHCH=NOCOCH_3$
	  $\|$
	  NO_2
		$C_6H_5N=CHCHCN$
		  $\|$
		  NO_2
			HCl, $(CH_3CO_2)O$, CH_3CO_2H	246

C_{10}	Citronellaldoxime	Citronellonitrile (72–86)	$(CH_3CO)_2O$	577
		Citronellamide (50)	Raney nickel	578
	syn-Mesitaldoxime	Formomesidide, mesitonitrile	PCl_5, $(C_2H_5)_2O$	221
	anti-Mesitaldoxime	Mesitonitrile	PCl_5, $(C_2H_5)_2O$	221
C_{12}	N-Glyoxyloximino-2-amino-5,6,7,8-tetrahydronaphthalene	![structures] (70) (30)	90% H_2SO_4	241
C_{15}	1,2,3,4-Tetrahydro-9-anthraldehyde oxime	Uncharacterized product reduced to 9-anthraldehyde with $SnCl_2$	PCl_5, $(C_2H_5)_2O$	579

Note: References 338 to 593 are on pp. 152–156.

TABLE XII
NITRONES

No. of C Atoms	Starting Material	Products (% Yield)	Catalysts and Experimental Conditions	References
C_8	Phenyl N-methyl nitrone	N-Methylbenzamide		270
	2-Nitrophenyl N-methyl nitrone	N-Methyl-2-nitrobenzamide	$(CH_3CO)_2O$	270
	3-Nitrophenyl N-methyl nitrone	N-Methyl-3-nitrobenzamide	$(CH_3CO)_2O$	270
	3-Nitrophenyl N-methyl nitrone hydrochloride	N-Methyl-3-nitrobenzamide		270
	4-Nitrophenyl N-methyl nitrone	$4\text{-}O_2NC_6H_4C(=NCH_3)OCH_3$	KCN, CH_3OH	269
		N-Methyl-4-nitrobenzamide	$(CH_3CO)_2O$	270
C_9	2-Anisyl N-methyl nitrone	N-Methyl-2-anisamide		270
	4-Anisyl N-methyl nitrone	N-Methyl-N-acetyl-4-anisamide	$(CH_3CO)_2O$	270
C_{10}	3,4-(Methylenedioxy)phenyl N-methyl nitrone	N-Methyl-3,4-(methylenedioxy)-benzamide		270
	Cinnamyl N-methyl nitrone	N-Methylcinnamide		265
C_{13}	Phenyl N-phenyl nitrone	Benzanilide	CH_3COCl	267
	2-Nitrophenyl N-phenyl nitrone	2-Nitrobenzanilide	$(CH_3CO)_2O$	267
		N-Acetyl-2-nitrobenzanilide	KCN, C_2H_5OH	269
		2-Nitrobenzanilide	$KCN; CH_3OH$	269
		$2\text{-}O_2NC_6H_4C(OCH_3)=NC_6H_5$	$(CH_3CO)_2O; KCN, C_2H_5OH$	267, 269
	3-Nitrophenyl N-phenyl nitrone	3-Nitrobenzanilide		
		$3\text{-}O_2NC_6H_4C(OCH_3)=NC_6H_5$	KCN, CH_3OH	269
	4-Nitrophenyl N-phenyl nitrone	$4\text{-}O_2NC_6H_4C(OC_2H_5)=NC_6H_5$	KCN, C_2H_5OH	269
		4-Nitrobenzanilide	$(CH_3CO)_2O$	267
	2,4-Dinitrophenyl N-phenyl nitrone	N-Acetyl-2,4-dinitrobenzanilide	$HCl, CH_3CO_2H, H_2O;$ CH_3COCl	266, 267
		N-Acetyl-2,4-dinitrobenzanilide	$(CH_3CO)_2O$	267

TABLE XII—Continued

NITRONES

No. of C Atoms	Starting Material	Products (% Yield)	Catalysts and Experimental Conditions	References
C_{13} (continued)	2,4,6-Trinitrophenyl N-phenyl nitrone	2,4,6-Trinitrobenzanilide	CH_3COCl	267
C_{14}	2-Hydroxyphenyl N-phenyl nitrone	Salicylanilide		265
	Phenyl N-benzyl nitrone	N-Benzylbenzamide, ammonium benzenesulfonate, N,N,N-tribenzylaminosulfonate	$C_6H_5SO_2Cl$, C_6H_6	274
		N-Benzylbenzamide	$C_6H_5SO_2Cl$, H_2O	274
	x-Methoxyphenyl N-phenyl nitrone	Anisanilide		265
	2,4-Dinitrophenyl N-2-tolyl nitrone	2,4-Dinitro-2'-methylbenzanilide	KOH, C_2H_5OH; CH_3COCl	268
		2,4-$(O_2N)_2C_6H_3CON(COCH_3)$-$C_6H_4CH_3$-2	$(CH_3CO)_2O$, CH_3CO_2Na	268
	2,4-Dinitrophenyl N-3-tolyl nitrone	2,4-Dinitro-3'-methylbenzanilide	CH_3COCl	268
		2,4-$(O_2N)_2C_6H_3CON(COCH_3)$-$C_6H_4CH_3$-3	$(CHCO)_2O$, CH_3CO_2Na	268
	2,4-Dinitrophenyl N-4-tolyl nitrone	2,4-Dinitro-4'-methylbenzanilide	CH_3COCl	267
		2,4-$(O_2N)_2C_6H_3CON(COCH_3)$-$C_6H_4CH_3$-2	$(CH_3CO)_2O$	267
	Diphenyl N-methyl nitrone	Benzanilide (27)	PCl_5, $POCl_3$	272
		$CH_3CON(CH_3)OCOCH_3$ (40)	$(CH_3CO)_2O$	272
		Benzophenone (24) and methylamine	$SbCl_5$, $CHCl_3$	272
	$\overset{O}{\underset{\uparrow}{C_6H_5N}}=CHCH=\overset{O}{\underset{\uparrow}{NC_6H_5}}$	N,N'-Diphenyloxamide	CH_3CO_2H; $(CH_3CO)_2O$	580

	Starting Material	Product	Reagent	Ref.
C_{15}	p-Anisyl N-benzyl nitrone	N-Benzyl-p-anisamide, sulfur dioxide, water, ammonium benzenesulfonate N-Benzyl-p-anisamide	$C_6H_5SO_2Cl$; C_6H_6 Phthaloyl chloride or picryl chloride, C_6H_6	274 274
	p-Nitrophenyl N-(4-dimethylamino)-phenyl nitrone	N-Acetyl-4-nitro-4'-dimethylamino-benzanilide	$(CH_3CO)_2O$	267
	2,4-Dinitrophenyl N-(4-dimethylamino)phenyl nitrone	2,4-$(O_2N)_2C_6H_3CON(COCH_3)$-$C_6H_4N(CH_3)_2$-4	$(CH_3CO)_2O$	268
	2,4-Dinitrophenyl N-(4-dimethylamino)phenyl nitrone	Unidentified product	CH_3COCl, PCl_5	268
C_{16}	Benzoyl N-(4-dimethylamino)phenyl nitrone	$C_6H_5COCONHC_6H_4N(CH_3)_2$-4 (88)	$(CH_3CO)_2O$	271
		C_6H_5COCCN $\parallel$ $NC_6H_4N(CH_3)_2$-4	Aq. NaCN	271
		N-Formyl-4'-dimethylamino-benzanilide (55)	Uv light, acetone	581
		4'-Dimethylaminobenzanilide (14, 25, —)	Air, 14 da.; aq. Na_2CO_3; uv light, pyridine	581
		Benzoic acid	NH_3 or aq. NaOH	581
C_{17}	Phenyl N-α-naphthyl nitrone	N-α-Naphthylbenzamide	$(C_6H_5CO)_2O$, C_6H_5COCl, CH_3COCl	275
C_{21}	1-Anthraquinoyl N-phenyl nitrone	Anthraquinone-1-carboxylic acid	H_2SO_4, CH_3CO_2H	582
C_{22}	2,3,5-Triphenyl-3-hydroxy-$\Delta^{3,5}$-pyrroline N-oxide	N-β-Benzoylstyrylbenzamide	$PCl_5, (C_2H_5)_2O$	545
C_{26}	Diphenyl N-benzhydryl nitrone	Benzophenone oxime O-benzhydryl ether (100)		273

Note: References 338 to 593 are on pp. 152–156.

REFERENCES FOR TABLES I-XII

[338] Todd, Kenner, and Webb, *J. Chem. Soc.*, **1956**, 1231.
[339] Bredt and Boeddinghous, *Ann.* **251**, 316 (1889).
[340] Hantsch, *Ber.*, **24**, 4018 (1891).
[341] Roberts and Chambers, *J. Am. Chem. Soc.*, **73**, 3176 (1951).
[342] (To Badische Anilin und Soda Fabrik), Ger. pat. 858,397 (1952) [*C.A.*, **48**, 12810 (1954)].
[343] Scholl, Weil, and Halderman, *Ann.*, **338**, 1 (1904).
[344] Schmidt and Furnee, *Arch. Pharm.*, **236**, 334 (1898) (*Chem. Zentr.*, **1898, II**, 629, 631).
[345] Nef, *Ann.*, **310**, 316 (1895).
[346] Meyer and Warrington, *Ber.*, **20**, 500 (1887).
[347] Hart and Curtis, *J. Am. Chem. Soc.*, **78**, 112 (1956).
[348] Drake, Kline, and Rose, *J. Am. Chem. Soc.*, **56**, 2076 (1934).
[349] Mayer and Scharvin, *Ber.*, **30**, 2862 (1897).
[350] Burrows and Eastman, *J. Am. Chem. Soc.*, **79**, 3756 (1957).
[351] Thoms, *Ber. deut. pharm. Ges.*, **11**, 3 (1900) (*Chem. Zentr.*, **1901, I**, 524).
[352] Houben, *Ber.*, **35**, 3587 (1902).
[353] Schmidt, *Arch. Pharm.*, **237**, 222 (1899) (*Chem. Zentr.*, **1900, II**, 581).
[354] Wallach, *Ann.*, **389**, 169 (1912).
[355] Rupe, Collin, and Schmiderer, *Helv. Chim. Acta*, **14**, 1340 (1931).
[356] Furukawa, *Sci. Papers Inst. Phys. Chem. Research (Tokyo)*, **20**, 71 (1933) [*C.A.*, **27**, 2131 (1933)].
[357] Blaise and Guerin, *Bull. soc. chim. France*, [3], **29**, 1211 (1903).
[358] Asahina and Hongo, *J. Pharm. Soc. Japan*, **506**, 227 (1924) [*C.A.*, **18**, 2510 (1924)].
[359] Shukoff and Schestakoff, *J. Russ. Phys.-Chem. Soc.*, **35**, 1 (1903) (*Chem. Zentr.*, **1903, I**, 825).
[360] Minunni, *Gazz. chim. ital.*, **27, II**, 263 (1899).
[361] Furukawa, *Sci. Papers Inst. Phys. Chem. Research (Tokyo)*, **19**, 27 (1932) [*C.A.*, **27**, 303 (1933)].
[362] Kohler and Heritage, *Am. Chem. J.*, **34**, 568 (1905).
[363] Asano and Takashi, *J. Pharm. Soc. Japan*, **65**, 811 (1945) [*C.A.*, **45**, 4303 (1951)].
[364] Tokura, Asami, and Todo, *J. Am. Chem. Soc.*, **79**, 3135 (1957).
[365] Stephen and Staskun, *J. Chem. Soc.*, **1956**, 4708.
[366] Richter, *J. Org. Chem.*, **21**, 619 (1956).
[367] von Auwers and Meyenburg, *Ber.*, **24**, 2370 (1891).
[368] Meisenheimer, Zimmermann, and Kummer, *Ann.*, **446**, 205 (1925).
[369] Korten and Scholl, *Ber.*, **34**, 1901 (1901).
[370] Callet, *Bull. soc. chim. France*, [3] **27**, 539 (1902).
[371] Jenkins and Richardson, *J. Am. Chem. Soc.*, **55**, 1618 (1933).
[372] Graziano, *Gazz. chim. ital.*, **45, II**, 390 (1915).
[373] Kissman and Williams, *J. Am. Chem. Soc.*, **72**, 5323 (1950).
[374] Buu-Hoï and Xuong, *J. Chem. Soc.*, **1953**, 386.
[375] Tshizaka, *Ber.*, **47**, 2460 (1914).
[376] Borsche and John, *Ber.*, **57**, 656 (1924).
[377] Buu-Hoï, Jacquignon, and Lavit, *J. Chem. Soc.*, **1956**, 2593.
[377a] Eckert, Royer, and Buu-Hoï, *Compt. rend.*, **233**, 1461 (1951).
[378] Bachmann and Dice, *J. Org. Chem.*, **12**, 876 (1947)
[379] Richtzenhain and Nippus, *Chem. Ber.*, **82**, 408 (1949).
[380] Ohara, *J. Pharm. Soc. Japan*, **71**, 1244 (1951) [*C.A.*, **46**, 5552 (1952)].
[381] Scharwin, *Ber.*, **35**, 2511 (1902).
[382] Anderson, Fritz, and Scotoni, *J. Am. Chem. Soc.*, **79**, 6511 (1957).
[383] Scharvin, *Ber.*, **30**, 2862 (1897).
[384] van der Zanden, de Vries, and Dijkstra, *Rec. trav. chim.*, **61**, 280 (1942).
[385] Jenkins, *J. Am. Chem. Soc.*, **55**, 703 (1933).
[386] Borsche and Pammer, *Ber.*, **54**, 102 (1921).
[387] Close, *J. Am. Chem. Soc.*, **79**, 1455 (1957).
[388] Arnold and Craig, *J. Am. Chem. Soc.*, **72**, 2728 (1950).

[389] Berezovskii, Kurdyukova, and Preobrazhenskii, *Zhur. Priklad. Khim.*, **22**, 533 (1949) [*C.A.*, **44**, 2463 (1950)].
[390] Buck and Ide, *J. Am. Chem. Soc.*, **53**, 1536 (1931).
[391] von Auwers and Seyfried, *Ann.*, **484**, 178 (1930).
[392] Henrich, *Ann.*, **351**, 172 (1907).
[393] Perold and Reiche, *J. Am. Chem. Soc.*, **79**, 465 (1957).
[394] Brewster and Kline, *J. Am. Chem. Soc.*, **74**, 5179 (1952).
[395] Vinkler and Autheried, *Acta Univ. Szegediensis, Chem. et Phys.*, **2**, 50 (1948) [*C.A.*, **45**, 580 (1951)].
[396] Buu-Hoï, Hoan, and Royer, *Bull. soc. chim. France*, **1950**, 489.
[397] Henrich and Wirth, *Ber.*, **37**, 731 (1904).
[398] Cain, Simonsen, and Sumter, *J. Chem. Soc.*, **103**, 1035 (1913).
[399] von Auwers, *Ber.*, **62**, 1320 (1929).
[400] Hasselstrom and Hampton, *J. Am. Chem. Soc.*, **61**, 3445 (1939).
[401] Buu-Hoï and Cagniant, *Rec. trav. chim.*, **62**, 719 (1943).
[402] Bruyn, *Compt. rend.*, **228**, 1953 (1949).
[403] Buu-Hoï, *J. Org. Chem.*, **19**, 721 (1954).
[404] Kohler, *Am. Chem. J.*, **31**, 642 (1904).
[405] Buu-Hoï, Hoan, and Xuong, *J. Am. Chem. Soc.*, **72**, 3992 (1950).
[406] Beckmann, *Ber.*, **19**, 988 (1886).
[407] Vargha and Gonczy, *J. Am. Chem. Soc.*, **72**, 2738 (1950).
[408] Lachmann, *J. Am. Chem. Soc.*, **46**, 1477 (1924).
[409] Hudlický, *Chem. Listy*, **51**, 470 (1957) [*C.A.*, **51**, 10443 (1957)].
[410] Turnbull, *Chem. & Ind. (London)*, **1956**, 350.
[411] Brodskii, *Izvest. Akad. Nauk S.S.S.R., Otdel Khim.*, **1949**, 3 [*C.A.*, **43**, 5011 (1949)].
[412] Meerwein, *Angew. Chem.*, **67**, 374 (1955).
[413] Exner, *Chem. Listy*, **48**, 1634 (1951) [*C.A.*, **49**, 14674 (1955)].
[414] Lampert and Bordwell, *J. Am. Chem. Soc.*, **73**, 2369 (1951).
[415] Amet, Bovin, and Dewar, *Can. J. Chem.*, **35**, 180 (1957) [*C.A.*, **51**, 11281 (1957)].
[416] Montagne, *Rec. trav. chim.*, **25**, 376 (1906).
[417] Brady and Mehta, *J. Chem. Soc.*, **125**, 2297 (1924).
[418] Meisenheimer and Gaiser, *Ann.*, **539**, 95 (1939).
[419] Sutton and Taylor, *J. Chem. Soc.*, **1931**, 2190.
[420] Meisenheimer, Hanssen, and Wachterowitz, *J. prakt. Chem.*, [2] **119**, 315 (1928).
[421] Shah and Ichaporia, *J. Univ. Bombay*, **3**, 172 (1934) [*C.A.*, **29**, 4755 (1935)].
[422] Meisenheimer, *Ann.*, **539**, 99 (1939).
[423] Scharwin and Schorigin, *Ber.*, **36**, 2025 (1903).
[424] Schnackenburg and School, *Ber.*, **36**, 654 (1903).
[425] Betti and Becciolini, *Gazz. chim. ital.*, **45**, II, 218 (1915).
[426] Poccianti, *Gazz. chim. ital.*, **45**, II, 111 (1915).
[427] Campbell and Woodham, *J. Chem. Soc.*, **1952**, 843.
[428] Munchmeyer, *Ber.*, **19**, 1845 (1886); **20**, 228 (1887).
[429] Baither, *Ber.*, **20**, 1731 (1887).
[430] Larner and Peters, *J. Chem. Soc.*, **1952**, 680.
[431] Ecary, *Ann. chim. (Paris)*, [12] **3**, 445 (1948).
[432] Lynch and Reid, *J. Am. Chem. Soc.*, **55**, 2515 (1933).
[433] Beckmann, Liesche, and Correns, *Ber.*, **56**, 341 (1923).
[434] Bradsher and Swerlick, *J. Am. Chem. Soc.*, **72**, 4189 (1950).
[435] Cass (to du Pont), U.S. pat. 2,221,369 (1940) [*C.A.*, **35**, 584 (1941)].
[436] Schlack (to Alien Property Custodian), U.S. pat. 2,313,026 (1943) [*C.A.*, **37**, 5078 (1943)].
[437] Wiest and Hopff (to I. G. Farbenind.), Ger. pat. 736,735 (1943) [*C.A.*, **38**, 2967 (1944)].
[438] (To Badische Anilin und Soda Fabrik), Ger. Pat. 863,657 (1957).
[439] Lowery, Ger. pat. appl. D 4,334 (1952).
[440] Stephen and Staskun, *J. Chem. Soc.*, **1956**, 4694.
[441] Van Krevelen and De Gelder (to Maatschappij voor Kohlenbeiverkung) U.S. pat. 2,769,000 (1956).

[442] Todd, Chase, Kenner, and Webb, *J. Chem. Soc.*, **1956**, 1371.
[443] Donaruma and England (to du Pont), U.S. pat. 2,769,807 (1956) [*C.A.*, **51**, 6685 (1957)]; Novotny, Czech. pats. 85,505 and 85,524 (1957) [*C.A.*, **51**, 463 (1957)].
[444] Hildebrand and Bogert, *J. Am. Chem. Soc.*, **58**, 650 (1936).
[445] Wallach, *Ann.*, **312**, 177 (1900).
[446] Oxley and Short, Brit. pat. 577,696 (1946).
[447] Marvel and Eck, *Org. Syntheses*, **17**, 60 (1937).
[448] Lukes and Smolek, *Collection Czechoslov. Chem. Communs.*, **11**, 506 (1939) [*C.A.*, **34**, 7868 (1940)].
[449] Mokudai and Oda, *Bull. Inst. Phys. Chem. Research (Tokyo)*, **20**, 343 (1941) [*C.A.*, **36**, 3049 (1942)].
[450] Ruzicka, *Helv. Chim. Acta*, **4**, 472 (1921).
[451] (To Dai Nippon Celluloid Co.), Jap. pat. 157,331 (1943) [*C.A.*, **44**, 1531 (1950)].
[452] Hopff and Wiest (to I. G. Farbenind.), Ger. pat. 755,944 (1944).
[453] Stickdorn and Hentrick (to Deutsche Hydrierwerke), Ger. pat. 919,047 (1954).
[454] Schlack (to I. G. Farbenind.), U.S. pat. 2,249,177 (1941) [*C.A.*, **35**, 6599 (1941)].
[455] (To Bata), Fr. pat. 903,687 (1945).
[456] (To. I. G. Farbenind.), Fr. pat. 1,001,570 (1952).
[457] Oxley and Short, (to Boots Pure Drug Co.), Brit. pat. 577,696 (1946) [*C.A.*, **41**, 2433 (1947)].
[458] Moller, Bayer, and Wilms (to Farbenfabriken Bayer), Ger. pat. 924,866 (1955).
[459] (To Bata), Belg. pat. 448,478 (1943) [*C.A.*, **41**, 6576 (1947)].
[460] Aikawa (to Oriental Rayon Co.), Jap. pat. 6,178 (1951) [*C.A.*, **50**, 3673 (1956)].
[461] (To Directie van de Staatsmijnen im Limburg), Fr. pat. 1,023,858 (1953).
[462] Taubock (to I. G. Farbenind.), Ger. pat. 686,902 (1939) [*C.A.*, **33**, 2663 (1944)].
[463] Tokura, Asami, and Tado, *Sci. Repts. Research Insts., Tohoku Univ.*, **A8**, 149 (1956) [*C.A.*, **51**, 4944 (1957)].
[464] (To Directie van de Staatsmijnen im Limburg), Dutch pat. 78,624 (1955) [*C.A.*, **50**, 13989 (1956)].
[465] Van Krevelin (to Stamicarbon N.V.), Ger. pat. appl. M 13,706 (1957).
[466] Novotny and Bauer (to Farbenfabriken Bayer), Ger. pat. 834,987 (1952) [*C.A.*, **51**, 14789 (1957)].
[467] Cohn (to Brit. Celanese Ltd.), Brit. pat. 750,222 (1956) [*C.A.*, **50**, 15579 (1956)].
[468] Blaser and Tischberek (to Henkel and Cie), Ger. pat. 944,730 (1956).
[469] Cohn, Groombridge, and Lincoln (to Brit. Celanese Ltd.), U.S. pat. 2,723,266 (1956) [*C.A.*, **50**, 15580 (1956)].
[470] Wagner and O'Hara (to Olin Matheson Corp.), U.S. pat. 2,797,216 (1957).
[471] (To Bata), Belg. pat. 447,859 (1942).
[472] Barnett, Cohn, and Lincoln (to Brit. Celanese Ltd.), U.S. pat. 2,754,298 (1956) [*C.A.*, **51**, 2583 (1957)].
[473] (To Stamicarbon), Belg. pat. 534,950 (1955).
[474] Coffman, Cox, Martin, Mochel, and van Natta, *J. Polymer Sci.*, **3**, 85 (1948).
[475] Donat and Nelson, *J. Org. Chem.*, **22**, 1107 (1957).
[476] Schaffler and Ziegenbein, *Chem. Ber.*, **88**, 1374 (1955).
[477] Wallach, *Ann.*, **346**, 249 (1906).
[478] Wallach, *Ann.*, **332**, 337 (1904).
[479] Knunyants and Fabrichnyi, *Doklady Akad. Nauk S.S.S.R.*, **85**, 793 (1952) [*C.A.*, **47**, 9945 (1953)].
[480] Schaffler, Kaufold, and Ziegenbein, *Chem. Ber.*, **88**, 1906 (1955).
[481] (To Inventa A. G.), Swiss pat. 287,863 (1953) [*C.A.*, **49**, 2490 (1955)].
[482] Kipping, *Proc. Chem. Soc.*, **1893**, 240.
[483] Peters, *Ber.*, **40**, 240 (1907).
[484] Pailer and Allmer, *Monatsh.*, **86**, 819 (1955).
[485] Mosher, Farker, Williams, and Oakwood, *J. Am. Chem. Soc.*, **74**, 4627 (1952).
[486] Morris (to Shell Development Co.), U.S. pat. 2,462,009 (1949) [*C.A.*, **43**, 3451 (1949)].
[487] Montgomery and Dougherty, *J. Org. Chem.*, **17**, 823 (1952).

[488] Read and Cook, *J. Chem. Soc.*, **127**, 2782 (1925).
[489] Wallach and Fritzsche, *Ann.*, **336**, 247 (1904).
[490] Cusmano, *Gazz. chim. ital.*, **42, I**, 1 (1919).
[491] Burrows and Eastman, *J. Am. Chem. Soc.*, **79**, 3756 (1957).
[492] Hardy, Ward, and Day, *J. Chem. Soc.*, **1956**, 1979.
[493] Knunyants and Fabrichnyi, *Doklady Akad. Nauk S.S.S.R.*, **68**, 523 (1949) [*C.A.*, **44**, 1469 (1950)].
[494] Wallach, *Ann.*, **408**, 163 (1914).
[495] Lloyd, Matternas, and Horning, *J. Am. Chem. Soc.*, **77**, 5932 (1955).
[496] Barnes and Beachem, *J. Am. Chem. Soc.*, **77**, 5388 (1955).
[497] Walbaum, *J. prakt. Chem.*, [2] **113**, 166 (1926).
[498] Ruzicka, *Helv. Chim. Acta*, **9**, 230 (1926).
[499] Cassaday and Bogert, *J. Am. Chem. Soc.*, **63**, 1452 (1941).
[500] Litvan and Robinson, *J. Chem. Soc.*, **1938**, 1997.
[501] Kaufmann, *Mem. Congr. cient. Mex., IV, Centenario Univ., Mexico*, **2**, 107 (1953) [*C.A.*, **49**, 7585 (1955)].
[502] Schmidt-Thomé, Ger. pat. 871,010 (1949) (*Chem. Zentr.*, **1953**, 6938).
[503] Rosenkrantz, Mancera, Sondheimer, and Djerassi, *J. Org. Chem.*, **21**, 520 (1956).
[504] Buckmuhl, Ehrhart, Ruschig, and Aumiller, Ger. pat. 723,615 (1942) [*C.A.*, **37**, 5558 (1943)].
[505] Schmidt-Thomé, *Angew. Chem.*, **67**, 715 (1955).
[506] Barnes, Barton, Fawcett, and Thomas, *J. Chem. Soc.*, **1952**, 2339.
[507] Schenck and Kirchof, *Z. physiol. Chem.*, **166**, 142 (1927).
[508] Borsche, *Nachr. kgl. Ges. Wiss. Göttingen*, **1920**, 192 (*Chem. Zentr.*, **1921, III**, 174).
[509] Schmidt-Thomé, *Ann.*, **603**, 43 (1957).
[510] Schenck, *Z. physiol. Chem.*, **128**, 53 (1923).
[511] Schenck and Kirchof, *Z. physiol. Chem.*, **180**, 107 (1929).
[512] Schenck, *Z. physiol. Chem.*, **89**, 360 (1914).
[513] Schenck, *Z. physiol. Chem.*, **211**, 88 (1932).
[514] Schenck, *Z. physiol. Chem.*, **263**, 55 (1940).
[515] Tsuda and Hayatsu, *J. Am. Chem. Soc.*, **78**, 4107 (1956).
[516] Djerassi and Hodges, *J. Am. Chem. Soc.*, **78**, 3534 (1956).
[517] Dickerman and Lindwell, *J. Org. Chem.*, **14**, 534 (1949).
[518] Cymerman-Craig and Willis, *J. Am. Chem. Soc.*, **1955**, 1071; Steinkopf, *Ann.*, **403**, 17 (1914).
[519] Foldi, *Acta Chim. Acad. Sci. Hung.*, **6**, 307 (1955) [*C.A.*, **51**, 1983 (1957)].
[520] Van der Meer, Koffmann, and Veldstino, *Rec. trav. Chim.*, **72**, 236 (1953).
[521] Rumpel, *Arch. Pharm.*, **237**, 233 (1899) (*Chem. Zentr.*, **1899, I**, 1284; **1900, II**, 581).
[522] Mameli, *Gazz. chim. ital.*, **56**, 759 (1926).
[523] Hennion and O'Brian, *J. Am. Chem. Soc.*, **71**, 2933 (1949).
[524] Truce and Simmons, *J. Org. Chem.*, **22**, 617 (1957).
[525] Ostrogovich and Tanislau, *Gazz. chim. ital.*, **66**, 672 (1936).
[526] Stark, *Ber.*, **40**, 3425 (1907).
[527] Etienne and Staehelin, *Compt. rend.*, **230**, 1960 (1950).
[528] Matsumura, *J. Am. Chem. Soc.*, **57**, 124 (1935).
[529] Galinawsky and Kainz, *Monatsh.*, **77**, 137 (1947).
[530] Villiani and Papa, *J. Am. Chem. Soc.*, **72**, 2722 (1950).
[531] Van Ark, *Arch. Pharm.*, **238**, 321 (1900) (*Chem. Zentr.*, **1900, II**, 581).
[532] Burger, Wartman, and Lutz, *J. Am. Chem. Soc.*, **60**, 2628 (1938).
[533] Nobis, Blaydinelli, and Blaney, *J. Am. Chem. Soc.*, **75**, 3384 (1953).
[534] Buu-Hoï, Lescot, and Xuong, *J. Chem. Soc.*, **1956**, 2408.
[535] Buu-Hoï and Hoan, *J. Org. Chem.*, **17**, 643 (1952).
[536] Buu-Hoï and Royer, *J. Org. Chem.*, **16**, 1198 (1951).
[537] Ghigi, *Ber.*, **75**, 1316 (1942).
[538] Benary and Psille, *Ber.*, **57**, 828 (1924).
[539] Schapf, *Ann.*, **452**, 211 (1927).
[540] Fuson and Kao, *J. Am. Chem. Soc.*, **54**, 313 (1932).

541 Königs, *Ber.*, **40**, 648, 2873 (1907).
542 Small and Mallanu, *J. Org. Chem.*, **5**, 286 (1940).
543 Kohler, *J. Am. Chem. Soc.*, **46**, 1733 (1924).
544 Borsche, *Ber.*, **40**, 737 (1907).
544a Beckmann and Koster, *Ann.*, **274**, 1 (1893).
545 Blatt, *J. Am. Chem. Soc.*, **56**, 2774 (1934).
546 Rydon, Williams, and Smith, *J. Chem. Soc.*, **1957**, 1900.
547 Badger and Seidler, *J. Chem. Soc.*, **1954**, 2329.
548 Graebe and Honigsberger, *Ann.*, **311**, 257 (1900).
549 Green and Saville, *J. Chem. Soc.*, **1956**, 3887.
550 Diels and Van der Leeden, *Ber.*, **38**, 3357 (1905).
550a Reissert, *Ber.*, **41**, 3810 (1908).
551 Goldschmidt, *Ber.*, **20**, 484 (1887).
552 Konowalov, *J. Russ. Phys.-Chem. Soc.*, **33**, 45 (1901) (*Chem. Zentr.*, **1901, II,** 1002).
553 Leuckart, *Ber.*, **20**, 110 (1887).
554 Nagata and Takeda, *J. Pharm. Soc. Japan*, **72**, 1566 (1952) [*C.A.*, **47**, 10504 (1953)].
555 Blaise and Blanc, *Compt. rend.*, **129**, 886 (1899).
556 Komatsu and Yamaguchi, *Mem. Coll. Sci., Kyoto Imp. Univ.*, **6**, 245 (1923) [*C.A.*, **17**, 3495 (1923)].
557 Wallach and Rojahn, *Ann.*, **313**, 367 (1900).
558 Diels and Sharkoff, *Ber.*, **46**, 1862 (1913).
559 Diels, Buddenberg, and Wang, *Ann.*, **451**, 223 (1927).
560 Haller and Bauer, *Compt. rend.*, **156**, 1503 (1913).
561 Rappoport and Pasky, *J. Am. Chem. Soc.*, **78**, 3788 (1956).
562 Blatt and Barnes, *J. Am. Chem. Soc.*, **58**, 1900 (1936).
563 Gheorghiu and Cozubschi-Sciurevici, *Ann. sci. univ. Jassy*, Sect. 1, **28**, 209 (1942) [*C.A.*, **38**, 3276 (1944)].
564 Matsumura, *J. Am. Chem. Soc.*, **57**, 955 (1935).
565 Rappoport and Lavigne, *J. Am. Chem. Soc.*, **75**, 5329 (1953).
566 Bently and Ringl, *J. Org. Chem.*, **22**, 424 (1957).
567 Findlay, *J. Org. Chem.*, **21**, 644 (1956).
568 Baiardo, *Gazz. chim. ital.*, **56**, 567 (1926).
569 Hauser and Hoffenberg, *J. Org. Chem.*, **20**, 1496 (1955).
570 Bamberger, *Ber.*, **33**, 1781 (1888).
571 Brady and Whitehead, *J. Chem. Soc.*, **1927**, 2933.
572 Meisenheimer, Theilacker, and Beisswenger, *Ann.*, **495**, 249 (1932).
573 Brady and Dunn, *J. Chem. Soc.*, **123**, 1783 (1923).
574 Ponzio and Avogadro, *Gazz. chim. ital.*, **53**, 311 (1923).
575 Martinet and Coisset, *Compt. rend.*, **172**, 1234 (1921).
576 Ressy and Ortodocsu, *Bull. soc. chim. France*, [4] **33**, 637 (1923).
577 Moluch and Minslow, *J. Org. Chem.*, **20**, 1311 (1955).
578 Coldwell and Jones, *J. Chem. Soc.*, **1946**, 599.
579 Coleman and Pyle, *J. Am. Chem. Soc.*, **68**, 2007 (1946).
580 Pechmann, *Ber.*, **30**, 2791, 2873 (1897).
581 Kröhnke, *Ann.*, **604**, 203 (1957).
582 Scholl and Donat, *Ber.*, **64**, 318 (1931).
583 Oddo and Curti, *Gazz. chim. ital.*, **54**, 577 (1924).
584 Blatt and Russell, *J. Am. Chem. Soc.*, **58**, 1903 (1936).
585 Fox, Dunn, and Stoddard, *J. Org. Chem.*, **6**, 410 (1914).
586 Terent'ev and Makarova, *Vestnik Moskov Univ.*, **1947**, No. 4,101 [*C.A.*, **42**, 1590 (1948)].
587 Buu-Hoï, Royer, Daudel, and Martin, *Bull. soc. chim. France*, **1948**, 329.
588 Hukki, *Acta Chem. Scand.*, **3**, 288 (1949) [*C.A.*, **43**, 7916 (1949)].
589 Kaslow, Genser, and Goodspeed, *Proc. Indiana Acad. Sci.*, **59**, 134 (1950) [*C.A.*, **45**, 8534 (1951)].
590 Bahner, Scott, Cate, Walden, and Baldridge, *J. Am. Chem. Soc.*, **73**, 4013 (1951).
591 v. Schickh (to Badische Anilin und Soda Fabrik), Ger. pat. 1,024,515 (1956).
592 (To Stamicarbon), Dutch pat. 81,037 (1956) [*C.A.*, **51**, 2853 (1957)].
593 Samlo, Lemetre, Giolitti, Marahini, and Pierucci, *Chim. e ind.* (*Milan*), **39**, 905 (1957).

CHAPTER 2

THE DEMJANOV AND TIFFENEAU-DEMJANOV RING EXPANSIONS

PETER A. S. SMITH AND DONALD R. BAER*

University of Michigan

CONTENTS

	PAGE
INTRODUCTION	158
MECHANISM	158
SCOPE AND LIMITATIONS	163
Ring Size	163
Unsaturated Rings	163
Heterocyclic Rings	164
Alkyl and Aryl Substitution	165
Bicyclic and Polycyclic Systems	169
Rings Substituted with Other Functional Groups	171
APPLICATION TO SYNTHESIS	172
EXPERIMENTAL CONDITIONS	175
EXPERIMENTAL PROCEDURES	178
Cycloheptanone	178
3-Hydroxydodecahydroheptalene and Decahydroheptalene by the Demjanov Rearrangement	178
Cycloheptanol by the Demjanov Rearrangement	178
Cycloöctanone by the Tiffeneau-Demjanov Rearrangement	179
TABULAR SURVEY	180
Table I. Mononuclear Carbocyclic Rings	181
Table II. Polynuclear Carbocyclic Systems with Fusion at a Single Side	185
Table III. Polynuclear Systems with Fusion at More than One Edge	187
Table IV. Heterocyclic Rings	187

* Present address: Jackson Laboratory, E. I. du Pont de Nemours and Company, Inc., Wilmington, Delaware.

INTRODUCTION

The reaction of aminomethylcycloalkanes with nitrous acid to produce cycloalkanols in which the ring is larger by one carbon atom is known as the Demjanov (Demianov, Demjanow, Dem'yanov) rearrangement. The first example of this type of ring expansion was encountered by Demjanov and Luschnikov in 1901,[1] but was not recognized until 1903 when cyclopentanol was identified as one of the products formed from cyclobutanemethylamine.[2] Since that time the reaction has been

$$\begin{array}{c} CH_2\!-\!CHCH_2NH_2 \\ | \qquad\qquad | \\ CH_2\!-\!CH_2 \end{array} \xrightarrow{HNO_2} \begin{array}{c} CH_2\!-\!CHOH \\ | \qquad\quad \diagdown \\ | \qquad\qquad CH_2 \\ | \qquad\quad \diagup \\ CH_2\!-\!CH_2 \end{array} + N_2 + H_2O$$

extended to rings of many sizes. Olefins almost invariably accompany the alcohols that are formed. The Demjanov rearrangement includes within its scope the rearrangements that occur when acyclic amines are treated with nitrous acid as well as the ring expansions considered in this chapter.

A highly useful extension of the Demjanov reaction, reported in 1937 by Tiffeneau, Weill, and Tchoubar,[3] consists of the treatment of 1-aminomethylcycloalkanols with nitrous acid, forming ring-enlarged ketones. Since Tiffeneau's name is associated with other reactions, the term

$$(CH_2)_n \!\!\begin{array}{c}\diagup CH_2NH_2 \\ C \\ \diagdown OH\end{array} \xrightarrow{HNO_2} (CH_2)_n \!\!\begin{array}{c} CH_2 \\ | \\ C\!=\!O \end{array} + N_2 + H_2O$$

Tiffeneau-Demjanov ring expansion will be used in this chapter to designate ring enlargements by pinacolic deamination.

Inasmuch as both alcohols and ketones can be converted readily to amines, and ketones can be converted to amino alcohols, the Demjanov or Tiffeneau-Demjanov ring expansion can be made the key step in the conversion of a cyclic alcohol or ketone into its next higher ring homolog.

MECHANISM

The Demjanov ring enlargement may be regarded as a special case of the rearrangement which so often accompanies the reaction of aliphatic

---

[1] Demjanov and Luschnikov, *J. Russ. Phys.-Chem. Soc.*, **33**, 279 (1901) (*Chem. Zentr.*, **1901, II,** 335).

[2] Demjanov and Luschnikov, *J. Russ. Phys.-Chem. Soc.*, **35**, 26 (1903) (*Chem. Zentr.*, **1903, I,** 828).

[3] Tiffeneau, Weill, and Tchoubar, *Compt. rend.*, **205**, 54 (1937).

primary amines with nitrous acid. Accordingly, information concerning its mechanism can be derived from investigations of analogous reactions of acyclic compounds. Similarly, the Tiffeneau-Demjanov ring expansion may be regarded as a special case of the semi-pinacol rearrangement, or pinacolic deamination.

Recent extensive kinetic investigations have established with high probability that the initial step of the reaction of most, if not all, amines with nitrous acid involves the free amine and a derivative of nitrous acid, such as N_2O_3, and results in the formation of a diazonium ion.[4-10] Such an ion is unstable in an aliphatic system, and may lose nitrogen by several possible paths or lose a proton from the α-carbon atom to give a diazo compound. Since the product formed by unimolecular elimination of nitrogen is a carbonium ion, the large body of information about the behavior of carbonium ions is applicable to nonreductive deaminations in general.

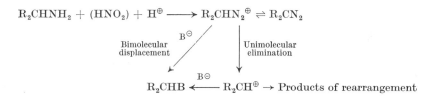

Both the Demjanov and the Tiffeneau-Demjanov ring expansions are commonly regarded as special cases of the rearrangement of a carbonium ion.[11-15] It is immediately seen that rearrangement is always competitive with a displacement reaction which precludes rearrangement, as well as with the possible combination of the unrearranged carbonium ion with a base. Consequently it is not surprising that rearrangement is generally only one of several reactions that take place.

These considerations are illustrated by the reaction of cyclohexanemethylamine with nitrous acid in dilute aqueous acetic acid.[16] The

---

[4] Austin, Hughes, Ingold, and Ridd, *J. Am. Chem. Soc.*, **74**, 555 (1952).
[5] Hughes, Ingold, and Ridd, *J. Chem. Soc.*, **1958**, 58.
[6] Hughes, Ingold, and Ridd, *J. Chem. Soc.*, **1958**, 65.
[7] Hughes, Ingold, and Ridd, *J. Chem. Soc.*, **1958**, 77.
[8] Hughes, Ingold and Ridd, *J. Chem. Soc.*, **1958**, 88.
[9] Hughes and Ridd, *J. Chem. Soc.*, **1958**, 70.
[10] Hughes and Ridd, *J. Chem. Soc.*, **1958**, 82.
[11] Hückel and Wilip, *J. prakt. Chem.*, [2] **158**, 21 (1941).
[12] Tchoubar, *Bull. soc. chim. France*, **1951**, C44.
[13] Wheland, *Advanced Organic Chemistry*, p. 512, John Wiley & Sons, New York, 1949.
[14] Alexander, *Principles of Ionic Organic Reactions*, pp. 49–51, John Wiley & Sons, New York, 1950.
[15] Fuson, *Advanced Organic Chemistry*, p. 523, John Wiley & Sons, New York, 1950.
[16] Smith and Baer, *J. Am. Chem. Soc.*, **74**, 6135 (1952).

products which result are cyclohexylcarbinol, 1-methylcyclohexanol, cycloheptanol, the acetates of these alcohols, and a mixture of isomeric olefins (cycloheptene[17] and presumably some methylenecyclohexane and 1-methylcyclohexene). Cycloheptanol and its acetate are the principal products. Rearrangement by migration of a hydride ion or a ring carbon

atom as shown is to be expected from the consideration that a secondary or tertiary carbonium ion is thereby produced from a primary one, in accord with the known relative stabilities of such species.[18] Predominance of ring expansion over the formation of tertiary alcohol is a fortunate circumstance arising from the higher entropy of activation required for hydrogen migration.[19]

The acetate esters are formed in amounts out of proportion to the stoichiometric concentration of acetic acid; the relative preferences of carbonium ions for the various nucleophilic species that may be available to them are governed by somewhat complex considerations which have not been completely elucidated.[11, 20]

[17] Ruzicka and Brugger, *Helv. Chim. Acta*, **9**, 399 (1926).
[18] Dostrovsky, Hughes, and Ingold, *J. Chem. Soc.*, **1946**, 173.
[19] Cannell and Taft, *J. Am. Chem. Soc.*, **78**, 5813 (1956).
[20] Hine, *Physical Organic Chemistry*, pp. 134–167, McGraw-Hill, New York, 1956.

In addition to the foregoing interpretation, aliphatic diazonium ions may be considered to yield a rearranged carbonium ion directly by internal displacement.[21] Furthermore this process may be considered to be a concerted one, leading then directly from a diazonium ion to a final

product of rearranged structure without passing through a carbonium ion stage. The evidence available at present does not permit one to specify with certainty the path or paths by which the Demjanov rearrangement

occurs. However, the carbonium ion path is supported by the success with which knowledge of the behavior of carbonium ions can be applied in interpreting amine-nitrous acid rearrangements. A general theoretical treatment of aliphatic deaminations, embracing the situations in which ring expansion is possible, has recently been given by Streitwieser.[22, 23]

Of historical interest is the earlier concept that a diazonium compound (usually written as a non-ionic hydroxide) may simultaneously lose water and nitrogen to form a cyclopropane derivative, which by cleavage of any of the three bonds of the cyclopropane ring might give rise to the observed products.[24, 25] One is reminded of this concept by the polycyclic, nonclassical carbonium ion proposed more recently to account for the singular behavior of cyclopropanemethylamine.[26] The facts that cyclobutylamine undergoes partial ring contraction[24] and cyclopropanemethylamine partial expansion,[27] to give in each case a nearly 1 : 1 mixture of cyclobutanol and cyclopropylcarbinol,[28] have been explained on the basis of

[21] Bernstein and Whitmore, *J. Am. Chem. Soc.*, **61**, 1324 (1939).
[22] Streitwieser and Schaeffer, *J. Am. Chem. Soc.*, **79**, 2888 (1957).
[23] Streitwieser, *J. Org. Chem.*, **22**, 861 (1957).
[24] Demjanov, *Ber.*, **40**, 4961 (1907).
[25] Wallach and Fleischer, *Ann.*, **353**, 318 (1907).
[26] Roberts and Mazur, *J. Am. Chem. Soc.*, **73**, 3542 (1951); Roberts, 16th Nationa lOrganic Chemistry Symposium, American Chemical Society, Seattle, June, 1959.
[27] Demjanov, *Ber.*, **40**, 4393 (1907); *J. Russ. Phys.-Chem. Soc.*, **39**, 1077 (1907).
[28] Skrabal, *Monatsh.*, **70**, 420 (1937).

a common set of intermediates deduced to have structures represented by I.[26]

The mechanisms of the Tiffeneau-Demjanov and the Demjanov ring expansions are fundamentally the same. However, two important effects are operative in the former that favor ring expansion. There is no hydrogen atom in the position from which it could migrate in competition

with a ring carbon atom; also, the ion resulting from rearrangement bears its positive charge on a protonated carbonyl group, an arrangement generally of much lower energy than a simple carbonium ion structure. As a result, ring expansion is more complete, and the product does not contain the substantial amount of olefins found in the Demjanov reaction.

In a consideration of the expansion of unsymmetrical rings, the question of "migration aptitudes" arises. The same circumstance introduces the possibility of diastereomeric aminomethylcycloalkanes, and with it the possibility of steric control of the direction of enlargement. Experimental evidence to resolve these questions is incomplete and in part contradictory.[29] However, there is partial evidence for steric control of the course of the Tiffeneau-Demjanov expansion in the steroid field.[30-33] Since steric control has been demonstrated in the analogous noncyclic pinacolic deamination,[34] and the pertinence of conformational factors has been justified in a general way,[22, 35] steric control in ring expansions seems probable.

SCOPE AND LIMITATIONS

Ring Size. All ring sizes from cyclopropane[27, 36] through cyclooctane[17] have been expanded by the Demjanov method with some degree of success. The ratio of the yield of the alcohol with one more carbon atom in the ring to the alcohol with the same carbon skeleton as the amine varies from 1 : 1 for cyclopropanemethylamine[36] through a maximum of > 3 : 1 for cyclobutane-[2] and cyclopentane-methylamines[37] to 2 : 3 for cyclooctanemethylamine.[17] The presence of substituents on the rings would be expected to change these ratios. It appears that the Demjanov expansion is most useful for the preparation of five-, six-, and seven-membered rings, and is of considerably less value for the preparation of smaller or larger rings.

The Tiffeneau-Demjanov expansion has been successfully applied to the preparation of five-,[38] six-, seven-, eight-, and nine-membered rings.[39] with a slight decrease in yield with increasing ring size.[40] It has not yet been applied to the expansion of three-membered rings. Whenever a comparison has been made, the Tiffeneau-Demjanov method has given a higher yield.

Unsaturated Rings. Two cycloalkenemethylamines have been studied, each one having a double bond on the carbon atom holding the

[29] Wendler, Taub, and Slates, *J. Am. Chem. Soc.*, **77**, 3559 (1955).
[30] Goldberg and Studer, *Helv. Chim. Acta*, **24**, 295E (1941).
[31] Heusser, Herzig, Fürst, and Plattner, *Helv. Chim. Acta*, **33**, 1093 (1950).
[32] Ramirez and Stafiej, *J. Am. Chem. Soc.*, **77**, 134 (1955).
[33] Ramirez and Stafiej, *J. Am. Chem. Soc.*, **78**, 644 (1956).
[34] Pollak and Curtin, *J. Am. Chem. Soc.*, **72**, 961 (1950).
[35] Cram and McCarty, *J. Am. Chem. Soc.*, **79**, 2866 (1957).
[36] Roberts and Mazur, *J. Am. Chem. Soc.*, **73**, 2509 (1951).
[37] Smith, Baer, and Ege, *J. Am. Chem. Soc.*, **76**, 4564 (1954).
[38] Roberts and Gorham, *J. Am. Chem. Soc.*, **74**, 2278 (1952).
[39] Ruzicka, Plattner, and Wild, *Helv. Chim. Acta*, **26**, 1631 (1943).
[40] Tchoubar, *Bull. soc. chim. France*, **1949**, 164.

aminomethyl group. Cyclohexene-1-methylamine (II) forms only the unrearranged alcohol,[41] and aminoterebenthene (III) undergoes an allylic rearrangement but not ring expansion. In the latter case the results must be interpreted with caution since uncertainties as to the structures of the starting material and product exist.

[cyclohexene-CH₂NH₂ (II) → cyclohexene-CH₂OH]

[structure III CH₂NH₂ → rearranged structure CH₂OH]

There are no data regarding the effect of an isolated double bond in a simple ring system, but expansion would be expected to be less affected in these cases.

Heterocyclic Rings. Of the small number of aminoheterocyclic compounds to which the Demjanov expansion has been applied, 2-aminomethylpyrrole (IV) and 2-aminomethylpyrrolidine (V) have given low

[IV: pyrrole-CH₂NH₂ → pyridine]   [V: pyrrolidine-CH₂NH₂ → tetrahydropyridine]

yields of pyridine and tetrahydropyridine, respectively.[42] It should be noted that the position of the nitrogen atom inevitably involves it in the structure of the carbonium ion formed in the rearrangement. The presence of a nitrogen (or other) atom further removed from the site of the expansion would be expected to have less effect on the course of the reaction. This presumption is supported by the success of the single reported example of the Tiffeneau-Demjanov expansion of a heterocyclic ring; 3-aminomethyl-3-tropanol (VI) gave R-homotropinone in good yield.[43]

[41] Jacquier and Zagdoun, *Bull. Soc. chim. France*, **1952**, 699.
[42] Putoshin, *J. Russ. Phys.-Chem. Soc.*, **62**, 2226 (1930) [*C.A.*, **25**, 3996 (1931)].
[43] Cope, Nace, and Estes, *J. Am. Chem. Soc.*, **72**, 1123 (1950).

One sulfur heterocycle, 2-thenylamine (VII), has been shown to give the unrearranged alcohol VIII and a small amount of what appears to be hydroxythiopyran (IX). Complete ring enlargment of 2-aminomethyl-furan to 2-hydroxypyran has been reported.[43a]

Alkyl and Aryl Substitution. Three cases of significantly different consequences can be distinguished: substitution on the aminomethyl carbon atom (R in the following formula); on the ring carbon atom attached to the aminomethyl group (R'), and elsewhere on the ring (R'').

Substitution of an aryl or alkyl group on the aminomethyl side chain (R) invariably hinders both the Demjanov and the Tiffeneau-Demjanov expansions. Thus α-cyclohexylethylamine (X)[32, 37] and its 4-methyl derivative[37, 44] do not give detectable amounts of cycloheptane derivatives, and α-cyclobutyl- and α-cyclopentyl-ethylamine give less expansion than retention of ring size.[37] The presence of a phenyl group introduces an even greater hindrance to ring expansion as evidenced by the fact that no Demjanov-type ring expansion occurs when α-cyclopentyl-[37] or α-cyclohexyl-benzylamine[45] is treated with nitrous acid. Only the unrearranged alcohols are obtained. Further proof of the stabilization of the benzyl cation is shown by the fact that 2-phenylcyclohexylamine

---

[43a] Colonge and Corbet, *Compt. rend.*, **247**, 2144 (1958).
[44] Wallach and Pohle, *Nachr. kgl. Ges. Wiss. Göttingen*, **1915**, 1–27 (16/1) (*Chem. Zentr.* **1915, II,** 828).
[45] Elphimoff-Felkin and Tchoubar, *Compt. rend.*, **233**, 799 (1951).

contracts its ring to form the same alcohol that arises from α-cyclopentylbenzylamine on treatment with nitrous acid.[46] The same results are obtained in the Tiffeneau-Demjanov expansion of three different α-(1-hydroxycyclohexyl)benzylamines (XI). Five-membered rings containing

$Ar = C_6H_5, p\text{-}CH_3C_6H_4, p\text{-}CH_3OC_6H_4.$

an aryl group on the aminomethyl side chain, in contrast to the six-membered rings, will enlarge under the Tiffeneau-Demjanov conditions. Thus α-(1-hydroxycyclopentyl)benzylamine (XII) produces about equal amounts of expanded and nonexpanded rings. Since both alkyl and aryl

substitution, particularly the latter, increase the stability of a carbonium ion, such substitution on the aminomethyl side chain saps the driving force of the ring expansion; only when additional driving force is available, such as by relief of ring strain or change to a more stable type of positive ion, does expansion occur when the side chain bears an alkyl group.

[46] Nightingale and Maienthal, *J. Am. Chem. Soc.*, **72**, 4823 (1950).

Thus 1-(α-aminoalkyl)cyclohexanols (XIII) rearrange readily to give 2-alkylcycloheptanones.[47]

[structure: cyclohexane with OH and CHNH₂-R substituents at same carbon → cycloheptanone with R substituent α to C=O]

XIII

In contrast, substitution at the ring carbon atom attached to the aminomethyl group (R') would be expected to favor expansion. Evidence on this point is confined to four examples, in which there are some uncertainties about the structures of the products. α-(1-Phenylcyclopentyl)ethylamine appears to undergo ring expansion without occurrence of side reactions to an appreciable extent, showing that a 1-phenyl group can completely override the hindrance to ring expansion due to a methyl substituent on the side chain.[37] Two cyclopentanemethylamine derivatives bearing 1-methyl groups and 1-methylcyclopropylmethylamine[26] have been found to give ring-enlarged alcohols,[48–50] indicating no adverse affect on ring expansion of the substitution of the 1-carbon atom.

Substitution on a ring carbon atom in a position other than the 1 position does not significantly affect the course of the expansion reactions if the substituent is symmetrically placed. Thus 4-methylcyclohexanemethylamine[51] and 4-methyl-1-hydroxycyclohexanemethylamine (XIV)[40,52] give good yields of 4-methylcycloheptanol and 4-methylcycloheptanone, respectively.

[structure: 4-methylcyclohexanol with CH₂NH₂ at 1-position → 4-methylcycloheptanone]

XIV

An unsymmetrically placed substituent on an aminomethylcycloalkane or aminomethylcycloalkanol gives rise to the possibility of alternative directions of expansion leading to products which are position isomers. In most cases of this type, mixtures have been obtained with one isomer usually predominating markedly over the other if the substituent was in

[47] Elphimoff-Felkin and Tchoubar, *Compt. rend.*, **233**, 964 (1951).
[48] Bredt, *J. prakt. Chem.*, [2], **95**, 70 (1917).
[49] Errera, *Gazz. chim. ital.*, **22**, II, 109 (1892).
[50] Rupe and Splittgerber, *Ber.*, **40**, 4311 (1907).
[51] Qudrat-i-Khuda and Ghosh, *J. Indian Chem. Soc.*, **17**, 19 (1940).
[52] F. F. Blicke, private communication.

the 2 position. Thus 1-aminomethyl-2-methylcyclohexanol (XV) gave a 66% yield of ketones consisting of 2- and 3-methylcycloheptanone in the proportion 1 : 9, while 1-aminomethyl-3-methylcyclohexanol gave 3- and

4-methylcycloheptanones in nearly equal amounts.[40] Other examples are encountered among the bicyclic compounds (see the next section) and in the tables. Information on the Demjanov expansion of unsymmetrically substituted rings is limited to the indication that mixtures are produced.[53, 54]

Since diastereomers of unsymmetrically substituted cyclic compounds are possible, the probable steric control of the direction of the expansion must be considered (see p. 163). The stereochemical nature of the amine to be subjected to ring expansion will depend on the method by which it was prepared. It is thus probable that the ratios of position isomers are determined at least in part by factors governing the reactions by which the amines were prepared, and that different routes for synthesizing an amine may result in different ratios of the position isomers of the product of ring expansion.

Since the amino alcohols required for the Tiffeneau-Demjanov expansion are usually produced by reduction of an addition product of a ketone (such as a cyanohydrin), a substituent in the 2 position has a much

[53] Barbier, *Helv. Chim. Acta*, **23**, 519 (1940).
[54] Barbier, *Helv. Chim. Acta*, **23**, 524 (1940).

greater influence than one further removed from the site of reaction, since it influences the stereochemistry of the addition product. Similar considerations presumably apply to the Demjanov expansion; however, the stereochemical nature of the amine is usually determined by a reductive step, such as the hydrogenation of an unsaturated nitrile.

Bicyclic and Polycyclic Systems. The principal synthetic application of the Demjanov and Tiffeneau-Demjanov ring expansions has been to polynuclear systems. Apart from the formation of position isomers when the aminomethyl group is unsymmetrically placed, ring expansion proceeds normally by both methods. Thus 5-aminomethylhydrindane has been converted to a mixture of isomeric bicyclo[5.3.0]-

decanols,[55,56] and 5-aminomethylhydrindan-5-ol (XVI) has been converted to a mixture of bicyclo[5.3.0]decanones (largely the 4-isomer) in useful yields. A mixture of 1-keto- and 2-keto-hexahydropentalene in the ratio 85 : 15 has been obtained from 6-aminomethylbicyclo[3.2.0]-2-hepten-6-ol (XVII).[38] Expansion is successful when one nucleus is aromatic, as shown by the conversion of β-aminomethyl-β-hydrindenol (XVIII) to β-tetralone.[40]

A number of steroids have been converted to ring homologs by the Tiffeneau-Demjanov method. The expanded ring was in all cases fused

[55] Arnold, *Ber.*, **76**, 777 (1943).
[56] Plattner, Fürst, and Studer, *Helv. Chim. Acta*, **30**, 1091 (1947).

to a saturated cyclohexane ring, but other portions of the molecules contained benzene nuclei, ethylenic double bonds, ester groups, hydroxyl groups, or an epoxide group. Of particular interest is the fact that the stereochemistry of the ring fusion of the expanded ring was apparently undisturbed.[57]. Throughout these examples the expanded ring was unsymmetrical and the formation of isomeric ketones was to be expected, but in practice one isomer always predominated. Thus from the hydrogenated cyanohydrin of *trans*-dehydroandrosterone acetate (XIX) there was obtained 37% of 3-β-acetoxy-17α-keto-D-homoandrostane (XX) and 5% of its 16-keto isomer (XXI).[58] However, when the diastereomeric cyanohydrins were separated beforehand, the major isomer gave only the 17a-ketone.[31]

Expansion of rings that are part of a cage structure has been accomplished by the Demjanov route. Thus 2,5-endomethylenehexahydrobenzylamine (XXII) gave bicyclo[3.2.1]octan-2-ol (XXIII) in good yield,[58a] and ω-aminoisocamphane gave an R-homocamphenilol of uncertain positional and stereochemical nature.[59] The opening of a bicyclic

[57] Goldberg and Studer, *Helv. Chim. Acta*, **25**, 1553 (1942).
[58] Goldberg and Wydler, *Helv. Chim. Acta*, **26**, 1142 (1943).
[58a] Kornblum and Iffland, *J. Am. Chem. Soc.*, **71**, 2137 (1949).
[59] Lipp, Dessauer, and Wolf, *Ann.*, **525**, 271 (1936).

structure is illustrated by the behavior of bornylamine (XXIV).[60] The major products, camphene and its hydrate, are the result of the usual Demjanov reaction; as a consequence of the bicyclic structure, the expansion of the ring not bearing the amino group simultaneously contracts the other ring. In addition, about 20% of (+)-α-terpineol (XXV) is formed; the opening of the transannular bridge can also be accounted for as a carbonium ion rearrangement. Isobornylamine gives only camphene and its hydrate.

XXIV   XXV

Rings Substituted with Other Functional Groups. The information about the effect of other functional groups on attempted ring expansion is limited to the several examples cited in the discussion of steroids, a few hydroxy compounds, and to two halogen compounds.

Compounds containing a hydroxyl group attached to the carbon atom bearing the aminomethyl group present the special case of the Tiffeneau-Demjanov ring expansion. A hydroxyl group in the 2 position of cyclohexanemethylamine has been reported to prevent ring expansion.[61] From the *trans* isomer XXVI a mixture of the corresponding glycol and

XXVI

2-methylcyclohexanone is obtained, and from the *cis* isomer cyclohexanecarboxaldehyde is also formed. *trans*-2-Hydroxycyclopentanemethylamine similarly gives 2-methylcyclopentanone and the unrearranged glycol. From 2-methyl-2-hydroxycyclohexanemethylamine only the glycol was obtained.

Halogenated rings show less tendency for ring enlargement. 2-Chlorocyclohexanemethylamine is reported to undergo no rearrangement.[62]

[60] Hückel and Nerdel, *Ann.*, **528**, 57 (1937).
[61] Mousseron, Jullien, and Winternitz, *Compt. rend.*, **226**, 1909 (1946).
[62] Mousseron, Jullien, and Winternitz, *Bull. soc. chim. France*, **1948**, 878.

Since 2,2,3,3-tetrafluorocyclobutanemethylamine gives the unrearranged alcohol as the sole product,[63] it appears that the presence of highly electronegative substituents such as fluorine inhibits ring expansion.

APPLICATION TO SYNTHESIS

The Demjanov ring expansion can be made the essential step in the conversion of a cyclic alcohol into its ring homolog when combined with one of several methods for preparing the aminomethyl compound from the alcohol. The obvious route via the cycloalkyl halide, cyanide, and reduction is not generally used because the reaction of a cycloalkyl halide with cyanide usually gives a poor yield of nitrile. Alternatively, the cyanide can be obtained via the Grignard reagent and the carboxylic acid. The alternative that often presents advantages consists of oxidation of the alcohol to a ketone, followed by preparation of the cyanohydrin, dehydration, and reduction.[17] In many cases direct reduction of the cyanohydrin is possible, and then the Tiffeneau-Demjanov expansion is used. Unsaturated nitriles can be reduced successfully by catalytic hydrogenation[17] or with sodium and alcohol.[17,51] A slightly longer route

$$(CH_2)_n\!\!-\!\!CHOH\!-\!\!CH_2 \rightarrow (CH_2)_n\!\!-\!\!C\!\!=\!\!O\!-\!\!CH_2 \rightarrow (CH_2)_n\!\!-\!\!C(OH)(CN)\!-\!\!CH_2 \rightarrow$$

$$(CH_2)_n\!\!-\!\!CCN\!\!=\!\!CH \rightarrow (CH_2)_n\!\!-\!\!CHCH_2NH_2\!-\!\!CH_2$$

$$(CH_2)_n\!\!-\!\!C\!\!=\!\!O + BrCH_2CO_2C_2H_5 \xrightarrow{Zn} (CH_2)_n\!\!-\!\!C(OH)(CH_2CO_2C_2H_5)$$

$$\downarrow$$

$$(CH_2)_n\!\!-\!\!CHCH_2NH_2 \leftarrow (CH_2)_n\!\!-\!\!CHCH_2CO_2H \leftarrow (CH_2)_n\!\!-\!\!C(OH)(CH_2CO_2H)$$

[63] Baer, *J. Org. Chem.*, **23**, 1560 (1958).

makes use of the Reformatskiĭ reaction,[64] followed by reduction to a cycloalkylacetic acid and degradation of the carboxyl group to an amino group.[65]

If ring expansion of an available cyclic alcohol is not the objective, other routes to aminomethylcycloalkanes may of course be used. The reduction of nitrosites, obtained by the addition of oxides of nitrogen to cycloalkenes with exocyclic double bonds, is a rare but applicable method.[66] The aminomethylcyclohexanes can be prepared by hydrogenation of the corresponding benzylamine or by the hydrogenation of an arylacetic acid[55] followed by any of the several methods for replacement of a carboxyl group by an amino group.[67-69]

The Tiffeneau-Demjanov expansion is somewhat more easily adapted to the preparation of the next higher ring homologs. A cyclic ketone may be converted in three steps, via its cyanohydrin and reduction to the aminocycloalkanol, to the next higher cyclic ketone. The reduction of cyanohydrins is usually successful by low-pressure hydrogenation with

$$(CH_2)_n\ C{=}O \rightarrow (CH_2)_n\ C{\overset{OH}{\underset{CN}{\diagup\diagdown}}} \rightarrow$$

$$(CH_2)_n\ C{\overset{OH}{\underset{CH_2NH_2}{\diagup\diagdown}}} \rightarrow (CH_2)_n{\overset{C=O}{\underset{CH_2}{\diagdown\diagup}}}$$

platinum oxide catalyst.[56, 70-73] Cyanohydrins vary in the ease with which they dissociate into ketone and hydrogen cyanide, and the occasionally poor results of catalytic hydrogenation have been attributed to the easy reversal and poisoning of the catalyst by the hydrogen cyanide

---

[64] Bachmann and Hoffman, in Adams, *Organic Rections*, Vol. I, pp. 224–262, John Wiley & Sons, New York, 1944.
[65] Wallach, *Ann.*, **353**, 284 (1907).
[66] Wallach and Isaac, *Ann.*, **346**, 243 (1906).
[67] Wallis and Lane, in Adams, *Organic Reactions*, Vol. III, pp. 267–306, John Wiley & Sons, New York, 1946.
[68] Wolf, in Adams, *Organic Reactions*, Vol. III, pp. 307–336, John Wiley & Sons, New York, 1946.
[69] Smith, in Adams, *Organic Reactions*, Vol. III, pp. 337–450, John Wiley & Sons, New York, 1946.
[70] Tchoubar, *Compt. rend.*, **212**, 1033 (1941).
[71] Gutsche, *J. Am. Chem. Soc.*, **71**, 3513 (1949).
[72] Goldberg and Kirchensteiner, *Helv. Chim. Acta*, **26**, 288 (1943).
[73] Tchoubar, *Bull. soc. chim. France*, **1949**, 160.

formed.[73] Cyclohexanone cyanohydrin presents such a case;[70, 72–74] consequently 1-aminomethylcyclohexanol is usually prepared either by reduction of the cyanohydrin with lithium aluminum hydride[74] or by electrolytic[75] or chemical[76] reduction of the nitromethane-cyclohexanone adduct. Reduction of some cyanohydrins with lithium aluminum hydride[74, 77, 78] also proceeds poorly, for the basic reagent appears to favor the reversal.[31, 38] However, the greater specificity of lithium aluminum hydride, which does not reduce unconjugated double bonds, makes it a desirable reagent for the reduction of cyanohydrins.[79] Thus dehydroepiandrosterone acetate was successfully expanded at ring D without disturbing the double bond in ring B; lithium aluminum hydride was used for the reduction of the cyanohydrin.[31] Dissociation of a cyanohydrin can be overcome by acetylation, and the route is then synthetically useful.[31, 38] However, acetylation of the cyanohydrin hydroxyl group does not appear to improve the yields in catalytic hydrogenation.[72] Dissociation of the cyanohydrin can also be prevented by temporarily converting the hydroxyl group to an ether with vinyl isopropyl ether[80] or dihydropyran.[81]

Cyclic ketones have occasionally been condensed with nitromethane to give 1-nitromethylcycloalkanols[76, 82] which can be reduced to 1-aminomethylcycloalkanols.[76] Such nitro alcohols appear to require rather

$$(CH_2)_n \,\, C{=}O + CH_3NO_2 \rightarrow (CH_2)_n \,\, C \!\! \begin{array}{c} OH \\ \diagdown \\ CH_2NO_2 \end{array} \rightarrow (CH_2)_n \,\, C \!\! \begin{array}{c} OH \\ \diagdown \\ CH_2NH_2 \end{array}$$

specific conditions for satisfactory reduction, but they have been reduced successfully both catalytically[76] and electrolytically.[75]

Amino alcohols for the Tiffeneau-Demjanov expansion have also been produced by the reaction of ammonia with epoxides,[3] but this route is not used much because the epoxides are relatively inaccessible.[40] Another route not involving reduction is the Reformatskiĭ reaction between a

[74] Nace and Smith, *J. Am. Chem. Soc.*, **74**, 1861 (1952).
[75] Blicke, Doorenbos, and Cox, *J. Am. Chem. Soc.*, **74**, 2924 (1952).
[76] Dauben, Ringold, Wade, and Anderson, *J. Am. Chem. Soc.*, **73**, 2359 (1951).
[77] Blicke, Azuara, Doorenbos, and Hotelling, *J. Am. Chem. Soc.*, **75**, 5418 (1953).
[78] Nystrom and Brown, *J. Am. Chem. Soc.*, **70**, 3738 (1948).
[79] Brown, in Adams, *Organic Reactions*, Vol. VI, pp. 409–571, John Wiley & Sons, New York, 1951.
[80] Tchoubar, *Compt. rend.*, **237**, 1006 (1953).
[81] Elphimoff-Felkin, *Compt. rend.*, **236**, 387 (1953).
[82] Nightingale, Erickson, and Shackelford, *J. Org. Chem.*, **17**, 1005 (1952).

cyclic ketone and ethyl bromoacetate, followed by conversion of the carboxylic ester to the amine.[83]

$$(CH_2)_n\ C{=}O + BrCH_2CO_2C_2H_5 \xrightarrow{Zn}$$

$$(CH_2)_n\ C{\diagup{OH}\atop\diagdown{CH_2CO_2C_2H_5}} \rightarrow (CH_2)_n\ C{\diagup{OH}\atop\diagdown{CH_2NH_2}}$$

EXPERIMENTAL CONDITIONS

The general procedure is to dissolve the amine in dilute aqueous acid, add excess aqueous sodium nitrite, and, when the evolution of nitrogen ceases, to isolate the product either by steam distillation or by extraction with an immiscible solvent. The optimum pH appears to be not far from 7, in agreement with the formulation of the reaction as one between the free base and nitrous acid. It has been shown that high acidity (pH 3) stops the reaction of aliphatic amines with nitrous acid.[84] At too low acidity (pH 7 or above), the reaction either does not occur or is impractically slow. The desired pH is readily provided by dissolving the amine or its acetate in excess dilute acetic acid.[2, 3, 25] Alternatively, the amine hydrochloride may be used with a few drops of excess acid (mineral or acetic).[1, 50, 85] Occasionally other salts, such as oxalates,[66] have been used. Hydrochloric,[77] sulfuric,[75] and perchloric[36] acids have been used successfully, but when acids of this strength are used the excess must be small. Sodium dihydrogen phosphate and phosphoric acid are quite satisfactory,[16, 37] but, owing to the weak acidity of the former reagent, reaction is slow.

Although the choice of acid is often dictated by convenience, the possible involvement of the anion of the acid in the reaction should not be overlooked. This does not appear to be important in the Tiffeneau-Demjanov expansion where the product results by elimination of a proton, even though halohydrins are by-products when the halide ion concentration is high.[86, 87] In the Demjanov expansion, the last step is a combination of an intermediate with a nucleophilic species, commonly water. It has been demonstrated that the alkyl group of an amine undergoing deamination with nitrous acid is ultimately found combined

[83] Bergmann and Sulzbacher, *J. Org. Chem.*, **16**, 84 (1951).
[84] Kornblum and Iffland, *J. Am. Chem. Soc.*, **71**, 2137 (1949).
[85] Alder and Windemuth, *Ber.*, **71**, 2404 (1938).
[86] Felkin, *Compt. rend.*, **226**, 819 (1948).
[87] Tchoubar, *Bull. soc. chim. France*, **1949**, 169.

to some extent with all anions present,[88] and that the relative amounts may not be in proportion to their concentrations.[11] Alcohols produced by the Demjanov expansion in acetic acid solution are usually heavily contaminated with their acetate esters.[16] It is for this reason that phosphate[16] and perchlorate[36] solutions have been used.

The temperature is usually adjusted to 0° at the start of the Demjanov or Tiffeneau-Demjanov reaction, allowed to rise slowly to room temperature, and finally raised to near 100°. The choice of an initially low temperature is perhaps in part due to the instability of free nitrous acid, and partly due to the very occasionally rapid evolution of nitrogen; nevertheless, it does not appear to be generally necessary. When gas evolution has subsided, heating is begun. Successful results have also been obtained without heating, when the reaction mixture was allowed to stand for several hours.[40, 43] The time and temperature required appear to depend as much on the acidity of the medium as on the nature of the amine.

The source of nitrous acid is almost invariably sodium or potassium nitrite, although in the older literature the use of silver nitrite with amine hydrochlorides is described.[49, 89] Excesses of nitrite as high as 50%[72] and 200%[40] have been used, although one equivalent is the common amount. Since some nitrous acid is almost invariably lost through disproportionation, the use of only one equivalent of nitrite usually leads to recovery of considerable amounts of unreacted amine.[16, 17, 56] Because nitrous acid may react with the olefinic products accompanying the Demjanov expansion and with the ketones from the Tiffeneau-Demjanov expansion, it is best to avoid an unnecessary excess. An effective scheme is to use at first one equivalent, remove the products which are formed (by steam distillation or extraction), and then treat the remaining aqueous solution with fresh portions of acid and nitrite.[85]

Moderately dilute solutions are usual, about 5–20% in amine and the same range of a weak acid, if one is employed; for strong acids, as has been mentioned, the total quantity is kept at little more than that equivalent to the amine, and the acid is usually diluted to a concentration of less than 10%.

Since the deamination products are usually not basic; they commonly separate from solution as the reaction proceeds. Solid products can, of course, be removed by filtration. Liquid products are commonly isolated by extraction with ether and fractional distillation of the dried extracts. Steam distillation from the reaction mixture[53, 54, 85] is occasionally employed; it has the advantage of freeing the product from the

[88] Whitmore and Langlois, *J. Am. Chem. Soc.*, **54**, 3441 (1932).
[89] Demjanov, *J. Russ. Phys.-Chem. Soc.*, **36**, 166 (1904) (*Chem. Zentr.*, **1904, I**, 1214).

nonvolatile tars which are so often formed, especially in the Demjanov expansion, and to some extent from the small amounts of glycols sometimes formed in the Tiffeneau-Demjanov expansion.[77, 87]

The products of a Demjanov expansion are easily separated into an olefin (lower boiling) and an alcohol fraction; either or both may, of course, be the desired product. Purification of the alcohol fraction is generally not practicable by distillation, owing to the similar boiling points of the isomeric alcohols. Where acetic acid solutions have been used, esters must first be saponified or cleaved with lithium aluminum hydride. Since the unrearranged alcohol is almost always primary, and the expanded alcohol is almost always secondary, either oxidation or differential esterification[17, 51, 54] may be used to separate the isomers. The small amounts of tertiary alcohols that are sometimes present may also often be eliminated by such procedures. Oxidation, usually with chromic acid, converts the expanded alcohol to a ketone and the primary alcohol either to an aldehyde or acid, allowing separation by obvious means.[17, 51] Esterification of primary alcohols with phthalic anhydride, usually in benzene solution, is fairly rapid; esterification of secondary alcohols is much slower and requires prolonged heating, and tertiary alcohols are either dehydrated or unaffected.[90] The alkyl hydrogen phthalates produced can be separated from unesterified material by extraction with very dilute alkali and then recrystallized.[51, 90] Regeneration of the alcohol by saponification presents no complications.[17, 54, 90] The olefins produced in the Demjanov reaction usually are not easily separated from each other, but oxidation to ketones, keto acids, or acids may elucidate their structures.[17]

The products of a Demjanov expansion usually include small amounts of nitrogen-containing compounds which often appear in the high-boiling residue. These substances are usually neglected. Those isolated have been identified as nitroalkanes,[60, 91] which presumably result from the action of oxides of nitrogen on the olefins formed.

The isolation of the ketones from Tiffeneau-Demjanov expansions is somewhat simpler, since the principal accompanying substances (other than unreacted amine) are glycols which are very much less volatile than the ketones. However, when it is not desirable to separate the ketone by distillation, as in the steroid field, it may be necessary to separate the ketone through the semicarbazone,[38, 72] by reaction with Girard's reagents, or by chromatography.[72, 92]

[90] Ingersoll, in Adams, *Organic Reactions*, Vol. II, p. 393, John Wiley & Sons, New York, 1944.
[91] Cook, Jack, and Loudon, *J. Chem. Soc.*, **1952**, 607.
[92] Goldberg and Studer, *Helv. Chim. Acta*, **24**, 478 (1941).

EXPERIMENTAL PROCEDURES

Cycloheptanone. Detailed directions for the preparation of cycloheptanone in a 40–42% over-all yield from cyclohexanone by the Tiffeneau-Demjanov rearrangement are given in *Organic Syntheses*.[93]

3-Hydroxydodecahydroheptalene and Decahydroheptalene by the Demjanov Rearrangement.[94] To a solution of 7.1 g. of 2'-aminomethylcyclohexanocycloheptane and 3.5 ml. of glacial acetic acid in 70 ml. of water is added a solution of 4.2 g. of sodium nitrite in 28 ml. of water, and the mixture is heated on a water bath until the evolution of nitrogen ceases. The oil that separates is extracted with ether and the extracts are washed with aqueous sodium hydroxide, dried, and distilled to give two main fractions: (a) 1.5 g. (23%) of crude olefins, b.p. 53–93°/0.5 mm.; and (b) 3.6 g. (51%) of crude alcohols, b.p. 105–115°/1 mm. Redistillation of fraction (a) over sodium gives decahydroheptalene as a colorless oil (1.0 g., 14%) b.p. 58–62°/0.5 mm. Fraction (b) consists mainly of 3-hydroxydodecahydroheptalene.

Cycloheptanol by the Demjanov Rearrangement.[16, 37] To a solution of 45 g. (0.038 mole) of syrupy, 85% orthophosphoric acid in about 300 ml. of water containing some ice is added 56.5 g. (0.5 mole) of cyclohexanemethylamine; a white, crystalline precipitate forms. A solution of 35 g. (0.5 mole) of sodium nitrite in about 60 ml. of water is added all at once to the cold amine solution, boiling chips are added, and the mixture is allowed to stand for one hour in a 1-l. round-bottomed flask. During this time the precipitate dissolves, nitrogen is slowly evolved, and an oil separates. The reaction mixture is then distilled; an efficient condenser should be used if it is desired to avoid loss of olefin. After most of the oil has come over (usually 200–250 ml. of distillate), distillation is stopped and the distilland is cooled somewhat; 7.5 g. (0.107 mole) of sodium nitrite in concentrated aqueous solution and 4.5 g. (0.04 mole) of 85% orthophosphoric acid are then added. Distillation is resumed until only isolated lumps of tar float in the distilland; usually 50–150 ml. of distillate is required. This second distillate is collected in a separate flask containing several grams of potassium carbonate in water solution to neutralize nitrous fumes.

The distillates are combined, the layers separated, and the aqueous layer extracted twice with 25-ml. portions of petroleum ether (b.p. 30–40°). The combined extracts are then dried over potassium carbonate, filtered through a filter paper moistened with petroleum ether, and distilled through an 18-inch Vigreux column or its equivalent until the solvent is

---

[93] Dauben, *Org. Syntheses*, **34**, 19 (1954).
[94] Aspinwall and Baker, *J. Chem. Soc.*, **1950**, 743.

removed (90°). If the distillation is carred further at this point, there are obtained 6–8 g. (12–17%) of mixed olefins, b.p. 95–125° (mostly 105–115°) and 25–30 g. (44–52%) of mixed alcohols, b.p. 125–185° (mostly 155–180°). It is usually desirable to purify the alcohol by chemical means, for which purpose the solvent-free but unfractionated material is suitable.

To remove cyclohexanemethanol from the product, the residue after removal of the solvent is mixed with 10 g. (0.07 mole) of phthalic anhydride and heated under reflux at 120–140° for one-half to one hour. The cooled mixture is shaken with 8.5 g. of sodium carbonate monohydrate in 350 ml. of water and 50 ml. of petroleum ether (b.p. 30–40°), and the layers are separated. The organic layer is washed with two 50-ml. portions of water.*

The combined petroleum ether solutions are dried over potassium carbonate and distilled through an 18-inch Vigreux column or its equivalent. There are obtained 5–6 g. (10–12%) of olefins, b.p. 103–127° (mostly 105–115°), and 22–23 g. (38–40%) of alcohol, b.p. 127–187°. Redistillation of the alcohol mixture gives about 20 g. (35%) of somewhat impure cycloheptanol, b.p. 150–180°. Further purification may be accomplished, if desired, by converting the crude cycloheptanol to its hydrogen phthalate, using the detailed directions given for 2-octyl hydrogen phthalate in an earlier volume of this series (Ref. 90, p. 400); pure cycloheptyl hydrogen phthalate melts at 100–102°.

Cycloöctanone by the Tiffeneau-Demjanov Rearrangement.[77] 1-Aminomethylcycloheptanol (124 g., 0.87 mole) is dissolved in 400 ml. of 10% hydrochloric acid and cooled to below 5°. A solution of 69 g. (1 mole) of sodium nitrite in 300 ml. of water is added slowly with stirring, and the resulting solution is allowed to stand for two hours, during which time it warms to room temperature. It is then heated on a steam bath for one hour, cooled, and the oily layer is separated. The aqueous layer is extracted with about 100 ml. of ether, and the combined extracts are dried over potassium carbonate and distilled under reduced pressure through a short column. There is obtained 67 g. (61%) of cycloöctanone, b.p. 85–87°/17 mm. The higher-boiling residue contains 2-hydroxymethylcycloheptanol, which may also be collected by distillation; the yield is 5 g. (4%), b.p. 142–147°/2 mm.

* To recover the unrearranged alcohol the combined aqueous layers are washed with 50 ml. of petroleum ether. The hexahydrobenzyl hydrogen phthalate is recovered by acidifying the aqueous solution with hydrochloric acid, allowing the precipitated oil to crystallize, and recrystallizing from ligroin or aqueous acetic acid. There is thus obtained 11–13 g. (8–10%) of a white solid whose melting point is usually in the range 110–120°.

TABULAR SURVEY OF THE DEMJANOV AND TIFFENEAU-DEMJANOV RING EXPANSIONS

In the tables are included all the examples of successful or attempted ring enlargements by the Demjanov and Tiffeneau-Demjanov methods that could be found through the year 1957. In addition some pertinent later examples are included.

The examples are given in four tables: mononuclear carbocyclic rings; polynuclear carbocyclic systems with one shared side; polynuclear carbocyclic systems with more than one shared side, and heterocyclic rings. Within each table the entries appear in the order of increasing ring size, and, for compounds of the same ring size, in the order of the number of carbon atoms. In many of the examples of the Demjanov reaction, the yields quoted do not represent pure substance isolated, but are calculated from the total yield and the composition analysis.

TABLE I
Mononuclear Carbocyclic Rings

Amine	Rearranged Alcohol or Ketone	Yield % (Ref.)	Olefin Yield % (Ref.)	Unrearranged Alcohol Yield % (Ref.)
Cyclopropanemethylamine	Cyclobutanol	50 (27), 17, (36)	(27, 36)	50 (27), 17 (36)
	Allylcarbinol	2 (36)		
α-Cyclopropylethylamine				— (95)
1-Methylcyclopropanemethylamine	1-Methylcyclobutanol	High (26)		0 (26)
Cyclobutanemethylamine	Cyclopentanol	— (2)	— (2)	— (2)
2,2,3,3-Tetrafluorocyclobutanemethylamine		0 (63)	— (63)	— (63)
α-Cyclobutylethylamine	1-Methylcyclopentanol, 1-ethylcyclobutanol, *trans*-2-methylcyclopentanol (trace)	46 (37)	14 (37)	— (37)
Cyclopentanemethylamine	Cyclohexanol	30 (25, 37), 7 (37)	— (25, 37)	3 (37)
1-Hydroxycyclopentanemethylamine	Cyclohexanone	75 (40, 96, 97)		
α-Cyclopentylethylamine	*trans*-2-Methylcyclohexanol	17 (37)	9 (37)	22 (37)
	1-Ethylcyclopentanol	16 (37)		
α-(1-Hydroxycyclopentyl)ethylamine	2-Methylcyclohexanone	70 (97)		
trans-2-Hydroxycyclopentanemethylamine	2-Methylcyclopentanone	— (61, 98)		— (61, 98)
2-Methyl-1-hydroxycyclopentanemethylamine	3-Methylcyclohexanone	80 (40, 99)		
3-Methyl-1-hydroxycyclopentanemethylamine	3-Methylcyclohexanone	35 (40)		
	4-Methylcyclohexanone	35 (40)		

Note: References 95 to 110 are on p. 188.

TABLE I—*Continued*
Mononuclear Carbocyclic Rings

Amine	Rearranged Alcohol or Ketone	Yield % (Ref.)	Olefin Yield % (Ref.)	Unrearranged Alcohol Yield % (Ref.)
1,2,2,3-Tetramethylcyclopentanemethylamine	1,3,3,4- or 1,2,2,3-Tetramethylcyclohexanol	— (48, 49)	— (49)	
1,2,2-Trimethyl-3-carboxycyclopentanemethylamine	A trimethylhydroxycyclohexanecarboxylic acid	— (48, 50)		
α-(1-Hydroxycyclopentyl)-benzylamine	2-Phenylcyclohexanone	50 (97)		
1-Phenylcyclopentyl-1′-ethylamine	2-Phenylcyclohexanol (*cis* and *trans*)	73 (37)	0 (37)	0 (37)
Cyclohexanemethylamine	Cycloheptanol	29 (16), 64 (17)	27 (16), 21 (17)	15 (16)
	1-Methylcyclohexanol	2 (16)	— (89)	
1-Hydroxycyclohexanemethylamine	Cycloheptanone	60 (3, 40)		— (40, 75)
		65 (76), 57 (75)		
α-Cyclohexylethylamine	1-Ethylcyclohexanol	16 (37, 76, 100)	3 (37)	23 (37)
β-Cyclohexylethylamine	α-Cyclohexylethanol	— (101)	Trace (101)	— (101)
cis-2-Hydroxycyclohexanemethylamine	Cyclohexanecarboxaldehyde	— (60, 98)		— (60, 98)
trans-2-Hydroxycyclohexanemethylamine	2-Methylcyclohexanone	— (60, 98)		— (59, 98)
2-Chlorocyclohexanemethylamine	None	0 (62)		— (62)
α-(1-Hydroxycyclohexyl)ethylamine	2-Methylcycloheptanone	60 (47), 55 (97)		
2-Methyl-1-hydroxycyclohexanemethylamine	2-Methylcycloheptanone	6 (40, 96)		
	3-Methylcycloheptanone	60 (40, 96)		
2-Hydroxy-2-methylcyclohexanemethylamine	None	0 (62)		— (40)

3-Methyl-1-hydroxycyclohex-anemethylamine	3-Methylcycloheptanone	40 (40, 96)	
4-Methyl-1-hydroxycyclohex-anemethylamine	4-Methylcycloheptanone	40 (40, 96)	
	4-Methylcycloheptanone	60 (52), 65 (40)	
4-Methylcyclohexanemethylamine	4-Methylcycloheptanol	55 (51)	20 (51)
α-(4-Methylcyclohexyl)ethylamine	None	0 (37, 44)	25 (37) 39* (37, 44)
3,5-Dimethylcyclohexanemethyl-amine	2,4-Dimethylcycloheptanol	— (44)	
3,3,5-Trimethylcyclohexane-methylamine	3,5,5-Trimethylcycloheptanol†	— (53)	— (53)
3,3,5-Trimethyl-1-hydroxycyclo-hexanemethylamine	3,5,5- and 3,3,5-Trimethylcyclo-heptanone	— (40)	
2,2,6-Trimethylcyclohexane-methylamine	2,2,6-Trimethylcycloheptanol†	— (54)	— (54)
1-Hydroxycyclohexane-1′-iso-butylamine	2-Isopropylcycloheptanone	50 (102)	— (102)
1-Hydroxycyclohexane-1′-neopentylamine	2-t-Butylcycloheptanone	30–40 (102)	§(102) 0 (102)
α-Cyclohexylbenzylamine	None	0 (45)	— (45)
2-Phenyl-1-hydroxycyclohex-anemethylamine	3-Phenylcycloheptanone	64‡ (71)	
α-(1-Hydroxycyclohexyl)benzyl-amine	None	0 (103)	— (103)

Note: References 95 to 110 are on p. 188.

* This figure includes some tertiary alcohol.
† The position of the hydroxyl group is uncertain.
‡ The yield is based on the cyanohydrin.
§ Appreciable amounts of cyclohexanone were formed in this experiment.

TABLE I—Continued
Mononuclear Carbocyclic Rings

Amine	Rearranged Alcohol or Ketone	Yield % (Ref.)	Olefin Yield % (Ref.)	Unrearranged Alcohol Yield % (Ref.)
α-(1-Hydroxycyclohexyl)-p-methylbenzylamine	None	— (103)		— (103)
α-(1-Hydroxycyclohexyl)hexahydrobenzylamine	2-Cyclohexylcycloheptanone	50 (102)		— (102)
Cycloheptanemethylamine	Cycloöctanol	— (25)		
1-Hydroxycycloheptanemethylamine	Cycloöctanone	61 (77), 70 (40)		4 (77)
Cycloöctanemethylamine	Cyclononanol	18 (17)	38 (17)	26 (17)
1-Hydroxycycloöctanemethylamine	Cyclononanone	50 (40), 57 (39)		

Note: References 95 to 110 are on p. 188.

TABLE II
Polynuclear Carbocyclic Systems with Fusion at a Single Side

Amine	Rearranged Alcohol or Ketone	Yield % (Ref.)	Olefin Yield % (Ref.)
6-Hydroxybicyclo[3.2.0]-2-heptene-6-methyl-amine	Bicyclo[3.3.0]-2-octen-6-one	47* (38)	
cis-2-Hydroxybicyclo[3.3.0]-octane-2-methylamine	Bicyclo[3.3.0]-2-octen-7-one Hydrindan-5-one	8* (38) 60 (104)	
2-Hydroxyindane-2-methylamine	β-Tetralone	— (40, 99)	
17-Aminomethylestradiol-3-acetate	D-Homoestrone acetate	38 (30, 92)	
3-trans-17-Dihydroxy-17-aminomethyl-androstane	3-trans-Hydroxy-D-homoandrostan-17a-one	51 (105)	
3-trans-Acetoxy-17-hydroxy-17-amino-methylandrostane	3-trans-Acetoxy-D-homoandrostan-17a-one	51 (105)	
3-epi-17-Hydroxy-17-aminomethyl-androstane	3-epi-Hydroxy-D-homoandrostan-17a-one	73 (105)	
3β-Acetoxy-17-hydroxy-17-aminomethyl-androstane	3β-Acetoxy-D-homoandrostan-17a-one	37 (58)	
Δ⁵,⁶-3β,17-Dihydroxy-17-aminomethyl-androstene	3β-Acetoxy-D-homoandrostan-17-one Δ⁵,⁶-3β-Hydroxy-17a-keto-D-homo-androstene	5 (58) 80 (31)	
3β-Acetoxy-5,6β-oxido-17-hydroxy-17-aminomethylandrostane	3β-Acetoxy-5,6β-oxido-D-homoandrostan-17a-one	26 (106)	
	3β-Acetoxy-5,6β-oxido-D-homoandrostan-17-one	2 (106)	

Note: References 95 to 110 are on p. 188.

* The yield is based on the acetylated cyanohydrin.

TABLE II—Continued
POLYNUCLEAR CARBOCYCLIC SYSTEMS WITH FUSION AT A SINGLE SIDE

Amine	Rearranged Alcohol or Ketone	Yield % (Ref.)	Olefin Yield % (Ref.)
Hydrindane-5-methylamine	4,5-Cyclopentanocycloheptanol	68 (55), 57 (56, 107)	20 (55), 15 (56)
5-Hydroxyhydrindane-5-methylamine	Bicyclo[5.3.0]decan-3-one	89 (56)	
5-Methylhydrindane-6-methylamine	2-Methyl-4,5-cyclopentanocycloheptanol + 6-methyl-3,4-cyclopentanocycloheptanol	54 (55)	— (55)
3,4-Cycloheptanocyclohexanemethylamine	3-Hydroxydodecahydroheptalene	50 (94)	24 (94)
9-Aminomethyl-9,10-dihydro-2,3,4,7-tetramethoxyphenanthrene	Deaminocolchinol methyl ether	— (91)	— (91)
17α-Hydroxy-17α-aminomethyl-D-homo-estrol-3-monoacetate	D-*bis*-Homoestrone acetate	76 (57)	
3-Aminomethyl-17-acetoxyandrostan-3-ol	A-Homo-17-acetoxyandrostan-4-one	— (72)	
3-Hydroxy-3-aminomethylcholestane	A-Homocholestanone	70 (72)	

Note: References 95 to 110 are on p. 188.

TABLE III
Polynuclear Systems with Fusion at More than One Edge

Amine	Rearranged Alcohol or Ketone	Yield % (Ref.)	Olefin Yield % (Ref.)
2,5-Endomethylenecyclohexanemethylamine	2,5-Endomethylenecycloheptanol	Good (85)	
ω-Aminoisocamphane	R-Homocamphenilol	45 (108)	16 (108)
ω-Aminotricyclene	Not identified	— (109)	
Bornylamine	(+)-Camphene hydrate	31 (60)	13* (60)
	(+)-α-Terpineol	26 (60)	
Isobornylamine	(−)-Camphene hydrate	— (60)	
ω-Aminopinene (aminoterebenthene)	p-Isopropyl-3,4-dihydrobenzyl alcohol	— (66)	— (60)*

Note: References 95 to 110 are on p. 188.

* The olefin was camphene

TABLE IV
Heterocyclic Rings

Amine	Rearranged Alcohol or Ketone	Yield % (Ref.)	Unrearranged Alcohol Yield % (Ref.)
2-Aminomethylfuran	2-Hydroxypyran	High (43a)	
Pyrrolidine-α-methylamine	Piperideine	Low (42)	
Pyrrole-α-methylamine	Pyridine	25 (42)	
3-Aminomethyl-3-tropanol	R-Homotropinone	57 (43)	
2-Thenylamine	Hydroxythiopyran	— (110)	— (110)

Note: References 95 to 110 are on p. 188.

REFERENCES FOR TABLES I–IV

[95] Demjanov and Pinegin, *J. Russ. Phys.-Chem. Soc.*, **46,** 58 (1914) (*C.A.*, **8,** 1965 (1914)].
[96] Tiffeneau and Tchoubar, *Compt. rend.*, **212,** 195 (1941).
[97] Elphimoff-Felkin and Tchoubar, *Compt. rend.*, **237,** 726 (1953).
[98] Mousseron, Jacquier, and Jullien, *Bull. soc. chim. France*, **1951,** C89.
[99] Tiffeneau and Tchoubar, *Compt. rend.*, **215,** 224 (1942).
[100] Jennen, *Compt. rend.*, **234,** 961 (1952).
[101] Wallach and Lorge, *Ann.*, **359,** 312 (1908).
[102] Elphimoff-Felkin and Gault, *Compt. rend.*, **246,** 1871 (1958).
[103] Elphimoff-Felkin and Tchoubar, *Compt. rend.*, **231,** 1314 (1950); **233,** 596 (1951).
[104] Granger and Nau, *Bull. soc. chim. France*, **1957,** 986.
[105] Goldberg and Monnier, *Helv. Chim. Acta*, **23,** 376 (1940).
[106] Goldberg, Sice, Robert, and Plattner, *Helv. Chim. Acta*, **30,** 1441 (1947).
[107] Plattner, Heilbronner, and Fürst, *Helv. Chim. Acta*, **30,** 1100 (1947).
[108] Lipp, Dessauer, and Wolf, *Ann.*, **525,** 271 (1936).
[109] Lipp, *Ber.*, **53,** 769 (1920).
[110] Putoshin and Egorova, *J. Gen. Chem. U.S.S.R.*, **10,** 1873 (1940) [*C.A.*, **35,** 4377 (1941)].

CHAPTER 3

ARYLATION OF UNSATURATED COMPOUNDS BY DIAZONIUM SALTS (THE MEERWEIN ARYLATION REACTION)

CHRISTIAN S. RONDESTVEDT, JR.*

CONTENTS

	PAGE
INTRODUCTION	190
MECHANISM	192
SCOPE AND LIMITATIONS	198
The Unsaturated Component	198
Quinones	201
Miscellaneous Unsaturated Compounds	203
Reactivities of Unsaturated Compounds	205
Decarboxylation during Arylation of Cinnamic and Maleic Acids	206
The Diazonium Salt	206
Factors Influencing Addition vs. Substitution	208
Side Reactions	209
COMPARISON WITH OTHER SYNTHETIC METHODS	211
EXPERIMENTAL CONDITIONS	214
Effects of Reaction Medium	215
Solvent	215
Anions	217
Catalysts	217
Acidity	218
EXPERIMENTAL PROCEDURES	219
1-p-Nitrophenylbutadiene	219
3-p-Nitrophenylcoumarin	219
trans-p-Chlorocinnamic Acid and p-Nitrocinnamic Acid	219
2-Methoxy-4'-phenylstilbene	220
p-Nitrocinnamonitrile	220

* Present address: Jackson Laboratory, E. I. du Pont de Nemours & Company, Inc. Wilmington 99, Delaware.

190   ORGANIC REACTIONS

PAGE

α-p-Chlorophenyl-N-isopropylmaleimide 220
p-Nitrophenylmaleic Anhydride 221
1,4-Bis(2'-chloro-2'-cyanoethyl)benzene (Use of a Diamine) . . . 221
2-o-Chlorophenylbenzoquinone 221

TABULAR SURVEY 222

    Table I. Nonconjugated Olefins and Acetylenes 224
    Table II. Conjugated Dienes and Arylolefins 225
    Table III. α,β-Unsaturated Aldehydes and Ketones 229
    Table IV. Aliphatic α,β-Unsaturated Monobasic Acids, Esters, Nitriles . 231
    Table V. Aromatic α,β-Unsaturated Acids, Esters, Nitriles . . . 235
    Table VI. Heterocyclic α,β-Unsaturated Acids 240
    Table VII. α,β-Unsaturated γ-Keto Acids 241
    Table VIII. Conjugated Dienoic Acids and Esters 242
    Table IX. Polybasic α,β-Unsaturated Acids, Nitriles, Esters, Imides . 243
    Table X. Quinones 248
    Table XI. Hydroquinone 254
    Table XII. Coumarins 254
    Table XIII. Miscellaneous 255
        A. Bisdiazonium Salts 255
        B. Nitroölefins 255
        C. Vinyl Ethers 255
        D. Active Methylene Compounds 256
        E. Oximes and Semicarbazones 256
        F. Furfural 258
        G. Furoic Acid 259

INTRODUCTION

The arylation of olefinic compounds by diazonium halides with copper salt catalysis was discovered by Hans Meerwein.[1,2] This reaction has been referred to as the Meerwein reaction despite the possibility of its being confused with the Meerwein-Ponndorf-Verley reduction or the Wagner-Meerwein rearrangement. The Meerwein arylation reaction proceeds best when the olefinic double bond is activated by an electron-attracting group Z, such as carbonyl, cyano, or aryl. The net result is the union of the aryl group from the diazonium salt with the carbon atom β to the activating group, either by substitution of a β-hydrogen atom or by addition of Ar and Cl to the double bond.

$$ArN_2Cl + RCH{=}CRZ \xrightarrow[\text{salt}]{\text{Copper}} ArCR{=}CRZ + ArCHRC(R)ClZ$$

[1] Meerwein, Büchner, and van Emster, *J. prakt. Chem.*, [2] **152**, 239 (1939); Schering-Kahlbaum, Brit. pat. 480,617 [*C.A.*, **32**, 6262⁶ (1938)]; Meerwein, U.S. pat. 2,292,461 [*C.A.*, **37**, 654⁹ (1943)].

[2] Franzen and Krauch, *Chemiker-Ztg.*, **79**, 101 (1955). These authors state that the original discovery is due to Curt Schuster, but his results were published only in internal reports of the I. G. Farbenindustrie.

The reaction is a valuable synthetic tool. Although the yields are often low (commonly 20–40%), such yields are offset by the availability at low cost of a wide variety of aromatic amines and unsaturated compounds, and by the ease and simplicity of performing the reaction. Furthermore, the polyfunctional product built up in a single operation from commercial chemicals is capable of undergoing many subsequent transformations.

The accompanying examples are typical of the scope of the reaction. They also show some of the realized and potential transformations of the products.

(1) $ArN_2Cl + CH_2=CHCO_2H \longrightarrow ArCH=CHCO_2H$

(2) $ArN_2Cl + CH_2=CHCN \longrightarrow ArCH_2CHClCN \longrightarrow ArCH_2CH_2CN \longrightarrow Ar(CH_2)_3NH_2$
$\downarrow \downarrow$
$ArCH=CHCN ArCH_2CH_2CO_2H$

(3) $ArN_2Cl + \begin{matrix} CHCO \\ \| \\ CHCO \end{matrix}\!\!>\!\!NR \longrightarrow \begin{matrix} ArC-CO \\ \| \\ CHCO \end{matrix}\!\!>\!\!NR + \begin{matrix} ArCHCO \\ | \\ ClCHCO \end{matrix}\!\!>\!\!NR \xrightarrow[\text{2. Heat}]{\text{1. Hydrol.}} \begin{matrix} ArC-CO \\ \| \\ CHCO \end{matrix}\!\!>\!\!O$

(4) $ArN_2Cl + CH_2=CHCH=CH_2 \longrightarrow ArCH_2CH=CHCH_2Cl \xrightarrow{\text{Base}} ArCH=CHCH=CH_2$
$[H]\downarrow \searrow LiAlH_4$
$ArCH_2CH_2CH_2CH_2Cl ArCH_2CH=CHCH_3$

(5) $ArN_2Cl + Ar'CH=CHCO_2H \longrightarrow Ar'CH=CHAr + CO_2$

(6) $ArN_2Cl + \underset{O}{\bigcirc}\!\!-CHO \longrightarrow Ar\underset{O}{\bigcirc}\!\!-CHO$

(7) $ArN_2Cl + \underset{O}{\overset{O}{\bigcirc}} \longrightarrow Ar\underset{O}{\overset{O}{\bigcirc}}$

This review will be confined to reactions in which a new carbon-carbon bond is formed between the aromatic ring of a diazonium salt and an aliphatic unsaturated compound, including olefins, acetylenes, quinones, oximes, and such heterocycles as furan and thiophene. The arylation of aromatic compounds by diazonium salts and related compounds (the Gomberg-Bachmann reaction) has been reviewed in Volume II of *Organic Reactions*.

MECHANISM

The mechanism of the Meerwein arylation reaction is not known with certainty, although some features have been established. The correct mechanism must account for the following facts. (1) The olefinic double bond must be activated by an electron-attracting group; the few reported exceptions[3,4] to this generalization have not been confirmed. (2) The incoming aryl group occupies the position β to the (stronger) activating group. (3) Diazonium salts bearing electron-attracting substituents usually give better results than those possessing electron-releasing substituents. (4) In most cases the reaction is specifically catalyzed by copper salts. (5) The rate of reaction (nitrogen evolution) appears to be markedly dependent on the structure of both the unsaturated compound and the diazonium salt. (6) The yields are dependent upon the pH, the nature of the solvent, and other components of the reaction medium; the presence of halide ion appears to be advantageous,[4] though not indispensable.[5]

Ionic Mechanism. Meerwein[1] proposed that the diazonium cation loses nitrogen to form an aryl cation as a result of "the polarizing influence of the unsaturated compound." The cation then adds to the double bond. He showed that iodonium salts, which he believed could react only by an ionic mechanism, likewise arylated unsaturated compounds. Recent work has shown that diaryliodonium salts also may react by radical mechanisms.[6,7] The ionic mechanism for the Meerwein arylation has been supported by other workers.[8-22]

---

[3] Müller, "Zetko Austausch," Dept. of Commerce, Office of Technical Services, P.B. No. 737.
[4] Müller, *Angew. Chem.*, **61**, 179 (1949).
[5] C. S. Rondestvedt, Jr., unpublished experiments.
[6] Sandin and Brown, *J. Am. Chem. Soc.*, **69**, 2253 (1947).
[7] Beringer, Geering, Kuntz, and Mausner, *J. Phys. Chem.*, **60**, 141 (1956).
[8] Brunner and Perger, *Monatsh.*, **79**, 187 (1948).
[9] Brunner and Kustatscher, *Monatsh.*, **82**, 100 (1951).
[10] Bergmann and Weinberg, *J. Org. Chem.*, **6**, 134 (1941).
[11] Bergmann, Weizmann, and Schapiro, *J. Org. Chem.*, **9**, 408 (1944).
[12] Bergmann and Weizmann, *J. Org. Chem.*, **9**, 415 (1944).
[13] Bergmann and Schapiro, *J. Org. Chem.*, **12**, 57 (1947).
[14] Bergmann, Dimant, and Japhe, *J. Am. Chem. Soc.*, **70**, 1618 (1948).
[15] Freund, Brit. pat. 670,317 [*C.A.*, **46**, 10201c (1952)]; Freund, U.S. pat. 2,710,874 [*C.A.*, **49**, 11705c (1955)]; Freund, Austral. pat. 147,045 [*C.A.*, **51**, 15595d (1957)].
[16] Freund, *J. Chem. Soc.*, **1951**, 1943.
[17] Freund, *J. Chem. Soc.*, **1952**, 1954.
[18] Freund, *J. Chem. Soc.*, **1952**, 3068.
[19] Freund, *J. Chem. Soc.*, **1952**, 3072.
[20] Freund, *J. Chem. Soc.*, **1952**, 3073.
[21] Freund, *J. Chem. Soc.*, **1953**, 2889.
[22] Freund, *J. Chem. Soc.*, **1953**, 3707.

THE MEERWEIN ARYLATION REACTION

$$ArN_2^{\oplus} \xrightarrow{(RCH=CRZ)} Ar^{\oplus} + N_2$$

$$Ar^{\oplus} + RCH=CRZ \longrightarrow ArCH(R)\overset{\oplus}{C}RZ$$

$$ArCH(R)\overset{\oplus}{C}RZ + Cl^{\ominus} \longrightarrow ArCH(R)CClRZ$$
$$\searrow ArC(R)=CRZ + H^{\oplus}$$

The cationic mechanism explains the effect of substituents in the diazonium salt (point 3 above): electron-attracting groups increase the electrophilicity of the cation. It also accounts for point 5, though "the polarizing influence of the olefin" is not a very specific explanation. However, a cationic mechanism fails to account for points 1 and 3, for the olefins most reactive toward arylation are those with double bonds rendered electron-deficient by the group Z. Yet these compounds are the least reactive in typical electrophilic additions (bromination, etc.). The normal ionic polarization of the olefins renders the β carbon positive

$$\left[{}^{\oplus}CH_2\overset{\ominus}{C}HC=O \leftrightarrow {}^{\oplus}CH_2CH=C-O^{\ominus} \atop R R \right],$$

as demonstrated by the following additions.

$$R\overset{\ominus}{O}\overset{\oplus}{H} + CH_2=CHCN \xrightarrow{\text{Base}} ROCH_2CH_2CN$$

$$\overset{\oplus}{H}\overset{\ominus}{Cl} + CH_2=CHCO_2H \rightarrow ClCH_2CH_2CO_2H$$

Alternatively one must invoke an abnormal polarization $\overset{\ominus}{C}H_2\overset{\oplus}{C}HCOR$ to explain why the hypothetical cation attacks the β-carbon atom. The alternative ionic mechanism involving an aryl anion is equally difficult to accept, for the existence of aryl anions in the aqueous acid medium is highly unlikely.

Finally, in reactions of diazonium salts with olefins that are certainly ionic, very different products are obtained, as shown in the ensuing equations.

$$C_6H_5N_2BF_4 + CH_2=CHCN \rightarrow [CH_2=\overset{\oplus}{C}H\overset{\ominus}{C}=NAr]BF_4 \xrightarrow{H_2O}$$
$$CH_2=CHCONHC_6H_5 \quad (\text{Refs. 23, 24})$$

[23] Makarova and Nesmeyanov, *Izvest. Akad. Nauk S.S.S.R., Otdel. Khim. Nauk*, **1954**, 1019; *Bull. Acad. Sci. U.S.S.R., Div. Chem. Sci.*, (*Engl. Transl.*), **1954**, 1109, [*C.A.*, **50**, 241a (1956)].

[24] Meerwein, Laasch, Mersch, and Spille, *Chem. Ber.*, **89**, 209 (1956).

Compare

$$[(C_2H_5)_3O]^{\oplus}BF_4^{\ominus} + RCN \rightarrow [RC{=}NC_2H_5]^{\oplus}BF_4^{\ominus} \xrightarrow{H_2O} RCONHC_2H_5$$
(Ref. 24)

$$C_6H_5N_2BF_4 + CH_2{=}CHCO_2CH_3 \rightarrow CH_2{=}C(C_6H_5)CO_2CH_3 \quad \text{(Ref. 25)}$$

Note the α-arylation, not β-arylation as obtained under Meerwein arylation conditions.

Free-Radical Mechanism. A radical mechanism was proposed by Koelsch and Boekelheide[26] and by Müller,[4] and supported by others.[2] At pH 3–5, the diazonium salt is in equilibrium with the covalent diazo acetate (from the acetate buffer) or diazo chloride, either of which may decompose to an aryl radical which then may add to the double bond. The alkyl radical is thought to be oxidized by cupric ion to a cation which then acquires chloride ion or loses a proton to give the product. The cuprous ion is reoxidized by the acetate (or chloride) radical to cupric ion.

$$ArN_2^{\oplus} + OCOCH_3^{\ominus} \rightarrow ArN{=}NOCOCH_3$$

$$ArN{=}NOCOCH_3 \rightarrow Ar\cdot\, +\, N_2\, +\, \cdot OCOCH_3$$

$$Ar\cdot\, +\, RCH{=}CRZ \rightarrow ArCH(R)CRZ$$

$$ArCH(R)CRZ + Cu^{++} \rightarrow ArCH(R)\overset{\oplus}{C}RZ + Cu^+$$

$$Cu^+ + \cdot OCOCH_3 \rightarrow Cu^{++} + OCOCH_3^{\ominus}$$

The radical mechanism explains the direction of addition to unsymmetrical olefins.* With a monosubstituted olefin, only one of the two possible intermediate radicals can be stabilized by resonance. With unsymmetrical 1,2-disubstituted ethylenes, such as β-substituted styrenes, resonance with the aryl group is more effective in controlling orientation than resonance with a carbonyl or cyano group. These principles are illustrated in the examples on p. 195.

Despite its success in accounting for the position occupied by the attacking group, the free-radical mechanism cannot be accepted without modification. Many of the olefins arylated in the Meerwein reaction are vinyl monomers which are readily polymerized by authentic radicals.

---

[25] Nesmeyanov, Makarova, and Tolstaya, *Tetrahedron*, **1**, 145 (1957).
[26] Koelsch and Boekelheide, *J. Am. Chem. Soc.*, **66**, 412 (1944).

* It was argued that the observed arylation of vinylacetic acid at the γ-carbon atom was possible only with an ionic mechanism.[13] Actually, this experiment provides no evidence for either mechanism, since both cations and radicals attack the γ-carbon atom; that is,

$$CH_3\overset{\oplus}{C}HCH_2CO_2H \xleftarrow{H^{\oplus}} CH_2{=}CHCH_2CO_2H \xrightarrow{Br\cdot} BrCH_2CHCH_2CO_2H$$

$$\text{Ar} \cdot\ +\ \text{CH}_2\!=\!\text{CHCN} \begin{array}{c} \nearrow \\ \xrightarrow{} \end{array} \begin{bmatrix} \text{ArCH}_2\text{CHCN} \longleftrightarrow \text{ArCH}_2\text{CH}\!=\!\text{C}\!=\!\text{N}\cdot \\ \text{ArCH(CN)CH}_2\cdot \\ \text{(less stable)} \end{bmatrix}$$

$$\text{Ar}\cdot\ +\cdot\text{C}_6\text{H}_5\text{CH}\!=\!\text{CHCHO} \begin{array}{c} \nearrow \\ \searrow \end{array} \begin{bmatrix} \text{C}_6\text{H}_5\text{CHCH(CHO)Ar} \longleftrightarrow \bigcirc\!=\!\text{CHCH(CHO)Ar} \\ \text{C}_6\text{H}_5\text{CH(Ar)CHCHO} \longleftrightarrow \text{C}_6\text{H}_5\text{CH(Ar)CH}\!=\!\text{CHO}\cdot \\ \text{(less stable)} \end{bmatrix}$$

It is known that many monomers are polymerized by diazonium salts in the absence of copper,[27] yet styrene,[9,28,29] acrylonitrile,[4,8,29] vinyl halides,[4,30] acrylic acid[31] and its esters,[32] and maleimide derivatives[33,34] give good yields of Meerwein products without appreciable formation of polymers other than "diazo resins." Probably the copper salt or another component of the medium functions as an efficient chain transfer agent to prevent the growth of the monomer radical ArCH_2CHZ, which is converted instead to $\text{ArCH}_2\text{CHClZ}$ or $\text{ArCH}\!=\!\text{CHZ}$. However, copper salts may also promote the polymerizing activity of diazonium salts under certain conditions.[35]

Other evidence suggests that the radical is different from the radicals which initiate vinyl polymerization. Diazonium salts under the conditions of the Meerwein reaction gave better yields in the arylation of coumarin and other selected olefins than the aryl radicals derived from aroyl peroxides, N-nitrosoacetanilides, and 1-aryl-3,3-dimethyltriazenes.[36] On the other hand, arylation of aromatic compounds by diazonium salts under Meerwein conditions proceeds in fair yields, in a few cases at least,[37,38] and arylation of aromatic compounds is normally a homolytic reaction.[39,40]

[27] Willis, Alliger, Johnson, and Otto, *Ind. Eng. Chem.*, **45**, 1316 (1953); Cooper, *Chem. & Ind. (London)*, **1953**, 407; Marvel, Friedlander, and Inskip, *J. Am. Chem. Soc.*, **75**, 3846 (1953); Horner and Stöhr, *Chem. Ber.*, **86**, 1066 (1953).
[28] Kochi, *J. Am. Chem. Soc.*, **77**, 5090 (1955).
[29] Kochi, *J. Am. Chem. Soc.*, **78**, 4815 (1956).
[30] Cristol and Norris, *J. Am. Chem. Soc.*, **76**, 3005 (1954).
[31] Rai and Mathur, *J. Indian Chem. Soc.*, **24**, 413 (1947).
[32] Koelsch, *J. Am. Chem. Soc.*, **65**, 57 (1943).
[33] Rondestvedt and Vogl, *J. Am. Chem. Soc.*, **77**, 2313 (1955).
[34] Rondestvedt, Kalm, and Vogl, *J. Am. Chem. Soc.*, **78**, 6115 (1956).
[35] Furukawa, Sasaki, and Murakami, *Chem. High Polymers (Tokyo)*, **11**, 77 (1954) [*C.A.*, **50**, 5548e (1956)].
[36] Vogl and Rondestvedt, *J. Am. Chem. Soc.*, **77**, 3067 (1955).
[37] Dickerman, Weiss, and Ingberman. *J. Org. Chem.*, **21**, 380 (1956).
[38] Dickerman and Weiss, *J. Org. Chem.*, **22**, 1070 (1957).
[39] Rondestvedt and Blanchard, *J. Org. Chem.*, **21**, 229 (1956).
[40] Augood and Williams, *Chem. Revs.*, **57**, 123 (1957).

Intermediate Complex Formation. Neither the simple ionic nor the radical mechanism accounts for the dependence of the reaction rate (nitrogen evolution) upon the structure of the olefin. For example, solutions of many diazonium chlorides in an acetate buffer containing cupric chloride are stable for some time. Addition of an olefin, such as acrylic acid,[31] initiates rapid nitrogen evolution. There is a wide range of temperatures at which nitrogen evolution begins, dependent upon the structure of the olefin.[41–44] These and other examples led to the proposal that a complex was formed between diazonium salt, olefin, and copper chloride which then decomposed by internal one-electron transfers to products.[26, 33] A tentative description of the complex has been given.[33]

Function of Catalyst. The copper salt is usually added as cupric chloride. However, it is known that cupric chloride reacts slowly with acetone to form cuprous chloride and chloroacetone.[37, 45] The cuprous chloride thus produced is a powerful catalyst for the Sandmeyer reaction and for the arylation of benzene by 2,4-dichlorobenzenediazonium chloride.[37] This cuprous chloride will also induce a Meerwein arylation of styrene or acrylonitrile by p-chlorobenzenediazonium chloride.[28, 29, 46] From these results it was concluded that the Meerwein reaction is catalyzed by univalent copper, not by divalent copper.[45] The following mechanism, reproduced in part, has been suggested.[37]

$$ArN_2^{\oplus} + CuCl_2^{\ominus} \rightarrow ArN{=}N\cdot + CuCl_2$$

$$ArN{=}N\cdot \rightarrow Ar\cdot + N_2$$

$$Ar\cdot + \overset{|}{\underset{|}{C}}{=}\overset{|}{\underset{|}{C}} \rightarrow Ar\overset{|}{\underset{|}{C}}{-}\overset{|}{\underset{|}{C}}\cdot$$

$$Ar\overset{|}{\underset{|}{C}}{-}\overset{|}{\underset{|}{C}}\cdot + CuCl_2 \rightarrow Ar\overset{|}{\underset{|}{C}}{-}\overset{|}{\underset{|}{C}}Cl + CuCl$$

$$Ar\cdot + CuCl_2 \rightarrow ArCl\ (\text{by-product}) + CuCl$$

$$Ar\cdot + CH_3COCH_3 \rightarrow ArH\ (\text{by-product}) + CH_3COCH_2\cdot$$

The mechanism involving cuprous catalysis is in harmony with some of the facts known about the Meerwein reaction, such as the formation

---

[41] L'Écuyer and Turcotte, *Can. J. Research*, **B25**, 575 (1947).
[42] L'Écuyer, Turcotte, Giguère, Olivier, and Roberge, *Can. J. Research*, **B26**, 70 (1948).
[43] L'Écuyer and Olivier, *Can. J. Research*, **B27**, 689 (1949).
[44] L'Écuyer and Olivier, *Can. J. Research*, **B28**, 648 (1950).
[45] Kochi, *J. Am. Chem. Soc.*, **77**, 5274 (1955).
[46] Kochi, *J. Am. Chem. Soc.*, **78**, 1228 (1956).

of chloroacetone,[1] the hydrocarbon, and the aryl halide, and it explains the generally beneficial effect of acetone and halide ions.[4,5] However, it is not compatible with other facts. Thus acetonitrile[1,36] (which does not reduce cupric chloride[45]), N-methylpyrrolidone,[5] dimethyl sulfoxide,[34] sulfolane,[5] and dimethylsulfolane[5] are fairly satisfactory solvents in the few cases studied. Furthermore, acetone is actually harmful in many reactions, as with acrylic acid,[31] maleic acid,[47] and furfural.[48-51] These compounds are better arylated in aqueous solution. Meerwein[1] and Terent'ev[52] commented that cuprous salts were poorer catalysts than cupric salts, or that they were ineffective, but they gave no experimental details in support of this statement. It may be mentioned that cuprous salt catalysis is strongly inhibited by oxygen,[46] yet a common experimental technique for the reaction involves vigorous stirring in contact with air, which oxidizes any cuprous copper as it is formed.

Recent experiments with methacrylonitrile[5] have shown that, when the diazonium salts bear electron-attracting groups, cupric copper gives better yields than cuprous copper. The reverse is true with diazonium salts lacking an electron-attracting group.

When considered together, all the facts suggest that there are at least two mechanisms of initiation of the Meerwein arylation. The rates of the reactions by the different mechanisms will probably be found to depend on the nature of the substituents in the diazonium salt and the character of the unsaturated compound. It is also likely that a variety of one-electron oxidation-reduction systems, such as ferrous-ferric or ferrocyanide-ferricyanide, can function as catalysts in selected examples. Indeed, if the olefin-diazonium salt combination possesses the proper one-electron oxidation-reduction potential, the reaction should proceed without a metallic catalyst. This has been realized with coumarin[53] and, especially, with quinones.[18, 20, 21, 54-71]

---

[47] Rai and Mathur, *J. Indian Chem. Soc.*, **24**, 383 (1947).
[48] Oda, *Mem. Fac. Eng. Kyoto Univ.*, **14**, 195 (1952) [*C.A.*, **48**, 1935c (1954)].
[49] Grummitt and Splitter, *J. Am. Chem. Soc.*, **74**, 3924 (1952).
[50] Kost and Terent'ev, *Zhur. Obshcheĭ Khim.*, **22**, 655 (1952) [*C.A.*, **47**, 2759c (1953)].
[51] Akashi and Oda, *J. Chem. Soc. Japan, Ind. Chem. Sect.*, **53**, 81 (1950) [*C.A.*, **47**, 2164e (1953)]; *Repts. Inst. Chem. Research, Kyoto Univ.*, **19**, 93 (1949) [*C.A.*, **45**, 7519h (1951)]; *Teijin Times*, **19**, No. 4, 7 (1949) [*C.A.*, **44**, 5314 (1950)].
[52] Dombrovskiĭ, Terent'ev, and Yurkevich, *Zhur. Obshcheĭ Khim.*, **26**, 3214 (1956); *J. Gen. Chem. U.S.S.R. (Engl. Transl.)*, **26**, 3585 (1956) [*C.A.*, **51**, 8038e (1957)].
[53] Rondestvedt and Vogl, *J. Am. Chem. Soc.*, **77**, 3401 (1955).
[54] Huisgen and Horeld, *Ann.*, **562**, 137 (1949).
[55] Borsche, *Ber.*, **32**, 2935 (1899); *Ann.*, **312**, 211 (1900).
[56] Günther, U.S. pat. 1,735,432 (*Chem. Zentr.*, **1930, II**, 137); Ger. pat. 508,395 (*Chem. Zentr.*, **1931, I**, 1676); Brit. pat. 390,029 [*C.A.*, **27**, 468² (1933)].
[57] Schimmelschmidt, *Ann.*, **566**, 184 (1950).
[58-71] See page 198.

Kinetic studies of the Meerwein arylation have suggested that it is mechanistically closely related to the Sandmeyer reaction.[29, 46, 72, 73] However, the rate expressions are too complicated to permit more than qualitative conclusions. These conclusions were based on the assumption that cuprous copper is the sole catalytic species, so that they do not apply to examples where cuprous copper cannot function.

SCOPE AND LIMITATIONS

The Unsaturated Component

Olefins ranging from simple to complicated have been arylated. For the most part, the ethylenic double bond is attached to an electron-attracting group such as carbonyl, cyano, halogen, aryl, or vinyl. Important examples are given in the accompanying equations, with selected references.

$ArN_2Br + CH_2=CHBr \rightarrow ArCH_2CHBr_2$ (Ref. 30)

$ArN_2Cl + Ar'CH=CH_2 \rightarrow ArCH_2CHClAr' + ArCH=CHAr'$
Ar' = phenyl, substituted phenyl, 2-pyridyl. (Refs. 9, 28, 46, 74, 75)

$ArN_2Cl + CH_2=CHCH=CH_2 \rightarrow ArCH_2CH=CHCH_2Cl$
(Refs. 3, 4, 49, 76–79)

$ArN_2Cl + CH_2=CHCO_2H \rightarrow ArCH=CHCO_2H$ (Refs. 4, 31, 80)

---

[58] Kvalnes, *J. Am. Chem. Soc.*, **56**, 2478 (1934).
[59] Marini-Bettòlo, *Gazz. chim. ital.*, **71**, 627 (1941).
[60] Marini-Bettòlo and Rossi, *Gazz. chim. ital.*, **72**, 208 (1942).
[61] Marini-Bettòlo, Polla, and Abril, *Gazz. chim. ital.*, **80**, 76 (1950).
[62] Neunhoeffer and Weise, *Ber.*, **71**, 2703 (1938).
[63] Kögl, Erxleben, and Janecke, *Ann.*, **482**, 119 (1930).
[64] Dobàš, *Chem. Listy*, **46**, 277 (1952) [*C.A.*, **47**, 8669d (1953)].
[65] Asano and Kameda, *J. Pharm. Soc. Japan*, **59**, 768 (1939) [*C.A.*, **34**, 2345⁶ (1940)].
[66] Malinowski, *Roczniki Chem.*, **29**, 47 (1955) [*C.A.*, **50**, 3364f (1956)].
[67] Fieser, Leffler, et al., *J. Am. Chem. Soc.*, **70**, 3203 (1948).
[68] Akagi and Hirose, *J. Pharm. Soc. Japan*, **62**, 191 (1942) [*C.A.*, **45**, 6169d (1951)].
[69] Akagi, *J. Pharm. Soc. Japan*, **62**, 195 (1942) [*C.A.*, **45**, 6169f (1951)].
[70] Akagi, *J. Pharm. Soc. Japan*, **62**, 199 (1942) [*C.A.*, **45**, 6169h (1951)].
[71] Akagi, *J. Pharm. Soc. Japan*, **62**, 202 (1942) [*C.A.*, **45**, 2898e (1951)].
[72] Kochi, *J. Am. Chem. Soc.*, **79**, 2942 (1957).
[73] Dickerman, Weiss, and Ingberman, *J. Am. Chem. Soc.*, **80**, 1904 (1958).
[74] Dale and Ise, *J. Am. Chem. Soc.*, **76**, 2259 (1954).
[75] Razumovskiĭ and Rychkina, *Doklady Akad. Nauk S.S.S.R.*, **88**, 839 (1953) [*C.A.*, **48**, 3311i (1954)]; cf. Dilthey, *J. prakt. Chem.*, **142**, 177 (1935).
[76] Ropp and Coyner, *Org. Syntheses*, **31**, 80 (1951).
[77] Coyner and Ropp, *J. Am. Chem. Soc.*, **70**, 2283 (1948).
[78] Ropp and Coyner, *J. Am. Chem. Soc.*, **72**, 3960 (1950).
[79] Braude and Fawcett, *J. Chem. Soc.*, **1951**, 3113.
[80] Krishnamurti and Mathur, *J. Indian Chem. Soc.*, **28**, 507 (1951).

THE MEERWEIN ARYLATION REACTION

$ArN_2Cl + CH_2{=}CHCN \rightarrow ArCH_2CHClCN$

(Refs. 4, 8, 15, 28, 32, 43, 46, 81–83)

$ClN_2ArN_2Cl + 2CH_2{=}CHCN \rightarrow Ar(CH_2CHClCN)_2$ (Refs. 4, 84)

$ArN_2Cl + CH_2{=}CHCOCH_3 \rightarrow ArCH_2CHClCOCH_3$ (Refs. 3, 4, 85)

Acetylenes will participate, but the examples are few.

$ArN_2Cl + CH{\equiv}CH \rightarrow ArCH{=}CHCl$ (Ref. 4)

(Diazonium salts react with cuprous acetylide to form mono- and di-arylacetylenes in low yield.[86])

$ArN_2Cl + C_6H_5C{\equiv}CH \rightarrow ArCH{=}CClC_6H_5$ (Ref. 5)

$ArN_2Cl + C_6H_5C{\equiv}CCO_2H \rightarrow C_6H_5CCl{=}C(Ar)CO_2H$ (Ref. 1)

The ethylenic bond may be substituted with two activating groups. If both are on the same carbon atom, the aryl group becomes attached to the other carbon atom. Symmetrical 1,2-disubstituted ethylenes can give only one orientation. If the activating groups on the α- and β-carbon atoms are different, the compound formed can be predicted from the rule that the product will be the one formed via the intermediate radical that is the more resonance stabilized.[26] The accompanying equations illustrate arylation of multiply activated olefins.

$ArN_2Cl + CHCl{=}CCl_2 \rightarrow ArCHClCCl_3$ (Refs. 3, 4)

$ArN_2Cl + Ar'CH{=}CHCO_2CH_3 \rightarrow Ar'CHClCH(Ar)CO_2CH_3$ (Refs. 1, 26)

$ArN_2Cl + C_6H_5(CH{=}CH)_2CO_2CH_3 \rightarrow C_6H_5CH{=}CHCH{=}C(Ar)CO_2CH_3$

(Ref. 26)

$ArN_2Cl + C_6H_5CH{=}CHCHO \rightarrow C_6H_5CH{=}C(Ar)CHO$ (Ref. 1)

ArN_2Cl + [chromen-2-one] $\rightarrow$ [3-aryl chromen-2-one] (Refs. 1, 15, 16, 53)

---

[81] Dhingra and Mathur, *J. Indian Chem. Soc.*, **24**, 123 (1947).

[82] Gaudry, *Can. J. Research*, **B23**, 88 (1945).

[83] Malinowski, *Roczniki Chem.*, **26**, 85 (1952) [*C.A.*, **48**, 620i (1954)].

[84] Malinowski and Benbenek, *Roczniki Chem.*, **27**, 379 (1953) [*C.A.*, **49**, 1034h (1955)].

[85] Malinowski, *Roczniki Chem.*, **29**, 37 (1955) [*C.A.*, **50**, 3292h (1956)].

[86] Sokol'skiĭ and Nikolenko, *Doklady Akad. Nauk S.S.S.R.*, **82**, 923 (1952) [*C.A.*, **47**, 2723b (1953)].

$ArN_2Cl + RO_2CCH{=}CHCO_2R \rightarrow$
  cis or *trans*

$RO_2CCH{=}C(Ar)CO_2R + RO_2CCHClCH(Ar)CO_2R$ (Refs. 1, 87, 88)

$$ArN_2Cl + \begin{array}{c}CHCO\\\|\quad\diagdown\\\quad\quad NR\\\diagup\\CHCO\end{array} \longrightarrow \begin{array}{c}ArC{-}CO\\\|\quad\diagdown\\\quad\quad NR\\\diagup\\CHCO\end{array} + \begin{array}{c}ArCHCO\\|\quad\diagdown\\\quad\quad NR\\\diagup\\ClCHCO\end{array} \quad \text{(Refs. 33, 34)}$$

Certain α,β-unsaturated acids, such as cinnamic acid and maleic acid, undergo arylation at the carbon atom bearing the carboxyl group. In these reactions decarboxylation accompanies arylation, the extent apparently depending upon the pH (see section on reaction conditions). Examples of this phenomenon follow.

$ArN_2Cl + Ar'CH{=}CHCO_2H \rightarrow ArCH{=}CHAr'$ (Refs. 1, 11–13, 16, 41)

$ArN_2Cl + HO_2CCH{=}CHCO_2H \rightarrow ArCH{=}CHCO_2H$ (Ref. 47)

$ArN_2Cl + C_6H_5COCH{=}CHCO_2H \rightarrow ArCH{=}CHCOC_6H_5$ (Refs. 89, 90)

$ArN_2Cl + RCH{=}CHCH{=}CHCO_2H \rightarrow RCH{=}CHCH{=}CHAr$
  $R = CH_3, C_6H_5.$ (Refs. 11–13, 26, 91)

Occasionally the reaction proceeds without decarboxylation. Thus maleic acid is arylated at a pH of about 2 in a reaction involving only addition.[92] Monoarylmaleic acids give α,β-diaryl-α-chlorosuccinic acids under these conditions.[92] Cinnamic acids are sometimes arylated without decarboxylation;[1] the resulting α-arylcinnamic acids are not further arylated.[14]

$ArN_2Cl + HO_2CCH{=}CHCO_2H \xrightarrow{pH\,2} ArCH(CO_2H)CHClCO_2H$ (Ref. 92)

$ArN_2Cl + HO_2CCH{=}C(Ar')CO_2H \rightarrow HO_2CCH(Ar)CCl(Ar')CO_2H$ (Ref. 92)

$ArN_2Cl + Ar'CH{=}CHCO_2H \rightarrow Ar'CH{=}C(Ar)CO_2H$ (Ref. 1)

There is one report of a nitro group being lost during arylation. The formation of benzyl *p*-nitrophenyl ketone from ω-nitrostyrene and

---

[87] Taylor and Strojny, *J. Am. Chem. Soc.*, **76**, 1872 (1954).
[88] Vogl and Rondestvedt, *J. Am. Chem. Soc.*, **78**, 3799 (1956).
[89] Mehra and Mathur, *J. Indian Chem. Soc.*, **32**, 465 (1955).
[90] Mehra and Mathur, *J. Indian Chem. Soc.*, **33**, 618 (1956).
[91] Fusco and Rossi, *Gazz. chim. ital.*, **78**, 524 (1948).
[92] Denivelle and Razavi, *Compt. rend.*, **237**, 570 (1954).

p-nitrobenezenediazonium chloride is believed to proceed via a Nef reaction on the supposed intermediate aci-nitro compound.[93]

$p\text{-}O_2NC_6H_4N_2Cl + C_6H_5CH=CHNO_2 \rightarrow$
$\qquad\qquad C_6H_5CH=CHC_6H_4NO_2\text{-}p + C_6H_5CH_2COC_6H_4NO_2\text{-}p$

Arylation of β-2-furyl- and β-2-thienyl-acrylic acid is complicated by the preferential or simultaneous occurrence of ring arylation at the 5 position. The high nuclear reactivity of furan derivatives in the Meerwein arylation has been demonstrated in arylations of furfural.[51] Since furan may also be arylated by diazonium salts under the conditions of the Gomberg-Bachmann free-radical biaryl synthesis,[94] its arylation under Meerwein conditions illustrates the similarity between these two reactions.

ArN₂Cl + [furan/thiophene]CH=CHCO₂H ⟶ Ar[furan/thiophene]CH=CHCO₂H +

X = O, S.

[furan/thiophene]CH=CHAr + Ar[furan/thiophene]CH=CHAr   (Refs. 15, 18, 21, 43, 48, 95)

ArN₂Cl + [furan]CH=CHCOCH₃ ⟶ Ar[furan]CH=CHCOCH₃   (Ref. 48)

Quinones. Apparently the first examples of quinone arylation were provided by Borsche,[55] who phenylated benzoquinone monoxime (p-nitrosophenol) and toluquinone monoxime in low yield. After a period of dormancy, the reaction was applied by Günther to the synthesis of arylbenzoquinones.[56] Subsequently others have shown the reaction to be general and to proceed according to the following equation.

O=⟨benzoquinone⟩=O $\xrightarrow{ArN_2X}$ O=⟨aryl-benzoquinone⟩=O

---

[93] Bergmann and Vromen, *Bull. Research Council Israel*, **3**, Nos. 1/2, 98 (1953) [*C.A.*, **49**, 1605f (1955)].
[94] Johnson, *J. Chem. Soc.*, **1946**, 895.
[95] Brown and Kon, *J. Chem. Soc.*, **1948**, 2147.

A large variety of quinones has been arylated by diazonium salts or by the related N-nitroso-N-arylacetamides. Methylated, halogenated, and arylated benzoquinones have been studied, although benzoquinone itself has been investigated most extensively. 1,4-Naphthoquinone has received some attention, though it is arylated much less readily than benzoquinone. An extensive series of 2-hydroxy-3-arylnaphthoquinones has been prepared by this reaction, though mostly in very poor yields.[66, 67]

Schimmelschmidt made a significant contribution to quinone arylation technique.[57] The reaction with benzoquinone could be run very efficiently in weakly alkaline medium *if a trace of hydroquinone was present*. Under these conditions, the diazonium salt reacts with the quinone with the speed of a titration. Pure benzoquinone did not react at all until a little hydroquinone was added. These conditions give very good (but unspecified) yields of arylquinones with a wide variety of diazonium salts, mostly the *ortho*-substituted ones which others had found to be recalcitrant.

Hydroquinone itself has been treated with diazonium salts, and it has been recommended as a reagent for the reductive removal of the diazo group.[96] Schimmelschmidt stated that diazonium salts and hydroquinone form an intractable tar, but other workers have had some success in preparing arylhydroquinones by this procedure.[61, 64] These compounds are probably better prepared by reduction of the quinones.

It is difficult to discuss the limitations of the arylation of quinones by the diazonium salt reaction because of the almost universal failure of authors in this field to report yields or exact reaction conditions. All that can be said is that most diazonium salts will give some product with a mononuclear quinone. The difficulty which many workers have experienced with *ortho*-substituted diazonium salts has been overcome by addition of a trace of hydroquinone.[57]

A number of different experimental conditions have been employed. Most authors have used an aqueous or ethanolic medium with the pH one or two units on either side of neutrality. Some have preferred a more strongly acidic medium with added copper powder or cupric chloride.[59-61, 66, 97] The only comparison of a variety of reaction conditions was made by Fieser and Leffler,[67] but they used the rather unreactive 2-hydroxy-1,4-naphthoquinone in their studies. They did not find that any one set of conditions consistently gave the best results. Since the best yields were reported by Schimmelschmidt, his conditions[57] are probably the most suitable for trial experiments with new examples of this reaction.

The use of N-nitroso-N-arylacetamides appears to be promising,

[96] Orton and Everatt, *J. Chem. Soc.*, **93**, 1021 (1908).
[97] Brassard and L'Écuyer, *Can. J. Chem.*, **36**, 700 (1958). See also refs. 158–160.

since these compounds are soluble in moderately polar or nonpolar solvents such as ethanol or ethanol-ether mixtures,[68-71] or benzene.[54]

Since the experimental conditions under which quinones may be arylated are so diverse, it appears that more than one mechanism may be operative. Schimmelschmidt[57] has proposed a scheme to account for the participation of hydroquinone. When the conditions approximate those of the Meerwein reaction, quinone arylation probably involves the same reaction path. In the absence of copper, or in neutral or alkaline solution, or with nitrosoacetanilides, the mechanism is doubtless similar to that for arylation of aromatic compounds.[39,40] Further study is required before the mechanism(s) of quinone arylation can be considered to be established.

Miscellaneous Unsaturated Compounds. Several examples of the C-arylation of aldoximes have been reported. Although this reaction has received only limited study, it appears to be a potentially useful way of synthesizing aromatic aldehydes and ketones.[98-103] Aldehyde semicarbazones react similarly.

$ArN_2Cl + RCH=NOH \rightarrow ArC(R)=NOH \rightarrow ArCOR$ (Refs. 100-103)
R = H, alkyl

$ArN_2Cl + Ar'COCH=NOH \rightarrow Ar'COC(Ar)=NOH$ (Refs. 98, 99, 103)

Malonic ester and nitromethane have been arylated, although the more usual reaction of active methylene compounds with diazonium salts is azo coupling followed by tautomerism to an arylhydrazone. Compare *Organic Reactions*, Volume 10, Chapter 1.

$ArN_2Cl + CH_2(CO_2C_2H_5)_2 \rightarrow ArCH(CO_2C_2H_5)_2$ (Ref. 104)

$ArN_2Cl + CH_3NO_2 \rightarrow ArCH_2NO_2$ (Ref. 105)

Despite the impressive array of examples of the Meerwein arylation reaction, there are numerous gaps. Any compound with olefinic unsaturation conjugated with another group should be a candidate for arylation, yet many important classes of such compounds have received little or no attention. Only one paper deals with arylation of acrolein

[98] Kanno, *J. Pharm. Soc. Japan*, **73**, 118 (1953) [*C.A.*, **47**, 11154b (1953)].
[99] Kanno, *J. Pharm. Soc. Japan*, **73**, 120 (1953) [*C.A.*, **47**, 11154e (1953)].
[100] Beech, *J. Chem. Soc.*, **1954**, 1297.
[101] Borsche, *Ber.*, **40**, 737 (1907).
[102] Philipp, *Ann.*, **523**, 285 (1936).
[103] Beech, *J. Chem. Soc.*, **1955**, 3094.
[104] Haginiwa and Murakoshi, *J. Pharm. Soc. Japan*, **73**, 1015 (1953) [*C.A.*, **48**, 10670d (1954)].
[105] Tsurata and Oda, *J. Chem. Soc. Japan, Ind. Chem. Sect.*, **53**, 16 (1950) [*C.A.*, **47**, 5909a (1953)]; cf. Busch and Schäffner, *Ber.*, **56**, 1613 (1923); Oda and Tsurata, *Repts. Inst. Chem. Research, Kyoto Univ.*, **19**, 89 (1940) [*C.A.*, **45**, 7541h (1951)].

and its derivatives,[106] and this reaction is worthy of more study as a new route to cinnamaldehydes. The only nitroölefin studied is β-nitrostryene,

$$ArN_2Cl + CH_2=C(R)CHO \rightarrow ArCH_2CCl(R)CHO \rightarrow$$
$$ArCH=C(R)CHO \quad (Ref. 106)$$

in which the phenyl group directs the incoming aryl group to the carbon atom holding the nitro group; the nitro group is lost.[93] Arylation of aliphatic nitroölefins has not been studied, but it would be expected to proceed as shown in the following equation.

$$ArN_2Cl + CH_2=CHNO_2 \rightarrow ArCH_2CHClNO_2 \rightarrow ArCH=CHNO_2$$

Vinyl esters have not been studied, while vinyl ethers reportedly give azo coupling in the absence of copper salts.[107] Both are worth examination as routes to arylacetaldehydes.

Simple dienes give 1-arylbutadienes after dehydrohalogenation. Further arylation of 1-arylbutadienes has been explored cursorily as a route to 1,4-diarylbutadienes.[108] The latter compounds can also be made by Meerwein arylation of cinnamylideneacetic acid.[26] Because arylation of anthracene is handicapped by its low solubility, its arylation has required very dilute solutions;[109, 110] discovery of a better solvent would enhance the attractiveness of this simple route to 9-aryl- and 9,10-diarylanthracenes. Phenanthrene has not been studied.

Unsaturated sulfur compounds have received little attention. The experiments with 2-phenylethene-1-sulfonic acid in aqueous solution gave no pure product; the sulfonic acid was not attacked at pH 3–6 by various diazonium salts, but at a more alkaline pH it was converted by p-nitrobenzenediazonium chloride to a neutral material (loss of the sulfo group) which was not p-nitrostilbene.[111, 112] Ethylenesulfonic acid has not been tested. A few unsaturated sulfides and sulfones have been tried.[5] There is no mention of the arylation of ethylenephosphonic acid or its derivatives. Enamines have not been studied.

Although unsaturated acids, esters, nitriles, and cyclic imides undergo the Meerwein reaction, amides appear not to react. It was observed that acrylamide, N-t-butylacrylamide, N,N'-methylenebisacrylamide, cinnamamide, and N-methylcinnamamide did not give detectable amounts

---

[106] Malinowski and Benbenek, *Roczniki Chem.*, **30**, 1121 (1956) [*C.A.*, **51**, 8688f (1957)].

[107] Terent'ev and Zagorevskiĭ, *Zhur. obshcheĭ Khim.*, **26**, 200 (1956); *J. Gen. Chem. U.S.S.R. (Engl. Transl.)*, **26**, 211 (1956) [*C.A.*, **50**, 13777i (1956)].

[108] Dombrovskiĭ, *Doklady Akad. Nauk S.S.S.R.*, **111**, 827 (1956); *Proc. Acad. Sci. U.S.S.R., Sect. Chem. (Engl. Transl.)*, **111**, 705 (1956) [*C.A.*, **51**, 9507f (1957).

[109] Étienne and Degent, *Compt. rend.*, **236**, 92 (1953); **238**, 2093 (1954).

[110] Dickerman, Levy, and Schwartz, *Chem. & Ind. (London)*, **1958**, 360.

[111] C. S. Rondestvedt, Jr., and C. D. Ver Nooy, unpublished results.

[112] C. S. Rondestvedt, Jr., and O. Vogl, unpublished results.

of arylated product by the customary procedure in acetone.[112] Acrylamide and methacrylamide were not arylated in aqueous solution in the presence of cuprous chloride.[5] It is not clear why amides should be so unreactive, particularly when contrasted with the high reactivity of maleimide derivatives.*

Reactivities of Unsaturated Compounds. In the absence of quantitative data concerning relative reactivities in the Meerwein arylation reaction, only a few qualitative trends based upon yields can be given (see, however, Ref. 73). Compounds with a terminal double bond usually give better results than compounds of the same type where the double bond is not terminal. Thus acrylic and methacrylic acids and their esters[31, 44, 80, 91] give much better yields than crotonic acid and its esters.[1, 26, 31] This may be due to steric factors, or it may reflect a lower degree of polarizability of the nonterminal double bond.[9] Parallel results have been obtained in polymerization studies.

Cinnamic acid appears to be less reactive than acrylic or maleic acid, since cinnamic acids can be prepared by the Meerwein arylation of acrylic and maleic acids.[31, 47] The difference is probably attributable to the energy barrier to decarboxylation which occurs during the reaction with cinnamic acids (see below), or to steric hindrance.

Activated cyclic double bonds are very reactive. The yields of 3-arylcoumarins[1] are high compared to the yields of products from benzalacetone[43] and methyl cinnamate.[1, 26] Maleimide and N-substituted maleimides[33, 34] generally give satisfactory yields of arylated products, while amides are quite unreactive.[5, 112] Quinones are sufficiently reactive to undergo arylation without a cupric catalyst. The possibility of arylating a double bond activated by a strained ring system, as in bicyclo[2.2.1]heptene, has not been tested.

A triple bond is less reactive than a double bond. One can arylate styrene in far higher yield than phenylacetylene.[5] Vinylacetylene is arylated in good yield at the double bond but not at the triple bond.[113] The difference can probably be ascribed to the greater rigidity of the intermediate radical, necessarily containing a double bond, or to the more strained geometry of the intermediate complex.

The relative efficiencies of various groups in directing the incoming aryl group should be noted. An aryl group is superior to vinyl, carboxyl, carbalkoxyl, cyano, aldehyde or ketone carbonyl, or nitro; no exceptions have been found to the generalization that the incoming aryl group always takes up the position β to the aryl group already present in the structure $\text{Ar}\overset{\alpha}{\text{CH}}=\overset{\beta}{\text{CH}}\text{Z}$. The other available comparison of directing

* A private communication from George Cleland indicates that conditions may be found in which amides will undergo the Meerwein arylation.

[113] Barney and Pinkney, U.S. pat. 2,657,244 [*C.A.*, **48**, 12800g (1954)].

power is that a benzoyl group is stronger than carboxyl; arylation of β-benzoylacrylic acid occurs β to the benzoyl group.[89, 90] These effects may be rationalized in terms of radical stabilities, as discussed above, or by the relative steric sizes of the directing groups.

More detailed comparisons of relative reactivities will require the results from competitive experiments or other quantitative studies.

Decarboxylation during Arylation of Cinnamic and Maleic Acids.
Cinnamic acids are decarboxylated during arylation. In only a few examples were small amounts of α-arylcinnamic acids isolated.[1, 15, 16] Likewise, when maleic, citraconic, and bromomaleic acids were arylated at the usual pH, monocarboxylic acids were the only acidic materials isolated.[31, 46, 80, 89, 114, 115]

Decarboxylation appears to depend on pH. By operating in somewhat more acidic solutions (about pH 2) than customary, maleic acid and arylmaleic acids were arylated without loss of carbon dioxide.[92] This information was utilized to prepare arylmaleic anhydrides by cyclizing the resulting α-aryl-β-chlorosuccinic acids with hot acetic anhydride. It has not been determined whether cinnamic acids may be arylated at a low pH without decarboxylation.

The mechanism of decarboxylation during arylation is obscure. One mechanism involves formation of the β-halo acid which then undergoes dehalogenative decarboxylation. This is unlikely at the pH commonly used, since dehalogenative decarboxylation is a reaction of the anion which occurs only in neutral or basic solution.[116] β-Lactone formation and decomposition are also unlikely.[116] Another mechanism was based on a study of the acid-catalyzed decarboxylation of cinnamic acids.[117] It proposes that the intermediate ion $Ar\overset{\oplus}{C}HCH(Ar')CO_2^{\ominus}$, which in the Meerwein arylation reaction could arise by oxidation of the free-radical intermediate,[26] undergoes scission to the olefin and carbon dioxide by a simple electron shift. The failure to decarboxylate at low pH is then attributable to the decreased dissociation of the carboxyl group.

The Diazonium Salt

A wide variety of diazotizable aromatic amines participate in the Meerwein arylation reaction. Thus halo-, nitro-, alkoxy-, acetamido-, sulfo-, arsono-, alkyl-, and aryl-anilines have been used, as well as α- and

[114] Rehan and Mathur, *J. Indian Chem. Soc.*, **28**, 540 (1951).
[115] Mathur, Krishnamurti, and Pandit, *J. Am. Chem. Soc.*, **75**, 3240 (1953).
[116] Vaughan and Craven, *J. Am. Chem. Soc.*, **77**, 4629 (1955).
[117] Johnson and Heinz, *J. Am. Chem. Soc.*, **71**, 2913 (1949).

β-naphthylamines. Disubstituted anilines, mostly dihaloanilines, and trisubstituted anilines have found occasional use. Diamines such as p-phenylenediamine and benzidine yield bis-products when tetrazotized and coupled with two equivalents of acrylonitrile.

No generalizations can be made about the effects of substituents that will be free from exceptions. However, several trends have been noticed that will be helpful in predicting whether a new example is likely to succeed. First, diazonium salts containing electron-attracting groups usually give better yields than does benzenediazonium chloride. Nitro groups and halogen atoms are often particularly beneficial. There are not enough comparisons with other electron-attracting substituents (such as carboxyl, cyano, acetyl, sulfo) to permit confident prediction, but they appear to lead to better yields. It also appears that the electron-attracting group must not be insulated from the ring by a methylene group; this statement is based on the report that p-carboxymethyl-, p-cyanomethyl-, and p-methoxymethylbenzenediazonium chloride fail to react with cinnamic acid.[118]

Alkyl groups, as in the toluidines and xylidines, are frequently harmful, and the yields from the alkylbenzenediazonium salts are usually inferior to those from nitro- and halo-benzenediazonium salts. An aryl group is usually helpful, unless condensed as in the naphthylamines.

The effect of a methoxyl group is ambiguous. Most of the data show that the yields from diazotized anisidines are better than with diazotized aniline, but not so good as with nitro- and halodiazonium salts. Occasionally the best yields (or the poorest) in a series are obtained from alkoxylated diazonium salts.

Second, the position of the substituent may be critical. The tables at the end of this chapter show that the best yields are usually obtained when the substituent is *para* to the diazonium function, poorest when it is *ortho*. This seems to be especially true of the more negative, bulkier groups such as nitro and carboxy and less true of methyl and methoxyl groups. One *ortho* halogen atom seems to have little effect, but two *ortho* halogen atoms sometimes completely prevent the reaction. Significant exceptions are found in the arylation of quinone,[57] butadiene,[79] benzalacetone,[43] and cinnamic acid,[1] where the yields of *ortho*- and *para*-nitro products are comparable.

Probably the position effect is not entirely steric. For example, in the arylation of acrylic acid, the yields of o-halocinnamic acid were not affected in the series o-chloro-, o-bromo-, and o-iodo-benzenediazonium chloride, being 26% in each case.[31] Even 2,6-dichlorobenzenediazonium chloride gave a 20% yield of 2,6-dichlorocinnamic acid. On the other

[118] Kon, *J. Chem. Soc.*, **1948**, 224.

hand, in the same reaction, the yields from o-, m-, and p-nitrobenzenediazonium chloride were 7, 29, and 60%, respectively.[31] Possibly the adverse effect of an o-nitro or o-carboxyl group is a result of formation of an internal complex between the diazonium group and the substituent, which does not readily accept an electron from the unsaturated compound. The yield differences also may result in part from the fact that among the three isomeric products, the *para* isomer is usually the easiest to purify because of its lower solubility and higher melting point.

In view of the numerous exceptions to these generalizations concerning the effects of substituents in the diazonium salt, the potential user of this reaction should not be deterred from attempting it with apparently unpromising diazonium salts.

Although the simple diazonium salts are well represented in the tables, less attention has been devoted to more complicated compounds. In view of the variety of aromatic amines commercially available as dye intermediates, it is surprising to find that investigation of the Meerwein arylation reaction with polysubstituted anilines has been limited almost entirely to the polyhaloanilines. One explanation may be that the more weakly basic amines require special techniques for diazotization. Since such procedures have been highly developed in the dye industry, their use should permit examination of many weakly basic amines. Heterocyclic primary amines comprise another large and neglected class. Quinoline-3-diazonium chloride reacted with methacrylonitrile in the expected manner.[5] 6-Methoxyquinoline-8-diazonium chloride gave only 6-methoxy-8-chloroquinoline on attempted reaction with cinnamic acid.[119] There is no reason to doubt that moderately stable heterocyclic diazonium salts will take part in the Meerwein arylation reaction. It is also possible that the less stable ones, such as those derived from 2- and 4-aminopyridine which commonly lose nitrogen to give 2- and 4-halopyridine, may be used in the Meerwein reaction by application of Malinowski's technique[84] of diazotizing the amine in the presence of the unsaturated compound and cupric chloride.

Factors Influencing Addition vs. Substitution

The Meerwein arylation reaction will in general give two products, one arising from substitution of a hydrogen on the β-carbon atom of the olefin by the aryl group, the other by addition of the aryl group and chlorine atom to the double bond. It would be helpful to be able to predict which product will be formed from a given reaction and what experimental conditions will favor one or the other product. (In many

[119] Cook, Heilbron, and Steger, *J. Chem. Soc.*, **1943**, 413.

cases this knowledge is not important, for the addition product can usually be converted to the substitution product by dehydrohalogenation with a tertiary amine or a stronger base such as potassium hydroxide.)

$$ArN_2Cl + RCH{=}CRZ \rightarrow \underset{A}{ArCR{=}CRZ} + \underset{B}{ArCH(R)C(R)ClZ}$$

However, no systematic study of this aspect of the reaction has been published. Therefore several tentative generalizations based upon a few scattered observations can serve only as rough guides.

The controllable factor which seems to influence the proportion of addition and substitution products is the pH of the reaction medium. The basis for this statement is the fact that arylation of maleic acid at the customary pH of 3 to 5 proceeds with decarboxylation,[47] while in more acidic medium the addition product is formed without decarboxylation.[92] If this is generally true, it is probable that the best yields of addition product will be obtained by operating in the most acidic medium that will permit the reaction to occur. The concentration of chloride ion probably also plays a role.

The most important factor, namely the structure of the olefin, cannot be controlled. It appears from the tables that most olefins give chiefly addition products. The exceptions are cinnamaldehyde, benzalacetone, acrylic acid, methacrylic acid, cinnamylideneacetic ester, coumarin, sometimes maleimides, and of course those compounds that undergo decarboxylation. It is likely that a careful examination of most of the reported reactions would disclose the presence of both types of product. One may tentatively conclude that, if the substitution product is extensively stabilized by resonance, as with the 3-arylcoumarins, such products will be formed, probably because an extended conjugated system is thereby formed. This explanation does not account for the fact that acrylic and methacrylic acids give the substitution product exclusively, whereas the corresponding esters give addition products. This situation may result from the use of sodium bicarbonate during the isolation of the products from the acids,[31, 47, 80, 81, 114, 115] since the addition product is dehydrohalogenated by this reagent, as shown by the presence of ionic halide after the bicarbonate treatment.

Side Reactions

The low yields often obtained in the Meerwein arylation reaction attest to the prominence of side reactions. This is not surprising in view of the wide variety of reactions that diazonium salts undergo. Those that have been identified as occurring during the Meerwein arylation reaction are replacement of the diazo group by a halogen atom, hydrolysis to the phenol, reductive loss of nitrogen (deamination), formation of symmetrical azo compounds, and decomposition to the inevitable tars (diazo resins) almost always associated with the reactions of diazonium salts.

The Sandmeyer reaction involving negatively substituted diazonium salts (precisely those diazonium salts that give the best results in the Meerwein reaction) proceeds well in the presence of cupric chloride,[120] although *cuprous* chloride formed in small amounts was the catalytic agent.[120] Cuprous chloride is also present in those Meerwein reactions conducted in the presence of acetone.[37, 45] The arylation reaction is much faster than the Sandmeyer reaction when the highly reactive olefin styrene is present.[46] With olefins less reactive than styrene, the Sandmeyer reaction competes very effectively.[119] Thus the reaction of *p*-nitrobenzenediazonium chloride with methyl vinyl ketone gave 20% of *p*-nitrochlorobenzene and 41% of the Meerwein product;[85] in the same reaction, diazotized anthranilic acid or ester gave no Meerwein product, but instead a large amount of *o*-chlorobenzoic acid or ester. It is doubtless true that most Meerwein reactions are accompanied by products of the Sandmeyer reaction, although most authors have not reported the presence of the latter, nor have they given yields.

It should be possible to suppress the competitive Sandmeyer reaction by control of the pH and halide ion concentration,[120] and by minimizing the quantity of *cuprous* halide by operating in a solvent other than acetone. Present knowledge suggests the desirability of changing the above variables and the reaction temperature before abandoning a new Meerwein arylation reaction because of the formation of the products of a Sandmeyer reaction.

Formation of phenols is not always detected: first, because the rapid hydrolysis of a diazonium salt usually requires temperatures much higher than those normally used in the Meerwein arylation reaction, and, second, because any phenol that is formed is likely to be consumed by azo coupling, especially if the medium is more alkaline than pH 3–4. Phenol formation seems to be seriously competitive with some *ortho*-substituted diazonium salts. Thus salicylic acid or ester was a major by-product in the reaction of *o*-carboxy- or carbalkoxy-benzenediazonium chloride with methyl vinyl ketone.[85] *o*-Cresol was the major product isolated from the reaction of *o*-toluenediazonium chloride with methacrylonitrile.[5]

Reductive replacement of the diazo group (deamination) is a much more serious side reaction. Meerwein traced it to a reaction with acetone.

$$ArN_2Cl + CH_3COCH_3 \rightarrow ArH + N_2 + ClCH_2COCH_3$$

[120] Pfeil, *Angew. Chem.*, **65**, 155 (1953), and papers cited therein.

In the reaction of p-chlorobenzenediazonium chloride with acetone without cupric salt and sodium acetate, about 14% of chloroacetone was produced. Cupric chloride and sodium acetate increased the yield of chloroacetone to 45%. Comparison of variously substituted diazonium salts showed that the yield of chloroacetone in the presence of cupric chloride and sodium acetate was greatest with negatively substituted diazonium salts; highest with 2,4-dichlorobenzenediazonium chloride (65%), and lowest with p-methoxybenzenediazonium chloride (18%). Unfortunately, although the deamination product was isolated in several cases, yields and reaction rates were not given. Therefore the data do not show what fraction of the chloroacetone arose from the reaction just written and what fractions came from the independent attack of cupric chloride on acetone.[28] This point deserves reinvestigation. The reduction may be explained as hydrogen transfer to the intermediate aryl radical from acetone.[37,121]

The symmetrical azo compound ArN=NAr is often one of the components of the tarry by-product that accompanies the Meerwein arylation reaction. In some reactions this azo compound has been isolated.[15,16,19,122] It not uncommonly accompanies the Sandmeyer reaction, especially in the presence of insufficient cuprous chloride.[123,124]

The most annoying and least understood side reaction is the formation of diazo resins. While these may be formed entirely from the diazonium salt, it is quite likely that some of the unsaturated compound is incorporated in the tar. Although the homopolymer of acrylonitrile could not be detected in a typical example,[36] it is known that diazonium salts may function as polymerization initiators.[27,35] If chain transfer is less than 100% efficient, the 1:1 radical intermediate may add a few more monomer molecules before its growth is stopped.

Further discussion of the decomposition of diazonium salts is given in the excellent monograph by Saunders.[123]

COMPARISON WITH OTHER SYNTHETIC METHODS

Despite the low yields often obtained in the Meerwein arylation reaction, an appreciation of its synthetic value is best obtained by surveying other methods that may be used for the preparation of the same compounds. The ensuing discussion is not intended to be an exhaustive survey of

[121] Waters, *J. Chem. Soc.*, **1937**, 2007; **1938**, 843.

[122] Nesmeyanov, Perevalova, and Golovnya, *Doklady Akad. Nauk S.S.S.R.*, **99**, 539 (1954) [*C.A.*, **49**, 15918c (1955)].

[123] Saunders, *The Aromatic Diazo Compounds*, 2nd ed., p. 228, Arnold, London, 1949.

[124] Holt and Hopson-Hill, *J. Chem. Soc.*, **1952**, 4251; Atkinson et al., *J. Am. Chem. Soc.*, **72**, 1397 (1950); **67**, 1513 (1945); and previous papers.

alternative routes. Rather, one or two of the more general alternative synthetic methods for the major classes of compounds available from the Meerwein arylation reaction will be considered.

The Meerwein reaction has been used most frequently for preparing stilbenes. One common alternative method involves the Perkin condensation of an arylacetic acid with an aromatic aldehyde, followed by decarboxylation of the resulting α-arylcinnamic acid—a two-step process. Except where the aldehyde and the arylacetic acid are commercially available, both must be synthesized. A second and more recent method[125] involves the self-condensation of benzyl halides in the presence of alkali metal amides. At present this method appears to be limited to symmetrical stilbenes and at least requires the synthesis of the substituted benzyl halide. In contrast, the Meerwein arylation requires the aromatic amines (more available than the corresponding aldehydes) and the cinnamic acids (or styrenes). The cinnamic acids usually may be prepared by a Meerwein arylation of acrylic or maleic acid. Thus, complicated stilbenes are available in two steps, and the starting materials are two aromatic amines and commercial acrylic or maleic acid.

Cinnamic acids may be prepared by the Reformatskiĭ, the Perkin, or the Doebner-Knoevenagel condensation.[126] The aromatic aldehyde is the required starting material and usually must be synthesized. The Meerwein procedure requires the aromatic amine and either acrylic or maleic acid. Though the yields may be low, the product is readily freed from tar by extraction of the acid with sodium bicarbonate.

The Meerwein arylation of acrolein and methacrolein, recently reported,[106] yields β-aryl-α-chloropropionaldehydes. If the yields could be improved, and if dehydrochlorination offered no difficulty, the reaction would constitute a valuable synthesis for ring-substituted cinnamaldehydes. These important compounds are usually prepared by a crossed aldol condensation between an aromatic and an aliphatic aldehyde.

3-Arylcoumarins are prepared by condensation of salicylaldehyde with ring-substituted phenylacetic acids.[127] Since the latter are more difficultly accessible than aromatic amines, the Meerwein reaction appears to be the method of choice for the synthesis of 3-arylcoumarins.

1-Arylbutadienes have been made by adding Grignard reagents to aldehydes and dehydrating the carbinols, for example, by adding allylmagnesium chloride to benzaldehydes or methylmagnesium iodide to cinnamaldehydes.[49, 77, 78] Again the aldehydes are the starting materials.

---

[125] Hauser, Brasen, Skell, Kantor, and Brodhag, *J. Am. Chem. Soc.*, **78**, 1653 (1956).

[126] Johnson, in Adams, *Organic Reactions*, Vol. I, p. 233, John Wiley & Sons, New York, 1942.

[127] von Walther and Wetzlich, *J. prakt. Chem.*, [2] **61**, 169 (1900).

The Meerwein reaction of aromatic amines with butadiene appears to be preferable, since the 1-aryl-4-chlorobutenes are readily dehydrochlorinated to 1-arylbutadienes.[76, 108, 128] 1,4-Diarylbutadienes can be prepared by successive Meerwein reactions, although this application has not been explored in detail.[108] At present, 1,4-diarylbutadienes are prepared by Grignard reactions or by the Meerwein arylation of cinnamylideneacetic acids.[26]

2-Aryl-1,4-quinones have been prepared in low yields by arylation of a quinone with a diaroyl peroxide,[67] the latter usually being made from the aromatic acid. The convenience of using an aromatic amine instead of a peroxide which usually must be synthesized, together with the better yields from the amine, suggests that the arylquinones are best prepared by the Schimmelschmidt,[57] Kvalnes,[58] or L'Écuyer[97] modification of the Meerwein reaction.

One important general method for coupling an aromatic ring to an aliphatic side chain is the Grignard reaction. It suffers from the serious limitation that arylmagnesium halides will react with functional groups other than the desired one. Thus one cannot prepare Grignard reagents from aryl halides containing nitro, cyano, sulfo, acyl, carboxy, or carbalkoxy groups, i.e., just those substituents which promote the Meerwein reaction.

Another method for attaching a functional aliphatic side chain to an aromatic nucleus is the Friedel-Crafts reaction.[129] For example, methacrylic acid condenses with toluene or p-xylene to form α-arylisobutyric acids.[130] Crotonic acid condenses with benzene to form, after cyclization, 3-methylhydrindanone.[131] Cinnamic acids react with aromatic compounds giving $β,β$-diarylpropionic acids,[132] although α-phenylacrylic acid is arylated at the α-carbon atom to give α,α-diarylpropionic acids.[132] In these examples, the orientation is the opposite of that obtained in the Meerwein reaction, and the acids obtained have saturated side chains. Furthermore the Friedel-Crafts reaction is hindered or prevented by strongly electron-attracting groups in the aromatic nucleus, again the same substituents which promote the Meerwein reaction.

In summary, the Meerwein reaction is no synthetic panacea. It occupies an important place among those reactions which form a new bond between an aromatic ring and a functionally substituted side chain.

[128] Dombrovskiĭ and Terent'ev, *Zhur. obshcheĭ Khim.*, **27**, 415 (1956); *J. Gen. Chem. U.S.S.R.* (*Engl. Transl.*), **27**, 469 (1956) [*C.A.*, **51**, 15454d (1957)].
[129] Kirk, U.S. pat. 2,497,673 [*C.A.*, **44**, 5389d (1950)].
[130] Colonge and Weinstein, *Bull. soc. chim. France*, **1951**, 820; Prijs, *Helv. Chim. Acta*, **35**, 780 (1952); Colonge and Pickat, *Bull. soc. chim. France*, **1949**, 177.
[131] Koelsch, *J. Am. Chem. Soc.*, **65**, 59 (1943).
[132] Dippy and Young, *J. Chem. Soc.*, **1955**, 3919; **1952**, 1817; **1951**, 1415.

It is particularly attractive because of the low cost and ready availability of aromatic amines and because of its experimental simplicity. Further study directed toward improving the yields obtainable by suppressing side reactions will increase its value still more.

EXPERIMENTAL CONDITIONS

The technique of a Meerwein reaction is usually very simple, requiring no elaborate apparatus. The diazonium salt is prepared from one equivalent of aromatic amine, dissolved in 2.5–3.0 equivalents of hydrochloric (or hydrobromic) acid, by the addition of sodium nitrite solution. The cold solution is filtered if necessary to remove any diazoamino compound. Although the excess nitrous acid may be removed with sulfamic acid or urea, it appears from qualitative experiments that the subsequent reaction proceeds faster in the presence of small amounts of nitrite ion.[5, 112] The cold mixture is then adjusted to about pH 3–4 by addition of concentrated sodium acetate or chloroacetate solution. A pH meter or short-range pH paper is helpful in the operation.

Meanwhile the unsaturated compound is dissolved in water, acetone, or other desired solvent. The two solutions are mixed and cupric chloride (or bromide) dihydrate (0.07–0.15 mole) is added. At this point, additional water or acetone may be needed to render the mixture homogeneous. Nitrogen evolution may begin immediately or after a short induction period. Otherwise, the solution is warmed slowly to the temperature at which nitrogen evolution begins; this is usually below 25°. Stirring is usually unnecessary. Once the reaction begins, some cooling may be necessary for control. Strong cooling may stop the reaction, and it is then difficult to initiate it again. Addition of 1–2% of nitrite ion is sometimes helpful to reinitiate reactions that have stopped.[112]

When nitrogen evolution is complete, the acetone, if present, is removed by distillation at ordinary or reduced pressure. Steam distillation is usually desirable since many of the by-products such as the chloro compound resulting from the Sandmeyer reaction, the phenol, the chloroacetone, the deamination product, and often the unreacted starting material are steam distillable. The product is separated from the aqueous phase by filtration or by extraction with methylene chloride, ether, or other solvent. The product may be freed from tar if the former is soluble in acid or base. Distillation of the product is recommended where feasible, since the tars are almost invariably nonvolatile.* If the product cannot be distilled, it often can be purified by dissolving it in petroleum ether, carbon tetrachloride, or benzene and passing the solution

---

* *Caution*: Distillation of nitro-containing tars may lead to explosions.

through a short column of alumina; the diazo resin is usually retained as a strongly adsorbed band at the top of the column. In favorable cases, the product may be crystallized from an appropriate solvent, often with the aid of activated charcoal.

Should the simple procedure just described be unsuccessful, the first variable to alter is the pH. It is probable that each combination of diazonium salt and unsaturated compound will have an optimum pH. For example, in the arylation of maleic acid, negatively substituted diazonium salts react at an appreciably lower pH than other diazonium salts.[47] The second variable to change is the solvent. As noted below, acetone is frequently harmful, and its use should probably be avoided when the unsaturated compound is sufficiently water-soluble.

In the event of continued failure, the experimenter should make at least one trial with 5–15% of *cuprous* chloride catalyst in the absence of oxygen before concluding that the reaction should be abandoned.

Difficulties in purification often arise because a mixture of substitution and addition products is formed (see above). When the substitution product is the one that is sought, the crude product may advantageously be treated with base to effect dehydrohalogenation. Treatment with hot or cold alcoholic alkali is doubtless the most rapid method. The use of tertiary amines such as dimethylaniline, 2,6-lutidine, *sym*-collidine, or triethylamine at temperatures from 25° to as high as 220° is recommended for products destroyed by stronger bases.

Effects of Reaction Medium

Solvent. When the unsaturated component is sufficiently soluble in water, an organic co-solvent is usually unnecessary. In the arylation of acrylic acid and maleic acid, the yields are considerably lower when acetone is present.[31, 47] The same is true in the arylation of furfural.[51] Ferrocene[122, 133–135] and quinones [57] do not require acetone, though comparisons of yields with and without acetone have not been made.

Acetone is by far the most popular organic solvent, though a few others have received some attention. Methyl ethyl ketone, acetonitrile, N-methylpyrrolidone, pyridine, dimethyl sulfoxide, sulfolane (tetrahydrothiophene-1,1-dioxide), and 2,4-dimethylsulfolane appear, from very limited data, to be useful. In the arylation of coumarin with p-chloro- or

[133] Weinmayr, *J. Am. Chem. Soc.*, **77**, 3012 (1955).

[134] Nesmeyanov, Perevalova, Golovnya, and Nesmeyanova, *Doklady Akad. Nauk S.S.S.R.*, **97**, 459 (1954) [*C.A.*, **49**, 9633f (1955)].

[135] Nesmeyanov, Perevalova, Golovnya, and Shilovtseva, *Doklady Akad. Nauk S.S.S.R.*, **102**, 535 (1955) [*C.A.*, **50**, 4925h (1956)].

p-nitro-benzenediazonium chloride, acetonitrile as the solvent gave yields comparable to acetone as the solvent. However, the yield in the p-chlorophenylation of methacrylonitrile was lower in acetonitrile and the reaction was slow.[5] Dimethyl sulfoxide gave fair results in the p-nitrophenylation of coumarin.[34] The two sulfolanes have been tried only in the p-chlorophenylation of methacrylonitrile, with excellent results, although the isolation of the products was more difficult because of the high boiling points of these solvents.[5] There are scattered reports of the use of pyridine as a buffering ingredient.[1,8] Pyridine also has been used as a constituent of a solvent mixture for difficulty soluble cinnamic acids.[136,137] Less satisfactory solvents are dimethylformamide, tetrahydrofuran, and ethylene glycol dimethyl ether, judging from results in the p-nitrophenylation of coumarin.[53] N-Methylpyrrolidone has been used in the p-chlorophenylation of methacrylonitrile with fair results.[5] Ethanol is definitely unsatisfactory.[28,53,138] Diethyl ether has been employed in the self-catalyzed (no copper salt) reaction of ferrocene with diazonium salts,[122] and ethanol-ether mixtures were satisfactory in the arylation of quinones by N-nitroso-N-arylacetamides.[68–71] However, these reactions are not typical Meerwein reactions.

As yet untried are esters such as methyl formate or butyrolactone. There is no report of attempts to conduct the reaction deliberately in a two-phase system with solvents such as chloroform, carbon tetrachloride, methylene chloride, or benzene. The two-phase technique might possess the same advantages it has in the related Gomberg-Bachmann arylation of aromatic compounds.[40]

Consideration of the structures of the useful solvents suggests that their beneficial effect is associated with the presence of easily polarized unsaturation electrons, which may assist in the transfer of an electron from the olefin to the diazonium salt. Alcohols and ethers, with merely unshared electrons, seem incapable of functioning as demanded. Furthermore, the latter solvents reduce (deaminate) diazonium salts.[138,139]

The state of the art does not permit a reliable prediction of the best solvent medium for a new Meerwein reaction. Initial experiments should be tried in aqueous solutions if the solubility of the olefin permits. Otherwise, acetone is probably the best cosolvent, considering cost, availability, and ease of subsequent removal. If acetone proves unsatisfactory, acetonitrile should be tried next. Not enough is known about the other solvents to provide a basis for comment.

[136] Drefahl, Seeboth, and Degen, *J. prakt. Chem.* [4] **4**, 99 (1956).
[137] Drefahl, Gerlach, and Degen, *J. prakt. Chem.* [4] **4**, 119 (1956).
[138] Meerwein, *Angew. Chem.*, **70**, 211 (1958).
[139] Dombrovskiĭ and Stadnichuk, *Zhur. Obshcheĭ Khim.*, **25**, 1737 (1955) [*C.A.*, **50**, 5548e (1956)].

Anions. Almost all studies of this reaction have been performed with diazonium chlorides. The few reported examples of the use of diazonium bromides have given roughly comparable yields.[1, 4, 30] On the other hand, some attempts to use the diazonium sulfates or nitrates have failed.[1] It has been stated (without specifying the particular examples) that no reaction (nitrogen evolution) took place between an olefin, a diazonium sulfate, and copper sulfate until hydrochloric or hydrobromic acid was added.[4]

This behavior is understandable for those reactions where halogen is incorporated into the product. Here the presence of a readily polarizable nucleophilic anion would be essential. It is not so clear why it should be true when the ionic halogen is not incorporated into the product, as is true with coumarin,[1] cinnamaldehyde,[1] cinnamic acid,[1] acrylic acid,[31] etc. In fact, it is not certain that halide ion is essential, since no specific examples have been cited in support of the claim that it is. Recent experiments have shown that halide ion is desirable but not indispensable.[5] Both the p-nitrophenylation of acrylic acid (no acetone) and the p-chlorophenylation of cinnamic acid (with acetone) proceed when the chloride ion is replaced by sulfate. However, the reactions had to be heated to 60° to produce a rate of nitrogen evolution equal to those from controls at room temperature containing a plentiful supply of chloride ion. The chloride-promoted reaction is thus about ten times faster. The yields without chloride were only about 60% of those with chloride. Other examples from the literature, such as arylation of quinones and ferricinium ion, are not typical Meerwein reactions.

One possible explanation for the function of halide is that a crucial stage in the reaction requires a covalent diazo compound $ArN\!=\!NX$. Anions such as bisulfate, sulfate, and nitrate do not readily form covalent bonds. A high concentration of acetate ions should then permit formation of a covalent diazo acetate in the absence of halide ions. A more plausible explanation is that a complex copper *anion* such as $CuCl_3^-$ or $CuCl_4^=$ is the effective catalyst. Such complex anions form readily with halides but not with nitrate, etc. If one accepts the postulate that *cuprous* salt is sometimes the active catalyst, halide is required both for the attack on acetone (see, however, Ref. 73) and for complexing and solubilizing the otherwise unstable and insoluble cuprous copper.

Further experimental evidence is necessary to clarify the function of the anion.

Catalysts. Apart from copper salts, which have been discussed above, only copper powder,[140] mercuric chloride, and zinc chloride exhibited a

---

[140] Dobáš, Marhan, Krejčí, and Pirkl, *Collection Czechoslov. Chem. Communs.*, **22,** 1473 (1957); *Chem. Listy*, **51,** 463 (1957) [*C.A.*, **51,** 10449 (1957)].

modest catalytic activity in the p-nitrophenylation of coumarin.[53] A wide variety of other transition metal salts was essentially inert, affording no better yield than that obtained in the absence of added catalyst (8%). More recent studies of various metal salts in other olefin-diazonium salt systems confirmed this observation. However, since the oxidation-reduction potential of each olefin-diazonium salt pair is different, it is probable that there exist systems in which other catalysts will be effective. The complexing ability of the metal salt is doubtless a significant factor, but it cannot be assessed at the present time.

Certain reactions proceed without a catalyst. None was employed in the arylation of ferrocene.[122, 133-135] Many satisfactory quinone arylations[57] require no copper salt; a trace of hydroquinone functions as the catalyst. In these typical cases, the unsaturated compound requires no added catalyst to transfer an electron to the diazonium salt. Furthermore, quinones are notably efficient radical traps. In a few reactions conducted near pH 6, nitrogen evolution was observed before the addition of a copper salt.[15-22] However, this observation was not followed up.

Acidity. Most Meerwein reactions have been conducted in the pH range 3-4, occasionally as low as pH 2[92] or as high as pH 6.[15-22] Control of the pH is important in minimizing side reactions. In the lower pH range, the Sandmeyer reaction consumes a large fraction of the diazonium salt,[1] and at high pH the formation of diazo resins is accelerated.[1] In the arylation of maleic acid the yields were poor if the mixture was too acidic.[47] However, maleic acid arylated at pH 2 gives α-aryl-β-chlorosuccinic acids in good (though unspecified) yields.[92] Acrylonitrile and methyl vinyl ketone have been arylated in unbuffered hydrochloric acid solution with good results.[84, 85, 119, 141] Müller apparently did not neutralize the free acid left over from diazotization in some reactions.[4] Ferricinium ion was arylated in strong aqueous sulfuric acid.[133]

A study of the effect of pH upon yield and quality of 3-p-nitrophenylcoumarin showed that best results were obtained in the range pH 2-4.[53] At pH 3, the nature of the buffering anion is important;[53] acetate and chloroacetate[1] are best, while succinate, phosphate, tartrate, and citrate are inferior. Pyridine usually, but not always, gives poorer results than acetate.

Deviations toward the alkaline side may result in azo coupling with some compounds. Thus 7-hydroxycoumarin and p-hydroxycinnamic acid were arylated in a chloroacetate buffer of unspecified pH, but underwent azo coupling if the medium became more alkaline.[1] The reverse of this pH effect was noted in the arylation of 2-hydroxy-1,4-naphthoquinone.[62]

[141] Malinowski, *Roczniki Chem.*, **27**, 54 (1953) [*C.A.*, **48**, 13678h (1954)].

THE MEERWEIN ARYLATION REACTION 219

Experiments with a new Meerwein reaction probably should begin in an acetate buffer at pH 3–4. Variations toward the acid side probably will be more fruitful than variations in the basic direction, but the optimum pH will have to be determined experimentally.

EXPERIMENTAL PROCEDURES

1-p-Nitrophenylbutadiene. The preparation of 1-p-nitrophenyl-4-chloro-2-butene from p-nitroaniline and butadiene (89% crude yield) and its dehydrohalogenation with methanolic potassium hydroxide to 1-p-nitrophenylbutadiene (57–61% based on p-nitroaniline) has been described in *Organic Synthesis*.[76]

3-p-Nitrophenylcoumarin.[1, 53] p-Nitroaniline (4.1 g., 0.03 mole) is diazotized by treatment with 25 ml. of 1:1 hydrochloric acid, 15 g. of ice, and 7.0 ml. of 30% aqueous sodium nitrite. The pH is brought to 3–4 by addition of saturated aqueous sodium acetate, and the filtered solution is added in one portion to a solution of 4.4 g. (0.03 mole) of coumarin in 75–90 ml. of acetone. Then 0.8 g. (0.0045 mole) of cupric chloride dihydrate is added, and the mixture is stirred at ambient temperature until nitrogen evolution is complete. Slight cooling may be necessary if the reaction becomes too vigorous. The mixture is then steam-distilled until no more organic material distills. The water-insoluble residue is collected by filtration, washed with water, triturated with several small portions of acetone to remove unchanged coumarin and diazo resins, and finally recrystallized from anisole (10–12 ml. per g.). Pure p-nitrophenylcoumarin melting at 264° is obtained in a yield of 2.8–3.6 g. (35–45%).

***trans*-p-Chlorocinnamic Acid.**[31] p-Chloroaniline (3.2 g., 0.025 mole) is diazotized as above. The filtered diazonium solution (22–25 ml.) is added to a solution of 1.8 g. (0.025 mole) of acrylic acid, 5.8 g. of sodium acetate, and 1 g. of cupric chloride dihydrate in 80 ml. of water. After the vigorous evolution of nitrogen ceases, the insoluble material is collected by filtration and extracted with 5% sodium bicarbonate solution. The insoluble tarry portion is discarded, and the aqueous filtrate is acidified with dilute sulfuric acid. The p-chlorocinnamic acid is collected and crystallized from aqueous methanol; yield, 1.3 g. (28%), m.p. 239–240°.

A similar procedure with p-nitroaniline yields 2.9 g. (60%) of p-nitrocinnamic acid, m.p. 285–286°.[31] The writer has confirmed this yield and has found that 2-methoxyethanol containing a little ethanol is a much better solvent than ethanol for the crystallization of p-nitrocinnamic acid. The Meerwein arylation is far more convenient than the nitration of cinnamic acid followed by separation of isomers.

2-Methoxy-4'-phenylstilbene.[42] *p*-Aminobiphenyl (16.9 g., 0.1 mole) is diazotized in hydrochloric acid in the usual manner. The diazonium solution is added to a solution of 17.8 g. (0.1 mole) of *o*-methoxycinnamic acid in 1 l. of acetone containing 25 g. of anhydrous sodium acetate and 4.2 g. of cupric chloride dihydrate. Nitrogen evolution is complete after 3 hours at 20–25°. The solid remaining after steam distillation is sublimed at 125°/1 μ and then crystallized from alcohol. Ten grams (35%) of the stilbene is obtained as small white prisms, m.p. 184–185°.

In the preparation of stilbenes substituted in both rings, it is highly desirable to use the more soluble of the two possible cinnamic acids and to supply the second aryl group via the amine.

trans-*p*-**Nitrocinnamonitrile.**[4] *p*-Nitroaniline (4.2 kg.) in 18 l. of hot 1:1 hydrochloric acid is cooled to 30–40°, mixed with 24 kg. of ice, and diazotized with 7.3 l. of 30% aqueous sodium nitrite. The filtered diazonium solution is added to 1.76 kg. of acrylonitrile in 15 l. of acetone. After addition of 0.6 kg. of cupric chloride dihydrate, nitrogen evolution sets in at 18°. (A sodium acetate buffer is not specified.) The temperature is maintained below 30° by cooling. After nitrogen evolution is complete, the product is collected and crystallized from methanol. The yield of α-chloro-*p*-nitrohydrocinnamonitrile, m.p. 110°, is 5.3 kg. (83%).

The chloronitrile (5.2 kg.) is dehydrohalogenated by boiling it for 10 hours with a solution of 4 kg. of sodium acetate in 20 l. of ethanol and 8 l. of water. The insoluble *p*-nitrocinnamonitrile which separates is collected, washed, and crystallized from chlorobenzene, m.p. 200°; yield, 3.6 kg. (79%).

α-*p*-Chlorophenyl-N-isopropylmaleimide.[33] *p*-Chlorobenzenediazonium chloride solution, prepared in the usual way from 0.1 mole of *p*-chloroaniline, is added to an ice-cold solution of 0.1 mole of N-isopropylmaleimide in 30 ml. of acetone. The *p*H is brought to 3 with aqueous sodium acetate, 0.015 mole of cupric chloride is added, then enough acetone or water to form a homogeneous solution. Nitrogen evolution begins immediately. The mixture is kept in an ice bath for $\frac{1}{2}$ hour, then warmed to 35–40° and maintained at that temperature with stirring for 3 hours. The acetone is then evaporated under reduced pressure, and the oily product is separated.

The oil is dissolved in 50 ml. of 2,6-lutidine, heated nearly to boiling, cooled, diluted with 75 ml. of benzene, and filtered. The filtrate is partitioned between ether and water, the organic layer is washed with dilute sulfuric acid and water, then dried and evaporated. The crystalline residue is recrystallized from ether-petroleum ether. Alternatively, the residue may be distilled at reduced pressure; the product is then more easily recrystallized. The yield of pure material, m.p. 102–104°, is 14.6 g. (51%).

p-Nitrophenylmaleic Anhydride.[92] A solution of p-nitrobenzene-diazonium chloride is prepared by diazotizing 27.6 g. (0.2 mole) of p-nitroaniline in the presence of sufficient hydrochloric acid to make the pH of the resulting solution about 2. It is then added with vigorous stirring to a solution of 23 g. (0.2 mole) of maleic acid in 80 ml. of acetone containing 8 g. of cupric chloride dihydrate in 14 ml. of water. The temperature is maintained between 12 and 18° for 2 hours, and the mixture is then allowed to stand for 24 hours at room temperature. The layers are separated, and the lower layer is concentrated under reduced pressure. The solid residue is crystallized from a mixture of ethanol and benzene, giving 27 g. (50%) of α-p-nitrophenyl-β-chlorosuccinic acid as micro crystals, m.p. 275° (dec.).

For the preparation of p-nitrophenylmaleic anhydride, 12 g. of the chlorosuccinic acid is dissolved in 24 g. of acetic anhydride and boiled under reflux for 6 hours. The solvent is then removed at reduced pressure, and the residue is crystallized from ligroin, giving 8.8 g. (92%) of p-nitrophenylmaleic anhydride, m.p. 127°.

1,4-Bis-(2'-chloro-2'-cyanoethyl)benzene (Use of a Diamine).[84] A solution of 21.2 g. (0.4 mole) of acrylonitrile in 100 ml. of acetone is added to a solution of 36 g. (0.2 mole) of p-phenylenediamine dihydrochloride, 100 ml. of water, 50 ml. of concentrated hydrochloric acid, and 10 g. of cupric chloride dihydrate. The mixture is cooled to −7° and slowly treated with 27.6 g. of sodium nitrite in water. During the course of 2 hours, about 1 mole of nitrogen is evolved. The end point is determined with starch-iodide paper.

The cold mixture (a dark bronze, oily liquid) is filtered and allowed to warm to 28° during the course of 1 hour. At this temperature nitrogen is evolved vigorously. On the following day, tarry particles are removed by filtration, and the filtrate is steam distilled. About 1 l. of distillate, containing about 3 ml. of a yellow immiscible liquid with an acrid odor, is collected. The distillation is then stopped despite the fact that the distillate is still cloudy.

The tarry residue solidifies on cooling. It is crystallized from 5 l. of methanol with 10 g. of decolorizing carbon. The product weighs 18 g. (36%) and melts at 178–180°. After two recrystallizations from ethanol, pure 1,4-bis(2'-chloro-2'-cyanoethyl)benzene, m.p. 184°, is obtained. A larger run gave a 45% yield.

2-o-Chlorophenylbenzoquinone.[57] A solution of 325 g. of o-chloroaniline in 500 ml. of water and 500 ml. of concentrated hydrochloric acid is prepared by warming, then cooled and mixed with 2 kg. of ice. Sodium nitrite (350 ml. of a 40% solution) is added with vigorous stirring and efficient cooling below the surface of the first solution as rapidly as possible. The mixture is filtered; the filtrate has a volume of 3.5–4.0 l. It must be acid to Congo red and contain free nitrous acid.

Meanwhile a suspension of p-benzoquinone is prepared by oxidizing 220 g. of hydroquinone in 2 l. of water with 121 g. of potassium bromate and 110 ml. of N sulfuric acid. The suspension is heated at 60–75° until all the dark quinhydrone crystals have disappeared. It is then cooled to 5°, and 350 g. of sodium bicarbonate is added just before the coupling reaction is started.

The quinone suspension is placed in a 10-l. flask and stirred vigorously while the diazonium solution is added below the surface of the suspension from a graduated dropping funnel at the rate of 25 ml. per minute. The temperature is maintained in the range 5–8° during addition. The mixture is tested periodically to be sure that it is still alkaline. It is also tested with cotton soaked in Naphthol-AS solution or paper soaked in the sodium salt of β-naphthol. If this test shows the presence of unreacted diazonium salt, a trace of hydroquinone is added.

Reaction stops abruptly when about 104% of the theoretical amount of diazonium solution has been added. The product is collected by filtration, washed with water, and dried. The crude product weighs 450 g. It is purified by distillation, giving 410 g., b.p. 160–162°/3 mm. The residue consists of the decomposition products of polyarylated benzoquinone. 2-*o*-Chlorophenylquinone may be recrystallized from methanol or ethanol; m.p. 82–83°. The yield is 90% based on amine or 94% based on hydroquinone.

TABULAR SURVEY OF THE MEERWEIN ARYLATION REACTION

In the following thirteen tables are collected the examples of the Meerwein reaction which could be found in the literature up to October, 1958. The search was conducted with *Chemical Abstracts Subject Indexes* through Vol. 50, 1956. More recent references were located by scanning titles in *Current Chemical Papers* for titles suggestive of the Meerwein reaction.

In each table, the unsaturated components are arranged in the following order: the parent compound of the series; its halogen derivatives in the order F, Cl, Br, I; its alkyl derivatives in the order of increasing size and complexity; its phenyl derivatives and its nuclear-substituted phenyl derivatives; and finally heterocyclic derivatives of the parent compound.

Under each unsaturated component the diazonium salts used are arranged in the following order: benzenediazonium chloride, then nuclear substitution products in the order F, Cl, Br, I, NO_2, OH, OCH_3, NH_2, $NHCOCH_3$, SO_3H, SO_2NH_2, AsO_3H_2, alkyl in the order of increasing size and complexity, aryl (including condensed aryl as in naphthalenediazonium chloride), CHO, CO_2H, CO_2R, COR, CN, and finally heterocyclic diazonium salts.

The individual diazonium salts are not entered in the tables since they are adequately identified by inspection of the products.

The practice has been followed of reporting the highest yield claimed in the literature for a particular reaction; that figure is given by the first reference cited, followed by the others in numerical order. The symbol (—) indicates that no yield was reported. Unsuccessful experiments have been included in the tables.

TABLE I
Nonconjugated Olefins and Acetylenes

Olefin or Acetylene	Product (Yield, %)	References
$CH_2=CH_2$	$p\text{-}O_2NC_6H_4CH_2CH_2Cl$ (< 20)	3, 4
$CH_2=CHBr$	$p\text{-}O_2NC_6H_4CH_2CHBr_2$ (77)	30
$CH_2=CHCH_2Cl$	$p\text{-}ClC_6H_4CH_2CHClCH_2Cl$ (—)	3, 4
$CHCl=CCl_2$	$p\text{-}ClC_6H_4CHClCCl_3$* (—)	3, 4
$CH_2=CHCH_2CO_2H$	$p\text{-}ClC_6H_4CH=CHCH_2CO_2H$ (5)	13
$CH≡CH$	$p\text{-}O_2NC_6H_4CH=CHCl$ (—)	3, 4
$CH_2=CHSi(C_2H_5)_3$	$p\text{-}HO_2CC_6H_4CH_2CHClSi(C_2H_5)_3$ (13)	142
$CH_2=CHSi(C_6H_5)_3$	$C_6H_5CH_2CHClSi(C_6H_5)_3$ (0)	142
	$p\text{-}ClC_6H_4CH_2CHClSi(C_6H_5)_3$ (37)	142
	$p\text{-}BrC_6H_4CH_2CHClSi(C_6H_5)_3$ (15)	142
	$m\text{-}O_2NC_6H_4CH_2CHClSi(C_6H_5)_3$ (16)	142
	$p\text{-}O_2NC_6H_4CH_2CHClSi(C_6H_5)_3$ (28)	142
	$p\text{-}CH_3OC_6H_4CH_2CHClSi(C_6H_5)_3$ (0)	142
	$p\text{-}CH_3C_6H_4CH_2CHClSi(C_6H_5)_3$ (0)	142
	$p\text{-}C_6H_5C_6H_4CH_2CHClSi(C_6H_5)_3$ (11)	142
	$p\text{-}HO_2CC_6H_4CH_2CHClSi(C_6H_5)_3$ (23)	142

Note: References 142 to 161 are on p. 260.

* This structure was assigned by analogy.

TABLE II
Conjugated Dienes and Acetylenes, Styrenes
A. Conjugated Dienes and Acetylenes

Diene	Product (Yield, %)	References
CH_2=CHCH=CH_2	$C_6H_5CH_2CH$=$CHCH_2Cl$ (70)	128, 108, 143, 144
	$C_6H_5CH_2CH$=$CHCH_2Br$ (33)	143
	p-$ClC_6H_4CH_2CH$=$CHCH_2Cl$ (67)	108, 143, 3, 4
	2,4-$Cl_2C_6H_3CH_2CH$=$CHCH_2Cl$ (64)	108, 143, 3, 4
	p-$BrC_6H_4CH_2CH$=$CHCH_2Cl$ (60)	108, 143, 78
	2,4-$Br_2C_6H_3CH_2CH$=$CHCH_2Cl$ (62)	108, 143
	p-$IC_6H_4CH_2CH$=$CHCH_2Cl$ (30)	108, 143
	o-$O_2NC_6H_4CH_2CH$=$CHCH_2Cl$ (76)	79, 108, 143
	m-$O_2NC_6H_4CH_2CH$=$CHCH_2Cl$ (57)	108, 143
	p-$O_2NC_6H_4CH_2CH$=$CHCH_2Cl$ (89)	49, 3, 4, 76, 77, 108, 143, 145
	p-$CH_3OC_6H_4CH_2CH$=$CHCH_2Cl$ (41)	108, 143
	o-$CH_3C_6H_4CH_2CH$=$CHCH_2Cl$ (52)	108, 143
	m-$CH_3C_6H_4CH_2CH$=$CHCH_2Cl$ (50)	108, 143
	p-$CH_3C_6H_4CH_2CH$=$CHCH_2Cl$ (52)	108, 143
CH$_2$=CHCCl=CH$_2$	$C_6H_5CH_2CH$=$CClCH_2Cl$* (57)	108, 3
	p-$ClC_6H_4CH_2CH$=$CClCH_2Cl$* (45)	3
	2,4-$Cl_2C_6H_3CH_2CH$=$CClCH_2Cl$* (ca. 70)	3, 4
	2,5-$Cl_2C_6H_3CH_2CH$=$CClCH_2Cl$* (68)	4
	3,4-$Cl_2C_6H_3CH_2CH$=$CClCH_2Cl$* (—)	3, 4
	o-$O_2NC_6H_4CH_2CH$=$CClCH_2Cl$* (—)	3, 4
	m-$O_2NC_6H_4CH_2CH$=$CClCH_2Cl$* (ca. 70)	3, 4
	p-$O_2NC_6H_4CH_2CH$=$CClCH_2Cl$* (—)	4
	m-$NCC_6H_4CH_2CH$=$CClCH_2Cl$* (ca. 70)	3

Note: References 142 to 161 are on p. 260.

* The structure is assigned by analogy; no conclusive structure proof is given.

TABLE II—*Continued*

CONJUGATED DIENES AND ACETYLENES, STYRENES

A. *Conjugated Dienes and Acetylenes—Continued*

Diene	Product (Yield, %)	References
CH$_2$=CHCH=CHCH$_3$	C$_6$H$_5$CH$_2$CH=CHCHClCH$_3$* (56)	108
	p-O$_2$NC$_6$H$_4$CH$_2$CH=CHCHClCH$_3$* (40)	108
CH$_2$=CHC(CH$_3$)=CH$_2$	C$_6$H$_5$CH$_2$CH=C(CH$_3$)CH$_2$Cl* (68)	108, 1, 3, 4
	o-O$_2$NC$_6$H$_4$CH$_2$CH=C(CH$_3$)CH$_2$Cl (—)	146
CH$_2$=CHC(CH$_3$)=CHCH$_3$	C$_6$H$_5$CH$_2$CH=C(CH$_3$)CHClCH$_3$* (48)	108
CH$_2$=C(CH$_3$)C(CH$_3$)=CH$_2$	C$_6$H$_5$CH$_2$C(CH$_3$)=C(CH$_3$)CH$_2$Cl (68)	108, 1, 3, 4
CH$_2$=CHCH=CHC$_6$H$_5$	C$_6$H$_5$CH=CHCH=CHC$_6$H$_5$ (80)	108
CH$_2$=CHCH=CHC$_6$H$_4$CH$_3$-p	C$_6$H$_5$CH=CHCH=CHC$_6$H$_4$CH$_3$-p (70)	108
CH$_2$=CHC(C$_6$H$_5$)=CH$_2$	C$_6$H$_5$CH=CHC(C$_6$H$_5$)=CH$_2$ (—)	108
CH$_2$=CHC≡CH	C$_6$H$_5$CH$_2$CHClC≡CH (40-45)	113
	2,5-Cl$_2$C$_6$H$_3$CH$_2$CHClC≡CH (—)	113
Anthracene (C$_{14}$H$_{10}$)	9-C$_6$H$_5$C$_{14}$H$_9$ (—)	110
	9,10-(C$_6$H$_5$)$_2$C$_{14}$H$_8$ (—)	109, 110
	9-p-ClC$_6$H$_4$C$_{14}$H$_9$ (7)	110
	9,10-(p-ClC$_6$H$_4$)$_2$C$_{14}$H$_8$ (37)	110, 109
	9,10-(o-O$_2$NC$_6$H$_4$)$_2$C$_{14}$H$_8$ (—)	109
	9-p-O$_2$NC$_6$H$_4$C$_{14}$H$_9$ (16)	110
	9,10-(p-O$_2$NC$_6$H$_4$)$_2$C$_{14}$H$_8$ (20)	109, 110
	9-p-CH$_3$OC$_6$H$_4$C$_{14}$H$_9$ (10)	110
	9,10-(p-CH$_3$OC$_6$H$_4$)$_2$C$_{14}$H$_8$ (9)	110
9-Phenylanthracene (9-C$_6$H$_5$C$_{14}$H$_9$)	9,10-(C$_6$H$_5$)$_2$C$_{14}$H$_8$ (18)	110
	9-C$_6$H$_5$-10-p-O$_2$NC$_6$H$_5$C$_{14}$H$_8$ (54)	110
Anthracene-9-carboxylic acid (C$_{14}$H$_9$CO$_2$H-9)	10-p-ClC$_6$H$_4$C$_{14}$H$_8$CO$_2$H-9 (8)	110
	10-p-O$_2$NC$_6$H$_4$C$_{14}$H$_8$CO$_2$H-9 (29)	110

Ferrocene (dicyclopentadienyliron, $C_{10}H_{10}Fe$)	$C_6H_5C_{10}H_9Fe$ (66)	147
	$(C_6H_5)_2C_{10}H_5Fe$* (42)	122
	m-$ClC_6H_4C_{10}H_9Fe$ (34)	147
	o-$O_2NC_6H_4C_{10}H_9Fe$† (5)	147
	m-$O_2NC_6H_4C_{10}H_9Fe$† (—)	135
	p-$O_2NC_6H_4C_{10}H_9Fe$† (64)	122, 134, 147
	p-$HOC_6H_4C_{10}H_9Fe$ (39)	135, 147
	p-$CH_3OC_6H_4C_{10}H_9Fe$ (40)	122, 147
	p-$HO_3SC_6H_4C_{10}H_9Fe$ (—)	147
	o-$CH_3C_6H_4C_{10}H_9Fe$ (43)	147
	p-$CH_3C_6H_4C_{10}H_9Fe$ (57)	122
	$(C_{10}H_7)xC_{10}H_{10-x}$*‡ (—)	122
	o-$HO_2CC_6H_4C_{10}H_9Fe$ (7)	147
	p-$CH_3COC_6H_4C_{10}H_9Fe$ (—)	148
Ferricinium ion	$C_6H_5C_{10}H_9Fe$ (17)	133
	$(C_6H_5)_2C_{10}H_8Fe$* (20)	133
	p-$ClC_6H_4C_{10}H_9Fe$ (—)	133
	p-$O_2NC_6H_4C_{10}H_9Fe$ (10)	133
	(p-$O_2NC_6H_4)_2C_{10}H_8Fe$* (60)	133
	p-$HOC_6H_4C_{10}H_9Fe$ (60)	133
	(p-$C_6H_5C_6H_4)_3C_{10}H_7Fe$* (50)	133
	(o-$HO_2CC_6H_4)_2C_{10}H_8Fe$* (15)	133
	8-HO_2C-1-$C_{10}H_6C_{10}H_9Fe$ (6)	133

Note: References 142 to 161 are on p. 260.

* The structure is assigned by analogy; no conclusive structure proof is given.
† The nitrobenzenediazonium salts oxidized some of the ferrocene to ferricinium ion; no product was obtained from 2,4-$(O_2N)_2C_6H_3N_2^+HSO_4^-$.[147]
‡ It was not specified whether the naphthyl group was α or β.

TABLE II—Continued
CONJUGATED DIENES AND ACETYLENES, STYRENES

B. Styrenes and Phenylacetylene

Unsaturated Compound	Product (Yield, %)	References
CH_2=CHC_6H_5	C_6H_5CH=CHC_6H_5 (23)	9
	p-ClC_6H_4CH=CHC_6H_5 (41)	9
	p-$ClC_6H_4CH_2CHClC_6H_5$ (75)	28
	2,4-$Cl_2C_6H_3CH_2CHClC_6H_5$ (—)	73
	p-$O_2NC_6H_4CH$=CHC_6H_5 (32)	9
	p-$CH_3OC_6H_4CH$=CHC_6H_5 (13)	9
CH_2=$CHC_6H_4NO_2$-p	p-$ClC_6H_4CH_2CHClC_6H_4NO_2$-$p$ (9)	74
	p-$O_2NC_6H_4CH_2CHClC_6H_4NO_2$-$p$ (4)	74
	p-$CH_3OC_6H_4CH_2CHClC_6H_4NO_2$-$p$ (4)	74
	2,4-$Cl_2C_6H_3CH_2CCl(CH_3)C_6H_5$ (—)	73
CH_2=$C(CH_3)C_6H_5$	$C_6H_5C(CH_3)$=$C(CH_3)C_6H_5$ (8)	9
CH_3CH=$C(CH_3)C_6H_5$	p-$O_2NC_6H_4C(CH_3)$=$C(CH_3)C_6H_5$ (36)	9
	p-$CH_3OC_6H_4C(CH_3)$=$C(CH_3)C_6H_5$ (0)	9
C_2H_5CH=$C(C_2H_5)C_6H_4OCH_3$-p	p-$CH_3OC_6H_4C(C_2H_5)$=$C(C_2H_5)C_6H_4OCH_3$-p (0.8)	9
CH_2=$C(C_6H_5)_2$	p-$O_2NC_6H_4CH$=$C(C_6H_5)_2$ (10)	75
2-Vinylpyridine	p-$ClC_6H_4CH_2CHClC_5H_4N$-2 (20)	74
	p-$O_2NC_6H_4CH_2CHClC_5H_4N$-2 (15)	74
	p-$CH_3OC_6H_4CH_2CHClC_5H_4N$-2 (54)	74
HC≡CC_6H_5	C_6H_5C=CC_6H_5 § (5)	5
	p-ClC_6H_4C=CC_6H_5 § (24)	5
	p-$O_2NC_6H_4C$=CC_6H_5 § (14)	5

§ The crude product was a mixture of ArC≡CC_6H_5 and ArCH=$CClC_6H_5$; it was dehydrohalogenated without purification to the diarylacetylene.

TABLE III
α,β-Unsaturated Aldehydes and Ketones

Unsaturated Carbonyl Compound	Product (Yield %)	References
CH_2=CHCHO	$C_6H_5CH_2CHClCHO$ (10)	106
	m-$ClC_6H_4CH_2CHClCHO$ (27)	106
	p-$ClC_6H_4CH_2CHClCHO$ (38)	106
	p-$O_2NC_6H_4CH_2CHClCHO$ (11)	106
	p-$ClC_6H_4CH_2CCl(CH_3)CHO$ (43)	106
CH_2=C(CH_3)CHO	p-$ClC_6H_4CH_2CCl(C_2H_5)CHO$ (33)	106
CH_2=C(C_2H_5)CHO	$C_6H_5CH_2CHClCOCH_3$ (18)	85
CH_2=CHCOCH$_3$	p-$ClC_6H_4CH_2CHClCOCH_3$ (22)	85
	2,5-$Cl_2C_6H_3CH_2CHClCOCH_3$ (—)	3, 4
	p-$O_2NC_6H_4CH_2CHClCOCH_3$ (41)	85
	o-$HO_2CC_6H_4CH_2CHClCOCH_3$ (—)	85
	p-$HO_2CC_6H_4CH_2CHClCOCH_3$ (26)	85
	o-$CH_3O_2CC_6H_4CH_2CHClCOCH_3$ (—)	85
	m-$NCC_6H_4CH_2CHClCOCH_3$ (—)	3
![cyclopentenone with CH3]	(10)* structure with OCH$_3$, CH$_3$O groups	149
C_6H_5CH=CHCHO	p-ClC_6H_4C(CHO)=CHC_6H_5 (33)	1

Note: References 142 to 161 are on p. 260.

* The starting amine was methyl 2-aminotrimethylgallate; the intermediate addition product underwent spontaneous hydrolysis and lactonization.

TABLE III—Continued

α,β-UNSATURATED ALDEHYDES AND KETONES

Unsaturated Carbonyl Compound	Product (Yield %)	References
C_6H_5CH=$CHCOCH_3$	o-$ClC_6H_4CH(COCH_3)CHClC_6H_5$ (20)	43
	m-$ClC_6H_4CH(COCH_3)CHClC_6H_5$ (39)	43
	p-$ClC_6H_4C(COCH_3)$=CHC_6H_5 (45)	1, 43
	p-$BrC_6H_4C(COCH_3)$=CHC_6H_5 (10)	43
	o-$O_2NC_6H_4CH(COCH_3)CHClC_6H_5$ (38)	43
	m-$O_2NC_6H_4C(COCH_3)$=CHC_6H_5 (18)	43
	p-$O_2NC_6H_4C(COCH_3)$=CHC_6H_5 (39)	43
	o-$CH_3C_6H_4CH(COCH_3)CHClC_6H_5$ (15)	43
	m-$CH_3C_6H_4CH(COCH_3)CHClC_6H_5$ (20)	43
	p-$C_6H_5C_6H_4CH(COCH_3)CHClC_6H_5$ (20)	43
m-$O_2NC_6H_4CH$=$CHCOCH_3$	p-$O_2NC_6H_4CH(COCH_3)CHClC_6H_4NO_2$-$m$† (—)	43
C_6H_5CH=$CHCOC_6H_5$	p-$O_2NC_6H_4C(COC_6H_5)$=CHC_6H_5 (20)	44
[furyl]CH=CHCOCH$_3$	5-p-$ClC_6H_4C_4H_2O(CH$=$CHCOCH_3)$-2 (30)	48

† This product could not be purified.

TABLE IV

ALIPHATIC α,β-UNSATURATED MONOBASIC ACIDS, ESTERS, NITRILES

Unsaturated Compound	Product (Yield, %)	References
CH_2=$CHCO_2H$	C_6H_5CH=$CHCO_2H$ (0)	31
	o-ClC_6H_4CH=$CHCO_2H$ (26)	31
	m-ClC_6H_4CH=$CHCO_2H$ (28)	31
	p-ClC_6H_4CH=$CHCO_2H$ (28)	31
	p-$ClC_6H_4CH_2CHClCO_2H$ (—)	3, 4
	2,6-$Cl_2C_6H_3CH$=$CHCO_2H$ (20)	31
	o-BrC_6H_4CH=$CHCO_2H$ (25)	31
	m-BrC_6H_4CH=$CHCO_2H$ (26)	31
	p-BrC_6H_4CH=$CHCO_2H$ (26)	31
	o-IC_6H_4CH=$CHCO_2H$ (26)	31
	o-$O_2NC_6H_4CH$=$CHCO_2H$ (7)	31
	m-$O_2NC_6H_4CH$=$CHCO_2H$ (29)	31
	p-$O_2NC_6H_4CH$=$CHCO_2H$ (60)	31
	p-$O_2NC_6H_4CH_2CHClCO_2H$ (—)	3, 4
	o-$CH_3OC_6H_4CH$=$CHCO_2H$ (0)	31
	p-$CH_3OC_6H_4CH$=$CHCO_2H$ (0)	31
	p-$CH_3COHNC_6H_4CH$=$CHCO_2H$ (0)	31
	p-$H_2NO_2SC_6H_4CH_2CHBrCO_2H$ (—)	3, 4
	3-O_2N-4-$H_3CC_6H_3CH$=$CHCO_2H$ (15)	31
	p-$CH_3C_6H_4CH$=$CHCO_2H$ (0)	31
	2,3-$(CH_3)_2C_6H_3CH$=$CHCO_2H$ (0)	31
	α-$C_{10}H_7CH$=$CHCO_2H$ (7)	31
	α-$C_{10}H_7CH_2CHClCO_2H$ (—)	3, 4
	β-$C_{10}H_7CH$=$CHCO_2H$ (10)	31

TABLE IV—*Continued*

Aliphatic α,β-Unsaturated Monobasic Acids, Esters, Nitriles

Unsaturated Compound	Product (Yield, %)	References
$CH_2=CHCO_2CH_3$	2,4-$Cl_2C_6H_3CH_2CHClCO_2CH_3$ (—)	3, 4
	p-$O_2NC_6H_4CH_2CHClCO_2CH_3$ (50)	150
	p-$CH_3C_6H_4CH_2CHClCO_2CH_3$ (23 crude)	32
	$C_6H_5CH_2CHClCN$ (81)	8, 32, 50, 82
	p-$ClC_6H_4CH_2CHClCN$ (85)	8, 3, 4, 28, 43, 46
$CH_2=CHCN$	2,4-$Cl_2C_6H_3CH_2CHClCN$ (—)	3, 4, 73
	3,4-$Cl_2C_6H_3CH_2CHClCN$ (—)	3, 4
	4-Cl-2-$HO_2CC_6H_3CH_2CHClCN$ (—)	3
	o-$O_2NC_6H_4CH_2CHClCN$ (—)	83
	m-$O_2NC_6H_4CH_2CHClCN$ (58)	8, 3, 4, 32
	p-$O_2NC_6H_4CH_2CHClCN$ (91)	8, 3, 4, 32, 43, 83
	5-O_2N-2-$HO_2CC_6H_3CH_2CHClCN$ (—)	83
	o-$CH_3OC_6H_4CH_2CHClCN$ (17)	8
	p-$CH_3OC_6H_4CH_2CHClCN$ (76)	8, 82
	p-$HO_3SC_6H_4CH_2CHClCN$ (93)	8
	p-$H_2NO_2SC_6H_4CH_2CHClCN$ (98 crude)	3, 4
	p-$H_2O_3AsC_6H_4CH_2CHClCN$ (—)	15
	p-$CH_3C_6H_4CH_2CHClCN$ (40)	32
	α-$C_{10}H_7CH_2CH_2CH_2CO_2H$* (45)	150
	β-$C_{10}H_7CH_2CH_2CO_2H$* (50)	150
	p-$NCC_6H_4CH_2CHClCN$ (—)	3, 4

CH$_2$=CHCONH$_2$	o-HO$_2$CC$_6$H$_4$CH$_2$CHClCN (—)	83
	p-HO$_2$CC$_6$H$_4$CH$_2$CHClCN (88)	8, 4, 83
	p-ClC$_6$H$_4$CH$_2$CHClCONH$_2$ (0)	112
CH$_2$=CHCONHC$_4$H$_9$-t	p-ClC$_6$H$_4$CH$_2$CHClCONHC$_4$H$_9$-t (0)	112
(CH$_2$=CHCONH)$_2$CH$_2$	(p-ClC$_6$H$_4$CH$_2$CHClCONH)$_2$CH$_2$ (0)	112
CH$_2$=C(CH$_3$)CO$_2$H	C$_6$H$_5$CH=C(CH$_3$)CO$_2$H (26)†	91
	p-ClC$_6$H$_4$CH=C(CH$_3$)CO$_2$H (12)	80
	m-O$_2$NC$_6$H$_4$CH=C(CH$_3$)CO$_2$H (12)	80
	p-O$_2$NC$_6$H$_4$CH=C(CH$_3$)CO$_2$H (20)	80
	o-CH$_3$OC$_6$H$_4$CH=C(CH$_3$)CO$_2$H (28)†	91
	p-CH$_3$OC$_6$H$_4$CH=C(CH$_3$)CO$_2$H (25)†	91
	o-CH$_3$C$_6$H$_4$CH=C(CH$_3$)CO$_2$H (23)†	91
	p-CH$_3$C$_6$H$_4$CH=C(CH$_3$)CO$_2$H (28)†	91
	β-C$_{10}$H$_7$CH=C(CH$_3$)CO$_2$H (9)	80
CH$_2$=C(CH$_3$)CO$_2$CH$_3$	C$_6$H$_5$CH$_2$CCl(CH$_3$)CO$_2$CH$_3$ (47)	150
	p-ClC$_6$H$_4$CH$_2$CCl(CH$_3$)CO$_2$CH$_3$‡ (57)	44
	2,4-Cl$_2$C$_6$H$_3$CH$_2$CCl(CH$_3$)CO$_2$CH$_3$ (—)	73
	p-BrC$_6$H$_4$CH$_2$CCl(CH$_3$)CO$_2$CH$_3$‡ (32)	44
	o-O$_2$NC$_6$H$_4$CH$_2$CCl(CH$_3$)CO$_2$CH$_3$‡ (27)	44
	m-O$_2$NC$_6$H$_4$CH$_2$CCl(CH$_3$)CO$_2$CH$_3$‡ (63)	44
	p-O$_2$NC$_6$H$_4$CH$_2$CCl(CH$_3$)CO$_2$CH$_3$ (72)	150, 44
	p-CH$_3$OC$_6$H$_4$CH$_2$CCl(CH$_3$)CO$_2$CH$_3$‡ (26)	44
	p-CH$_3$C$_6$H$_4$CH$_2$CCl(CH$_3$)CO$_2$CH$_3$ (57)	44

Note: References 142 to 161 are on p. 260.

* The intermediate product C$_{10}$H$_7$CH$_2$CHClCN was not isolated as such, but was reduced and hydrolyzed directly to C$_{10}$H$_7$CH$_2$CH$_2$CO$_2$H.

† This was the yield of a mixture of stereoisomers whose separation was attended by great loss of material.

‡ The low halogen content of the product suggests that partial dehydrochlorination occurred on distillation.

TABLE IV—Continued

ALIPHATIC α,β-UNSATURATED MONOBASIC ACIDS, ESTERS, NITRILES

Unsaturated Compound	Product (Yield, %)	References
$CH_2=C(CH_3)CN$	$C_6H_5CH_2CCl(CH_3)CN$ (42)	5
	$p\text{-}ClC_6H_4CH_2CCl(CH_3)CN$ (66)	5
	$2,4\text{-}Cl_2C_6H_3CH_2CCl(CH_3)CN$ (56)	5
	$3,4\text{-}Cl_2C_6H_3CH_2CCl(CH_3)CN$ (58)	5
	$m\text{-}BrC_6H_4CH_2CCl(CH_3)CN$ (42)	5
	$m\text{-}O_2NC_6H_4CH_2CCl(CH_3)CN$ (59)	5
	$p\text{-}O_2NC_6H_4CH_2CCl(CH_3)CN$ (64)	5
	$p\text{-}CH_3OC_6H_4CH_2CCl(CH_3)CN$ (40)	5
	$2\text{-}CH_3O\text{-}5\text{-}ClC_6H_3CH_2CCl(CH_3)CN$ (0)	5
	$o\text{-}CH_3C_6H_4CH_2CCl(CH_3)CN$ (27)	5
	$p\text{-}CH_3C_6H_4CH_2CCl(CH_3)CN$ (53)	5
	$m\text{-}CF_3C_6H_4CH_2CCl(CH_3)CN$ (68)	5
	$2,6\text{-}(C_2H_5)_2C_6H_3CH_2CCl(CH_3)CN$ (0)	5
	![quinoline-CH2CCl(CH3)CN] (59)	5
$CH_3CH=CHCO_2H$	$p\text{-}O_2NC_6H_4CH(CH_3)CHClCO_2H$ (9)	26, 1, 31
$CH_3CH=CHCO_2CH_3$	$2,4\text{-}Cl_2C_6H_3CH(CH_3)CHClCO_2CH_3$ (20)	26, 1
$CH_3CH=CHCO_2C_2H_5$	$C_6H_5CH(CH_3)CHClCO_2C_2H_5$ (8)	26
	$p\text{-}ClC_6H_4CH(CH_3)CHClCO_2C_2H_5$ (34)	26

TABLE V

AROMATIC α,β-UNSATURATED ACIDS, ESTERS, NITRILES

Unsaturated Compound	Product (Yield, %)	References
$C_6H_5CH=CHCO_2H$	$C_6H_5CH=CHC_6H_5$ (36)*	1, 12
	$o\text{-}ClC_6H_4CH=CHC_6H_5$ (9)	11
	$m\text{-}ClC_6H_4CH=CHC_6H_5$ (16)	11
	$p\text{-}ClC_6H_4CH=CHC_6H_5$ (69)*	1, 11
	$2,4\text{-}Cl_2C_6H_3CH=CHC_6H_5$ (34)	1, 13
	$2,5\text{-}Cl_2C_6H_3CH=CHC_6H_5$ (28)	13
	$2,6\text{-}Cl_2C_6H_3CH=CHC_6H_5$ (0)	13
	$3,4\text{-}Cl_2C_6H_3CH=CHC_6H_5$ (40)	13
	$2\text{-}Cl\text{-}5\text{-}H_3CC_6H_3CH=CHC_6H_5$ (27)	13
	$o\text{-}BrC_6H_4CH=CHC_6H_5$ (8)	11
	$m\text{-}BrC_6H_4CH=CHC_6H_5$ (17)	11
	$p\text{-}BrC_6H_4CH=CHC_6H_5$ (23)	11
	$5\text{-}Br\text{-}2\text{-}CH_3OC_6H_3CH=CHC_6H_5$ (28)	13
	$o\text{-}O_2NC_6H_4CH=CHC_6H_5$ (44)*	1
	$m\text{-}O_2NC_6H_4CH=CHC_6H_5$ (33)	13
	$p\text{-}O_2NC_6H_4CH=CHC_6H_5$ (58)*	1
	$2\text{-}O_2N\text{-}4\text{-}CH_3OC_6H_3CH=CHC_6H_5$ (18)	13
	$4\text{-}O_2N\text{-}1\text{-}C_{10}H_6CH=CHC_6H_5$ (12)	10
	$o\text{-}CH_3OC_6H_4CH=CHC_6H_5$ (—)	13
	$p\text{-}CH_3OC_6H_4CH=CHC_6H_5$ (49)	1
	$p\text{-}HO_3SC_6H_4CH=CHC_6H_5$ (78)*	1
	$o\text{-}H_2O_3AsC_6H_4CH=CHC_6H_5$ (0)	16
	$m\text{-}H_2O_3AsC_6H_4CH=CHC_6H_5$ (—)	22, 16
	$p\text{-}H_2O_3AsC_6H_4CH=CHC_6H_5$ (30–35)	16, 15
	$o\text{-}CH_3C_6H_4CH=CHC_6H_5$ (12)	13

* This yield has been corrected to allow for recovered starting acid.

TABLE V—Continued

AROMATIC α,β-UNSATURATED ACIDS, ESTERS, NITRILES

Unsaturated Compound	Product (Yield, %)	References
$C_6H_5CH=CHCO_2H$ (continued)	$m\text{-}CH_3C_6H_4CH=CHC_6H_5$ (14)	13
	$p\text{-}CH_3C_6H_4CH=CHC_6H_5$ (40)	13, 1
	$p\text{-}CH_3OCH_2C_6H_4CH=CHC_6H_5$ (0)	118
	$p\text{-}CH_3OCCH_2C_6H_4CH=CHC_6H_5$ (0)	118
	$p\text{-}HO_2CCH_2C_6H_4CH=CHC_6H_5$ (0)	118
	$p\text{-}C_2H_5O_2CCH_2C_6H_4CH=CHC_6H_5$ (0)	118
	$p\text{-}NCCH_2C_6H_4CH=CHC_6H_5$ (0)	12
	$p\text{-}C_6H_5C_6H_4CH=CHC_6H_5$ (12)	13
	$\alpha\text{-}C_{10}H_7CH=CHC_6H_5$ (trace)	13
	$\beta\text{-}C_{10}H_7CH=CHC_6H_5$ (5)	41
	$o\text{-}C_6H_5CH=CHC_6H_4CH=CHC_6H_5$ (15)	41
	$m\text{-}C_6H_5CH=CHC_6H_4CH=CHC_6H_5$ (20)	41
	$p\text{-}C_6H_5CH=CHC_6H_4CH=CHC_6H_5$ (35)	151
	$p\text{-}OCHC_6H_4CH=CHC_6H_5$ (20)	152
	$o\text{-}HO_2CC_6H_4CH=CHC_6H_5$ (0)	152
	$m\text{-}HO_2CC_6H_4CH=CHC_6H_5$ (good)	152, 118
	$p\text{-}HO_2CC_6H_4CH=CHC_6H_5$ (60)	153
	$p\text{-}CH_3O_2CC_6H_4CH=CHC_6H_5$ (52)	153
	$p\text{-}C_2H_5O_2CC_6H_4CH=CHC_6H_5$ (36)	118
	$p\text{-}CH_3COC_6H_4CH=CHC_6H_5$ (45)	154
	$p\text{-}C_2H_5COC_6H_4CH=CHC_6H_5$ (22)	154
	$p\text{-}C_6H_5COC_6H_4CH=CHC_6H_5$ (25)	
	![structure with OCH3, N, CH=CHC6H5] (0)	119

o-ClC$_6$H$_4$CH=CHCO$_2$H	o-ClC$_6$H$_4$CH=CHC$_6$H$_4$Cl-o (12)	42
	p-ClC$_6$H$_4$CH=CHC$_6$H$_4$Cl-o (28)	42
	p-BrC$_6$H$_4$CH=CHC$_6$H$_4$Cl-o (17)	42
	o-O$_2$NC$_6$H$_4$CH=CHC$_6$H$_4$Cl-o (8)	42
	m-O$_2$NC$_6$H$_4$CH=CHC$_6$H$_4$Cl-o (25)	42
	p-O$_2$NC$_6$H$_4$CH=CHC$_6$H$_4$Cl-o (26)	42
	o-CH$_3$OC$_6$H$_4$CH=CHC$_6$H$_4$Cl-o (—)	42
	p-CH$_3$OC$_6$H$_4$CH=CHC$_6$H$_4$Cl-o (12)	42
	p-C$_6$H$_5$C$_6$H$_4$CH=CHC$_6$H$_4$Cl-o (12)	42
	p-RO$_2$CC$_6$H$_4$CH=CHC$_6$H$_4$Cl-o† (low)‡	153
p-ClC$_6$H$_4$CH=CHCO$_2$H	p-ClC$_6$H$_4$CH=CHC$_6$H$_4$Cl-p (8)‡	1
	p-H$_2$O$_3$AsC$_6$H$_4$CH=CHC$_6$H$_4$Cl-p (poor)‡	15
m-O$_2$NC$_6$H$_4$CH=CHCO$_2$H	o-ClC$_6$H$_4$CH=CHC$_6$H$_4$NO$_2$-m (17)	42
	p-ClC$_6$H$_4$CH=CHC$_6$H$_4$NO$_2$-m (12)	42
	p-BrC$_6$H$_4$CH=CHC$_6$H$_4$NO$_2$-m (10)	42
	o-O$_2$NC$_6$H$_4$CH=CHC$_6$H$_4$NO$_2$-m (12)	41
	m-O$_2$NC$_6$H$_4$CH=CHC$_6$H$_4$NO$_2$-m (18)	41
	p-O$_2$NC$_6$H$_4$CH=CHC$_6$H$_4$NO$_2$-m (25)	41
	o-CH$_3$OC$_6$H$_4$CH=CHC$_6$H$_4$NO$_2$-m (8)	42
	p-CH$_3$OC$_6$H$_4$CH=CHC$_6$H$_4$NO$_2$-m (12)	42
	o-CH$_3$C$_6$H$_4$CH=CHC$_6$H$_4$NO$_2$-m (5)	42
	p-CH$_3$C$_6$H$_4$CH=CHC$_6$H$_4$NO$_2$-m (10)	42
p-O$_2$NC$_6$H$_4$CH=CHCO$_2$H‡	o-ClC$_6$H$_4$CH=CHC$_6$H$_4$NO$_2$-p (5)	42
	p-ClC$_6$H$_4$CH=CHC$_6$H$_4$NO$_2$-p (12)	42, 1
	p-BrC$_6$H$_4$CH=CHC$_6$H$_4$NO$_2$-p (8)	42
	o-O$_2$NC$_6$H$_4$CH=CHC$_6$H$_4$NO$_2$-p (5)	42

Note: References 142 to 161 are on p. 260.

† The group R was not specified.
‡ The low yields probably resulted from the sparing solubility of the cinnamic acid. The better yields reported with o-chlorocinnamic acid in Ref. 42 were obtained by the use of a large volume of acetone.

238 ORGANIC REACTIONS

TABLE V—Continued

AROMATIC α,β-UNSATURATED ACIDS, ESTERS, NITRILES

Unsaturated Compound	Product (Yield, %)	References
p-$O_2NC_6H_4CH$=$CHCO_2H$‡ (continued)	p-$O_2NC_6H_4CH$=$CHC_6H_4NO_2$-p (11)	42
	o-$CH_3OC_6H_4CH$=$CHC_6H_4NO_2$-p (8)	42
	p-$CH_3OC_6H_4CH$=$CHC_6H_4NO_2$-p (10)	42
	p-$CH_3C_6H_4CH$=$CHC_6H_4NO_2$-p (14)	42
	p-$C_6H_5C_6H_4CH$=$CHC_6H_4NO_2$-p (12)	42
	p-$H_2O_3AsC_6H_4CH$=CHC_6H_3OH-4-NO_2-3 (2)	19, 15, 16
3-O_2N-4-HOC_6H_3CH=$CHCO_2H$		1
p-HOC_6H_4CH=$CHCO_2H$	p-ClC_6H_4CH=CHC_6H_4OH-p (56)	16, 15
	p-$H_2O_3AsC_6H_4CH$=CHC_6H_4OH-p (31)	19, 15
	3-O_2N-4-$H_2O_3AsC_6H_3CH$=CHC_6H_4OH-p (7)	42
o-$CH_3OC_6H_4CH$=$CHCO_2H$	p-ClC_6H_4CH=$CHC_6H_4OCH_3$-o (8)	42
	p-BrC_6H_4CH=$CHC_6H_4OCH_3$-o (13)	42
	o-$O_2NC_6H_4CH$=$CHC_6H_4OCH_3$-o (5)	42
	m-$O_2NC_6H_4CH$=$CHC_6H_4OCH_3$-o (25)	42
	p-$O_2NC_6H_4CH$=$CHC_6H_4OCH_3$-o (8)	42
	o-$CH_3OC_6H_4CH$=$CHC_6H_4OCH_3$-o (8)	42
	p-$CH_3OC_6H_4CH$=$CHC_6H_4OCH_3$-o (21)	42
	p-$C_6H_5C_6H_4CH$=$CHC_6H_4OCH_3$-o (35)	42
p-$CH_3OC_6H_4CH$=$CHCO_2H$	o-ClC_6H_4CH=$CHC_6H_4OCH_3$-p (14)	13
	p-ClC_6H_4CH=$CHC_6H_4OCH_3$-p (61)	1
	p-$H_2O_3AsC_6H_4CH$=$CHC_6H_4OCH_3$-p (ca. 20)	16, 15
p-$CH_3COHNC_6H_4CH$=$CHCO_2H$	p-$HO_3SC_6H_4CH$=$CHC_6H_4NHCOCH_3$-p (—)	137
	p-$H_2NO_2SC_6H_4CH$=$CHC_6H_4NHCOCH_3$-p (20)	137
	p-$H_2O_3AsC_6H_4CH$=$CHC_6H_4NHCOCH_3$-p (25)	137
	p-$OHCC_6H_4CH$=$CHC_6H_4NHCOCH_3$-p (30)	136
	p-$HO_2CC_6H_4CH$=$CHC_6H_4NHCOCH_3$-p (35)	136
p-$CH_3C_6H_4CH$=$CHCO_2H$	p-$H_2O_3AsC_6H_4CH$=$CHC_6H_4CH_3$-p (30)	16, 15

THE MEERWEIN ARYLATION REACTION

Starting material	Product	Ref.
$C_6H_5C(CH_3)=CHCO_2H$	$C_6H_5CH=C(CH_3)C_6H_5$ § (36)	44
	$p\text{-}ClC_6H_4CH=C(CH_3)C_6H_5$ (35)	44
	$p\text{-}BrC_6H_4CH=C(CH_3)C_6H_5$ § (23)	44
	$o\text{-}O_2NC_6H_4CH=C(CH_3)C_6H_5$ (18)	44
	$m\text{-}O_2NC_6H_4CH=C(CH_3)C_6H_5$ § (28)	44
	$p\text{-}O_2NC_6H_4CH=C(CH_3)C_6H_5$ (32)	44
	$p\text{-}CH_3OC_6H_4CH=C(CH_3)C_6H_5$ § (11)	44
	$p\text{-}CH_3C_6H_4CH=C(CH_3)C_6H_5$ § (11)	44
	$p\text{-}C_6H_5C_6H_4CH=C(CH_3)C_6H_5$ § (21)	44
$(C_6H_5)_2C=CHCO_2H$	$p\text{-}O_2NC_6H_4CH=C(C_6H_5)_2$ (48)	14
	$p\text{-}CH_3C_6H_4CH=C(C_6H_5)_2$ (11)	14
	$p\text{-}O_2NC_6H_4CH=C(C_6H_5)C_6H_4F\text{-}p$ (35)	14
$p\text{-}FC_6H_4C(C_6H_5)=CHCO_2H$	$p\text{-}O_2NC_6H_4CH=C(C_6H_4F\text{-}p)_2$ (50)	14
$(p\text{-}FC_6H_4)_2C=CHCO_2H$	$p\text{-}O_2NC_6H_4CH=C(C_6H_5)C_6H_4Br\text{-}p$ (30, 11)‖	14
$p\text{-}BrC_6H_4C(C_6H_5)=CHCO_2H$	$p\text{-}O_2NC_6H_4CH=C(C_6H_4OCH_3\text{-}p)_2$ (28)¶	14
$(p\text{-}CH_3OC_6H_4)_2C=CHCO_2H$	$p\text{-}CH_3C_6H_4CH=C(C_6H_4OCH_3\text{-}p)_2$ (small)¶	14
$(p\text{-}H_3CC_6H_4)_2C=CHCO_2H$	$m\text{-}O_2NC_6H_4CH=C(C_6H_4CH_3\text{-}p)_2$ (30)*	14
$C_6H_5CH=C(C_6H_5)CO_2H$	$p\text{-}O_2NC_6H_4C(C_6H_5)=CHC_6H_5$ (0)¶	14
$C_6H_5CH=CHCO_2CH_3$	$C_6H_5CHClCH(CO_2CH_3)C_6H_4Cl\text{-}p$ (30)	1, 26
	$C_6H_5CHBrCH(CO_2CH_3)C_6H_4Cl\text{-}p$ (26)	1
	$C_6H_5CH=C(CN)C_6H_4Cl\text{-}p$ (76)	1
$C_6H_5CH=CHCN$	$C_6H_5CH=C(CN)C_6H_4AsO_3H_2\text{-}p$ (ca. 20)	16, 15
$p\text{-}O_2NC_6H_4CH=CHCN$	$p\text{-}O_2NC_6H_4CH=C(CN)C_6H_4NO_2\text{-}p$ (12)	43

* This yield has been corrected to allow for recovered starting acid. The better yields reported with o-chlorocinnamic acid in Ref. 42 were obtained by the use of a large volume of acetone.
‡ The low yields probably resulted from the sparing solubility of the cinnamic acid. The better yields reported with o-chlorocinnamic acid in Ref. 42 were obtained by the use of a large volume of acetone.
§ The analysis of the product suggests the presence of some hydrogen chloride addition product.
‖ The 30% yield was obtained from the cinnamic acid of m.p. 175°; the 11% yield from acid of m.p. 169–170°.
¶ The starting acid decarboxylates extensively under the reaction conditions.

TABLE VI
Heterocyclic α,β-Unsaturated Acids

Acid	Products (Yield, %)	References
furan-CH=CHCO$_2$H	p-ClC$_6$H$_4$CH=CHC$_4$H$_3$O (—),	18
	5-p-ClC$_6$H$_4$C$_4$H$_2$O(CH=CHCO$_2$H)-2 (—, 26*),	18
	5-p-ClC$_6$H$_4$C$_4$H$_2$O(CH=CHC$_6$H$_4$Cl-p)-2 (—)	
	o-O$_2$NC$_6$H$_4$C(CO$_2$H)=CHC$_4$H$_3$O† (21)	18
	p-O$_2$NC$_6$H$_4$CH=CHC$_4$H$_3$O (23, 30*)	43, 95
	5-p-O$_2$NC$_6$H$_4$C$_4$H$_2$O(CH=CHCO$_2$H)-2 (12),	18
	5-p-O$_2$NC$_6$H$_4$C$_4$H$_2$O(CH=CHC$_6$H$_4$NO$_2$-p)-2 (36)	
	5-p-HO$_3$SC$_6$H$_4$C$_4$H$_2$O(CH=CHC$_6$H$_4$SO$_3$H-p)-2 (—),	18
	5-p-HO$_3$SC$_6$H$_4$C$_4$H$_2$O(CH=CHCO$_2$H)-2 (4)	
	5-p-H$_2$O$_3$AsC$_6$H$_4$C$_4$H$_2$O(CH=CHCO$_2$H)-2 (—),	18, 15
	5-p-H$_2$O$_3$AsC$_6$H$_4$C$_4$H$_2$O(CH=CHC$_6$H$_4$AsO$_3$H$_2$-p)-2 (—)	
	5-p-C$_2$H$_5$O$_2$CC$_6$H$_4$C$_4$H$_2$O(CH=CHCO$_2$H)-2 (14),	18
	5-p-C$_2$H$_5$O$_2$CC$_6$H$_4$C$_4$H$_2$O(CH=CHC$_6$H$_4$CO$_2$C$_2$H$_5$-p)-2 (—)	
thiophene-CH=CHCO$_2$H	p-ClC$_6$H$_4$CH=CHC$_4$H$_3$S (35)	21
	p-O$_2$NC$_6$H$_4$CH=CHC$_4$H$_3$S (36),	21
	5-p-O$_2$NC$_6$H$_4$C$_4$H$_3$S(CH=CHC$_6$H$_4$NO$_2$-p)-2 (8)	
	p-H$_2$O$_3$AsC$_6$H$_4$CH=CHC$_4$H$_3$S (30)	20
	p-HO$_2$CC$_6$H$_4$CH=CHC$_4$H$_3$S (22)	21

* This yield refers to an article by Oda,[48] who was probably describing the product in question. The original article was not available, and the nomenclature used in the abstract is ambiguous.
† The structure of the product was not proved.

TABLE VII

α,β-Unsaturated γ-Keto Acids

Acid	Product (Yield, %)	References
$C_6H_5COCH=CHCO_2H$	$C_6H_5CH=CHCOC_6H_5$ (trace)	90
	o-$ClC_6H_4CH=CHCOC_6H_5$ (5)	89
	m-$ClC_6H_4CH=CHCOC_6H_5$ (5)	89
	p-$ClC_6H_4CH=CHCOC_6H_5$ (27–29)	89
	p-$BrC_6H_4CH=CHCOC_6H_5$ (17–18)	89
	o-$O_2NC_6H_4CH=CHCOC_6H_5$ (10–14)	89
	m-$O_2NC_6H_4CH=CHCOC_6H_5$ (12–14)	89
	p-$O_2NC_6H_4CH=CHCOC_6H_5$ (16–19)	89
p-$CH_3OC_6H_4COCH=CHCO_2H$	$C_6H_5CH=CHCOC_6H_4OCH_3$-p (7)	90
	o-$ClC_6H_4CH=CHCOC_6H_4OCH_3$-$p$ (12)	90
	p-$ClC_6H_4CH=CHCOC_6H_4OCH_3$-$p$ (13)	90
	o-$BrC_6H_4CH=CHCOC_6H_4OCH_3$-$p$ (9)	90
	p-$BrC_6H_4CH=CHCOC_6H_4OCH_3$-$p$ (10)	90
	o-$O_2NC_6H_4CH=CHCOC_6H_4OCH_3$-$p$ (8)	90
	m-$O_2NC_6H_4CH=CHCOC_6H_4OCH_3$-$p$ (10)	90
	p-$O_2NC_6H_4CH=CHCOC_6H_4OCH_3$-$p$ (23)	90
$3,4$-$(CH_3O)_2C_6H_3COCH=CHCO_2H$	$C_6H_5CH=CHCOC_6H_3(OCH_3)_2$-$3,4$ (8)	90
	2-$O_2NC_6H_4CH=CHCOC_6H_3(OCH_3)_2$-$3,4$ (23)	90
	3-$O_2NC_6H_4CH=CHCOC_6H_3(OCH_3)_2$-$3,4$ (22)	90
	4-$O_2NC_6H_4CH=CHCOC_6H_3(OCH_3)_2$-$3,4$ (21)	90

TABLE VIII
Conjugated Dienoic Acids and Esters

Unsaturated Compound	Product (Yield, %)	References
$CH_3CH=CHCH=CHCO_2H$	$CH_3CH=CHCH=CHC_6H_5$ (26)	26
$C_6H_5CH=CHCH=CHCO_2H$	$C_6H_5CH=CHCH=CHC_6H_5$ (28)	26, 12
	$C_6H_5CH=CHCH=CHC_6H_4Cl\text{-}o$ (10)	11
	$C_6H_5CH=CHCH=CHC_6H_4Cl\text{-}m$ (29)	11
	$C_6H_5CH=CHCH=CHC_6H_4Cl\text{-}p$ (33)	11
	$C_6H_5CH=CHCH=CHC_6H_4NO_2\text{-}o$ (10)	155, 13
	$C_6H_5CH=CHCH=CHC_6H_4NO_2\text{-}m$ (12)	13, 41
	$C_6H_5CH=CHCH=CHC_6H_4NO_2\text{-}p$ (25)	10
	$C_6H_5CH=CHCH=CHC_6H_4OCH_3\text{-}o$ (18)	41
	$C_6H_5CH=CHCH=CHC_6H_4OCH_3\text{-}p$ (22)	41
	$C_6H_5CH=CHCH=CHC_6H_4C_6H_5\text{-}p$ (20)	12
	$C_6H_5CH=CHCH=CHC_6H_4CH=CHC_6H_5\text{-}m$ (—)	41
	$C_6H_5CH=CHCH=C(CO_2H)C_6H_5$* (19)	26
	$C_6H_5CH=CHCH=C(CO_2H)C_6H_4Cl\text{-}p$* (37)	26
$C_6H_5CH=CHCH=CHCO_2CH_3$		

Note: References 142 to 161 are on p. 260.

* The intermediate ester was saponified directly.

TABLE IX

POLYBASIC α,β-UNSATURATED ACIDS, NITRILES, ESTERS, IMIDES

Unsaturated Compound	Product (Yield, %)	References
Maleic acid (Reactions conducted at pH 3–4, in the absence of acetone)	$C_6H_5CH\!=\!CHCO_2H$ (0)	47
	$o\text{-}ClC_6H_4CH\!=\!CHCO_2H$ (28)	47
	$m\text{-}ClC_6H_4CH\!=\!CHCO_2H$ (28)	47
	$p\text{-}ClC_6H_4CH\!=\!CHCO_2H$ (28)	47
	$2,6\text{-}Cl_2C_6H_3CH\!=\!CHCO_2H$ (21)	47
	$o\text{-}BrC_6H_4CH\!=\!CHCO_2H$ (23)	47
	$m\text{-}BrC_6H_4CH\!=\!CHCO_2H$ (26)	47
	$p\text{-}BrC_6H_4CH\!=\!CHCO_2H$ (29)	47
	$o\text{-}IC_6H_4CH\!=\!CHCO_2H$ (23)	47
	$o\text{-}O_2NC_6H_4CH\!=\!CHCO_2H$ (7)	47
	$m\text{-}O_2NC_6H_4CH\!=\!CHCO_2H$ (24)	47
	$p\text{-}O_2NC_6H_4CH\!=\!CHCO_2H$ (58)*	47
	$2,4\text{-}(O_2N)_2C_6H_3CH\!=\!CHCO_2H$ (7)	47
	$3\text{-}O_2N\text{-}4\text{-}H_3CC_6H_3CH\!=\!CHCO_2H$ (14)	47
	$o\text{-}CH_3OC_6H_4CH\!=\!CHCO_2H$ (0)	47
	$p\text{-}CH_3OC_6H_4CH\!=\!CHCO_2H$ (0)	47
	$p\text{-}CH_3COHNC_6H_4CH\!=\!CHCO_2H$ (0)	47
	$2,3\text{-}(CH_3)_2C_6H_3CH\!=\!CHCO_2H$ (0)	47
	$\alpha\text{-}C_{10}H_7CH\!=\!CHCO_2H$ (7)	47
	$\beta\text{-}C_{10}H_7CH\!=\!CHCO_2H$ (8)	47

* The author of this chapter was unable to duplicate this yield in several attempts. The average yield in his experiments was 30%.

TABLE IX—Continued

POLYBASIC α,β-UNSATURATED ACIDS, NITRILES, ESTERS, IMIDES

Unsaturated Compound	Product (Yield, %)	References
Maleic acid (Reactions conducted at pH 1–2, in the presence of acetone)	$HO_2CCHClCH(CO_2H)C_6H_5$† (—)	92
	$HO_2CCHClCH(CO_2H)C_6H_4NO_2$-$p$ (50)	92
	$HO_2CCHClCH(CO_2H)C_6H_4CH_3$-$p$† (—)	92
	$HO_2CCHClCH(CO_2H)C_{10}H_7$-$\alpha$† (—)	92
	$HO_2CCHClCH(CO_2H)C_{10}H_7$-$\beta$† (—)	92
Dimethyl maleate	$HO_2CCH=C(CO_2H)C_6H_5$‡ (18)§	87
	$HO_2CCH=C(CO_2H)C_6H_4Cl$-p‡ (47)§	1
Dimethyl fumarate	$CH_3O_2CCH=C(CO_2CH_3)C_6H_4Cl$-$p\|$ (26)§	88
	$HO_2CCH=C(CO_2H)C_6H_4Cl$-p‡ (70)	1
	$CH_3O_2CCH=C(CO_2CH_3)C_6H_4Cl$-$p\|$ (48)§	88
Di-n-butyl maleate	n-$C_4H_9O_2CCH=C(CO_2C_4H_9$-$n)C_6H_4Cl$-$p\|$ (40)§	88
Di-n-butyl fumarate	n-$C_4H_9O_2CCH=C(CO_2C_4H_9$-$n)C_6H_4Cl$-$p\|$ (62)§	88
Maleonitrile	$NCCH=C(CN)C_6H_4Cl$-$p\|$ (45)§	88
Fumaronitrile	$NCCH=C(CN)C_6H_4Cl$-$p\|$ (52)§	88
	$NCCH=C(CN)C_6H_3Cl_2$-2,4$\|$ (good)§	112
	$NCCH=C(CN)C_6H_4NO_2$-p¶ (36)	88
	$NCCH=C(CN)C_6H_4OCH_3$-p (— crude)	112
Maleimide ($C_4H_2O_2NH$)**	$C_6H_5C_4HO_2NH$ (21)	33
	o-$ClC_6H_4C_4HO_2NH$ (—)	34
	α-m-ClC_6H_4-β-$ClC_4H_2O_2NH$ (41)	34
	p-$ClC_6H_4C_4HO_2NH$ (> 50)	112, 33
	2,4-$Cl_2C_6H_3C_4HO_2NH$ (56)	34
	2,5-$Cl_2C_6H_3C_4HO_2NH$ (51 crude)	34
	m-$BrC_6H_4C_4HO_2NH$ (— crude)	34
	p-$BrC_6H_4C_4HO_2NH$ (47)	33
	o-$O_2NC_6H_4C_4HO_2NH$ (0)	112
	p-$O_2NC_6H_4C_4HO_2NH$ (36)	33

		Ref.
	o-CH$_3$OC$_6$H$_4$C$_4$HO$_2$NH (—)††	34
	m-CH$_3$OC$_6$H$_4$C$_4$HO$_2$NH (—)††	34
	p-CH$_3$OC$_6$H$_4$C$_4$HO$_2$NH (45)‡‡	33
	(p-CH$_3$OC$_6$H$_4$)$_2$C$_4$O$_2$NH (35)§§	33
	p-CH$_3$OC$_6$H$_4$C$_4$HO$_2$NH (28)	33
	α-C$_{10}$H$_7$C$_4$HO$_3$†† (—)	33
	β-C$_{10}$H$_7$C$_4$HO$_3$†† (—)	33
N-Ethylmaleimide	p-ClC$_6$H$_4$C$_4$HO$_2$NC$_2$H$_5$ (27)	33
N-Isopropylmaleimide	C$_6$H$_5$C$_4$HO$_2$NCH(CH$_3$)$_2$ (27)	33
	o-ClC$_6$H$_4$C$_4$HO$_2$NCH(CH$_3$)$_2$ (36)	34
	m-ClC$_6$H$_4$C$_4$HO$_2$NCH(CH$_3$)$_2$ (36)	34
	p-ClC$_6$H$_4$C$_4$HO$_2$NCH(CH$_3$)$_2$ (51)	34
	2,4-Cl$_2$C$_6$H$_3$C$_4$HO$_2$NCH(CH$_3$)$_2$ (26)	34
	2,5-Cl$_2$C$_6$H$_3$C$_4$HO$_2$NCH(CH$_3$)$_2$ (54)	34
	o-BrC$_6$H$_4$C$_4$HO$_2$NCH(CH$_3$)$_2$ (51)	34
	m-BrC$_6$H$_4$C$_4$HO$_2$NCH(CH$_3$)$_2$ (— crude)	34
	o-O$_2$NC$_6$H$_4$C$_4$HO$_2$NCH(CH$_3$)$_2$ (0)	34
	p-O$_2$NC$_6$H$_4$C$_4$HO$_2$NCH(CH$_3$)$_2$ (19)	34
	p-CH$_3$OC$_6$H$_4$C$_4$HO$_2$NCH(CH$_3$)$_2$ (41)	34

† The product isolated was the substituted maleic anhydride, obtained by heating the chlorosuccinic acid with acetic anhydride.
‡ The intermediate ester was saponified without purification.
§ This is the combined yield of a mixture of stereoisomers.
‖ The crude product was dehydrohalogenated by treatment with a tertiary amine.
¶ The structure of this product is not certain.
** The yields could doubtless be improved in most of these reactions if the product were dehydrohalogenated before, rather than after, purification.
†† The intermediate imide was saponified and cyclized to the anhydride.
‡‡ Excess maleimide was used in this reaction.
§§ Excess diazonium salt was used in this reaction.

TABLE IX—Continued
Polybasic α,β-Unsaturated Acids, Nitriles, Esters, Imides

Unsaturated Compound	Product (Yield, %)	References
N-n-Hexylmaleimide	o-ClC$_6$H$_4$C$_4$HO$_2$NC$_6$H$_{13}$-n (48)	156
N-Phenylmaleimide	o-ClC$_6$H$_4$C$_4$HO$_2$NC$_6$H$_5$ (—)	156
	p-ClC$_6$H$_4$C$_4$HO$_2$NC$_6$H$_5$ (33)	33
Maleic hydrazide	p-ClC$_6$H$_4$C$_4$HO$_2$N$_2$H$_2$ (0)	34
Bromomaleic acid	HO$_2$CCBr=CHC$_6$H$_4$Cl-o (11)	114
	HO$_2$CCBr=CHC$_6$H$_4$Cl-m (20)	114
	HO$_2$CCBr=CHC$_6$H$_4$Cl-p (20)	114
	HO$_2$CCBr=CHC$_6$H$_4$Br-p (27) §	114
	HO$_2$CCBr=CHC$_6$H$_4$NO$_2$-o (5)	114
	HO$_2$CCBr=CHC$_6$H$_4$NO$_2$-m (21)	114
	HO$_2$CCBr=CHC$_6$H$_4$NO$_2$-p (15)	114
	HO$_2$CCBr=CHC$_{10}$H$_7$-α (4)	114
	HO$_2$CCBr=CHC$_{10}$H$_7$-β (3)	114
Dibromomaleic acid	HO$_2$CCBr=CBrC$_6$H$_4$Cl-p (0)	114
HO$_2$CC(CH$_3$)=CHCO$_2$H (cis)	HO$_2$CC(CH$_3$)=CHC$_6$H$_5$ (0)	81
	HO$_2$CC(CH$_3$)=CHC$_6$H$_4$Cl-p (34)	80
	HO$_2$CC(CH$_3$)=CHC$_6$H$_4$Br-p (10)	81
	HO$_2$CC(CH$_3$)=CHC$_6$H$_4$NO$_2$-o (—)	81
	HO$_2$CC(CH$_3$)=CHC$_6$H$_4$NO$_2$-m (—)	81
	HO$_2$CC(CH$_3$)=CHC$_6$H$_4$NO$_2$-p (14)	81
	HO$_2$CC(CH$_3$)=CHC$_6$H$_4$CO$_2$H-p (0)	81
	HO$_2$CC(CH$_3$)=CHC$_6$H$_4$SO$_3$H-p (0)	81

THE MEERWEIN ARYLATION REACTION

Reactant	Product (Yield, %)	Ref.
$HO_2CC(CH_3)=CHCO_2H$ (trans)	$HO_2CC(CH_3)=CHC_{10}H_7$-α (0)	81
	$HO_2CC(CH_3)=CHC_{10}H_7$-β (—)	81
	$HO_2CC(CH_3)=CHC_6H_4NO_2$-$p$ (10)	81
N-Isopropylcitraconimide	α-Methyl-α-chloro-α'-(p-chlorophenyl)-N-isopropylsuccinimide (56)	33
$CH_2=C(CO_2H)CH_2CO_2H$	$C_6H_5CH_2C(CO_2H)=CH_2$ (few drops)	115
	o-$ClC_6H_4CH_2C(CO_2H)=CH_2$ (16)	115
	p-$ClC_6H_4CH_2C(CO_2H)=CH_2$ (10)	115
	p-$BrC_6H_4CH_2C(CO_2H)=CH_2$ (20)	115
	p-$O_2NC_6H_4CH_2C(CO_2H)=CH_2$ (18)	115
$HO_2CCH=C(CO_2H)CH_2CO_2H$	$C_6H_5CH=C(CO_2H)CH_2CO_2H$ (0)	115
	m-$ClC_6H_4CH=C(CO_2H)CH_2CO_2H$ (0)	115
	p-$ClC_6H_4CH=C(CO_2H)CH_2CO_2H$ (25)	115
	o-$BrC_6H_4CH=C(CO_2H)CH_2CO_2H$ (0)	115
	m-$BrC_6H_4CH=C(CO_2H)CH_2CO_2H$ (8–16)	115
	p-$BrC_6H_4CH=C(CO_2H)CH_2CO_2H$ (8–16)	115
	m-$O_2NC_6H_4CH=C(CO_2H)CH_2CO_2H$ (0)	115
	p-$O_2NC_6H_4CH=C(CO_2H)CH_2CO_2H$ (8–16)	115
Phenylmaleic acid	Diphenylmaleic anhydride‖ (—)	92
	Phenyl-p-tolylmaleic anhydride‖ (—)	92
	Phenyl-β-naphthylmaleic anhydride‖ (28)	92
p-Nitrophenylmaleic acid	Di-(p-nitrophenyl)maleic anhydride‖ (—)	92

Note: References 142 to 161 are on p. 260.

§ This is the combined yield of a mixture of stereoisomers.

‖ This experiment was conducted at pH 1–2 in the presence of acetone. The crude product was cyclized to the substituted maleic anhydride by heating with acetic anhydride.

TABLE X

QUINONES

A. Benzoquinone Derivatives

Starting Quinone	Substituent(s) in Product Benzoquinone (Yield, %)	References
p-Benzoquinone	C_6H_5*† (84)	97, 54, 56, 58, 65, 68, 157
	o-ClC_6H_4 (90–94)	57, 97
	m-ClC_6H_4 (90)	97
	p-ClC_6H_4 (88)	97, 56
	2,3-$Cl_2C_6H_3$ (—)	57
	2,4-$Cl_2C_6H_3$ (—)	57
	2,5-$Cl_2C_6H_3$ (—)	57
	2,6-$Cl_2C_6H_3$ (73)	97, 57
	o-BrC_6H_4 (75)	97
	p-BrC_6H_4 (—)	61
	o-$O_2NC_6H_4$ (76)	97, 56, 57, 59–61
	m-$O_2NC_6H_4$ (—)	58, 59
	p-$O_2NC_6H_4$ (89)	97, 56, 58, 59
	o-HOC_6H_4 (77)	97
	p-HOC_6H_4 (59)	97
	o-$CH_3OC_6H_4$ (81)	97
	p-$CH_3OC_6H_4$*† (93)	97, 58, 63, 65, 68
	2-Cl-4-$CH_3OC_6H_3$ (—)	57
	2-Cl-5-$CH_3OC_6H_3$ (—)	57
	2-Cl-6-$CH_3OC_6H_3$ (—)	57
	3,4-$(CH_3O)_2C_6H_3$ (84)	97, 71
	p-$CH_3COHNC_6H_4$ (good)	56

p-H$_2$NO$_2$SC$_6$H$_4$ (—)	59
p-(p-HO$_3$SC$_6$H$_4$N=N)C$_6$H$_4$ (—)	56
o-CH$_3$C$_6$H$_4$ (62)	97
m-CH$_3$C$_6$H$_4$ (81)	97
p-CH$_3$C$_6$H$_4$† (—)	65, 58, 68
2-Cl-3-CH$_3$C$_6$H$_3$ (—)	57
2-Cl-4-CH$_3$C$_6$H$_3$ (—)	57
2-Cl-5-CH$_3$C$_6$H$_3$ (—)	57
2-Cl-6-CH$_3$C$_6$H$_3$ (—)	57
2-Br-4,5-(CH$_3$)$_2$C$_6$H$_2$ (—)	57
4-Br-2,6-(CH$_3$)$_2$C$_6$H$_2$ (73)	97
o-C$_6$H$_5$C$_6$H$_4$ (88)	97
p-C$_6$H$_5$C$_6$H$_4$ (—)	56, 58
α-C$_{10}$H$_7$ (78)	97
β-C$_{10}$H$_7$ (—)	58
2-Cl-1-C$_{10}$H$_6$ (—)	57
1-Cl-2-C$_{10}$H$_6$ (—)	57
3-Br-2-C$_{10}$H$_6$ (—)	57
o-HO$_2$CC$_6$H$_4$ (good)	56
o-CH$_3$O$_2$CC$_6$H$_4$ (81)	97
p-HO$_2$CC$_6$H$_4$ (—)	58
p-C$_2$H$_5$O$_2$CC$_6$H$_4$ (—)	58
m-CH$_3$COC$_6$H$_4$ (84)	97
p-CH$_3$COC$_6$H$_4$ (—)	58
2-Chlorobenzoquinone	
2-Cl, 6-C$_6$H$_5$ (54); 2-Cl, 3-C$_6$H$_5$ (30)	158
2-Cl, 5-C$_6$H$_5$ (—)	56
2-Cl, 6-p-ClC$_6$H$_4$ (66); 2-Cl, 3-p-ClC$_6$H$_4$ (18)	158

Note: References 142 to 161 are on p. 260.

* The product was accompanied by diaryl and/or polyaryl quinones.
† This product was prepared by the action of an N-nitrosoacetanilide upon the quinone.

TABLE X—Continued

QUINONES

A. Benzoquinone Derivatives—Continued

Starting Quinone	Substituent(s) in Product Benzoquinone (Yield, %)	References
2,3-Dichlorobenzoquinone	2,3-Cl$_2$, 5-C$_6$H$_5$ (69)	158
	2,3-Cl$_2$, 5-p-ClC$_6$H$_4$ (81)	158
2,5-Dichlorobenzoquinone	2,5-Cl$_2$, 3-C$_6$H$_5$*† (17)	69, 58
	2,5-Cl$_2$, 3-p-CH$_3$OC$_6$H$_4$*† (—)	69, 71
	2,5-Cl$_2$, 3-p-CH$_3$O$_2$CC$_6$H$_4$*† (—)	69
	2,5-Cl$_2$, 3-p-ClC$_6$H$_4$*† (—)	69
2,6-Dichlorobenzoquinone	2,6-Cl$_2$, 3-p-ClC$_6$H$_4$ (72)	158
2,5-Dihydroxybenzoquinone	2,5-(HO)$_2$, 3-C$_6$H$_5$, 6-C$_6$H$_5$N=N (28)	159
	2,5-(HO)$_2$, 3-m-CH$_3$C$_6$H$_4$, 6-m-CH$_3$C$_6$H$_4$N=N (48)	159
	2,5-(HO)$_2$, 3,6-(o-CH$_3$C$_6$H$_4$)$_2$ (32)	159
2,5-Dimethylbenzoquinone	2,5-(CH$_3$)$_2$, 3-m-O$_2$NC$_6$H$_4$ (—)	58
2-Chloro-6-phenylbenzoquinone	2-Cl, 3,6-(C$_6$H$_5$)$_2$ (20)	158
o-Chlorophenylbenzoquinone	2,5-(o-ClC$_6$H$_4$)$_2$ (42)	160
m-Chlorophenylbenzoquinone	2,5-(m-ClC$_6$H$_4$)$_2$ (46)	160
p-Chlorophenylbenzoquinone	2,5-(p-ClC$_6$H$_4$)$_2$ (43)	160
2-Chloro-6-p-chlorophenylbenzoquinone	2-Cl, 3,6-(p-ClC$_6$H$_4$)$_2$ (—)	158
o-Bromophenylbenzoquinone	2,5-(o-BrC$_6$H$_4$)$_2$ (40)	160
‡‡	2-C$_6$H$_5$, 5-m-BrC$_6$H$_4$ (32)	160
‡‡	2-C$_6$H$_5$, 5-o-CH$_3$OC$_6$H$_4$ (20)	160
‡‡	2-C$_6$H$_5$, 5-p-CH$_3$OC$_6$H$_4$ (32)	160
‡‡	2-p-CH$_3$C$_6$H$_4$, 5-p-CH$_3$OC$_6$H$_4$† (—)	68
‡‡	2-C$_6$H$_5$, 5-p-CH$_3$OC$_6$H$_4$† (—)	68
‡‡	2-C$_6$H$_5$, 5-β-C$_{10}$H$_7$ (36)	160
o-Carbomethoxyphenylbenzoquinone	2,5-(o-CH$_3$O$_2$CC$_6$H$_4$)$_2$ (38)	160
¶¶	2,5-Cl$_2$, 3-C$_6$H$_5$, 6-p-CH$_3$OC$_6$H$_4$ (—)	70
¶¶	2,5-Cl$_2$, 3-C$_6$H$_5$, 6-p-CH$_3$C$_6$H$_4$ (—)	70
¶¶	2,5-Cl$_2$, 3-p-CH$_3$OC$_6$H$_4$, 6-p-CH$_3$C$_6$H$_4$ (—)	70
2,5-Dichloro-3-p-tolylbenzoquinone	2,5-Cl$_2$, 3-p-CH$_3$C$_6$H$_4$, 6-[3,4-(CH$_3$O)$_2$C$_6$H$_3$] (29)	71

B. Naphthoquinones

Starting Naphthoquinone	Substituent(s) in Product Naphthoquinone (Yield, %)	References
1,2-Naphthoquinone ($C_{10}H_6O_2$)	3,4-(p-$HO_2CC_6H_4$)$_2$ (—)	58
1,4-Naphthoquinone ($C_{10}H_6O_2$)	2-C_6H_5 (poor)*†	58, 54
	2-o-$O_2NC_6H_4$ (0)	58
	2-m-$O_2NC_6H_4$ § (—)	58
	2-p-$O_2NC_6H_4$ § (50)	60, 56
	2-p-$HO_2CC_6H_4$ (—)	58
	2-[2,6-(CH_3)$_2C_6H_3$] (0)	58
	2-α-$C_{10}H_7$ (0)	58
2-Hydroxy-1,4-naphthoquinone	3-C_6H_5, 2-HO (—)	62, 67
	3-p-FC_6H_4, 2-HO (18)	67
	3-o-ClC_6H_4, 2-HO (low)	67
	3-m-ClC_6H_4, 2-HO (20)	67
	3-p-ClC_6H_4, 2-HO (30)	67
	3-(2,4-$Cl_2C_6H_3$), 2-HO (20)	67
	3-(2,5-$Cl_2C_6H_3$), 2-HO (20)	67
	3-o-BrC_6H_4, 2-HO (20)	67
	3-m-BrC_6H_4, 2-HO (20)	67
	3-p-BrC_6H_4, 2-HO (18-31)	67

Note: References 142 to 161 are on p. 260.

* The product was accompanied by diaryl and/or polyaryl quinones.
† This product was prepared by the action of an N-nitrosoacetanilide upon the quinone.
‡ The authors did not specify which of the two possible pairs of starting compounds (monoarylquinone and diazonium salt) was employed to prepare this product.
§ Copper powder was beneficial in this reaction.
¶ A monoaryl-2,5-dichloroquinone and a nitrosoacetanilide were used in this reaction. The author did not specify which aryl group in the product came from the quinone and which from the nitrosoacetanilide.

TABLE X—Continued

QUINONES

B. Naphthoquinones—Continued

Starting Naphthoquinone	Substituent(s) in Product Naphthoquinone (Yield, %)	References
2-Hydroxy-1,4-naphthoquinone (continued)	3-p-IC$_6$H$_4$, 2-HO (11)	67
	3-m-O$_2$NC$_6$H$_4$, 2-HO (low)‖	67
	3-p-O$_2$NC$_6$H$_4$, 2-HO (low)‖	67
	3-o-CH$_3$OC$_6$H$_4$, 2-HO (0)	67
	3-p-CH$_3$OC$_6$H$_4$, 2-HO (6)	67, 62
	3-p-C$_2$H$_5$OC$_6$H$_4$, 2-HO (9)	67
	3-p-HO$_3$SC$_6$H$_4$, 2-HO (—)	62
	3-p-H$_2$NO$_2$SC$_6$H$_4$, 2-HO (27)	67
	3-p-(2-Pyridyl)HNO$_2$SC$_6$H$_4$, 2-HO (20)	67
	3-p-H$_2$NC(=NH)HNO$_2$SC$_6$H$_4$, 2-HO (20)	67
	3-p-(2-Thiazolyl)HNO$_2$SC$_6$H$_4$, 2-HO (20)	67
	3-p-(2-Pyrimidyl)HNO$_2$SC$_6$H$_4$, 2-HO (20)	67
	3-p-H$_2$O$_3$AsC$_6$H$_4$, 2-HO (0)	67
	3-p-C$_6$H$_5$N=NC$_6$H$_4$, 2-HO (0)	67
	3-o-CH$_3$C$_6$H$_4$, 2-HO (66)	62, 67
	3-m-CH$_3$C$_6$H$_4$, 2-HO (10)	67
	3-p-CH$_3$C$_6$H$_4$, 2-HO (—)	62
	3-[2,4-(CH$_3$)$_2$C$_6$H$_3$], 2-HO (11)	67
	3-[2'-CH$_3$-5-i-C$_3$H$_7$C$_6$H$_3$], 2-HO (0)	67
	3-(p-t-C$_5$H$_{11}$C$_6$H$_4$), 2-HO (0)	67
	3-(2-CH$_3$-4-ClC$_6$H$_3$), 2-HO (7)	67
	3-(2-CH$_3$-4-BrC$_6$H$_3$), 2-HO (21)	67
	3-p-C$_6$H$_5$C$_6$H$_4$, 2-HO (20)	67

3-α-$C_{10}H_7$, 2-HO (10)	67
3-β-$C_{10}H_7$, 2-HO (—)	62
3-(4-Br-1-$C_{10}H_6$), 2-HO (0)	67
3-(2-Fluorenyl), 2-HO (trace)	67
3-(3-Acenaphthenyl), 2-HO (0)	67
3-(2-Dibenzofuranyl), 2-HO (0)	67
3-(2-CH_3-1-anthraquinonyl), 2-HO (0)	67
3-(o-$HO_2CC_6H_4$), 2-HO (—)	62
3-(p-$HO_2CC_6H_4$), 2-HO (—)	62
3-(o-$CH_3O_2CC_6H_4$), 2-HO (0)	67
3-(p-$CH_3COC_6H_4$), 2-HO (20)	67

2-Methoxy-1,4-naphthoquinone**

3-C_6H_5, 2-CH_3O (0)	66, 58
3-p-ClC_6H_4, 2-CH_3O (—)	66
3-p-$O_2NC_6H_4$, 2-CH_3O (41)	66
3-p-$CH_3OC_6H_4$, 2-CH_3O (0)	66
3-p-$HO_2CC_6H_4$, 2-CH_3O (—)	66
3-(3-HO-4-$HO_2CC_6H_3$), 2-CH_3O (—)	66

2-Methyl-1,4-naphthoquinone**

3-m-$O_2NC_6H_4$, 2-CH_3 (—)	60
3-p-$O_2NC_6H_4$, 2-CH_3 (—)	60
3-p-$CH_3OC_6H_4$, 2-CH_3 (—)	60
3-p-$CH_3C_6H_4$, 2-CH_3 (—)	60

2,6-Dimethyl-1,4-naphthoquinone

3-C_6H_5, 2,6-$(CH_3)_2$ (poor)	58

‖ The arylating agent was a diaroyl peroxide.
** Attempts to arylate this quinone with tetrazotized benzidine yielded only polymer.[66]

TABLE XI
Hydroquinone

Substituent in Hydroquinone Product (Yield, %)	References
2-Phenyl (0)	61
2-p-Bromophenyl* (—)	61
2-o-Nitrophenyl† (26)	64
2-m-Nitrophenyl (12–15)	61
2-p-Nitrophenyl (89)	64, 61
2-(2′,4′-Dinitrophenyl) (87 crude)	64
2-p-Carbethoxyphenyl* (50–55)	61

* The quinhydrone was also formed.
† This product was accompanied by a 23% yield of diarylhydroquinones.

TABLE XII
Coumarins

Coumarin	Arylated Coumarin	Substituents in Coumarin Product (Yield, %)	References
Coumarin		3-Phenyl (60)*	1
		3-p-Chlorophenyl (78)*	1
		3-o-Nitrophenyl (11)	1
		3-m-Nitrophenyl (—)	41
		3-p-Nitrophenyl (50)	1, 53
		3-p-Anisyl (66)*	1
		3-p-Acetamidophenyl (28)	1
		3-p-Sulfophenyl (58)*	1
		3-p-Arsonophenyl (55)	16, 15
		3-p-Arsenosophenyl (—)	15
		3-β-Naphthyl (30)*	1
		3-p-Carboxyphenyl (82)*	1
7-Hydroxycoumarin		3-p-Chlorophenyl-7-hydroxy (48)	17
4-Methyl-7-hydroxycoumarin		3-p-Chlorophenyl-4-methyl-7-hydroxy (small)	17
		3-p-Bromophenyl-4-methyl-7-hydroxy (—)	17
		3-p-Anisyl-4-methyl-7-hydroxy (very poor)	17

TABLE XIII
MISCELLANEOUS

Unsaturated Compound	Product (Yield, %)	References
A. Bisdiazonium Salts*†		
CH_2=CHCN	$p\text{-}C_6H_4(CH_2CHCN)_2$ (36)	84
	$CH_2(C_6H_4CH_2CHClCN\text{-}p)_2$ (—)	4
B. Nitroölefins		
C_6H_5CH=$CHNO_2$‡	$C_6H_5CH_2COC_6H_4NO_2\text{-}p$ (—), C_6H_5CH=$CHC_6H_4NO_2\text{-}p$ (—)	93
C. Vinyl Ethers§		
(pyranone with CH_2OH, HO, Ar substituents)	Ar = $p\text{-}ClC_6H_4$ (—)	161
	$p\text{-}O_2NC_6H_4$ (—)	
	$2,4\text{-}(O_2N)_2C_6H_3$ (0)	

Note: References 142 to 161 are on p. 260.

* Müller[3, 4] refers to the reaction of tetrazotized benzidine. dichlorobenzidine, 4,4′-diaminodiphenylmethane, diaminodimethyldiphenylmethane, and 4,4′-diaminodiphenylsulfone with acrylonitrile, acrylic acid, and methyl vinyl ketone. No details of the reactions or properties of the products are given.

† The reactions of tetrazotized 2,2′-diaminobiphenyl with maleimide and tetrazotized benzidine with N-isopropylmaleimide gave products that could not be purified.[34]

‡ On treatment with $p\text{-}O_2NC_6H_4N_2Cl$, the aliphatic nitro group was lost, perhaps as a result of a Nef reaction.

§ Exposure of alkyl vinyl ethers to diazonium salts in the absence of copper salts led to azo coupling.[107]

256

ORGANIC REACTIONS

TABLE XIII—Continued

MISCELLANEOUS

Unsaturated Compound	Product (Yield, %)	References
	D. Active Methylene Compounds	
CH$_3$NO$_2$‖	C$_6$H$_5$CH$_2$NO$_2$ (—)	105
	p-CH$_3$OC$_6$H$_4$CH$_2$NO$_2$ (—)	105
	p-CH$_3$C$_6$H$_4$CH$_2$NO$_2$ (—)	105
CH$_2$(CO$_2$C$_2$H$_5$)$_2$	C$_6$H$_5$CH$_2$CO$_2$C$_2$H$_5$¶ (—)	104
	E. Oximes and Semicarbazones	
CH$_2$=NOH**	C$_6$H$_5$CHO (40)	100
	o-ClC$_6$H$_4$CHO (52)	100
	m-ClC$_6$H$_4$CHO (50)	100
	p-ClC$_6$H$_4$CHO (60)	100
	o-O$_2$NC$_6$H$_4$CHO (33)	100
	o-HOC$_6$H$_4$CHO (9)	100
	o-CH$_3$OC$_6$H$_4$CHO (34)	100
	p-CH$_3$OC$_6$H$_4$CHO (42)	100
	o-CH$_3$C$_6$H$_4$CHO (46)	100
	m-CH$_3$C$_6$H$_4$CHO (41)	100
	p-CH$_3$C$_6$H$_4$CHO (46)	100
	o-C$_6$H$_5$C$_6$H$_4$CHO (very poor)	100
	β-C$_{10}$H$_7$CHO (25)	100
	3-Pyridylcarboxaldehyde (14)	100
	o-C$_2$H$_5$O$_2$CC$_6$H$_4$CHO (0)	100
	p-C$_2$H$_5$O$_2$CC$_6$H$_4$CHO (20)	100
	o-NCC$_6$H$_4$CHO (0)	100
	p-OHCC$_6$H$_4$C$_6$H$_4$CHO-p (very poor)	100
	p-OHCC$_6$H$_4$OC$_6$H$_4$CHO-p (very poor)	100

THE MEERWEIN ARYLATION REACTION

Compound	Product	Ref.
$CH_3CH=NOH$**	$o\text{-}ClC_6H_4COCH_3$ (43)	100
	$p\text{-}ClC_6H_4COCH_3$ (35–45)	100
	$3,4\text{-}(HO_2C)_2C_6H_3COCH_3$ (27)	100
	$m\text{-}C_6H_4(COCH_3)_2$†† (27)	100
	$p\text{-}C_6H_4(COCH_3)_2$†† (33)	100
	$4\text{-}ClC_6H_3(COCH_3)_2\text{-}1,3$†† (15)	100
$CH_3CH=NNHCONH_2$	$p\text{-}ClC_6H_4C(CH_3)=NNHCONH_2$ (40)	100
$CH_3CH_2CH=NOH$**	$p\text{-}ClC_6H_4COCH_2CH_3$ (30)	100
$C_6H_5CH=NOH$	$p\text{-}ClC_6H_4COC_6H_5$ (0)	100
$CH_3COCH=NOH$	$CH_3COC(=NOH)C_6H_5$ (82)	103, 101, 102
	$CH_3COC(=NOH)C_6H_4Cl\text{-}m$ (—)	103
	$CH_3COC(=NOH)C_6H_4Cl\text{-}p$ (—)	103, 102
	$CH_3COC(=NOH)C_6H_4NO_2\text{-}p$ (—)	103
	$CH_3COC(=NOH)C_6H_4OCH_3\text{-}o$ (22)	101
	$CH_3COC(=NOH)C_6H_4OCH_3\text{-}p$ (50)	102, 101
	$CH_3COC(=NOH)C_6H_4OC_2H_5\text{-}p$ (65)	102
	$CH_3COC(=NOH)C_6H_4NHCOCH_3\text{-}p$ (—)	103
	$CH_3COC(=NOH)C_6H_3OCH_3\text{-}3\text{-}NHCOCH_3\text{-}4$ (—)	103
	$CH_3COC(=NOH)C_6H_4CH_3\text{-}o$ (—)	103
	$CH_3COC(=NOH)C_6H_4CH_3\text{-}p$ (60)	102, 101
	$CH_3COC(=NOH)C_6H_3(CH_3)_2\text{-}2,4$ (—)	101
	$CH_3COC(=NOH)C_{10}H_7\text{-}\beta$ (40)	102
	1-Oximino-1-(3′-pyridyl)-2-propanone (60)	103
	$CH_3COC(=NOH)C_6H_4CO_2C_2H_5\text{-}p$ (70)	103

|| With $p\text{-}ClC_6H_4N_2Cl$ or $p\text{-}O_2NC_6H_4N_2Cl$, the product is the hydrazone resulting from conventional azo coupling.
¶ The product was isolated by hydrolysis, decarboxylation, and re-esterification.
** The product was isolated as the aldehyde or ketone after hydrolysis of the oxime. Control experiments[125] showed a loss of about 15% during hydrolysis and purification.
†† The appropriate aminoacetophenone was diazotized and allowed to react with acetaldoxime.

TABLE XIII—Continued

MISCELLANEOUS

E. Oximes and Semicarbazones—Continued

Unsaturated Compound	Product (Yield, %)	References
$C_6H_5COCH=NOH$	$C_6H_5COC(=NOH)C_6H_5$ (—)	98, 101
	$C_6H_5COC(=NOH)C_6H_4NO_2$-$p$ (—)	98
	$C_6H_5COC(=NOH)C_6H_4OC_2H_5$-$p$ (—)	98
p-$O_2NC_6H_4COCH=NOH$**	p-$O_2NC_6H_4COCOC_6H_4NO_2$-p (—)	98
	p-$O_2NC_6H_4COCOC_6H_4OC_2H_5$-$p$ (—)	98
3-Pyridylglyoxal monoxime	3-$C_5H_4NCOC(=NOH)C_6H_5$ (—)	99
	3-$C_5H_4NCOC(=NOH)C_6H_4NO_2$-$p$ (—)	99
	3-$C_5H_4NCOC(=NOH)C_6H_4OC_2H_5$-$p$ (—)	99
	3-$C_5H_4NCOC(=NOH)C_6H_4CH_3$-$p$ (—)	99
4-Pyridylglyoxal monoxime	4-$C_5H_4NCOC(=NOH)C_6H_4CH_3$-$p$ (—)	99

F. Furfural

Substituents in Furfural (Yield, %)	References
5-Phenyl (49)	48, 51
5-p-Chlorophenyl (90)	48, 51, 141
5-o-Nitrophenyl (—)	51
5-p-Nitrophenyl (96)	48, 51, 141

5-*p*-Sulfophenyl (44)	48, 51
5-*p*-Anisyl (—)	51
5-*p*-Aminophenyl (—)	51
5-*p*-Acetamidophenyl (—)	51
5-*p*-Dimethylaminophenyl (—)	51
5-*p*-Tolyl (—)	51
5-*p*-Carboxyphenyl (—)	141
5-(2-Anthraquinonyl) (—)	51
5-(2-Hydroxy-1-naphthyl-4-sulfonic acid) (—)	51

OHC—⟨furan⟩—C₆H₄—CH=CH—C₆H₄—⟨furan⟩—CHO (—)   51

G. *Furoic Acid*

Product (Yield, %)	References
5-Phenylfuroic Acid (60)	48
5-*p*-Chlorophenylfuroic acid (93)	48, 141
5-*p*-Nitrophenylfuroic acid (96)	48, 141

** The product was isolated as the aldehyde or ketone after hydrolysis of the oxime. Control experiments showed a loss of about 15% during hydrolysis and purification.

REFERENCES FOR TABLES I–XIII

[142] Benkeser, Bennett, and Hickner, *J. Am. Chem. Soc.*, **79**, 6253 (1957); Benkeser and Hickner, *ibid.*, **80**, 5298 (1958).
[143] Dombrovskiĭ and Terent'ev, *Zhur. obshcheĭ Khim.*, **26**, 2776 (1956); *J. Gen. Chem. U.S.S.R. (Engl. Transl.)*, **26**, 3091 (1956) [*C.A.*, **51**, 7337 (1957)].
[144] Dombrovskiĭ and Terent'ev, *Zhur. obshcheĭ Khim.*, **27**, 2000 (1957); *J. Gen. Chem. U.S.S.R. (Engl. Transl.)*, **27**, 2058 (1957).
[145] Ropp and Raaen, *J. Am. Chem. Soc.*, **76**, 4484 (1954).
[146] Taylor and Strojny, *J. Am. Chem. Soc.*, **78**, 5104 (1956).
[147] Broadhead and Pauson, *J. Chem. Soc.*, **1955**, 367.
[148] Rosenblum, Abstr. Am. Chem. Soc. Meeting, San Francisco, April 1958, p. 5N.
[149] Bernauer and Schmidt, *Ann.*, **591**, 153 (1955).
[150] Dombrovskiĭ, Terent'ev, and Yurkevich, *Zhur. Obshcheĭ Khim.*, **27**, 419 (1957); *J. Gen. Chem. U.S.S.R. (Engl. Transl.)*, **27**, 473 (1957) [*C.A.*, **51**, 15454g (1957)].
[151] Drefahl and Hartrodt, *J. prakt. Chem.* [4] **4**, 124 (1956).
[152] Bell and Waring, *J. Chem. Soc.*, **1948**, 1024.
[153] Fuson and Cooke, *J. Am. Chem. Soc.*, **62**, 1180 (1940).
[154] Drefahl, Hochbarth, and Möller, *J. prakt. Chem.* [4] **4**, 130 (1956).
[155] Bachman and Hoaglin, *J. Org. Chem.*, **8**, 300 (1943).
[156] A. Johnson, private communication.
[157] Wieland et al., *Ann.*, **514**, 148 (1934); Betterton and Waters, *J. Chem. Soc.*, **1953**, 329.
[158] Brassard and L'Écuyer, *Can. J. Chem.*, **36**, 814 (1958).
[159] Brassard and L'Écuyer, *Can. J. Chem.*, **36**, 1346 (1958).
[160] Brassard and L'Écuyer, *Can. J. Chem.*, **36**, 709 (1958).
[161] Quilico and Musante, *Gazz. chim. ital.*, **74**, 26 (1944).

CHAPTER 4

THE FAVORSKIĬ REARRANGEMENT OF HALOKETONES

ANDREW S. KENDE

Lederle Laboratories Division, American Cyanamid Co.

CONTENTS

	PAGE
NATURE OF THE REACTION	262
MECHANISM AND STEREOCHEMISTRY	263
Unsymmetrical Mechanisms	263
Symmetrical Mechanisms	265
Stereospecificity	267
SCOPE AND LIMITATIONS	270
Acyclic Monohaloketones	270
Alicyclic Monohaloketones	271
Aralkyl Monohaloketones	274
Steroid Monohaloketones	275
Dihaloketones	277
Trihaloketones	282
EXPERIMENTAL CONDITIONS	283
Side Reactions	283
Nature of the Halogen	285
Choice of Base and Solvent	286
Table I. Reaction of $(CH_3)_2CBrCOCH_3$	286
Table II. Yields of Rearrangement Acid Using Various Alkoxide-Solvent Pairs	287
Reaction Time and Temperature	288
EXPERIMENTAL PROCEDURES	289
Methyl Cyclopentanecarboxylate	289
Ethyl Trimethylacetate	289
Ethyl 3,3-Diphenylpropionate	289
Cyclohexanecarboxylic Acid	290
Methyl 3-Methyl-2-butenoate	290
20-Bromo-17(20)-pregnen-3β-ol-21-oic Acid	290

		PAGE
Tabular Survey		291
Table III.	Acyclic Monohaloketones	292
Table IV.	Alicyclic Monohaloketones	294
Table V.	Aralkyl Monohaloketones	302
Table VI.	Steroid Monohaloketones	307
Table VII.	Dihaloketones	310
Table VIII.	Trihaloketones	315

NATURE OF THE REACTION

The Favorskiĭ rearrangement is the skeletal rearrangement of α-halogenated ketones in the presence of certain nucleophilic bases, such as hydroxides, alkoxides, or amines, to give carboxylic acid salts, esters, or amides, respectively. Monohaloketones undergo the reaction to yield derivatives of saturated acids having the same number of carbon atoms.

$$(CH_3)_2CBrCOCH_3 + {}^\ominus OCH_3 \rightarrow (CH_3)_3CCO_2CH_3 + Br^\ominus$$

In a similar manner, suitable dihaloketones produce unsaturated carboxylic acids.

$$CH_3CCl_2COCH_3 + 2OH^\ominus \rightarrow CH_2{=}C(CH_3)CO_2H + 2Cl^\ominus$$

Analogous rearrangement of trihaloketones can give rise to unsaturated halo acids.

$$(CH_3)_2CBrCOCHBr_2 + 2OH^\ominus \rightarrow (CH_3)_2C{=}CBrCO_2H + 2Br^\ominus$$

Since the description of this rearrangement by Favorskiĭ[1] in 1894, successive investigations have largely clarified its scope, mechanism, and, more recently, its stereochemistry. Accordingly, the Favorskiĭ rearrangement has become an increasingly reliable and specialized instrument of organic synthesis. The reaction has found application for the preparation of highly branched acyclic carboxylic acids. It is a preferred route to various 1-substituted cycloalkanecarboxylic acids, and provides a direct method for ring contraction in simple alicyclic systems and in the steroids. Other typical applications include its use in the modification of the ring-D side chain of steroids and in the stereospecific synthesis of 8-methyl-1-hydrindone.

A review of the Favorskiĭ rearrangement, covering the literature through 1949, has been published.[2]

---

[1] Favorskiĭ, *J. Russ. Phys.-Chem. Soc.*, **26**, 559 (1894); *J. prakt. Chem.*, [2] **51**, 533 (1895).
[2] Jacquier, *Bull. soc. chim. France*, [5] **17**, D35 (1950).

MECHANISM AND STEREOCHEMISTRY

Five fundamental mechanisms have been advanced to account for the Favorskiĭ rearrangement. These are discussed here with immediate reference to the action of alkoxides on α-monohaloketones, but their extension to other bases or to polyhaloketones will be evident.

Unsymmetrical Mechanisms

The rearrangement was considered by Favorskiĭ[3] to proceed by addition of alkoxide to the carbonyl carbon, with concomitant ejection of halide ion, to produce an epoxyether (I), followed by rearrangement to product.

$$\begin{matrix} \overset{\ominus}{O}R \\ R_1-C=O \\ R_2-CH-X \end{matrix} \xrightarrow{-X^{\ominus}} \begin{bmatrix} R_1-C\overset{OR}{\underset{O}{\diagup}} \\ R_2-CH \end{bmatrix} \longrightarrow \begin{matrix} OR \\ | \\ C=O \\ | \\ R_1-CH-R_2 \end{matrix}$$

$$\text{I}$$

Although the isolation of epoxyethers from the action of alkoxides on certain α-haloketones is well established, the postulated rearrangement of the epoxyether I into product is inherently improbable.* Such a transformation is experimentally precluded by failure to effect this rearrangement starting with pure epoxyethers under a variety of conditions. Thus the epoxyether intermediate is clearly not involved in the main course of the Favorskiĭ reaction, although it plays a central role in the formation of certain by-products.

A second mechanism, that of Richard,[4] envisions the action of base on α-haloketones to involve abstraction of hydrogen halide, either by simultaneous α-elimination[5] or by loss of halide from a mesomeric enolate anion. The resulting species II would rearrange directly to the ketene

$$\begin{matrix} R_1-C=O \\ | \\ R_2-CHX \end{matrix} \xrightarrow{-HX} \begin{bmatrix} R_1-C-O^{\ominus} & R_1-C=O \\ \parallel & | \\ R_2-C^{\oplus} & R_2-C: \end{bmatrix} \rightarrow \begin{matrix} O \\ \parallel \\ C \\ \parallel \\ R_1-C-R_2 \end{matrix}$$

$$\text{II} \qquad\qquad\qquad \text{III}$$

---

[3] Favorskiĭ, *J. prakt. Chem.*, [2] **88**, 641 (1913).

* The formation and reactions of these epoxyethers are outlined in the discussion of side reactions.

[4] Richard, *Compt. rend.*, **197**, 1432 (1933).

[5] Hine, *Physical Organic Chemistry*, pp. 131–133, 188, McGraw-Hill, New York, 1956.

III, which would rapidly react with the nucleophile to give product.[6] This mechanism fails to accommodate those numerous examples of the Favorskiĭ rearrangement that produce esters of the trialkylacetic type, which cannot arise from a ketene precursor.

A third mechanism has seemed particularly attractive because of its analogy to the benzilic acid rearrangement. This semibenzilic mechanism

$$\begin{array}{c} {}^{\ominus}\text{OR} \\ R_1-C{=}O \\ R_2-CH-X \end{array} \longrightarrow \left[\begin{array}{c} \text{OR} \\ R_1-C-O^{\ominus} \\ R_2-CH-X \end{array} \right] \longrightarrow \begin{array}{c} \text{OR} \\ C{=}O \\ R_1-CH-R_2 \end{array}$$

features addition of alkoxide to the carbonyl carbon atom of the haloketone, followed by a concerted displacement of halide ion by the 1,2-migration of an alkyl group with its electron pair.[7]

A common feature of each of the three preceding mechanisms is their prediction that the rearrangement product of a given α-haloketone would be different from that derived from its α'-halogenated isomer.* For example, 1-chloro-3-phenylacetone (IV) should, according to any of the above pathways, give rise to 3-phenylpropionic acid (V), while 1-chloro-1-phenylacetone (VI) should rearrange exclusively to 2-phenylpropionic acid (VII). It is found, however, that both haloketones IV and VI yield

$$\begin{array}{ccc} C_6H_5CH_2COCH_2Cl & \longrightarrow & C_6H_5CH_2CH_2CO_2H \\ \text{IV} & & \text{V} \\ \\ C_6H_5CHClCOCH_3 & \xrightarrow{\times} & C_6H_5CH(CH_3)CO_2H \\ \text{VI} & & \text{VII} \end{array}$$

the same acid, V, and that such a result normally occurs.[8] Evidently the preceding mechanisms, which would maintain a given positional asymmetry from starting haloketone to product, are untenable without appropriate modification.

[6] Horner, Spietschka, and Gross, *Ann.*, **573**, 17 (1951); *Ber.*, **85**, 225 (1952).
[7] Tchoubar and Sackur, *Compt. rend.*, **208**, 1020 (1939).
* The prefixes α and α' will be used to differentiate the two carbon atoms which are adjacent to the carbonyl function of a haloketone. The halogen substituent of a monohaloketone is regarded as being on the α-carbon atom.
[8] McPhee and Klingsberg, *J. Am. Chem. Soc.*, **66**, 1132 (1944).

Symmetrical Mechanisms

One rationalization of the above observations would require halogen migration from the α- to the α'-carbon atom.[9,10] Relevant here are such reactions as the solvolysis of 3-bromo-1,1-diphenylacetone to 1-hydroxy-1,1-diphenylacetone,[11] the reaction of α-chloroacetoacetic ester with ethanolic potassium cyanide to form both α- and γ-cyanoacetic esters,[12] and the conversion of 2α-bromocholestan-3-one to both the 2α- and 4α-acetoxycholestan-3-ones by potassium acetate in acetic acid.[13] Alternatively, McPhee and Klingsberg postulate a carbonium ion mechanism in which a haloketone such as VI undergoes unimolecular dissociation (a) to a carbonium ion VIII which can tautomerize (b) through a common enol IX to the isomeric carbonium ion X.[8] The latter can then undergo rearrangement (c) to the acid V. The carbonium ion mechanism largely

(a) $C_6H_5CHClCOCH_3 \rightarrow [C_6H_5\overset{\oplus}{C}HCOCH_3]$
              VI   VIII

(b) $[C_6H_5\overset{\oplus}{C}HCOCH_3] \rightleftharpoons [C_6H_5\overset{\oplus}{C}HC(OH)=CH_2] \rightleftharpoons [C_6H_5CH_2\overset{\oplus}{C}OCH_2]$
            VIII   IX   X

(c) $[C_6H_5CH_2\overset{\oplus}{C}OCH_2] \rightarrow [C_6H_5CH_2CH_2\overset{\oplus}{C}O] \rightarrow C_6H_5CH_2CH_2CO_2H$
              X   V

lacks analogy and has the drawback that no key role is assigned to the base which is a normal requisite of the Favorskiĭ rearrangement.

The generality of any of the preceding mechanisms was disproved in 1950 by the elegant work of Loftfield.[14] A study was made of the rearrangement of C^{14}-labeled 2-chlorocyclohexanone, a structure which did not preclude the operation of any of the postulated mechanisms. The rearrangement of this chloroketone in dilute ethanolic sodium ethoxide was shown to follow essentially first-order kinetics with respect to both haloketone and alkoxide. When 2-chlorocyclohexanone-1,2-C^{14}, in which the isotope was equally distributed between carbon atoms 1 and 2, was treated with less than one equivalent of sodium isoamyloxide in isoamyl alcohol, the principal product was isoamyl cyclopentanecarboxylate, accompanied by some recovered chloroketone. Careful stepwise

[9] Richard, *Compt. rend.*, **200**, 1944 (1935).
[10] Wendler, Graber, and Hazen, *Chem. & Ind. (London)*, **1956**, 847; *Tetrahedron*, **3**, 144 (1958).
[11] Stevens and Lenk, *Org. Chem. Abstr., XIIth Congr. Intern. Union Pure and Appl. Chem.*, 1951, p. 470.
[12] Hantzsch and Schiffer, *Ber.*, **25**, 728 (1892).
[13] Fieser and Romero, *J. Am. Chem. Soc.*, **75**, 4716 (1953).
[14] Loftfield, *J. Am. Chem. Soc.*, **72**, 632 (1950); **73**, 4707 (1951).

degradation of both the ester and the haloketone established that the recovered chloroketone had the same isotope distribution as starting material, and that the radiocarbon in the ester fraction was distributed 50% on the carboxyl carbon atom, 25% on the ring α-carbon atom, and 25% on the two ring β-carbon atoms.

The preceding facts clearly exclude any reversible halogen migration in a rearrangement of this type, and necessarily rule out significant participation by any of the mechanisms so far discussed. The data are compatible, however, with any reaction intermediate in which, by reason of symmetry, the α- and α'-carbon atoms of the cyclohexanone are formally equivalent. This criterion is satisfied by a mechanism that involves a cyclopropanone intermediate. (The concept of cyclopropanone intermediates in the reactions of α-haloketones with bases was well established in the German chemical literature prior to 1900.[12,15–17]) According to this view, the initial step is the removal of a proton from the α'-carbon atom to give the haloketone enolate anion XI. Concerted or subsequent ejection of halide ion leads to a cyclopropanone which is rapidly cleaved by alkoxide to give the rearrangement product. In the Loftfield experiment, random cleavage of the cyclopropanone XII, having radiocarbon

distributed as marked, would lead to the isotope distribution observed in the ester fraction.

The Loftfield mechanism resembles the pathways suggested for the rearrangement of α-halosulfones,[18] α-haloacetanilides,[19] and oxime *p*-toluenesulfonates.[20] It is consistent with the known behavior of cyclopropanone derivatives[21,22] and in good agreement with the observed effect of various substituents on the facility and course of the Favorskiĭ

---

[15] Wolff, *Ann.*, **260**, 79 (1890); *Ber.*, **26**, 2220 (1893).
[16] Conrad, *Ber.*, **32**, 1005 (1899).
[17] Pauly and Rossbach, *Ber.*, **32**, 2000 (1899).
[18] Bordwell and Cooper, *J. Am. Chem. Soc.*, **73**, 5187 (1951).
[19] Sarel and Greenberger, *J. Org. Chem.*, **23**, 330 (1958).
[20] Hatch and Cram, *J. Am. Chem. Soc.*, **75**, 38 (1953).
[21] Lipp, Buchkremer, and Seeles, *Ann.*, **499**, 1 (1932).
[22] R. B. Woodward and A. S. Kende, unpublished observations; A. S. Kende, Ph.D. thesis, Harvard University, 1956.

rearrangement. In particular, it leads to the correct prediction that rearrangement of unsymmetrical α-haloketones leads to the product formed through cleavage of the cyclopropanone intermediate so as to give the more stable of the two possible transient carbanions. Stabilities of unconjugated carbanions increase in the order tertiary < secondary < primary < benzyl.[23-25] Thus the cyclopropanone XIII derived from

$$\begin{bmatrix} CH_3 & O \\ & \triangle \\ CH_3 & CH_2 \end{bmatrix} \xrightarrow{ROH} (CH_3)_3CCO_2R$$

XIII

3-bromo-3-methylbutan-2-one opens to the tertiary trimethylacetic ester, forming a transient primary rather than tertiary carbanion.[26] Similarly, the cyclopropanone from 1-chloro-1-phenylacetone opens by way of a benzylic carbanion to give 3-phenylpropionic acid derivatives.[8]

On the basis of the evidence at hand, it is likely that the Favorskiĭ rearrangement normally proceeds by a cyclopropanone mechanism. The few rearrangements which for structural reasons cannot utilize this pathway require special reaction conditions and probably take place through a variant of the semibenzilic mechanism.[27] A "push-pull" modification of the latter has been proposed for the quasi-Favorskiĭ rearrangement of such haloketones on treatment with silver salts.[28]

Stereospecificity

Although the cyclopropanone mechanism has received general acceptance and can often predict the formation of a preferred position isomer, its stereochemical implications are less firmly established. The Loftfield thesis implies that cyclopropanone formation is synchronous with an internal S_N2-type displacement on the halogen-bearing carbon atom with consequent inversion at that center.

[23] Haubein, *Iowa State Coll. J. Science*, **18**, 48 (1943) [*C.A.*, **38**, 716 (1944)].
[24] Bartlett, Friedman, and Stiles, *J. Am. Chem. Soc.*, **75**, 1771 (1953).
[25] G. S. Hammond, in Newman, *Steric Effects of Organic Chemistry*, pp. 439–441, John Wiley & Sons, New York, 1956.
[26] Aston and Greenburg, *J. Am. Chem. Soc.*, **62**, 2590 (1940).
[27] Stevens and Farkas, *J. Am. Chem. Soc.*, **74**, 5352 (1952).
[28] Cope and Graham, *J. Am. Chem. Soc.*, **73**, 4702 (1951).

This view has been questioned by Burr and Dewar on quantum mechanical grounds.[29] The latter suggest that the geometry of the enolate π-orbital is not suitable for effective $S_N 2$-type overlap with the σ-orbital of the halogen-bearing α-carbon atom. Rather, they agree with Aston and Newkirk[30] that loss of halide from the enolate anion precedes cyclopropanone formation, and involves the generation of a species variously represented as a mesomeric zwitterion[30] (XIV) or as a "no-bond" canonical form (XV) of a cyclopropanone.[29, 31] Subsequent collapse of this species to the more stable cyclopropanone would lead to the product.

$$\begin{array}{c}\text{O}\\\|\\\text{C}\\/\quad\backslash\\{}^{\ominus}\text{CH}\quad\text{CHX}\end{array}\xrightarrow{-\text{X}^{\ominus}}\begin{array}{c}\text{O}^{\ominus}\\|\\\text{C}\\/\!/\quad\backslash\\\text{HC}\quad\overset{\oplus}{\text{CH}}\\|\quad\quad|\\\text{XIV}\end{array}\quad\text{or}\quad\begin{array}{c}\text{O}\\\|\\\text{C}\\/\quad\backslash\\\text{HC}\cdot\quad\cdot\text{CH}\end{array}\rightarrow\begin{array}{c}\text{O}\\\|\\\text{C}\\/\quad\backslash\\\text{HC}\text{———}\text{CH}\\|\quad\quad|\\\text{XV}\end{array}$$

The synchronous and nonsynchronous mechanisms are not kinetically distinguishable if enolate formation is rate-determining, but they clearly differ in stereochemical implications. The synchronous process would entail steric inversion with the maintenance of essentially sp^3 hybridization at the halogen-bearing carbon. However, the intermediacy of a discrete species XIV or XV of high resonance energy would predict racemization of the α-carbon atom. The pathways could thus be differentiated by the rearrangement of a suitable optically active haloketone, such as XVI, into a trialkylacetic acid which would indicate by its optical purity the degree of participation of the synchronous as against the nonsynchronous mechanism.

$$\begin{array}{c}\text{C}_4\text{H}_9\text{-}n\\|\\\text{C}_2\text{H}_5\text{—C—COCH}_3\\|\\\text{Cl}\\\text{XVI}\end{array}$$

For some years there has been only meager evidence on this point.[32, 33] Wendler has shown that 17α-bromo-3α-acetoxypregnane-11,20-dione (XVII) of proven configuration gives on rearrangement a

---

[29] Burr and Dewar, *J. Chem. Soc.*, **1954**, 1201.

[30] Aston and Newkirk, *J. Am. Chem. Soc.*, **73**, 3900 (1951).

[31] J. G. Burr, Jr., private communication; Burr, *Tracer Applications for the Study of Organic Reactions*, Interscience, New York, 1957.

[32] Heusser, Engel, and Plattner, *Helv. Chim. Acta*, **33**, 2237 (1950).

[33] W. S. Johnson, M. M. Roth, and D. D. Cameron, unpublished observations; M. M. Roth, Ph.D. thesis, 1951, and D. D. Cameron, Ph.D. thesis, 1953, University of Wisconsin.

3 : 2 mixture of the epimeric 17-methyl-17-carboxylic esters XVIII and XIX, respectively.[10] This result, inexplicable by the synchronous mechanism, was rationalized by invoking bromine migration to C-21 prior to rearrangement, although independent evidence for such a shift was not adduced.

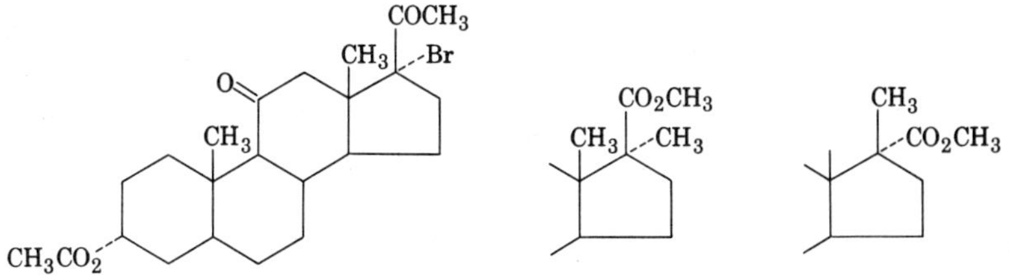

XVII XVIII XIX

A clearcut case of stereospecific rearrangement has recently been demonstrated using the pair of epimeric 1-chloro-1-acetyl-2-methyl-cyclohexanes XX and XXII of proven configuration.[34] Rearrangement of XX with sodium benzyloxide gave a benzyl ester converted by hydrogenolysis into a single 1,2-dimethylcyclohexanecarboxylic acid, XXI.

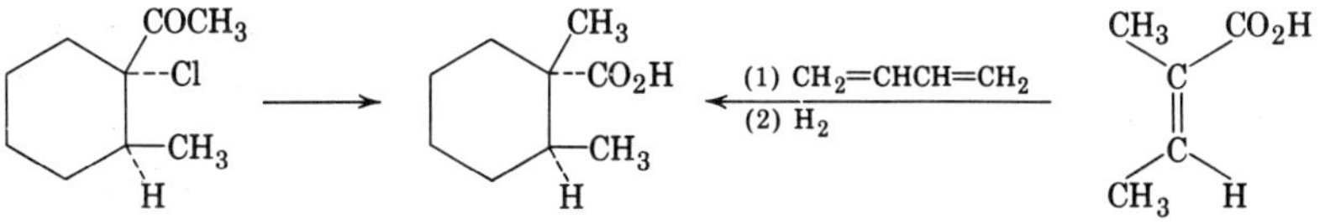

XX XXI

The stereochemistry of this acid was demonstrated by independent synthesis involving the stereospecific Diels-Alder addition of butadiene to tiglic acid.

Rearrangement of the epimeric chloroketone XXII gave in turn exclusively the benzyl ester of the diastereomeric acid XXIII. In addition, the chloroketone XXIV was shown to rearrange to the ester of

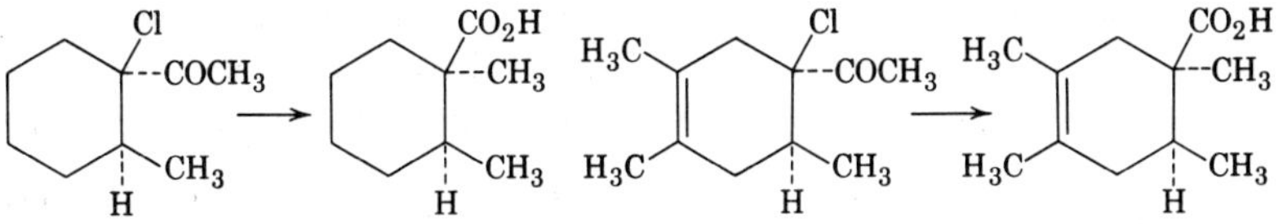

XXII XXIII XXIV XXV

[34] G. Stork and I. Borowitz, *J. Am. Chem. Soc.*, **82** (1960), in press; I. Borowitz, Ph.D. thesis Columbia University, 1956.

XXV, proven to have carboxyl and methyl *cis* by its nonidentity with the adduct of tiglic acid and 2,3-dimethylbutadiene.

These results are consistent with the Loftfield mechanism and suggest that cyclopropanone formation and halide loss are synchronous or very nearly so; as a minimum they would require that any intermediate XIV or XV, if formed, should collapse stereospecifically to a cyclopropanone before the departing halide recedes beyond "shielding" range.[35] However, the zwitterion mechanism may have significance for systems wherein steric barriers retard ring closure in the normal direction and thus allow the halide anion to travel beyond the range of stereoselective electrostatic interaction before the new bond is formed.

SCOPE AND LIMITATIONS

Acyclic Monohaloketones

The Favorskiĭ rearrangement of acyclic α-monohaloketones is particularly sensitive to both structural factors and reaction conditions. Because some of the acyclic haloketones reported in the literature are of uncertain structure, and because of reaction conditions that are not comparable, precise evaluation of the scope of the reaction in the acyclic series is difficult. Certain general structural correlations are nevertheless possible. In accord with the cyclopropanone mechanism, it is observed that the rearrangement becomes more difficult as the rate of proton release from the α'-carbon atom is reduced by increasing alkyl substitution.[14, 36, 37] For example, in the series $(CH_3)_2CBrCOR$, the yield of rearrangement product where R is methyl, ethyl, or *n*-propyl ranges from 39% to 69% (dry alkoxides in ether being used); where R is isopropyl the yield is at most 29%, while where R is *t*-butyl (no α'-hydrogen atom) rearrangement is not observed.[26,38]

Alkyl substituents on the halogen-bearing carbon atom, on the other hand, promote the rearrangement. This has been ascribed to steric hindrance toward competing bimolecular substitution or addition reactions.[39] For this reason, rearrangement of halomethyl alkyl ketones is unfavorable, whereas a number of α-haloisopropyl alkyl ketones do rearrange to give, as a rule, alkyldimethylacetic acids in good yields.

Although the formation of the more fully substituted acetic acids from the above rearrangements is generally observed, instances are known in

---

[35] Ingold, *Structure and Mechanism in Organic Chemistry*, pp. 382–384, Cornell Univ. Press, Ithaca, 1953.

[36] Pearson and Dillon, *J. Am. Chem. Soc.*, **75**, 2439 (1953).

[37] Cardwell, *J. Chem. Soc.*, **1951**, 2442.

[38] Sacks and Aston, *J. Am. Chem. Soc.*, **73**, 3902 (1951).

[39] Aston, Clarke, Burgess, and Greenburg, *J. Am. Chem. Soc.*, **64**, 300 (1942).

which the product formed is the unexpected, less-branched isomer. Thus rearrangement of the bromination product of 2,2,5-trimethylhexan-3-one (XXVI) leads to 93% of the ester XXVII, rather than to the isomer XXVIII.[38] Possibly the steric hindrance to solvation of the carbanion intermediate leading to XXVIII, in which the negative charge is on a particularly hindered neopentyl-type carbon atom, is greater than that required by the intermediate leading to the observed XXVII.

$$CH_3-CH-CO-CH_2-\underset{\underset{CH_3}{|}}{\overset{\overset{CH_3}{|}}{C}}-CH_3 \rightarrow CH_3-CH-CH-\underset{\underset{CH_3}{|}}{\overset{\overset{CO_2CH_3}{|}}{C}}\underset{CH_3}{\overset{CH_3}{\diagdown}}$$
$$\underset{CH_3}{|} \qquad\qquad\qquad\qquad \underset{CH_3}{|}$$
$$\text{XXVI} \qquad\qquad\qquad\qquad \text{XXVII}$$

$$CH_3-\underset{\underset{CH_3}{|}}{\overset{\overset{CO_2CH_3}{|}}{C}}-CH_2-\underset{CH_3}{\overset{CH_3}{\diagdown}}C$$
$$\text{XXVIII}$$

Alicyclic Monohaloketones

The ring contraction of α-halocyclanones to carboxylic acid derivatives of the next lower cycle is an important application of the Favorskiĭ reaction. (Ring contraction of cyclic ketones to carboxylic acids has also been directly achieved in 23–34% yields by use of hydrogen peroxide in the presence of selenium dioxide.[40]) Such rearrangements are usually less sensitive to variations of structure and reaction conditions than in the acyclic series, and thus prove a valuable synthetic route to certain alicyclic intermediates. The reaction is reasonably general for α-halocyclanones in rings of from six to ten carbon atoms. Under appropriate conditions, yields ranging from 40% to 75% can be obtained from the unsubstituted as well as from the majority of alkyl-substituted α-haloketones that have been studied.

A possible limitation would seem to be rearrangement of 2-halo-2-alkylcyclohexanones, two examples of which reportedly fail to undergo the reaction.[34,41] In contrast, 2-chloro-2-methylcycloheptanone gives the expected 1-methylcyclohexanecarboxylic acid in 41% yield.[34]

The rearrangement of 2-bromocyclodecanone in over 75% yield provides a preferred synthesis of cyclononanecarboxylic acid.[42]

[40] Payne and Smith, *J. Org. Chem.*, **22**, 1680 (1957).
[41] Mousseron and Granger, *Bull. soc. chim. France*, [5] **10**, 428 (1943).
[42] Schenker and Prelog, *Helv. Chim. Acta*, **36**, 896 (1953).

A number of α-halogenated acylcycloalkanes undergo rearrangement to derivatives of the corresponding 1-alkylcycloalkanecarboxylic acids. With these haloketones, the position of the halogen has a characteristic effect on the yield of the rearrangement product. The 1-halo-1-acylcycloalkanes (XXIX) tend to rearrange smoothly, while the isomeric halomethyl cycloalkyl ketones (XXXI) do so in lower yield. A striking illustration arises from the set of bromoketones derived from acetylcyclohexane itself. The bromoketone XXIX (X = Br) gives the methyl ester XXX in 79% yield, whereas the isomer XXXI (X = Br) leads only to a side reaction under identical conditions.[33, 43, 44] This difference, which is

    XXIX   XXX   XXXI

less pronounced in the chloro analogs, has been attributed to the relatively slow rate-determining ionization of the tertiary proton in XXXI, which allows competing side reactions to predominate.[14, 43] Of interest in this connection is the rearrangement of the comparatively acidic β-keto ester 6-bromo-2-carbethoxycyclohexanone, which furnishes cyclopentane-1,2-*trans*-dicarboxylic acid in high yield.[44a]

The rearrangement has been adapted to a reaction sequence which serves as a model for the stereospecific synthesis of the steroid D ring.[33, 34, 45] The

   XXXII   XXXIII   XXXIV

XXXV

[43] Loftfield and Schaad, *J. Am. Chem. Soc.*, **76**, 35 (1954).
[44] Wagner and Moore, *J. Am. Chem. Soc.*, **72**, 2884 (1950).
[44a] E. E. van Tamelen and J. E. Brenner, unpublished observations; J. E. Brenner, Ph.D thesis, University of Wisconsin, 1958.
[45] G. Stork and W. S. Worrall, unpublished observations.

epoxynitrile ester **XXXII**, obtained by Darzens condensation of 2-chloropropionitrile with the appropriate keto ester, was treated with hydrogen chloride followed by dilute base to give the chloroketone **XXXIII**. Rearrangement of this chloroketone with sodium benzyloxide led to the diester **XXXIV** (R=$C_6H_5CH_2$ or C_2H_5) which on Dieckmann cyclization and hydrolysis gave 8-methyl-*trans*-1-hydrindone (**XXXV**). The rearrangement proceeded in 21–25% yield.

A lower homolog of **XXXIV**, the diester **XXXVI**, was obtained in about 15% yield by stereospecific rearrangement of the chloroketone **XXXVII**, which in turn was prepared by sulfuryl chloride chlorination

XXXVI   XXXVII

of the corresponding δ-ketoester. Although the yields in the rearrangement of the chloroketones **XXXIII** and **XXXVII** were low, the stereospecificity of the reaction can make this a preferred route of synthesis for such intermediates.

The rearrangement of an α-chlorodicycloalkyl ketone, **XXXVIII**, to the difficultly accessible acid **XXXIX** has found synthetic utility.[46]

XXXVIII   XXXIX

Limited data on α-haloketones in fused bicyclic systems suggest that their behavior parallels the monocyclic as well as the more complex polycyclic analogs. The rearrangement of 4-chloro-*cis*-5-hydrindone (**XL**) led to a 65% yield of a mixture of the bicyclo[3.3.0]octane-2- and -3-carboxylic acids **XLI** and **XLII**.[47] The rearrangement of 3-chloro-*trans*-2-decalone to hydrindane derivatives has been reported.[48,49]

[46] Kopp and Tchoubar, *Bull. soc. chim. France*, [5] **19**, 84 (1952); **22**, 1363 (1955).
[47] Granger, Nau, and Corbier, *Bull. soc. chim. France*, [5] **22**, 5, 479 (1955); **23**, 247 (1956).
[48] Cauquil and Tsatsas, *Bull. soc. chim. France*, [5] **10**, 47 (1943).
[49] Mousseron, Granger, et al., *Bull. soc. chim. France*, [5] **10**, 42 (1943); **14**, 606 (1947).

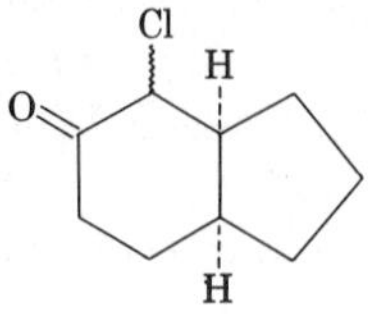

[Structures XL, XLI, XLII]

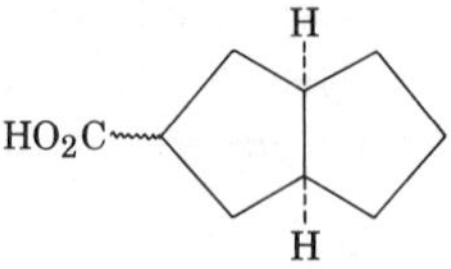

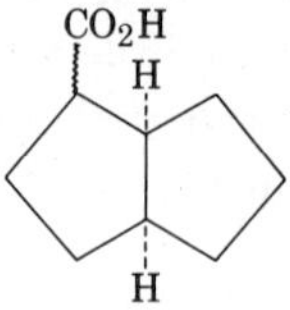

The 1-bromo-bicyclo[3.3.1]nonan-9-one system (XLIII) is readily transformed by a variety of reagents, such as silver or mercuric salts, sodium amide, or potassium hydroxide in ether, into derivatives of bicyclo-[3.3.0]octane-1-carboxylic acid (XLIV).[28, 50] These quasi-Favorskiĭ rearrangements are believed to proceed by a special "push-pull" mechanism related to the benzilic acid rearrangement.

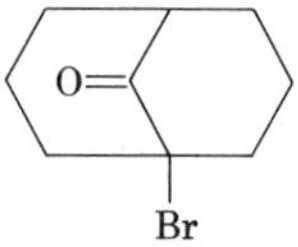

[Structures XLIII, XLIV]

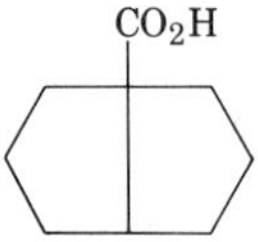

Aralkyl Monohaloketones

The labilizing effect of an aryl group leads to particularly facile rearrangement for haloketones of the type $ArCH_2COCHXR$. Yields of the order of 80% are obtained in the conversion of 1-chloro-3-arylacetones to the corresponding 3-arylpropionic esters.[8, 51] When two aryl groups activate the α'-carbon atom, rearrangement is very rapid, so that even the highly nucleophilic dialkylamines can serve as the basic reagents. Thus the dihydroanthracene ketones XLV (X = Cl, Br) on treatment with diethylamine give the diethylamide rearrangement product XLVI in about 40% yield.[14, 52, 53]

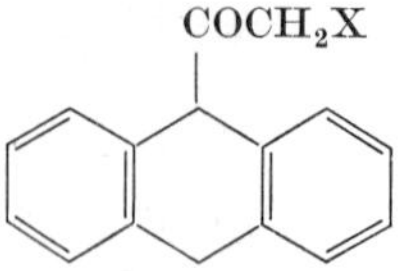

[Structures XLV, XLVI]

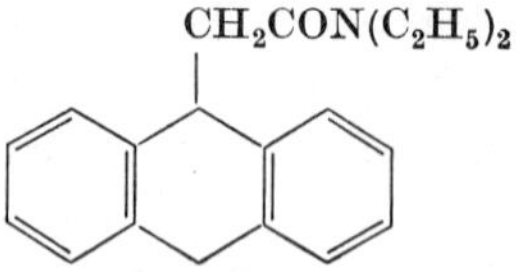

[50] Cope and Synerholm, *J. Am. Chem. Soc.*, **72**, 5228 (1950).
[51] Eastham, Fisher, Kulka, and Hibbert, *J. Am. Chem. Soc.*, **66**, 26 (1944).
[52] Dauben, Hiskey, and Muhs, *J. Am. Chem. Soc.*, **74**, 2082 (1952).
[53] May and Mosettig, *J. Am. Chem. Soc.*, **70**, 1077 (1948).

The presence of an enolizable α'-hydrogen atom remains a requirement for rearrangement under normal conditions. Haloketones lacking this feature, such as 1-chloro-1-benzoylcyclohexane or 2-chloro-1-tetralone, do not give rearrangement products on treatment with alkoxides.[54, 55] However, the use of silver salts or solid alkali-metal hydroxides can sometimes effect a quasi-Favorskiĭ rearrangement of these systems,[27, 28, 56] as illustrated by the nonstereospecific conversion of the levorotatory chloroketone XLVII to the racemic acid XLVIII by the action of sodium hydroxide in boiling xylene.[57]

$$\underset{\text{XLVII}}{\begin{array}{c}\text{piperidine ring with } COC_6H_5 \text{ and Cl}\\ \text{N-CH}_3\end{array}} \longrightarrow \underset{\text{XLVIII}}{\begin{array}{c}\text{piperidine ring with } C_6H_5 \text{ and } CO_2H\\ \text{N-CH}_3\end{array}}$$

Aryl substitution on the halogen-bearing carbon atom appears to have a favorable effect on the rearrangement. Thus 1-chloro-1-phenylacetone reacts with methanolic methoxide to give rearrangement products in 69% yield,[8] and the tertiary haloketone XLIX rearranges to give ethyl 3,3-diphenylpropionate in 85% yield.[58]

$$(C_6H_5)_2\overset{\overset{\displaystyle Cl}{|}}{C}COCH_3$$
XLIX

Steroid Monohaloketones

The Favorskiĭ rearrangement has found synthetic utility in the steroids as a direct route to A-norsteroids and in transformations leading to 17-methyletianic acid derivatives.

Reaction of 2-halocholestanones (L) with alkoxides has been studied in several laboratories.[59–62] Two esters, LI and LII, can be isolated, the

---

[54] Stevens, Malik, and Pratt, *J. Am. Chem. Soc.*, **72**, 4758 (1950).
[55] Stevens, Beereboom, and Rutherford, *J. Am. Chem. Soc.*, **77**, 4590 (1955).
[56] Tchoubar, *Compt. rend.*, **228**, 580 (1949); **235**, 720 (1952).
[57] Smissman and Hite, *J. Am. Chem. Soc.*, **81**, 1201 (1959); Abstracts, Medicinal Chemistry Section, 135th Meeting, Am. Chem. Soc., Boston, 1959, p. 18N.
[58] Stevens and Sherr, *J. Org. Chem.*, **17**, 1228 (1952).
[59] Winternitz and de Paulet, *Bull. soc. chim. France*, [5] **21**, 288 (1954); **22**, 1393 (1955).
[60] Evans, de Paulet, Shoppee, and Winternitz, *Chem. & Ind. (London)*, **1955**, 355; *J. Chem. Soc.*, **1957**, 1451.
[61] Smith and Nace, *J. Am. Chem. Soc.*, **76**, 6119 (1954).
[62] A. S. Kende, unpublished observations.

former predominating. The position of the carboxyl group was demonstrated in each product by Barbier-Wieland degradation to the corresponding A-norcholestan-2-one and A-norcoprostan-3-one, respectively. The reaction of 4β-bromocoprostan-3-one proceeds along similar lines to give approximately 25% each of the A-norcoprostane-2- and -3-carboxylates.

In contrast to the above instances, the reaction with methoxide ion of 17-brominated D-homoandrostan-17a-one LIII, which lacks an α'-hydrogen atom, gave only traces of the ester LIV.[60, 63]

Halogenated 20-ketosteroids undergo rearrangement very readily. A number of 17α-bromo-20-ketosteroids (LV) are transformed by methanolic bicarbonates in high yield to 17-methyletianic esters. The 17α-methyl ester LVI is invariably the principal product, but it is usually accompanied by a significant amount of the 17β-epimer LVII.[10, 32, 64]

[63] Prins and Shoppee, *J. Chem. Soc.*, **1946**, 494.
[64] Engel, *J. Am. Chem. Soc.*, **78**, 4727 (1956).

The action of potassium methoxide on 21-chloro-5-pregnen-3β-ol-20-one (LVIII) proceeds comparably to give 63% and 24%, respectively, of the 17α- and 17β-methyletianic esters described above.[32] The rearrangement of a 21-fluoro-20-ketosteroid takes a similar course.[64a]

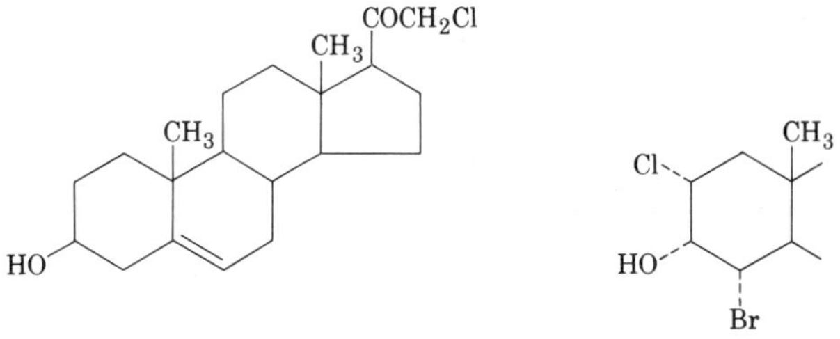

LVIII   LIX

The reaction of 2α-chloro-4α-bromocholestan-3α-ol (LIX) with ethanolic potassium hydroxide appears to involve a Favorskiĭ transformation.[65] The $C_{27}H_{46}O_2$ acid product, obtained in high yield, was assigned an A-norcholestane structure corresponding to the Favorskiĭ ester LI or LII, and could arise by rearrangement of an intermediate halocholestan-3-one.

Dihaloketones

In 1894 Favorskiĭ reported that several aliphatic dichloroketones were rearranged in refluxing potassium carbonate solution into unsaturated acids.[1] Subsequent studies by Wagner have shown that the rearrangement of a number of α,α'- or α,β-dihaloketones can be effected smoothly with sodium alkoxides. It was established that the primary product from an α,α'-dihaloketone (LX) is an α,β-unsaturated ester (LXI), while

LX   LXI

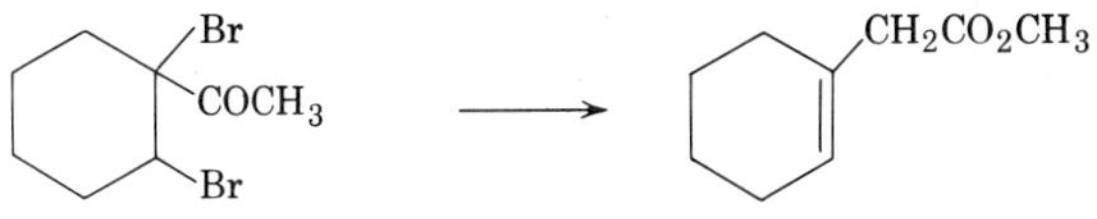

LXII   LXIII

[64a] Kende, *Chem & Ind.* (*London*), **1959**, 1346.
[65] Beereboom and Djerassi, *J. Org. Chem.*, **19**, 1196 (1954).

the product from an α,β-dihaloketone (LXII) is a β,γ-olefinic ester (LXIII).[66]

In practice this product specificity is not always observed because prototropic equilibration between an α,β- and a β,γ-isomer can occur.[67, 68] However, the above primary course of the reaction is well accommodated by the cyclopropanone mechanism which, moreover, is consistent with the stereochemistry found for some of the olefinic rearrangement products. Thus it has been pointed out that the dibromoketone LXV, derived from what is most probably *trans*-3-methyl-3-penten-2-one (LXIV), gives solely the *trans*-pentenoate LXVI on rearrangement.[14, 68]

Likewise, rearrangement of the dibromoketone LXVII should proceed through both cyclopropanones LXVIII and LXIX.[14] The observed yields of 29% *cis*-pentenoate LXX and 22% *trans*-pentenoate LXXI are in accord with this reasoning.[66]

[66] Wagner and Moore, *J. Am. Chem. Soc.*, **72**, 974 (1950).
[67] Marker, Wagner, and Wittbecker, *J. Am. Chem. Soc.*, **64**, 2093 (1942).
[68] Wagner, *J. Am. Chem. Soc.*, **71**, 3214 (1949).

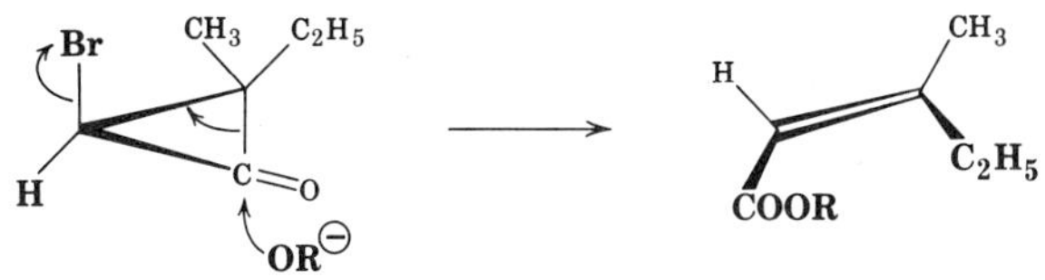

THE FAVORSKIĬ REARRANGEMENT OF HALOKETONES

[Structures LXVIII → LXX and LXIX → LXXI]

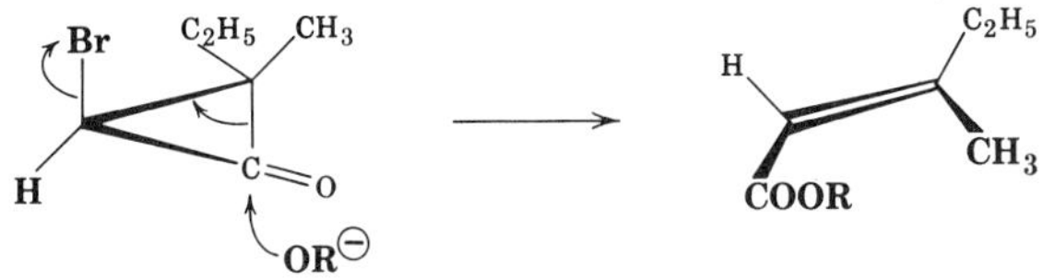

Yields of 51–84% are reported by Wagner for the alkoxide-catalyzed rearrangement of several aliphatic α,α'- and α,β-dibromoketones.[66,68] The principal side reaction is the addition of alcohol to the α,β-olefinic esters, which gives rise to β-alkoxy esters.[68] The rearrangement of the endocyclic dibromoketone LXXII to derivatives of 2-methylcyclohexene-1-carboxylic acid is effected by sodium benzyloxide.[34]

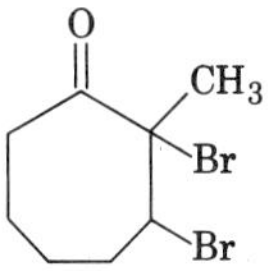

[Structure LXXII]

Certain dibromoketones are rearranged by the action of amines. Of particular interest is the heterocyclic dihaloketone LXXIII, which reacts with ammonia or primary amines to give the Δ^3-pyrroline derivatives LXXIV.[17,69]

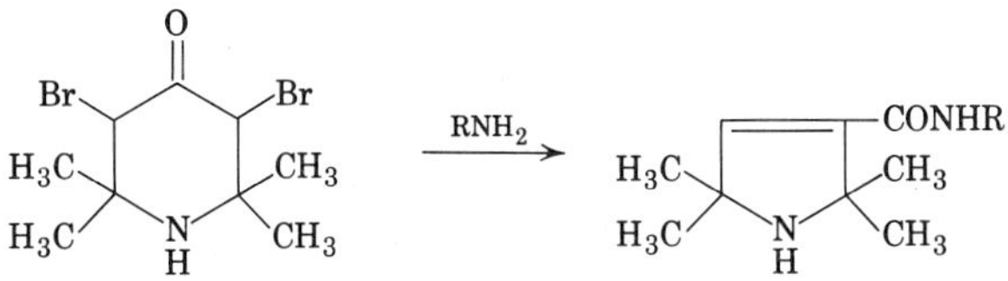

[Structures LXXIII → LXXIV]

[69] Pauly, *Ber.*, **31**, 668 (1898).

The use of a tertiary amine is illustrated by the transformation of 5α,7α-dibromo-3β-acetoxycholestan-6-one (LXXV) into the olefinic acid LXXVII by refluxing pyridine.[70] The acylpyridinium salt LXXVI has been suggested as an intermediate in this reaction.

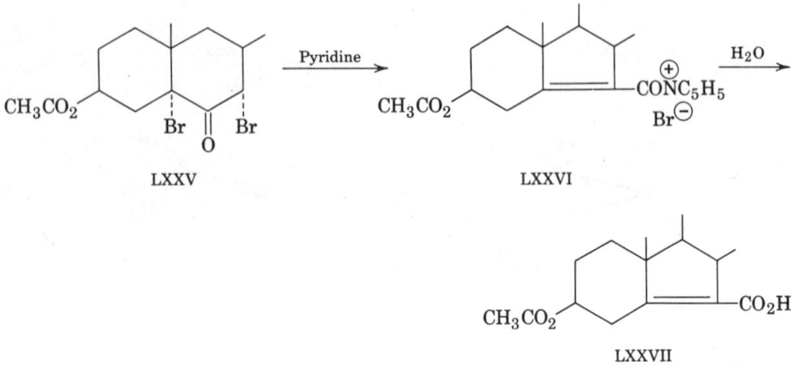

LXXV   LXXVI

LXXVII

On treatment with hot dimethylaniline or potassium hydroxide solution, the dibromination product of cyclononanone undergoes a transannular reaction to give the bicyclic ketone LXXVIII.[42]

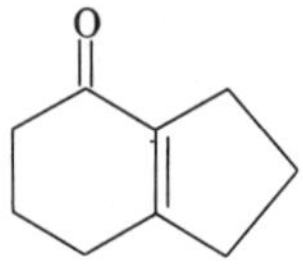

LXXVIII

Steroidal 17,21-dihalo-20-ketones (LXXIX, X = Br, I) and 16,17-dibromo-20-ketones (LXXX) are smoothly converted by methanolic potassium hydroxide into the corresponding $\Delta^{17(20)}$-21-carboxylic acids. The rearrangement of 17α-bromo-21-iodopregn-5-ene-3β-ol-20-one acetate has been shown to give both the *trans* and *cis* acids, LXXXI and LXXXII respectively, the former predominating.[71]

The rearrangement of certain terpene dibromoketones by aqueous base is a feature of the "Wallach degradation."[72] An illustration is the transformation of pulegone dibromide (LXXXIII) to "pulegenic acid," a mixture from which a 2-isopropylidene-5-methylcyclopentanecarboxylic

[70] Woodward and Clifford, *J. Am. Chem. Soc.*, **63**, 1123, 2727 (1941).
[71] Romo and Romo de Vivar, *J. Am. Chem. Soc.*, **79**, 1118 (1957).
[72] Wallach, *Ann.*, **414**, 271 (1918).

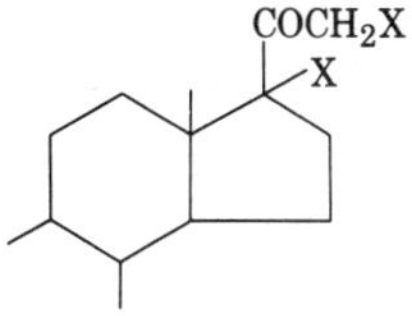

LXXIX 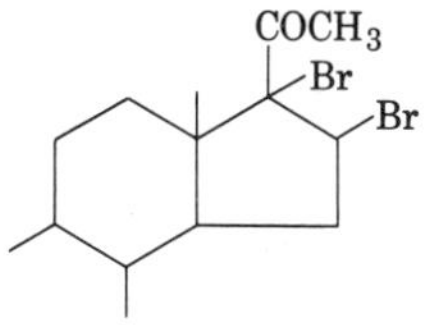   LXXX

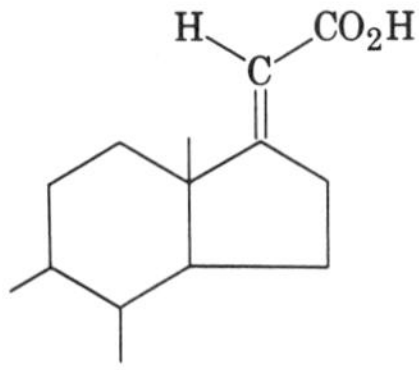

LXXXI 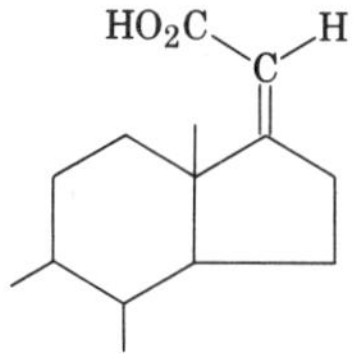   LXXXII

acid (LXXXIV) has been characterized.[73] Many of Wallach's dibromoketones and their transformation products are of uncertain purity and structure. Phenols, α-hydroxy acids, and substances resulting from ring cleavage are frequently produced in preference to the Favorskiĭ product.

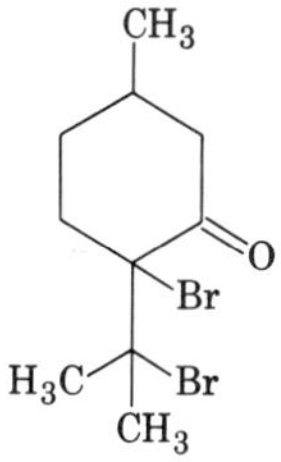

LXXXIII 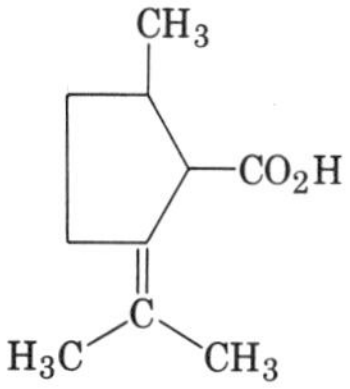   LXXXIV

The conversion of the β-keto ester LXXXV, R = CH_3, to mesaconic acid (LXXXVI, R = CH_3) may be regarded as the earliest example of the Favorskiĭ rearrangement.[74,75] Although the generality of the

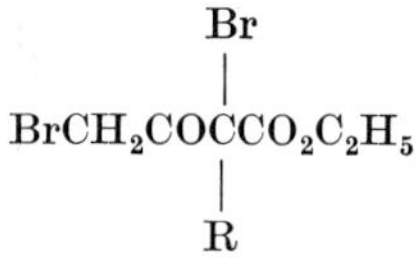

$BrCH_2COCCO_2C_2H_5$ with Br and R substituents
LXXXV

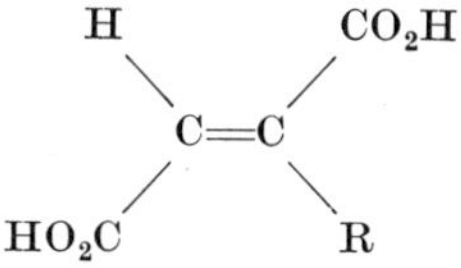

HO_2C, H, R, CO_2H on C=C
LXXXVI

[73] Wallach, Ann., **327**, 125 (1903); **414**, 233 (1918)
[74] Demarcay, Ann. chim. et phys., [5] **20**, 433 (1880).
[75] Cloez, Bull. soc. chim. France, [3] **3**, 602 (1890).

reaction had not been established, its course was clearly discussed by Wolff four years before Favorskiĭ's initial paper appeared.[15] Subsequently Conrad showed that the acetylsuccinic ester LXXXV, $R = CH_2CO_2C_2H_5$, behaves similarly, giving aconitic acid (LXXXVI, $R = CH_2CO_2C_2H_5$).[16]

Trihaloketones

The reaction of several α,α,α'-trihaloketones with alkaline reagents has been examined. The aliphatic tribromoketone LXXXVII reacts with aqueous base to give β,β-dimethylglyceric acid (LXXXVIII).[3] (The formation of LXXXVIII is analogous to the production of mandelic acid from the action of alkali on α,α-dibromoacetophenone.[76]) However, ethanolic potassium hydroxide converts LXXXVII to the Favorskiĭ product LXXXIX in low yield.[77]

$$\underset{\text{LXXXVII}}{(CH_3)_2\overset{\overset{Br}{|}}{C}COCHBr_2} \qquad \underset{\text{LXXXVIII}}{(CH_3)_2\overset{\overset{OH}{|}}{C}\text{——}\overset{\overset{OH}{|}}{C}HCO_2H} \qquad \underset{\text{LXXXIX}}{(CH_3)_2C\text{=}CBrCO_2H}$$

Similarly, dibromomethyl α-bromocyclohexyl ketone (XC) gives α-bromocyclohexylideneacetic acid (XCI).[77]

The cyclic trihaloketone XCII reacts with sodium acetate in aqueous ethanol to give the Favorskiĭ product 2-chloro-1-cyclohexene carboxylic acid (XCIII); the 2,2,8-trihalocycloöctanones undergo rearrangement with comparable ease.[78]

---

[76] Neville, *J. Am. Chem. Soc.*, **70**, 3499 (1948).
[77] Wagner and Moore, *J. Am. Chem. Soc.*, **72**, 3655 (1950).
[78] Hesse and Krehbiel, *Ann.*, **593**, 42 (1955); Hesse and Urbanek, *Chem. Ber.*, **91**, 2733, (1958).

In the steroids, rearrangements of the tribromoketone system XCIV to the corresponding bromo acids XCV are effected in 57–72% yield by ethanolic potassium hydroxide.[71, 77]

XCIV (with COCHBr$_2$ and Br substituents on decalin)

XCV (with =CBrCO$_2$H substituent on decalin)

EXPERIMENTAL CONDITIONS

Side Reactions

The principal side reactions encountered in the rearrangement of α-haloketones by alkoxides give rise to epoxyethers (XCVI), α-hydroxy ketals (XCVII) and α-hydroxy ketones (XCVIII) having the same carbon skeleton as the original haloketone. Less frequent by-products are α-alkoxyketones, unsaturated ketones, and acids resulting from secondary cleavage reactions.

The main side reaction competing with rearrangement proceeds through nucleophilic addition of alkoxide to the carbonyl group, with the formation of a labile epoxyether (XCVI). This intermediate can react further with alcohols or water to form hydroxy ketal or hydroxy ketone, respectively.[14, 43, 54, 79–83]

$$\begin{array}{c}
X \\
| \\
-C-C-C- \\
| \| | \\
  O
\end{array}
\xrightarrow[-X^{\ominus}]{RO^{\ominus}}
\begin{array}{c}
OR \\
| \\
C-C-C- \\
\diagdown O \diagup |
\end{array}
\xrightarrow{ROH}
\begin{array}{c}
  OR \\
  | \\
-C-C-C- \\
| | | \\
HO OR \\
\text{XCVII}
\end{array}$$

XCVI

$$\xrightarrow{H_2O}
\begin{array}{c}
HO O \\
| \| | \\
-C-C-C- \\
| | \\
\text{XCVIII}
\end{array}$$

---

[79] Ward, *J. Chem. Soc.*, **1929**, 1541.
[80] Mousseron, Jacquier, and Fontaine, *Bull. soc. chim. France*, [5] **19**, 767, (1952).
[81] Stevens and Farkas, *J. Am. Chem. Soc.*, **74**, 618 (1952).
[82] Stevens and Tazuma, *J. Am. Chem. Soc.*, **76**, 715 (1954).
[83] Bergmann and Miekeley, *Ber.*, **64**, 802 (1931).

Pure epoxyethers have been obtained by action of ethereal alkoxides on α-halopropiophenones and α-halocyclohexyl phenyl ketones.[54, 81, 84] These well-characterized epoxyethers reacted rapidly with methanol or methanolic methoxide to form α-hydroxy ketals, and with aqueous acid or base to give α-hydroxy ketones, but no rearrangement to esters was observed. Because of their lability, α-epoxyethers are not normally isolated as such from Favorskiĭ reaction mixtures.* In the presence of alcohols during reaction or isolation of the products, the principal byproduct is the expected hydroxy ketal,[26, 44] or an epoxyether dimer believed to be formed by reaction of hydroxy ketal with the epoxyether.[43]

Hydroxy ketones result on treatment of α-haloketones by hydroxides,[26, 38] or through hydrolysis of epoxyethers during reaction or isolation of the products.[43, 77] Such α-hydroxy ketones may undergo subsequent hydrolytic or oxidative cleavage to give carboxylic acids. The formation of 21% of cyclohexanecarboxylic acid from chloromethyl cyclohexyl ketone and sodium methoxide has been ascribed to hydrolysis of the intermediate hydroxymethyl ketone, since formation of the acid was largely eliminated under rigorously anhydrous reaction conditions.[43]† Formally similar reactions in the steroids have been attributed to the reaction of hydroxy ketone intermediates with oxygen in the presence of alkoxide.[60, 61, 85]

The extent to which side reactions such as the above interfere with the normal Favorskiĭ reaction must depend on the rate of epoxyether formation compared to the rate of rearrangement. This ratio is a function of several factors, primarily the structure of the haloketone and the nature of the halogen. With a given haloketone, there appears to be a dependence on the polarity of the reaction medium and possibly the nature of the alkoxide.[14, 43] The effects of these experimental variables are discussed in the following sections.

Other side reactions include direct substitution of certain α-haloketones by alkoxides, particularly methoxide ion, to form α-alkoxy ketones.[26, 38, 86] The use of amines as Favorskiĭ reagents gives rise to α-amino ketones.[52, 53, 87, 88] In some instances, dehydrohalogenation to unsaturated ketones may occur.[80]

[84] Temnikova and Kropacheva, *J. Gen. Chem. U.S.S.R.*, **19**, 1917 (1949) [*C.A.*, **44**, 1929 (1950)].

* The reported[80] formation of α-epoxyethers from the action of alcoholic alkoxides on alicyclic α-haloketones has been questioned by Stevens, who has identified several such products as α-hydroxy ketals.[82]

† Hydroxymethyl cyclohexyl ketone and 1-hydroxy-1-benzoylcyclohexane are known to cleave in base to give cyclohexanecarboxylic acid and benzoic acid, respectively.[27, 33, 43]

[85] Stoll and Hulstkamp, *Helv. Chim. Acta*, **30**, 1815 (1947).
[86] Barnes, Pausacker, and Badcock, *J. Chem. Soc.*, **1951**, 730.
[87] Jullien and Fauche, *Bull. soc. chim. France*, [5] **20**, 374 (1953).
[88] Dodson, Morello, and Dauben, *J. Am. Chem. Soc.*, **76**, 606 (1954).

The cationoid character of halogen in bromoketones renders the latter liable to reduction or disproportionation in the presence of strong bases.[46, 89] The reaction of 2-bromocyclohexanone and related substances with alkali is accompanied by formation of α-hydroxy acids having a rearranged carbon skeleton.[14, 72, 89] These are considered to arise through disproportionation to dibromoketones followed by hydrolysis to α-diketones, which undergo the benzilic acid rearrangement.[90, 91]

Nature of the Halogen

Chloroketones are normally preferable to bromoketones as reactants in the Favorskiĭ rearrangement. For example, chloromethyl cyclohexyl ketone (XCIX) reacts with sodium methoxide to give 38% of Favorskiĭ ester, whereas the corresponding bromoketone (C) under these conditions gives exclusively side-reaction products[43] Comparable differences have been observed for the 2-halocyclohexanones[14] and the α-halodicyclohexyl ketones.[47]

$$\underset{\substack{\text{XCIX} \ (X = Cl) \\ C \ (X = Br)}}{\overset{H}{\underset{|}{\bigcirc}}\overset{O}{\underset{\|}{-C}}CH_2X}$$

Loftfield has pointed out that, although the rates of rearrangement for haloketones XCIX and C are probably comparable, the rate of the main competing side reaction, epoxyether formation, is much greater for the bromoketone C than for the chloro compound XCIX.[43] (The consequent suggestion that α-fluoroketones might serve as superior starting materials for the rearrangement awaits experimental verification.[64a]) Extension of this principle to aliphatic α-monohaloketones has not been investigated in detail but is probably valid.[26, 92] In the rearrangement of 2-halo-3-ketosteroids[62] or 21-halo-20-ketosteroids[93] the chloro compound offers only minor advantages over the bromoketone. Data are lacking concerning the rearrangement of simple α-iodoketones. The reaction of α-p-toluenesulfonyloxyketones with alkoxides can proceed with elimination of p-toluenesulfinate anion to give α-diketones.[94]

[89] Lyle and Covey, *J. Am. Chem. Soc.*, **75**, 4973 (1953).
[90] Schwarzenbach and Wittwer, *Helv. Chim. Acta*, **30**, 663 (1947).
[91] Buchman and Sargent, *J. Org. Chem.*, **7**, 148 (1952).
[92] Delbaere, *Bull. soc. chim. Belges*, **51**, 1 (1942).
[93] Plattner, Heusser, and Boyce, *Helv. Chim. Acta*, **31**, 603 (1948).
[94] R. B. Woodward and S. Levine, unpublished observations; S. Levine, Ph.D. thesis, Harvard University, 1953.

Choice of Base and Solvent

The choice of base and solvent can profoundly affect the yield of a Favorskiĭ reaction. This is particularly clear-cut in the aliphatic series, as is illustrated by the data in Table I on the rearrangement of the bromoketone CI, in which the Favorskiĭ ester CII, the hydroxy ketal CIII, and the ketol CIV may be formed.

$$(CH_3)_2CBrCOCH_3 \xrightarrow{RO^\ominus} (CH_3)_3CCO_2R + (CH_3)_2C(OR)(OH)CCH_3 + (CH_3)_2\overset{OH}{\underset{}{C}}COCH_3$$

| CI | CII | CIII | CIV |

TABLE I

REACTION OF $(CH_3)_2CBrCOCH_3$ (CI) UNDER CONDITIONS OF THE FAVORSKIĬ REACTION

Base	Solvent	Yield (%) of CII	Yield (%) of By-products	Reference
Sodium isopropoxide	Diethyl ether	64	0	26
Sodium ethoxide	Diethyl ether	61	0	26
Sodium methoxide	Diethyl ether	39	20 CIII	26
Sodium isopropoxide	Isopropyl alcohol	20	8 CIII	26
Sodium ethoxide	Ethanol	14	32 CIII	26
Sodium methoxide	Methanol	0	77 CIII	26
Barium carbonate	Water	3		95
Potassium hydroxide	Water	0	76 CIV	95

Base and solvent effects on the rearrangement of 2-chlorocyclohexanone and 1-chloro-1-acetylcyclohexane have been studied in detail by Stork and Borowitz.[34] No correlation is found between yield and the pK_a of the alcohol, nor is there observed a simple dependence of yield on the size of the alkoxide ion, as earlier data seem to suggest.[26,80]* The use of excess alkoxide (2 to 4 equivalents) and high base concentrations leads to significantly higher yields in the homogeneous reactions. Rigorously anhydrous conditions are not essential for these haloketones, although traces of water have a deleterious effect in the reaction of other haloketones.[43,60] Yields obtained with given solvent-alkoxide combinations are listed in Table II.

[95] Venus-Danilova, *J. Gen. Chem. U.S.S.R.*, **11**, 847, (1941) [*C.A.* **36**, 4094 (1942)].

* Potassium *t*-butoxide gave a poor yield in the rearrangement of 2-chlorocyclohexanone.[34]

TABLE II
Yields of Rearrangement Acid Using Various Alkoxide-Solvent Pairs[34]

Base	Solvent	Yield (%) from 2-Chlorocyclo-hexanone	Yield (%) from 1-Chloro-1-acetylcyclo-hexane
Sodium ethoxide	Ethanol	60 (64)*	41
Sodium ethoxide	Diethyl ether		56
Sodium methoxide	Methanol	44	
Sodium isopropoxide	Isopropyl alcohol	36	
Sodium isopropoxide	Diethyl ether		45
Sodium isoamyloxide	Isoamyl alcohol	(47)*	
Sodium benzyloxide	Benzyl alcohol	75	57
Sodium benzyloxide	Diethyl ether	57	72

* Data of Loftfield.[14]

Survey of the literature reveals no single alkoxide-solvent combination as clearly superior for α-monohaloketones in general. The use of diethyl ether as solvent is indicated for the simpler haloketones, and theoretical considerations suggest that solvents of low polarity might have a generally favorable effect.[14] Sodium benzyloxide, used under a nitrogen atmosphere, and sodium ethoxide are among the more consistently successful reagents. The optimum choice of base and solvent appears to vary with the structure of the individual haloketone.

The use of hydroxides or carbonates generally leads to extensive hydroxyketone formation. Significant exceptions include the conversion of 2-chlorocycloheptanone to cyclohexanecarboxylic acid (69% yield) on treatment with hot aqueous potassium carbonate.[96] Similarly high yields are obtained in the rearrangement of 17-bromo-20-ketosteroids with refluxing methanolic bicarbonates.[64, 97] Sodium hydroxide in an inert solvent is moderately effective with some aralkyl ketones,[8, 57, 58, 98] and appears to be the reagent of choice in the quasi-Favorskiĭ rearrangement of 1-chloro-1-benzoylcyclohexane.[7, 27]

The use of secondary amines has limited scope and offers no advantages over alkoxides.[53, 87, 88] Sodium salts of various bifunctional alcohols and of alicyclic alcohols, such as menthol, also appear relatively unpromising.[80] Phenoxides and thiophenoxides lead primarily to substitution products.[80, 99]* Relatively non-nucleophilic bases, such as

[96] Gutsche, J. Am. Chem. Soc., **71**, 3513 (1949).
[97] Heusser, Engel, Herzig, and Plattner, Helv. Chim. Acta, **33**, 2229 (1950).
[98] Richard, G., Thèse Sciences, Univ. Nancy, 1936.
[99] Mousseron and Jacquier, Bull. soc. chim. France, [5] **16**, 689 (1949).
* Kopp-Mayer has claimed high yields of esters on treatment of aralkyl chloroketones with sodium phenoxide in dioxane.[100]
[100] Kopp-Mayer, Compt. rend., **240**, 1115 (1955).

sodium hydride or sodium triphenylmethide, do not effect rearrangement of 2-chloro-2-methylcycloheptanone.[34]

Rearrangement of the dibromoketone CV using sodium methoxide in diethyl ether proceeds in 48% yield;[66] the yield drops to 20% and 7% with the use of aqueous potassium hydroxide and carbonate, respectively.[101] Steroidal 17,21-dibromo-20-ketones, however, show relatively little sensitivity to such variations in reaction conditions.[64,102]

$$\text{cyclohexyl}(Br)-COCH_2Br$$

CV

Reaction Time and Temperature

Rearrangement of an α-monohaloketone is effected by adding the ketone to a fairly concentrated solution or suspension of the alkoxide at $-20°$ to $+30°$. Rapid addition of the ketone to an excess of the base is recommended. A mildly exothermic reaction usually results; short-term variations in reaction temperature normally have no effect on yield.[34]

Under the above conditions, homogeneous reactions of simple α-haloketones are generally complete within 10–30 minutes at room temperature.[14,34,58] With α-haloketones requiring ionization of a hindered proton, or with heterogeneous reactions, e.g., sodium alkoxides in ether, considerably longer reaction times may be required.[26,34,43] The reaction rate may be followed by determining the hydrogen halide liberated, through titration as acid or ionic halogen.[14,58,80]

Reaction temperatures above 50° are rarely necessary for rearrangements using alkoxides and, if maintained, may reduce the yield.[61] On the other hand, reactions in which a weak base such as methanolic bicarbonate is employed usually require 2 to 4 hours of heating under reflux.[64,93]

In the rearrangement of aliphatic dibromoketones, minimum reaction time and temperature, together with inverse addition of base to haloketone, are advisable to reduce the formation of β-alkoxy esters and resins.[66,68] For example, reaction of the dibromoketone CVI with ethereal sodium methoxide for 2.5 hours gives 64% of the olefinic ester CVII and 2% of the alkoxy ester CVIII, whereas a 30-hour reaction period leads to 42% of CVII and 16% of CVIII.

[101] Wagner and Moore, *J. Am. Chem. Soc.*, **72**, 1873 (1950).
[102] Koechlin and Reichstein, *Helv. Chim. Acta*, **27**, 549 (1944).

$$\underset{\text{CVI}}{\underset{|}{\overset{\overset{\text{Br}}{|}}{\text{BrCH}_2\text{CCOCH}_3}}\atop\text{CH}_3} \quad \underset{\text{CVII}}{(\text{CH}_3)_2\text{C}=\text{CHCO}_2\text{CH}_3} \quad \underset{\text{CVIII}}{\underset{|}{\overset{\overset{\text{OCH}_3}{|}}{(\text{CH}_3)_2\text{CCH}_2\text{CO}_2\text{CH}_3}}}$$

Experimental Procedures

Methyl Cyclopentanecarboxylate. Detailed directions for the preparation of methyl cyclopentanecarboxylate in 56–61% yield from 2-chlorocyclohexanone and sodium methoxide in diethyl ether are given in *Organic Syntheses*.[102a]

Ethyl Trimethylacetate.[26] (Rearrangement of a Bromoketone with Sodium Ethoxide in Diethyl Ether). To a dry 1-l. three-necked flask equipped with a dropping funnel and an efficient reflux condenser, each protected by a drying tube, is added 500 ml. of anhydrous diethyl ether. Into the ether is placed finely sliced sodium (11.5 g., 0.5 mole), which is followed by the addition of 29.2 ml. (0.5 mole) of absolute ethanol. The mixture is held at reflux for 48 hours to ensure reaction of the metal.*

The suspension is cooled in ice and 82.5 g. (0.5 mole) of 3-bromo-3-methyl-2-butanone is added over a period of 2 hours. The reaction mixture is heated under reflux for 3 hours, then water is added to dissolve the precipitated sodium bromide. The layers are separated and the ether dried over sodium sulfate. Fractionation gives 39.8 g. (61%) of ethyl trimethylacetate, b.p. 116°/725 mm., n_D^{20} 1.3912.

Ethyl 3,3-Diphenylpropionate.[58] (Rearrangement of a Chloroketone with Sodium Ethoxide in Ethanol). In a 100-ml. round-bottomed flask fitted with a calcium chloride tube is placed 5.4 g. (0.022 mole) of 1-chloro-1,1-diphenylacetone in 40 ml. of absolute ethanol. To this solution is added 9.2 ml. of freshly prepared ethanolic sodium ethoxide containing 2.42 millimoles of sodium ethoxide per milliliter of solution. During the addition, heat is evolved and the reaction mixture turns brown. After 1 minute, titration of an aliquot of the solution with hydrochloric acid shows that 89% of the sodium ethoxide has been consumed. The solution is poured onto ice, the water layer neutralized with dilute hydrochloric

---

[102a] Goheen and Vaughan, *Org. Syntheses*, **39**, 37 (1959).

* Preparation of the alkoxide is facilitated by equipping the flask with a sealed stirrer and replacing the sodium metal with 12.0 g. of sodium hydride powder (Metal Hydrides Inc., Beverly, Mass.). The ethanol is slowly added to the stirred hydride suspension at a rate that maintains steady hydrogen evolution. After the reaction has largely subsided, a 1-hour reflux period completes formation of the ethoxide.[62]

acid, and the organic material extracted with several portions of ether. The combined ether layers are dried over sodium sulfate, and the solvent is removed at room temperature with a water aspirator. The residue, 4.75 g. of a dark yellow oil, is distilled to give 4.5 g. (85%) of ethyl 3,3-diphenylpropionate, b.p. 129–133°/0.3 mm., m.p. 19–22°, n_D^{25} 1.4850.

Cyclohexanecarboxylic Acid.[96] (Rearrangement of a Chloroketone Using Aqueous Potassium Carbonate). A mixture of 5.0 g. of 2-chlorocycloheptanone, 15 g. of potassium carbonate, and 20 ml. of water is stirred vigorously at the reflux temperature for 6 hours. The reaction mixture is cooled and extracted with ether to remove neutral by-products (0.76 g.). The aqueous layer is acidified and is re-extracted with ether to isolate the acid fraction. Evaporation of the dried extract gives 3.0 g. (69%) of cyclohexanecarboxylic acid, m.p. 22–26°.

Methyl 3-Methyl-2-butenoate.[66] (Rearrangement of a Dibromoketone Using Inverse Addition of Sodium Methoxide in Diethyl Ether). A 2-l. three-necked flask is equipped with a sealed stirrer, thermometer, and a 5-l. separatory funnel. The funnel is equipped with a sealed stirrer and a wide-bore stopcock. A solution of 244 g. (1 mole) of 1,3-dibromo-3-methyl-2-butanone in 250 ml. of absolute diethyl ether is placed in the flask and cooled in a salt-ice bath. In the separatory funnel is placed 111.5 g. (2 moles) of freshly opened sodium methoxide powder (95% assay, Mathieson Alkali Works) suspended in 500 ml. of ether. The slurry of sodium methoxide is kept stirred and is added in small portions, over a 4-hour period, to the stirred reaction mixture at a temperature of 0–5°. After stirring for an additional 30 minutes, an aliquot of the reaction mixture is titrated with standard acid and it is found that less than 4% of the sodium methoxide remains. The reaction mixture is poured onto ice, the layers are separated, and the water layer is extracted with ether. The combined ether extracts are dried over anhydrous potassium carbonate and the ether is removed by distillation. The concentrate is rapidly distilled through a Claisen flask under reduced pressure to free it from any high-boiling and bromine-containing material. The crude distillate is carefully fractionated through a column packed with glass helices and the methyl 3-methyl-2-butenoate collected at 60°/50 mm. The product weighs 66 g. (58%) and has n_D^{20} 1.4382.

20-Bromo-17(20)-pregnen-3β-ol-21-oic Acid.[77] (Rearrangement of a Tribromoketone Using Potassium Hydroxide in Ethanol). To a solution of 3.0 g. of 17,21,21-tribromopregnan-3β-ol-20-one acetate in 600 ml. of boiling ethanol is added a solution of 12.0 g. of potassium hydroxide in 40 ml. of aqueous ethanol. The solution is refluxed for 2 hours, and the ethanol is then distilled under reduced pressure until solid material separates. The mixture is diluted with water and extracted with several

THE FAVORSKIĬ REARRANGEMENT OF HALOKETONES   291

portions of ether to remove neutral products. The aqueous layer, containing the sparingly soluble potassium salt of the acid, is treated with an excess of dilute sulfuric acid, and the organic acid is then extracted with ether. The ether extracts are washed with water, dried over sodium sulfate, and concentrated. When the volume is reduced to 100 ml., crystals begin to appear. After further concentration of the solution, the crystals are filtered and dried. The yield of bromo acid, m.p. 264–265°, is 1.27 g. (61%).

TABULAR SURVEY OF FAVORSKIĬ REARRANGEMENTS

Tables III–VIII list those haloketones from which products of the Favorskiĭ reaction have been isolated. In addition, characteristic examples of unsuccessful Favorskiĭ reactions have been included. The haloketones are tabulated in the order acyclic monohaloketones, alicyclic monohaloketones (except steroids), aralkyl monohaloketones, steroid monohaloketones, dihaloketones, trihaloketones. Since halogenation of unsymmetrical ketones can give rise to position isomers, a question mark following the position of the halogen is used to indicate doubt as to the identity or purity of a claimed structure. The yields given refer to the stated rearrangement product, except that when the yield figure is in parentheses it refers to yield of free acid derived from the primary rearrangement product.

The survey covers the literature available to the author through September 1958. A few later references are included.

TABLE III
ACYCLIC MONOHALOKETONES

Formula	Haloketone	Base	Solvent	Rearrangement Product	Yield (%)	References
C_3H_5OCl	Chloroacetone	KOH	$(C_2H_5)_2O$	Propionic acid		98, 103
C_3H_5OBr	Bromoacetone	$KOCH_3$	CH_3OH	*		83
C_4H_7OCl	1-Chloro-2-butanone	KOH	$(C_2H_5)_2O$	Butyric acid		98
	3-Chloro-2-butanone	KOH	$(C_2H_5)_2O$	Butyric and isobutyric acids		98
C_4H_7OBr	3-Bromo-2-butanone	$NaOCH_3$	$(C_2H_5)_2O$	†		39
C_5H_9OCl	3-Chloro-3-methyl-2-butanone	NaOH	H_2O	Trimethylacetic acid		92
C_5H_9OBr	2-Bromo-3-pentanone	$BaCO_3$	H_2O	3-Methylbutyric acid	2	95
	3-Bromo-3-methyl-2-butanone	$NaOCH(CH_3)_2$	$(CH_3)_2CHOH$	Isopropyl trimethylacetate	20	26
		$NaOCH(CH_3)_2$	$(C_2H_5)_2O$	Isopropyl trimethylacetate	64	26
		$NaOC_2H_5$	C_2H_5OH	Ethyl trimethylacetate	14	26
		$NaOC_2H_5$	$(C_2H_5)_2O$	Ethyl trimethylacetate	61	26
		$NaOCH_3$	CH_3OH	*		26
		$NaOCH_3$	$(C_2H_5)_2O$	Methyl trimethylacetate	39	26
		KOH	C_2H_5OH	†		26
		KOH	C_2H_5OH	Trimethylacetic acid	33	95
		$BaCO_3$	H_2O	Trimethylacetic acid	3	95
$C_6H_{11}OBr$	2(?)-Bromo-2-methyl-3-pentanone	$NaOCH_3$	$(C_2H_5)_2O$	Methyl 2,2-dimethyl-butyrate	57	26
$C_7H_{13}OCl$	2-Chloro-3-heptanone	$NaOCH_3$	$(C_2H_5)_2O$	Methyl 2-ethylvalerate	65	30
	4-Chloro-3-heptanone	$NaOCH_3$	$(C_2H_5)_2O$	Methyl 2-ethylvalerate	77	30
$C_7H_{13}OBr$	3-Bromo-4-heptanone	K_2CO_3	H_2O	2-Ethylvaleric acid	10	95
		$CaCO_3$	H_2O	2-Ethylvaleric acid	9	95
		$BaCO_3$	H_2O	2-Ethylvaleric acid	2	95

	2(?)-Bromo-2-methyl-3-hexanone	NaOCH$_3$	(C$_2$H$_5$)$_2$O	Methyl 2-ethyl-3-methylbutyrate	69	38
	2-Bromo-2,4-dimethyl-3-pentanone	NaOCH$_2$C$_6$H$_5$	(C$_2$H$_5$)$_2$O	Benzyl 2,2,3-trimethylbutyrate	29	30
		NaOCH(CH$_3$)$_2$	(C$_2$H$_5$)$_2$O	Isopropyl 2,2,3-trimethylbutyrate †	17	30
		NaOCH$_3$	(C$_2$H$_5$)$_2$O			30
C$_8$H$_{15}$OBr	3(?)-Bromo-4,4-dimethyl-2-pentanone	NaOCH$_3$	(C$_2$H$_5$)$_2$O	Methyl 2,3,3-trimethylbutyrate	73	39
	2(?)-Bromo-2-methyl-3-heptanone	NaOCH$_3$	(C$_2$H$_5$)$_2$O	Methyl 2,2-dimethylhexanoate	73	38
	3(?)-Bromo-3-methyl-4-heptanone	NaOCH(CH$_3$)$_2$	(C$_2$H$_5$)$_2$O	Isopropyl 2-methyl-2-ethylvalerate	41	38
		NaOCH$_3$	(C$_2$H$_5$)$_2$O	Methyl 2-methyl-2-ethylvalerate		39
	2(?)-Bromo-2,5-dimethyl-3-hexanone	NaOCH$_3$	(C$_2$H$_5$)$_2$O	Methyl 2,4-dimethylpentane-3-carboxylate †	83	38
	2-Bromo-2,4,4-trimethyl-3-pentanone	NaOCH(CH$_3$)$_2$	(C$_2$H$_5$)$_2$O			38
C$_9$H$_{17}$OBr	2(?)-Bromo-2-methyl-3-octanone	NaOCH$_3$	(C$_2$H$_5$)$_2$O	Methyl 2,2-dimethylheptanoate	83	38
	2(?)-Bromo-2,5,5-trimethyl-3-hexanone	NaOCH$_3$	(C$_2$H$_5$)$_2$O	Methyl 2,2,4-trimethylpentane-3-carboxylate	93	38
	3(?)-Bromo-3,5,5-trimethyl-2-hexanone	NaOCH$_3$	(C$_2$H$_5$)$_2$O	Methyl 2,2,4,4-tetramethylvalerate	78	38

Note: References 103 to 127 are on p. 316.

* Only hydroxy ketal was isolated.
† No rearrangement product was isolated.

TABLE IV
ALICYCLIC MONOHALOKETONES

Formula	Haloketone	Base	Solvent	Rearrangement Product	Yield (%)	References
C_5H_7OCl	2-Chlorocyclopentanone	$NaOCH_3$	CH_3OH	*		80
		KOH	C_2H_5OH	*		104
C_6H_9OF	2-Fluorocyclohexanone	$NaOC_2H_5$	C_2H_5OH	Ethyl cyclopentanecarboxylate	6	64a
		$NaOCH_3$	$(C_2H_5)_2O$	Methyl cyclopentanecarboxylate	40	64a
C_6H_9OCl	2-Chlorocyclohexanone†	$NaOCH_2C_6H_5$	$C_6H_5CH_2OH$	Benzyl cyclopentanecarboxylate	30	80
		$NaOCH_2C_6H_5$	$C_6H_5CH_2OH$	Benzyl cyclopentanecarboxylate	75	34
		$NaOCH_2C_6H_5$	$(C_2H_5)_2O$	Benzyl cyclopentanecarboxylate	(53, 57)	34
		$NaOCH_2CH_2-CH(CH_3)_2$	$(CH_3)_2CHCH_2CH_2OH$	Isoamyl cyclopentanecarboxylate	47	14
		$NaOCH(CH_3)_2$	$(CH_3)_2CHOH$	Isopropyl cyclopentanecarboxylate	(25, 36)	34
		$NaOCH(CH_3)_2$	$(CH_3)_2CHOH$	Isopropyl cyclopentanecarboxylate	55–60	80
		$NaOC_2H_5$	C_2H_5OH	Ethyl cyclopentanecarboxylate	(64)	14
		$NaOC_2H_5$	C_2H_5OH	Ethyl cyclopentanecarboxylate	(42–60)	34
		$NaOC_2H_5$	C_2H_5OH	Ethyl cyclopentanecarboxylate	(53)	105
		$NaOC_2H_5$	C_2H_5OH	Ethyl cyclopentanecarboxylate	(45–50)	80

		NaOCH$_3$	CH$_3$OH	Methyl cyclopentane-carboxylate	(44)	34
		NaOCH$_3$	(C$_2$H$_5$)$_2$O	Methyl cyclopentane-carboxylate	(56–61)	102a
		KOH	C$_2$H$_5$OH	Cyclopentanecarboxylic acid		106
C$_6$H$_9$OBr	2-Bromocyclohexanone	NaOC$_2$H$_5$	C$_2$H$_5$OH	Ethyl cyclopentane-carboxylate	(10)	14
		NaOC$_2$H$_5$	(C$_2$H$_5$)$_2$O	Ethyl cyclopentane-carboxylate	(21)	14
C$_7$H$_{11}$OCl	Chloromethyl cyclopentyl ketone	NaOCH$_3$	CH$_3$OH	Methyl 1-methylcyclo-pentanecarboxylate	50	80
	2-Chloro-2-methylcyclo-hexanone	NaOCH$_2$C$_6$H$_5$	C$_6$H$_5$CH$_2$OH	*		34
		NaOC$_2$H$_5$	C$_2$H$_5$OH	*		34, 107
		NaOCH$_3$	CH$_3$OH	*		80
	2-Chloro-4-methylcyclo-hexanone	NaOCH$_3$	CH$_3$OH	3-Methylcyclopentane-carboxylic acid	40–45	80, 108
	2-Chloro-5(?)-methylcyclo-hexanone	NaOCH$_3$	CH$_3$OH	3-Methylcyclopentane-carboxylic acid	40–45	108
		KOH	CH$_3$OH	3-Methylcyclopentane-carboxylic acid	50	106
		KOH	C$_2$H$_5$OH	3-Methylcyclopentane-carboxylic acid	43	107
	2-Chloro-6-methylcyclo-hexanone	NaOCH$_3$	CH$_3$OH	2(?)-Methylcyclopentane-carboxylic acid		87
	2-Chlorocycloheptanone	NaOC$_2$H$_5$	Not given	Ethyl cyclohexane-carboxylate	58	104
		KOH	C$_2$H$_5$OH	Cyclohexanecarboxylic acid	50	109
		KOH	C$_2$H$_5$OH	Cyclohexanecarboxylic acid	53	

Note: References 103 to 127 are on p. 316.
* No rearrangement product was isolated.
† A number of base-solvent combinations applied to this ketone have not been tabulated because of space limitations. The reader is referred to the original work.[80]

TABLE IV—Continued

ALICYCLIC MONOHALOKETONES

Formula	Haloketone	Base	Solvent	Rearrangement Product	Yield (%)	References
$C_7H_{11}OCl$ (continued)	2-Chlorocycloheptanone (continued)	NaOH	Not given	Cyclohexanecarboxylic acid	69	87
		K_2CO_3	H_2O	Cyclohexanecarboxylic acid	80	96
		Na_2CO_3	Not given	Cyclohexanecarboxylic acid	(20)	87
		$(CH_2)_5NH$	Not given	N,N-Pentamethylenecyclohexanecarboxamide	(20)	87
		$(CH_3)_2NH$	Not given	N,N-Dimethylcyclohexanecarboxamide	(20)	87
$C_8H_{11}OCl$	Chloromethyl cyclohexenyl ketone	$NaOCH_3$	CH_3OH	Methyl cyclohexenyl-1-acetate	20	80
$C_8H_{13}OCl$	Chloromethyl cyclohexyl ketone	$NaOC_2H_5$	C_2H_5OH	Ethyl 1-methylcyclohexanecarboxylate	20	43
		$NaOCH_3$	$(C_2H_5)_2O$	Methyl 1-methylcyclohexanecarboxylate	9	33
		$NaOCH_3$	CH_3OH	Methyl 1-methylcyclohexanecarboxylate	15, 35	33, 43
		$NaOCH_3$	CH_3OH	Methyl 1-methylcyclohexanecarboxylate and methyl cyclohexylacetate‡	50	80, 110
		$NaOCH_3$	CH_3OH-pet. ether	Methyl 1-methylcyclohexanecarboxylate	38	43
	1-Chloro-1-acetylcyclohexane	$NaOCH_2C_6H_5$	$C_6H_5CH_2OH$	Benzyl 1-methylcyclohexanecarboxylate	50, 57	34
		$NaOCH_2C_6H_5$	$(C_2H_5)_2O$	Benzyl 1-methylcyclohexanecarboxylate	72	34

Starting material	Reagent	Solvent	Product	Yield	Reference(s)
	$NaOCH(CH_3)_2$	$(C_2H_5)_2O$	Isopropyl 1-methylcyclohexanecarboxylate	45	34
	$NaOC_2H_5$	C_2H_5OH	Ethyl 1-methylcyclohexanecarboxylate	41	34
	$NaOC_2H_5$	$(C_2H_5)_2O$	Ethyl 1-methylcyclohexanecarboxylate	56	34
	$NaOCH_3$	CH_3OH	Methyl 1-methylcyclohexanecarboxylate	30	80, 110
	KOH	$(C_2H_5)_2O$	1-Methylcyclohexanecarboxylic acid		7
2-Chloro-2-methylcycloheptanone	$AgNO_3$	Aq. dioxane	*		56
2-Chlorocycloöctanone	$NaOCH_2C_6H_5$	$C_6H_5CH_2OH$	Benzyl 1-methylcycloheptanecarboxylate	(41)	34
$C_8H_{13}OBr$ Bromomethyl cyclohexyl ketone	NaOH	C_2H_5OH	Cycloheptanecarboxylic acid		109
	$NaOC_2H_5$	C_2H_5OH	*		43
1-Bromo-1-acetylcyclohexane	$NaOCH_3$	$(C_2H_5)_2O$			44
	$NaOCH_3$	$(C_2H_5)_2O$	Methyl 1-methylcyclohexanecarboxylate	79	44
2-Bromocycloöctanone	NaOH	H_2O	Cycloheptanecarboxylic acid	68	78
$C_9H_{13}OCl$ 4-Chloro-cis-5-hydrindone	$NaOCH_3$	CH_3OH	Methyl cis-bicyclo[3.3.0]octane-2- and 3-carboxylates	65	47

Note: References 103 to 127 are on p. 316.

* No rearrangement product was isolated.

† The report of the formation of the methyl cyclohexylacetate in this reaction has been shown to be erroneous.[43]

TABLE IV—Continued
Alicyclic Monohaloketones

Formula	Haloketone	Base	Solvent	Rearrangement Product	Yield (%)	References
$C_9H_{13}OBr$	1-Bromo-bicyclo[3.3.1]-nonan-9-one	KOH	$(C_2H_5)_2O$	Bicyclo[3.3.0]octane-1-carboxylic acid	34	28
		$Hg(OCOCH_3)_2$	C_2H_5OH	Ethyl bicyclo[3.3.0]octane-1-carboxylate	71	28
		$AgNO_3$	Aq. alcohols	Bicyclo[3.3.0]octane-1-carboxylic acid (and corresponding esters)		28
		Na or $NaNH_2$	Liq. NH_3	Bicyclo[3.3.0]octane-1-carboxamide	65–70	50
$C_9H_{13}O_3Br$	6-Bromo-2-carbethoxy-cyclohexanone	NaOH	Aq. C_2H_5OH	Cyclopentane-1,2-*trans*-dicarboxylic acid	91	44a
$C_9H_{15}OCl$	[structure: cyclohexane with COCH₃, Cl, CH₃, H substituents]	$NaOCH_2C_6H_5$	$(C_2H_5)_2O$	[structure: cyclohexane with CO₂CH₂C₆H₅, CH₃, CH₃, H]	(44, 53)	34
	[structure: cyclohexane with Cl, COCH₃, CH₃, H substituents]	$NaOCH_2C_6H_5$	$(C_2H_5)_2O$	[structure: cyclohexane with CO₂CH₂C₆H₅, CH₃, CH₃, H]	(29)	34
$C_9H_{15}OBr$	2-Bromocyclononanone	$C_6H_5N(CH_3)_2$	None	Cycloöctanecarboxylic acid		42
$C_{10}H_{15}OCl$	3-Chloro-*trans*-2-decalone	$NaOCH_3$	CH_3OH	Methyl *trans*-2-hydrindane-carboxylic acid	(20)	49
		KOH	C_2H_5OH	Two unidentified acids		48

THE FAVORSKIĬ REARRANGEMENT OF HALOKETONES

Formula	Compound	Base	Solvent	Product	(Yield %)	Ref.
$C_{10}H_{17}OCl$	2-Chloro-2-isopropyl-5-methylcyclohexanone	$NaOCH_3$	CH_3OH	*		41
$C_{10}H_{17}OBr$	2-Bromocyclodecanone	$NaOCH_3$	CH_3OH	Methyl cyclononane-carboxylate	(75)	42
	2-Bromo-3-methyl-6-isopropylcyclohexanone	$NaOCH_3$	CH_3OH	Methyl 2-methyl-5-isopropylcyclopentane-carboxylate		41
$C_{11}H_{17}OCl$	(structure: H₃C, H₃C-substituted cyclohexene with Cl, COCH₃, CH₃, H)	$NaOCH_2C_6H_5$	$(C_2H_5)_2O$	(structure: H₃C, H₃C-substituted cyclohexene with CO₂CH₂C₆H₅, CH₃, CH₃, H)	(44)	34
$C_{11}H_{17}O_3Cl$	(cycloheptanone with CH₃, Cl, CH₂CO₂CH₃)	$NaOCH_2C_6H_5$	$(C_2H_5)_2O$	(cyclohexane with CH₃, CO₂CH₂C₆H₅, CH₂CO₂CH₂C₆H₅, H)	(15)	34
$C_{12}H_{19}O_3Cl$	(cyclohexane with COCH₂Cl§, CH₂CH₂CO₂CH₃, H)	$NaOCH(CH_3)_2$	$(CH_3)_2CHOH$	*		33

* No rearrangement product was isolated.
§ A number of base-solvent combinations applied to this ketone have not been tabulated because of space limitations. The reader is referred to the original work.[33]

TABLE IV—Continued
ALICYCLIC MONOHALOKETONES

Formula	Haloketone	Base	Solvent	Rearrangement Product	Yield (%)	References
$C_{12}H_{19}O_3Cl$ (continued)	cyclohexane with —COCH$_2$Cl and —CH$_2$CH$_2$CO$_2$CH$_3$ substituents (H, H) §	NaOCH$_3$	$(C_2H_5)_2O$	cyclohexane with CH$_3$, CO$_2$CH$_3$, CH$_2$CH$_2$CO$_2$CH$_3$, H — CIX	(11)	33
	(continued)			cyclohexane with CO$_2$CH$_3$, CH$_3$, CH$_2$CH$_2$CO$_2$CH$_3$, H — CX	6‖	
	cyclohexane with Cl, —COCH$_3$ and —CH$_2$CH$_2$CO$_2$CH$_3$ (H)	NaOCH$_3$	CH$_3$OH	Diester CIX Diester CX	(27) 23‖	33
		NaOCH$_3$	$(C_2H_5)_2O$	Diester CIX Diester CX	(9) 5‖	33
		NaOCH$_3$	CH$_3$OH	*		33
$C_{12}H_{19}O_3Br$	cyclohexane with H, —COCH$_2$Br and —CH$_2$CH$_2$CO$_2$CH$_3$ (H)	NaOCH$_3$	CH$_3$OH or $(C_2H_5)_2O$	*		33

THE FAVORSKIĬ REARRANGEMENT OF HALOKETONES 301

Formula	Haloketone	Base	Solvent	Product	Yield (%)	Ref.
$C_{13}H_{21}OCl$	(cyclohexane with Br, COCH₃, CH₂CH₂CO₂CH₃, H)	$NaOCH_3$	$(C_2H_5)_2O$	Diester CIX / Diester CX	(3) / Trace ‖	33
		$NaOCH_3$	CH_3OH	Diester CIX	6	33
$C_{13}H_{21}OCl$	(dicyclohexyl ketone with Cl)	KOH	Dioxane	(dicyclohexyl-CO_2H)	50	46, 56
$C_{13}H_{21}OBr$	(dicyclohexyl ketone with Br)	KOH	Dioxane	*		46, 56
$C_{13}H_{21}O_3Cl$	(cyclohexane with COCH₃, Cl, CH₂CH₂CO₂C₂H₅)	$NaOCH_2C_6H_5$	$(C_2H_5)_2O$	(cyclohexane with CH₃, CO₂R, CH₂CH₂CO₂R, H) (R = $CH_2C_6H_5$ or C_2H_5)	21–25	34

* No rearrangement product was isolated.
§ A number of base-solvent combinations applied to this ketone have not been tabulated because of space limitations. The reader is referred to the original work.
‖ This was the yield of 8-methyl-*cis*-1-hydrindone from pyrolysis of the rearrangement acid.

TABLE V
Aralkyl Monohaloketones

Formula	Haloketone	Base	Solvent	Rearrangement Product	Yield (%)	References
C_9H_9OCl	1-Chloro-1-phenylacetone	NaOCH$_3$	CH$_3$OH	Methyl 3-phenylpropionate	60	8, 98
				3-Phenylpropionic acid	9	
		KOH	(C$_2$H$_5$)$_2$O	3-Phenylpropionic acid		98
		NaOH	CH$_3$OH	3-Phenylpropionic acid	48	8
		NaOC$_6$H$_5$	C$_6$H$_5$OH	Phenyl 3-phenylpropionate	65	100
		NaOC$_6$H$_5$	Dioxane	Phenyl 3-phenylpropionate	100	100
	1-Chloro-3-phenylacetone	NaOCH$_3$	CH$_3$OH	Methyl 3-phenylpropionate	80	8
	α-Chloropropiophenone	NaOCH$_3$	(C$_2$H$_5$)$_2$O	*		54
$C_{10}H_9OCl$	2-Chloro-1-tetralone	NaOCH$_3$	CH$_3$OH	*		55
	3-Chloro-2-tetralone	NaOCH$_3$	CH$_3$OH	Methyl 1-indanecarboxylate		80, 111
		NaOCH$_3$	CH$_3$OH	Methyl 2-indanecarboxylate		80, 111
$C_{10}H_9OBr$	2-Bromo-1-tetralone	NaOCH$_3$	CH$_3$OH	*		55
$C_{10}H_{11}OCl$	1-Chloro-1-phenyl-2-butanone	NaOCH$_3$	CH$_3$OH	2-Benzylpropionic acid		98
		KOH	(C$_2$H$_5$)$_2$O	2-Benzylpropionic acid		98
		NaOC$_6$H$_5$	C$_6$H$_5$OH	Phenyl 2-benzylpropionate	30	100
		NaOC$_6$H$_5$	Dioxane	Phenyl 2-benzylpropionate	50	100
	2-Chloro-1-phenyl-3-butanone	NaOCH$_3$	CH$_3$OH	4-Phenylbutyric acid		98
		NaOH	CH$_3$OH	Unidentified acid		8
		KOH	(C$_2$H$_5$)$_2$O	4-Phenylbutyric acid		9, 98
	1-Chloro-4-phenyl-2-butanone	NaOH	CH$_3$OH	4-Phenylbutyric acid		8
		KOH	(C$_2$H$_5$)$_2$O	4-Phenylbutyric acid		98

THE FAVORSKIĬ REARRANGEMENT OF HALOKETONES

Formula	Compound	Reagent	Solvent	Product	Yield	Ref.
$C_{10}H_{11}OBr$	α-Bromoisobutyrophenone	$NaOCH_3$	$(C_2H_5)_2O$	†		39
		KOH	$(C_2H_5)_2O$	†		28
		$AgNO_3$	Aq. C_2H_5OH	2-Methyl-2-phenylpropionic acid		28
$C_{11}H_{13}O_3Cl$	1-Chloro-3-(3,4-dimethoxyphenyl)acetone	$NaOC_2H_5$	C_2H_5OH	Ethyl 3-(3,4-dimethoxyphenyl)propionate		8, 51
		$NaOCH_3$	CH_3OH	Methyl 3-(3,4-dimethoxyphenyl)propionate		8, 51
		KOH	CH_3OH	Methyl 3-(3,4-dimethoxyphenyl)propionate	80	8, 51
$C_{13}H_{15}OCl$	1-Chloro-1-benzoylcyclohexane	$NaOCH_3$	$(C_2H_5)_2O$	†		54
		NaOH	Xylene	1-Phenylcyclohexanecarboxylic acid	53	27
		NaOH	Toluene	1-Phenylcyclohexanecarboxylic acid	51	27
		NaOH	$(C_2H_5)_2O$	1-Phenylcyclohexanecarboxylic acid	8	27
		KOH	$(C_2H_5)_2O$	1-Phenylcyclohexanecarboxylic acid	30–40	7
		KOH	Aq. dioxane	1-Phenylcyclohexanecarboxylic acid		27
		$AgNO_3$	Aq. dioxane	†	40	56
$C_{13}H_{15}OBr$	1-Bromo-1-benzoylcyclohexane	$NaOCH_3$	CH_3OH	†		81
		NaOH	Xylene	1-Phenylcyclohexanecarboxylic acid	39	27

Note: References 103 to 127 are on p. 316.
* Only hydroxy ketal was isolated.
† No rearrangement product was isolated.

TABLE V—*Continued*

ARALKYL MONOHALOKETONES

Formula	Haloketone	Base	Solvent	Rearrangement Product	Yield (%)	References
$C_{13}H_{15}OBr$ *(continued)*	1-Bromo-1-benzoylcyclo-hexane *(continued)*	NaOH	Toluene	1-Phenylcyclohexane-carboxylic acid	34	27
		NaOH	$(C_2H_5)_2O$	1-Phenylcyclohexane-carboxylic acid	6	27
		$AgNO_3$	C_2H_5OH	1-Phenylcyclohexane-carboxylic acid	18	27
		$AgNO_3$	Aq. dioxane	1-Phenylcyclohexane-carboxylic acid	30	56
		None	Aq. dioxane	1-Phenylcyclohexane-carboxylic acid	2	27
$C_{13}H_{16}ONCl$	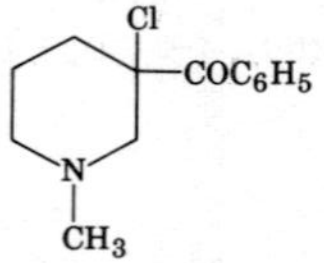 (structure: Cl, COC$_6$H$_5$ on piperidine N–CH$_3$)	NaOH	Xylene	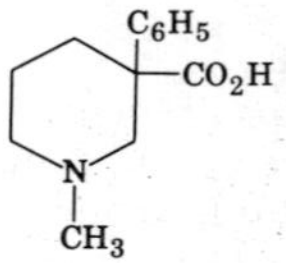 (structure: C$_6$H$_5$, CO$_2$H on piperidine N–CH$_3$)	8	57
	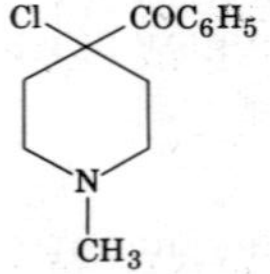 (structure: Cl, COC$_6$H$_5$ on piperidine N–CH$_3$)	NaOH	Xylene	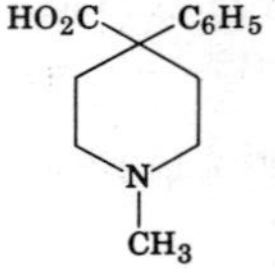 (structure: HO$_2$C, C$_6$H$_5$ on piperidine N–CH$_3$)	25	57

THE FAVORSKIĬ REARRANGEMENT OF HALOKETONES

$C_{15}H_{13}OCl$	1-Chloro-1,1-diphenylacetone	$NaOC_2H_5$	C_2H_5OH	Ethyl 3,3-diphenylpropionate	85	58
		$NaOCH_3$	CH_3OH	3,3-Diphenylpropionic acid		98
		NaOH	$(C_2H_5)_2O$	3,3-Diphenylpropionic acid	55	58
		KOH	$(C_2H_5)_2O$	3,3-Diphenylpropionic acid	40	9, 98
	1-Chloro-1,3-diphenylacetone	$NaOC_2H_5$	Not given	Ethyl 2,3-diphenylpropionate		87
		NaOH	Not given	2,3-Diphenylpropionic acid		87
		KOH	$(C_2H_5)_2O$	2,3-Diphenylpropionic acid		98
		Piperidine	Not given	$C_6H_5CH_2CH(C_6H_5)CONC_5H_{10}$ †	20	87
		$(CH_3)_2NH$	Not given			87
	1-Chloro-3,3-diphenylacetone	$NaOC_2H_5$	C_2H_5OH	Ethyl 3,3-diphenylpropionate	69	58
		$NaOCH_3$	CH_3OH	3,3-Diphenylpropionic acid	12	52
				Methyl 3-3-diphenylpropionate	77	
		$NaOCH_3$	$(C_2H_5)_2O$	3,3-Diphenylpropionic acid	3.5	52
				Methyl 3,3-diphenylpropionate	43	
$C_{15}H_{13}OBr$	1-Bromo-3,3-diphenylacetone	$NaOCH_3$	CH_3OH	3,3-Diphenylpropionic acid	7	52
				Methyl 3,3-diphenylpropionate	71	
		$NaOCH_3$	$(C_2H_5)_2O$	3,3-Diphenylpropionic acid	6	52
				Methyl 3,3-diphenylpropionate	31	
		$(C_2H_5)_2NH$	$(C_2H_5)_2O$	N,N-Diethyl-3,3-diphenylpropionamide	15	88

† No rearrangement product was isolated.

TABLE V—Continued
ARALKYL MONOHALOKETONES

Formula	Haloketone	Base	Solvent	Rearrangement Product	Yield (%)	References
$C_{16}H_{13}OCl$	9-(COCH$_2$Cl)-9,10-dihydroanthracene	$(C_2H_5)_2NH$	$(C_2H_5)_2O$	9-[CH$_2$CON(C$_2$H$_5$)$_2$]-9,10-dihydroanthracene	45	53
$C_{16}H_{13}OBr$	9-(COCH$_2$Br)-9,10-dihydroanthracene	R_2NH (R = C_2H_5, n-C_3H_7, n-C_5H_{11})	$(C_2H_5)_2O$	9-(CH$_2$CONR$_2$)-9,10-dihydroanthracene	37–45	53
$C_{19}H_{19}OBr$	2-Bromo-7,7-diphenyl-cycloheptanone	KOH	C_2H_5OH	‡		89
$C_{21}H_{27}O_4Cl$	(structure with COCH$_2$Cl, CH$_2$CH$_2$CO$_2$CH$_3$, CH$_3$O on phenanthrene skeleton)	CH_3OH		(rearranged phenanthrene with CO$_2$CH$_3$, CH$_3$, CH$_2$CH$_2$CO$_2$CH$_3$, H, CH$_3$O)	44§	33

‡ No normal Favorskii product was isolated.
§ This was the yield of estrone-c methyl ether obtained from the rearrangement product by Dieckmann cyclization and subsequent hydrolysis.

TABLE VI
Steroid Monohaloketones

Formula	Haloketone	Base	Solvent	Rearrangement Product	Yield (%)	References
$C_{21}H_{29}O_2Cl$	21-Chloro-4-pregnen-3,20-dione	$KOCH_3$	CH_3OH	Methyl 3-oxo-17β-methyl-4-etienate	58	112
				Methyl 3-oxo-17α-methyl-4-etienate	25	32, 93
$C_{21}H_{31}O_2F$	21-Fluoro-5-pregnen-3β-ol-20-one	$NaOCH_3$	CH_3OH	Methyl 3β-hydroxy-17α-methyl-5-etienate	ca. 20	64a
				Methyl 3β-hydroxy-17β-methyl-5-etienate	ca. 10	
$C_{21}H_{31}O_2Cl$	21-Chloro-5-pregnen-3β-ol-20-one	$KOCH_3$	CH_3OH	Methyl 3β-hydroxy-17α-methyl-5-etienate	63	32, 93
				Methyl 3β-hydroxy-17β-methyl-5-etienate	24	
$C_{21}H_{31}O_2Br$	21-Bromo-5-pregnen-3β-ol-20-one	$KOCH_3$	CH_3OH	Methyl 3β-hydroxy-17α-methyl-5-etienate		32, 93
				Methyl 3β-hydroxy-17β-methyl-5-etienate		
$C_{22}H_{33}O_3Br$	17-Bromo-D-homoandrostan-3β-ol-17a-one acetate	$NaOCH_3$	Dioxane	Methyl 3β-hydroxy-allo-etianate	0.3	63
$C_{23}H_{33}O_3Br$	17α-Bromo-5-pregnen-3β-ol-20-one acetate	$NaHCO_3$	H_2O-CH_3OH	3β-Hydroxy-17α-methyl-5-etienic acid and methyl ester	85*	97

Note: References 103 to 127 are on p. 316.

* This was the yield of the methyl ester acetate; its stereochemical homogeneity (about C-17) is uncertain.

TABLE VI—Continued

STEROID MONOHALOKETONES

Formula	Haloketone	Base	Solvent	Rearrangement Product	Yield (%)	References
$C_{23}H_{33}O_4Br$	17α-Bromopregnan-3α-ol-11,20-dione acetate	NaOCH$_3$	CH$_3$OH	Methyl 3α-hydroxy-11-oxo-17α-methyletianate	60	10
				Methyl 3α-hydroxy-11-oxo-17β-methyletianate	40	
		KHCO$_3$	CH$_3$OH	3α-Hydroxy-11-oxo-17α-methyletianic acid and methyl ester	77 (crude)	64, 113
				Methyl 3α-hydroxy-11-oxo-17β-methyletianate	20 (crude)	
$C_{23}H_{35}O_3Br$	17α-Bromo-allopregnane-3β-ol-20-one acetate	KHCO$_3$	CH$_3$OH	Methyl 3β-hydroxy-17-methylalloetianate	37*	93
	17α-Bromopregnane-3β-ol-20-one acetate	KHCO$_3$	CH$_3$OH	Methyl 3β-hydroxy-17-methyletianate	49*	114
$C_{27}H_{45}OCl$	2α-Chlorocholestan-3-one	NaOCH$_3$	CH$_3$OH	Methyl A-norcholestane-2-carboxylate	30	62

$C_{27}H_{45}OBr$	2α-Bromocholestan-3-one	$NaOC_2H_5$	C_2H_5OH	Ethyl A-norcholestane-2-carboxylate	14–30	60, 61
				Ethyl A-norcholestane-3-carboxylate	12–20	
		$NaOCH_3$	CH_3OH-$(C_2H_5)_2O$	Methyl A-norcholestane-2-carboxylate	25	59, 60
				Methyl A-norcholestane-3-carboxylate	1	
	4β-Bromocoprostan-3-one	$NaOCH_3$	CH_3OH-$(C_2H_5)_2O$	Methyl A-norcoprostane-2-carboxylate	24	60
				Methyl A-norcoprostane-3-carboxylate	24	
		$NaOCH_3$	CH_3OH	Methyl A-norcoprostane-2-carboxylate and methyl A-norcoprostane-3-carboxylate	61	60
		$NaOCH_3$	Aq. CH_3OH	A-Norcoprostane 2-carboxylic acid and A-norcoprostane 3-carboxylic acid	18	60

Note: References 103 to 127 are on p. 316.

* This was the yield of the methyl ester acetate; its stereochemical homogeneity (about C-17) is uncertain.

TABLE VII
DIHALOKETONES

Formula	Haloketone	Base	Solvent	Rearrangement Product	Yield (%)	References
$C_3H_4OCl_2$	1,1-Dichloroacetone	K_2CO_3	H_2O	Acrylic acid		1
$C_4H_6OCl_2$	3,3-Dichloro-2-butanone	K_2CO_3	H_2O	α-Methylacrylic acid		1
$C_5H_8OCl_2$	Mixture of 3,3-dichloro-2-pentanone and 2,2-dichloro-3-pentanone	K_2CO_3	H_2O	Angelic acid 2-Ethylacrylic acid		1
$C_5H_8OBr_2$	1,3-Dibromo-3-methyl-2-butanone	KOH	C_2H_5OH	3-Methyl-2-butenoic acid Ethyl 3-methyl-2-butenoate		3
		$NaOCH_3$	$(C_2H_5)_2O$	Methyl 3-methyl-2-butenoate	58	66
	1,2-Dibromo-2-methyl-3-butanone	$NaOCH_3$	$(C_2H_5)_2O$	Methyl 3-methyl-2-butenoate	42, 64	68
$C_6H_8OBr_2$	2,6-Dibromocyclohexanone	$NaOCH_3$	CH_3OH	Methyl cyclopentene-1-carboxylate	(5)	110
$C_6H_{10}OCl_2$	Mixture of 3,3-dichloro-2-hexanone and 2,2-dichloro-3-hexanone	KOH K_2CO_3	H_2O H_2O	* 2-n-Propylacrylic acid 2-Methyl-2-pentenoic acid		72 1
$C_6H_{10}OBr$	1,3-Dibromo-3-methyl-2-pentanone	$NaOCH_3$	$(C_2H_5)_2O$	Methyl cis-2-methyl-2-pentenoate Methyl trans-2-methyl-2-pentenoate	29 22	66
	3,4-Dibromo-3-methyl-2-pentanone	$NaOCH_3$	$(C_2H_5)_2O$	Methyl trans-2-methyl-3-pentenoate	55	68

THE FAVORSKIĬ REARRANGEMENT OF HALOKETONES 311

Formula	Compound	Base	Solvent	Product	Yield	Ref.
$C_7H_{10}OBr$	2,3-Dibromo-2-methyl-cyclohexanone	$NaOCH(CH_3)_2$	$(CH_3)_2CHOH$	*		34
$C_7H_{10}O_3Br$	$BrCH_2COCO_2C_2H_5$ $\underset{Br}{|}$ CH_3	KOH	C_2H_5OH	$HO_2C\underset{H_3C}{\overset{H}{\diagup}}C{=}C\underset{CO_2H}{\diagdown}$		74, 75
$C_7H_{12}OCl_2$	Mixture of 3,3-dichloro-4-heptanone and 4,4-dichloro-3-heptanone	K_2CO_3	H_2O	Unidentified unsaturated acid		1
$C_8H_{12}OCl_2$	2,8-Dichlorocycloöctanone	NaOH	Aq. C_2H_5OH	Cycloheptene-1-carboxylic acid	85	78
$C_8H_{12}OBr_2$	1-Bromo-1-bromoacetyl-cyclohexane	$NaOCH_3$	$(C_2H_5)_2O$	Methyl cyclohexylidene-acetate	48	66
		KOH	C_2H_5OH	Cyclohexylideneacetic acid	20	115
		KOH	H_2O	Cyclohexylideneacetic acid	7	101
		K_2CO_3	H_2O	Cyclohexylideneacetic acid	34	101
	1,2-Dibromo-1-acetylcyclohexane	$NaOCH_3$	$(C_2H_5)_2O$	Methyl cyclohexenyl-1-acetate		103
	2,3-Dibromo-2-methyl cycloheptanone	$NaOCH_2C_6H_5$	$C_6H_5CH_2OH$	Benzyl 1-methylcyclohexene-2-carboxylate and benzyl 1-methyl-cyclohexene-6-carboxylate	(69)	34
	2,8-Dibromocycloöctanone	NaOH	C_2H_5OH	Cycloheptene-1-carboxylic acid	87	78
		NaOH	H_2O	Cycloheptene-1-carboxylic acid	96	78

Note: References 103 to 127 are on p. 316.
* No normal Favorskiĭ product was isolated.

TABLE VII—Continued
DIHALOKETONES

Formula	Haloketone	Base	Solvent	Rearrangement Product	Yield (%)	References
$C_8H_{14}OBr_2$	2,4-Dibromo-2,5-dimethyl-3-hexanone	$NaOCH_3$	$(C_2H_5)_2O$	$(CH_3)_2C=C(CO_2CH_3)CH(CH_3)_2$	84	66
$C_9H_{14}OBr_2$	2,2(?)-Dibromocyclononanone	$C_6H_5N(CH_3)_2$	None	*		42
$C_9H_{15}ONBr_2$	3,5-Dibromo-2,2,6,6-tetramethyl-4-piperidone	NH_3	H_2O	2,2,6,6-Tetramethyl-3-pyrroline-3-carboxamide		26, 69
		CH_3NH_2	H_2O	2,2,6,6-Tetramethyl-3-pyrroline-3-N-methyl-carboxamide		26, 116
		RNH_2	None	2,2,6,6-Tetramethyl-3-pyrroline-3-N-alkyl-carboxamides *		26, 116
$C_{10}H_{14}OCl_2$	3,3-Dichloro-*trans*-2-decalone	Na_2CO_3	H_2O	*		117
$C_{10}H_{14}O_5Br_2$	BrCH$_2$COCCH$_2$CO$_2$C$_2$H$_5$ | CO$_2$C$_2$H$_5$ / Br	$BaCO_3$	H_2O	$HO_2C\diagdown\diagup H$ $C=C$ $HO_2CCH_2\diagup\diagdown CO_2H$		16
$C_{10}H_{16}OBr_2$	Dibromopulegone	KOH	H_2O	2-Methyl-5-isopropylidene-cyclopentanecarboxylic acid (and unidentified congeners)		73
	2,2(?)-Dibromocyclodecanone	$C_6H_5N(CH_3)_2$	None	*		42

Formula	Compound	Reagent	Product	Yield	Ref.
$C_{21}H_{30}OBr_2$	17α,21-Dibromo-allopregnan-20-one	KOH	17(20)-Allopregnen-21-oic acid	30	118
$C_{21}H_{30}O_2F_2$	21,21-Difluoro-5-pregnen-3β-ol-20-one	NaOCH₃	Methyl 5,17(20)-*trans*-pregnadien-3β-ol-21-oate	55	64a
$C_{21}H_{32}O_2Br_2$	17α,21-Dibromopregnan-3β-ol-20-one	KOH	17(20)-Pregnen-3β-ol-21-oic acid	85, 100 (crude)	119
$C_{23}H_{32}O_3BrI$	17α-Bromo-21-iodo-5-pregnen-3β-ol-20-one acetate	KOH	5,17(20)-Pregnadien-3β-ol-21-oic acid

![CXI structure: H–C(CO₂R)=C(CH₃)–ring with CH₃] (R = H and CH₃)

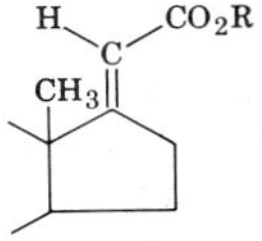

CXI

![CXII structure: RO₂C–C(H)=C(CH₃)–ring with CH₃] (R = H and CH₃)

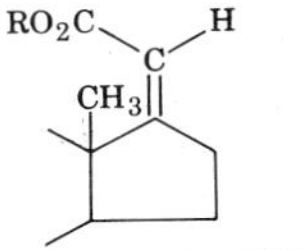

CXII | ca. 25

ca. 15 | 71 |
| $C_{23}H_{32}O_4Br_2$ | 17α,21-Dibromopregnan-3α-ol-11,20-dione acetate | Aq. CH₃OH | 17(20)-Pregnen-3α-ol-11-one-21-oic acid | 70 | 10, 64 |

Note: References 103 to 127 are on p. 316.

* No normal Favorskiĭ product was isolated.

TABLE VII—Continued
DIHALOKETONES

Formula	Haloketone	Base	Solvent	Rearrangement Product	Yield (%)	References
$C_{23}H_{34}O_3Br_2$	16,17-Dibromopregnan-3β-ol-20-one acetate	KOH	CH_3OH	17(20)-Pregnen-3β-ol-21-oic acid	60	67
				Methyl 17(20)-pregnen-3β-ol-21-oate	8	
	17α,21-Dibromo-allopregnan-3β-ol-20-one acetate	KOH	CH_3OH	17(20)-Allopregnen-3β-ol-21-oic acid	83	115
		KOH or aq. $KHCO_3$	CH_3OH	17(20)-Allopregnen-3β-ol-21-oic acid and methyl ester		93, 102
$C_{25}H_{30}O_5Br_2$	21,21-Dibromo-21-ethoxy-oxalyl-1,4-pregnadien-3,20-dione	$NaOCH_3$	CH_3OH	Methyl 1,4,17(20)-pregnatrien-3-one-21-oate†	40	123
$C_{25}H_{30}O_6Br_2$	21,21-Dibromo-21-ethoxy-oxalyl-1,4-pregnadien-11α-ol-3,20-dione	$NaOCH_3$	CH_3OH	Methyl 1,4,17(20)-pregnatrien-11α-ol-3-one-21-oate†		123
$C_{25}H_{32}O_6Br_2$	21,21-Dibromo-21-ethoxy-oxalyl-4-pregnen-3,11,20-trione	$NaOCH_3$	CH_3OH	Methyl 4,17(20)-pregnadien-3,11-dione-21-oate†	60	124
$C_{25}H_{34}O_6Br_2$	21,21-Dibromo-21-ethoxy-oxalyl-4-pregnen-11α-ol-3,20-dione	$NaOCH_3$	CH_3OH	Methyl 4,17(20)-pregnadien-11α-ol-3-one-21-oate†		124
$C_{25}H_{36}O_5Br_2$	17,21-Dibromopregnan-3α,12β-diol-20-one diacetate	KOH	CH_3OH	17(20)-Pregnen-3α,12β-diol-21-oic acid		102
$C_{29}H_{46}O_3Br_2$	5α,7α-Dibromocholestan-3β-ol-6-one-acetate	C_5H_5N	None	B-Nor-5(6)-cholestene-3β-ol-6-carboxylic acid acetate	23	70

Note: References 103 to 127 are on p. 316.

† In the absence of excess base the thermodynamically less stable $\Delta^{17(20)}$-*cis* ester (partial structure CXII) is obtained.[123–125]

TABLE VIII
TRIHALOKETONES

Formula	Haloketone	Base	Solvent	Rearrangement Product	Yield (%)	References
$C_5H_7OBr_3$	1,1,3-Tribromo-3-methyl-2-butanone	KOH	Aq. C_2H_5OH	2-Bromo-3-methyl-2-butenoic acid	10	77
		KOH	H_2O	*		3
$C_7H_9OClBr_2$	2-Chloro-2,7-dibromocycloheptanone	CH_3CO_2Na	Aq. C_2H_5OH	2-Chloro-1-cyclohexene-carboxylic acid	43–55	78
$C_8H_{11}OCl_3$	2,2,8-Trichlorocyclooctanone	NaOH	Aq. C_2H_5OH	2-Chlorocycloheptene-1-carboxylic acid	68	78
$C_8H_{11}OBr_3$	1-Bromo-1-dibromoacetylcyclohexane	KOH	C_2H_5OH	α-Bromocyclohexylideneacetic acid	33	77
	2,2,8-Tribromocyclooctanone	$NaOC_2H_5$	C_2H_5OH	Ethyl 2-bromocycloheptene-1-carboxylate	83	78
		NaOH	H_2O	2-Bromocycloheptene-1-carboxylic acid	83	78
		CH_3CO_2Na	C_2H_5OH	Ethyl 2-bromocycloheptene-1-carboxylate	83	78
		CH_3CO_2Na	CH_3CO_2H	2-Bromocycloheptene-1-carboxylic acid	83	78
		HCO_2Na	C_2H_5OH	Ethyl 2-bromocycloheptene-1-carboxylate		78
$C_{23}H_{31}O_3Br_3$	17α,21,21-Tribromo-5-pregnen-3β-ol-20-one acetate	KOH	Aq. CH_3OH	20-Bromo-5,17(20)-pregnadien-3β-ol-21-oic acid	72	71
$C_{23}H_{33}O_3Br_3$	17α,21,21-Tribromopregnan-3β-ol-20-one acetate	KOH	C_2H_5OH	20-Bromo-17(20)-pregnen-3β-ol-21-oic acid	57, 61	77
$C_{25}H_{35}O_5Br_3$	17,21,21-Tribromo-3α,12α-diacetoxypregnen-20-one	KOH	Aq. C_2H_5OH	20-Bromo-17(20)-pregnen-3α,12α-diol-21-oic acid	58	126
$C_{29}H_{33}O_9Br_3$	2,21,21-Tribromo-2,21-bis-ethoxyoxalyl-4-pregnen-3,11,20-trione	$NaOCH_3$	CH_3OH	Methyl 2-bromo-4,17(20)-pregnadiene-3,11-dione-21-carboxylate		127

Note: References 103 to 127 are on p. 316.
* No normal Favorskiĭ product was isolated.

REFERENCES FOR TABLES III–VIII

[103] Richard, *Bull. soc. chim. France*, [5] **5**, 286 (1938).
[104] Favorskiĭ and Bozhovskiĭ, *J. Russ. Phys.-Chem. Soc.*, **50**, 582 (1920) [*C.A.*, **18**, 1476 (1924)].
[105] Jackman, Bergman, and Archer, *J. Am. Chem. Soc.*, **70**, 497 (1948).
[106] Favorskiĭ and Bozhovskiĭ, *J. Russ. Phys.-Chem. Soc.*, **46**, 1097 (1914) [*C.A.*, **9**, 1900 (1915)].
[107] Mousseron, Winternitz, and Jacquier, *Bull. soc. chim. France*, [5] **14**, 83 (1947).
[108] Mousseron, Richaud, and Granger, *Bull. soc. chim. France*, [5] **13**, 625 (1946).
[109] Steadman, *J. Am. Chem. Soc.*, **62**, 1606 (1940).
[110] Mousseron, Jacquier, Fontaine, and Canet, *Compt. rend.*, **232**, 1562 (1951).
[111] Mousseron and Phuoc Du, *Compt. rend.*, **218**, 281 (1944).
[112] Engel and Just, *J. Am. Chem. Soc.*, **76**, 4909 (1954).
[113] Engel, *J. Am. Chem. Soc.*, **77**, 1064, (1955).
[114] Marker and Wagner, *J. Am. Chem. Soc.*, **64**, 216 (1942).
[115] Marker, Crooks, Wagner, and Wittbecker, *J. Am. Chem. Soc.*, **64**, 2089 (1942).
[116] Pauly and Boehm, *Ber.*, **33**, 919 (1900).
[117] Lehmann and Krätschell, *Ber.*, **68**, 360 (1935).
[118] Marker et al., *J. Am. Chem. Soc.*, **64**, 822 (1942).
[119] Marker, Crooks, and Wagner, *J. Am. Chem. Soc.*, **64**, 817 (1942).
[120] Julian and Karpel, *J. Am. Chem. Soc.*, **72**, 362 (1950).
[121] Marker, Crooks, Jones, and Shabica, *J. Am. Chem. Soc.*, **64**, 1276 (1942).
[122] Sondheimer, Mancera, Urquiza, and Rosenkranz, *J. Am. Chem. Soc.*, **77**, 4145 (1955).
[123] Magerlein and Hogg, *J. Am. Chem. Soc.*, **80**, 2220 (1958).
[124] Hogg et al., *J. Am. Chem. Soc.*, **77**, 4436 (1955).
[125] J. Hogg, private communication.
[126] Adams, Patel, Petrow, and Stuart-Webb, *J. Chem. Soc.*, **1954**, 1825.
[127] Hogg et al., *J. Am. Chem. Soc.*, **77**, 4438 (1955).

CHAPTER 5

OLEFINS FROM AMINES: THE HOFMANN ELIMINATION REACTION AND AMINE OXIDE PYROLYSIS

ARTHUR C. COPE

Massachusetts Institute of Technology

ELMER R. TRUMBULL

Colgate University

CONTENTS

	PAGE
INTRODUCTION	319
THE HOFMANN EXHAUSTIVE METHYLATION	320
MECHANISM	322
DIRECTION OF ELIMINATION	331
Aliphatic Amines	332
Alicyclic Amines	336
Heterocyclic Amines	338
The Hofmann Rule	348
REACTION WITH DIAMINES	349
SIDE REACTIONS: ALKYLATIONS BY QUATERNARY COMPOUNDS	350
Alcohol Formation	350
Ethers and Epoxides	352
ISOMERIZATION OF OLEFINS FORMED	355
MOLECULAR REARRANGEMENTS	356
ANALOGOUS "ONIUM" COMPOUNDS	357
EXPERIMENTAL CONSIDERATIONS	357
NATURE OF THE BASE	357

	PAGE
PYROLYSIS OF AMINE OXIDES	361
Mechanism	362
DIRECTION OF ELIMINATION	363
Acyclic Amines	363
Alicyclic Amines	365
Heterocyclic Amines	368
Side Reactions	368
DECOMPOSITION OF AMINE PHOSPHATES	371
DECOMPOSITION OF ACYL DERIVATIVES OF AMINES	371
REACTION OF QUATERNARY SALTS WITH ORGANOMETALLIC COMPOUNDS OR ALKALI METAL AMIDES	373
COMPARISON OF METHODS	374
EXPERIMENTAL CONDITIONS AND PROCEDURES	376
Isolation of Products	377
Preparation of Amine Oxides	378
Phosphoric Acid Deamination	379
Cycloheptyltrimethylammonium Iodide. Alkylation with Methyl Iodide.	379
n-Propyltrimethylammonium Iodide. Alkylation of Trimethylamine	380
Di-n-butyldiisoamylammonium Iodide. Alkylation of a Hindered Amine	380
Preparation of Silver Oxide	380
Di-n-butyldiisoamylammonium Hydroxide. Use of Silver Oxide	381
Decomposition of Di-n-butyldiisoamylammonium Hydroxide	381
1-Hexene. Methylation with Dimethyl Sulfate and Decomposition of the Sulfate	381
des-N-Methylaphylline. Decomposition of a Quaternary Hydroxide under Reduced Pressure	382
Dihydro-des-N-dimethylcytisine. Decomposition Followed by Hydrogenation	382
Decomposition of Cyclopropyltrimethylammonium Hydroxide. High Temperature Decomposition	383
1-Benzoyl-7-propionylheptatriene. Decomposition of a β Amino Ketone	383
N-Uramidohomomeroquinene. Decomposition of a Quaternary Iodide with Excess Base	384
Preparation and Decomposition of Cyclohexylphenethyldimethylammonium Hydroxide. Use of Thallous Hydroxide	384
trans-1,2-Octalin. Use of Silver Sulfate and Barium Hydroxide	385
des-N-Methyldihydro-β-erythroidinol. Use of an Ion Exchange Resin	385
Cularinemethine. Decomposition in Aqueous Solution with Added Base	386
Methylenecyclohexane and N,N-Dimethylhydroxylamine Hydrochloride	386
N,N-Dimethylcyclooctylamine Oxide	386
cis-Cyclooctene	387

	PAGE
TABULAR SURVEY	387
Table XI. Epoxides from β Amino Alcohols	389
Table XII. Pyrolysis of Amines with Phosphoric Acid or Phosphorus Pentoxide	391
Table XIII. Olefins from Acetyl Derivatives of Amines	393
Table XIV. Pyrolysis of Amine Oxides	394
Table XV. Pyrolysis of Oxides of Tertiary Amines without N-Methyl Groups	400
Table XVI. Decomposition of Quaternary Ammonium Compounds	401
Table XVII. Quaternary Compounds that Contain No N-Methyl Groups	430
Table XVIII. Hofmann Elimination Reaction with Alkaloids	433
Table XIX. List of Alkaloids by Type	487

INTRODUCTION*

The conversion of an amine to an olefin by elimination of the nitrogen atom and an adjoining hydrogen atom is a useful procedure for degradation and synthesis.

$$-\underset{H}{\overset{|}{C}}-\underset{N}{\overset{|}{C}}- \rightarrow \hspace{1em} \text{C}=\text{C}$$

The Hofmann exhaustive methylation method has been used most often to bring about this change, but other methods such as the thermal decomposition of amine oxides and the pyrolysis of amine phosphates or acetyl or benzoyl derivatives have often been employed to advantage.

$$-\underset{H}{\overset{|}{C}}-\underset{\overset{\oplus}{N}(CH_3)_3\,\overset{\ominus}{O}H}{\overset{|}{C}}- \rightarrow \text{C}=\text{C} + (CH_3)_3N + H_2O$$

$$-\underset{H}{\overset{|}{C}}-\underset{\underset{O\quad CH_3}{N-CH_3}}{\overset{|}{C}}- \rightarrow \text{C}=\text{C} + (CH_3)_2NOH$$

$$-\underset{H}{\overset{|}{C}}-\underset{\underset{\underset{R}{C=O}}{NH}}{\overset{|}{C}}- \rightarrow \text{C}=\text{C} + RCN + H_2O$$

* The authors are indebted to Robert W. Gleason for checking the literature referred to in the final draft of this chapter.

In this chapter the Hofmann elimination will be reviewed first because of its extensive history. This will be followed by a consideration of the alternative methods and a comparison of these reactions as a means of converting amines to olefins.

THE HOFMANN EXHAUSTIVE METHYLATION*

Decomposition of a quaternary ammonium hydroxide with the formation of a tertiary amine, an olefin, and water was reported by Hofmann in 1851.[1, 2] However, it was only with his application of the reaction to the study of the structure of piperidines in 1881[3, 4] that the utility of this method in the investigation of nitrogenous bases was appreciated. Since then it has become a routine step in the study of alkaloids. Since a methyl group cannot be eliminated as an olefin, cleavage must take place to free another group from the nitrogen atom. If the original amine is

$$\begin{bmatrix} H & H & CH_3 \\ | & | & \oplus \diagup \\ CH_3-C-C-N-CH_3 \\ | & | & \diagdown \\ H & H & CH_3 \end{bmatrix} OH^{\ominus} \rightarrow \begin{matrix} CH_3 & & H \\ \diagdown & & \diagup \\ & C=C & \\ \diagup & & \diagdown \\ H & & H \end{matrix} + (CH_3)_3N + H_2O$$

heterocyclic, this cleavage gives rise to a compound containing both an olefinic and a tertiary amino group. Repetition of the procedure yields a diene and trimethylamine. The degradation of N-methylpyrrolidine[5] (I) may be used to illustrate these steps.

---

* The term "Hofmann degradation" is often used to describe the reaction sequence under discussion but may be confusing because it is also used to designate the Hofmann hypobromite reaction (*Organic Reactions*, Vol. III, Chapter 7). Furthermore, some authors distinguish between the pyrolysis of a quaternary ammonium hydroxide itself and the pyrolysis of the same compound in the presence of excess alkali hydroxide, calling only the latter a "Hofmann degradation." Recently it has been proposed to restrict the phrase "exhaustive methylation" to those instances in which the procedure of methylation and pyrolysis is carried through enough stages to eliminate the nitrogen atom from the original molecule. However, most authors seem to use the phrase "exhaustive methylation" to designate an elimination reaction which involves the preparation of a quaternary ammonium compound by methylation and pyrolysis of this compound in the presence of base or pyrolysis of the corresponding quaternary hydroxide. It is in this sense that "Hofmann exhaustive methylation" is used in this chapter. The more general phrases "decomposition of quaternary salts" and "decomposition of quaternary hydroxides" will be used to denote reactions that do not fit the foregoing definition.

[1] Hofmann, *Ann.*, **78**, 253 (1851).
[2] Hofmann, *Ann.*, **79**, 11 (1851).
[3] Hofmann, *Ber.*, **14**, 494 (1881).
[4] Hofmann, *Ber.*, **14**, 659 (1881).
[5] Ciamician and Magnaghi, *Ber.*, **18**, 2079 (1885).

OLEFINS FROM AMINES

[Reaction scheme I: N-methylpyrrolidine ($C_5H_{11}N$) → [N,N-dimethylpyrrolidinium]$^+$ I^- → [N,N-dimethylpyrrolidinium]$^+$ OH^- → N,N-dimethylaminobutene ($C_6H_{13}N$, II) + H_2O]

[Reaction scheme II → III: N,N-dimethylaminobutene (II) → trimethylammonium iodide intermediate I^- → trimethylammonium hydroxide intermediate OH^- → butadiene (C_4H_6, III) + $(CH_3)_3N$ + H_2O]

In compounds like quinolizidine derivatives in which the nitrogen atom is located at a bridgehead, three such steps would be necessary to eliminate it as trimethylamine.

Quinolizidine

Thus the degradation not only introduces a new functional group, the olefinic double bond, which allows further degradation, but the number of steps required to liberate the nitrogen atom as trimethylamine is an indication of its situation in the original compound. In some instances the course of the reaction has been cited as evidence for a particular stereochemical assignment in the original amine.[6,7]

In order to describe these reaction products in cases in which the structure of the parent amine is still unknown, or systematic nomenclature would be too cumbersome, two systems are in common use. According to the "methine" system, the Hofmann product is called the methine or methine base of the parent alkaloid; so II would be pyrrolidinemethine. The product obtained by repeating the process of methylation and pyrolysis would be the *bis*-methine and that obtained after three steps, a *tris*-methine. This nomenclature is used widely in naming degradation products of morphine and its derivatives and some other alkaloids. The

---

[6] Findlay, *J. Am. Chem. Soc.*, **76**, 2855 (1954).
[7] Goutarel, Janot, Prelog, and Sneeden, *Helv. Chim. Acta*, **34**, 1962 (1951).

alternative "des" system takes advantage of the fact that, after each step of the Hofmann degradation, one more methyl group has been added to the nitrogen atom. When the amino group is finally eliminated, the resulting compound may be described as the "des aza" derivative. Thus II would be des-N-dimethylpyrrolidine and III would be des-aza-pyrrolidine. The product is called the "des" base of the parent amine with a prefix to indicate the number of methyl groups which have been added to the nitrogen atom.

In addition to its value in alkaloid studies, the Hofmann elimination reaction has been useful in the preparation of certain cyclic olefins such as cyclopropene[8] and *trans*-cycloöctene.[9] It may be useful also in preparing other olefins of known configuration although little advantage has been taken of this possibility.

MECHANISM

The decomposition of quaternary ammonium compounds was described as belonging to that class of bimolecular elimination reactions called E2 reactions by Hughes, Ingold, and Patel in 1933.[10] Subsequent work has served to confirm the opinion that this is the usual course of the reaction, but it has also revealed cases in which this mechanism is not correct. In some instances the nature of the alternative mechanism seems clear, while in others a choice cannot be made at present. In this section consideration will be given first to the E2 process and then to the other possibilities. It may be well to point out here, however, that the fact that mechanisms other than E2 are known to prevail in some Hofmann eliminations and that these do not require *trans* elimination means that it is not safe to assign stereochemical configuration to an amine on the basis of this reaction alone.

The general requirements of the Hofmann elimination reaction suggest that a moderately strong base, a β hydrogen atom, and a positively charged nitrogen center are involved since all of these are usually necessary. Most quaternary salts do not undergo elimination in the presence of phenoxide or acetate ions[11] or amines;[12] quaternary salts derived from phenethylamines do. Elimination proceeds without difficulty in many compounds that do not have an α hydrogen atom. Several examples of this type can be found in the tables at the end of this chapter. These observations are in accord with either a concerted process (E2) or a stepwise reaction (Elcb, El

---

[8] Schlatter, *J. Am. Chem. Soc.*, **63**, 1733 (1941).
[9] Cope, Pike, and Spencer, *J. Am. Chem. Soc.*, **75**, 3212 (1953).
[10] Hughes, Ingold, and Patel, *J. Chem. Soc.*, **1933**, 526.
[11] Hanhart and Ingold, *J. Chem. Soc.*, **1927**, 997.
[12] Hunig and Baron, *Chem. Ber.*, **90**, 395 (1957).

elimination in the conjugate base) in which the β hydrogen atom is removed first, forming a carbanion intermediate. Actually, as Ingold pointed out in 1933[10] and as has been restated recently,[13] these mechanisms may be taken as extremes which merge as the lifetime of the carbanion is considered to become shorter in the stepwise reaction or as the degree of carbon to hydrogen bond breaking in the transition state becomes greater in the concerted process.

Concerted: E2

$$R_2\overset{H}{\underset{\underset{\oplus}{NR_3}}{C}}\text{---}CR_2 + B^{\ominus} \rightleftharpoons \underset{R_2}{\overset{B}{\underset{\diagdown}{\overset{H}{\underset{\diagdown}{C}}}}}\text{---}\underset{NR_3}{CR_2} \rightarrow R_2C{=}CR_2 + BH + NR_3$$

Stepwise: E1cb

$$R_2\overset{H}{\underset{\underset{\oplus}{NR_3}}{C}}\text{---}CR_2 + B^{\ominus} \rightleftharpoons R_2\overset{\ominus}{\underset{\underset{\oplus}{NR_3}}{C}}\text{---}CR_2 + BH$$

$$R_2\overset{\ominus}{\underset{\underset{\oplus}{NR_3}}{C}}\text{---}CR_2 \rightarrow R_2C\text{---}\underset{NR_3}{CR_2} \rightarrow R_2C{=}CR_2 + R_3N$$

A choice between these mechanisms cannot be made on the basis of kinetic order, since both require second order behavior. The two extremes in mechanism do, however, lead to different predictions about the stereochemistry of the process. One of the requirements of the E2 mechanism is that the hydrogen atom and the nitrogen group involved in the elimination process be coplanar and in the *trans* conformation. This arrangement is shown using Newman's convention.[14] (It must be

---

[13] Saunders and Williams, *J. Am. Chem. Soc.*, **79**, 3712 (1957).
[14] Newman, *Steric Effects in Organic Chemistry*, John Wiley and Sons, New York, 1956, Chap. 1.

emphasized that *trans* as used in the phrase "*trans* elimination" is employed in this sense and does not refer to the geometrical isomer of the olefin produced in such an elimination; with suitable starting materials either a *cis* or a *trans* olefin can be prepared by a stereospecific elimination.) If the substituent groups are properly chosen, it is possible to test the *trans* nature of an elimination reaction. This criterion has been applied very convincingly to the Hofmann elimination in 1,2-diphenylpropylamines by treating the quaternary iodides with ethoxide ion in ethanol.[15] The *erythro* and *threo* isomers were studied separately and found to undergo stereospecific *trans* elimination. The *erythro* form gives *cis*-1,2-diphenylpropene, while the *threo* compound gives the *trans* olefin. It is also a consequence of the relatively rigid geometrical requirements of the transition state in these isomers that the *threo* form should react more rapidly than the *erythro* form, and this prediction was verified.

There can be no clearer demonstration of the E2 mechanism; the only question that may arise is how far these results can be extrapolated to other compounds and to other reaction conditions.

A study of the elimination reaction with the same compounds using *t*-butoxide ion in *t*-butyl alcohol[15] provides an example of the consequences of the stepwise mechanism and emphasizes the risk of extrapolation from one set of conditions to another. In this instance both the *erythro* and *threo* forms gave the same *trans* olefin and the isomers reacted at virtually the same rate. The *cis* olefin was shown to be stable under the reaction conditions, so it cannot have been formed and then isomerized. These

---

[15] Cram, Greene, and Depuy, *J. Am. Chem. Soc.*, **78**, 790 (1956).

are the results to be expected of the two-step reaction if the carbanion has an appreciable lifetime. Presumably the change from the E2 mechanism to the stepwise mechanism is due to the greater basicity of the t-butoxide ion which favors removal of the β hydrogen atom to a greater degree than does ethoxide ion. The carbanion then equilibrates so that the species obtained from either the *erythro* or the *threo* compound is the same and must go through the rate- and product-determining steps in the same way. In this instance these steps lead to the formation of the *trans* isomer, presumably because the transition state from carbanion to *trans* product involves less steric interaction than the one leading to *cis* olefin.

$$C_6H_5-\overset{H}{\underset{CH_3}{C}}-\overset{H}{\underset{C_6H_5}{C}}-\overset{\oplus}{N}R_3 \rightarrow C_6H_5-\overset{\ominus}{\underset{CH_3}{C}}-\overset{H}{\underset{C_6H_5}{C}}-\overset{\oplus}{N}R_3$$

$$C_6H_5-\overset{\ominus}{\underset{CH_3}{C}}-\overset{H}{\underset{C_6H_5}{C}}-\overset{\oplus}{N}R_3 \rightarrow \underset{CH_3}{\overset{C_6H_5}{}}C\!=\!\!C\underset{C_6H_5}{\overset{H}{}}-NR_3 \rightarrow \underset{CH_3}{\overset{C_6H_5}{}}C\!=\!C\underset{C_6H_5}{\overset{H}{}}$$

Other evidence for the *trans* nature of the Hofmann elimination reaction is provided by a study of the olefins produced from the N,N,N-trimethylammonium hydroxides of menthyl- and neomenthyl-amine.[16, 17] With neomenthylamine there is a hydrogen atom in the *trans* relationship to the amino group on both β carbon atoms, and elimination can give either 2-menthene or 3-menthene. The predominance of the latter isomer is taken to indicate that, given suitable geometry, the hydrogen atom at the 4 position is removed preferentially. The course of the reaction of menthylamine that yields 2-menthene as the major product must be governed by the fact that in menthylamine the only *trans* hydrogen atom suitable for elimination is the one located on the 2 carbon atom. The change in product composition is some measure of the preference for *trans* elimination in this series. The 3-menthene produced from menthylamine must be formed by some other reaction path. (See equation on p. 326.)

Similar evidence for *trans* elimination in alicyclic amines is provided by certain 3-amino steroids in the 5α-cholestane and 5α-pregnane (A-B *trans*) series.[18] In these compounds conversion of one chair form to

[16] Cope and Acton, *J. Am. Chem. Soc.*, **80**, 355 (1958).
[17] McNiven and Read, *J. Chem. Soc.*, **1952**, 153.
[18] Haworth, McKenna, and Powell, *J. Chem. Soc.*, **1953**, 1110.

Neomenthylamine → 3-Menthene (92%) + 2-Menthene (8%) (94%)

Menthylamine → (14%) + (86%) (80%)

another whereby all axial positions become equatorial and vice-versa is prohibited by the fused ring system. Consequently the equatorial β amino isomers have no hydrogen atom in the coplanar *trans* orientation but the axial α isomers do. Only the α forms undergo elimination in

(β form)   (α form)

reasonable yield. A similar illustration is provided by the 6-aminocholestanes, except that in this system the 6β amine has the axial conformation.[19] However, with a double bond in the 5 position, the stereospecificity is lost and the 3β amino compounds give the 3,5-diene.[18]

Evidence for the E2 mechanism instead of the two-step process in a simple alkyl ammonium compound is provided by the studies of Shiner and Smith,[20] who found that hydrogen atoms in the position β to the amino group were not exchanged for deuterium atoms during reaction although α hydrogen atoms were exchanged. Furthermore, by comparing the rate of decomposition of ethyl-2,2,2-d_3-trimethylammonium hydroxide

[19] Gent and McKenna, *J. Chem. Soc.*, **1959**, 137.
[20] Shiner and Smith, *J. Am. Chem. Soc.*, **80**, 4095 (1958).

with that of ethyltrimethylammonium hydroxide, it was found that replacement of hydrogen by deuterium caused roughly a four-fold decrease in rate. This isotope effect shows that a β hydrogen atom is involved in the rate-determining step, and lack of exchange at the β position shows that any intermediate carbanion that may be postulated collapses to olefin much more rapidly than it is neutralized by solvent, indicating that the elimination reaction is of the E2 type.

Evidence for the E2 mechanism is provided by kinetic, stereochemical, and isotope exchange data for aliphatic and alicyclic amines. Yet, one instance has already been discussed[15] in which use of t-butoxide ion as the base caused a change to a non-stereospecific reaction, presumably proceeding through the intermediate carbanion. Usually the E1cb mechanism requires a higher free energy of activation than the E2 process, but conditions may be found in which this relationship is reversed. The reaction of cis- and trans-2-phenylcyclohexylammonium compounds may provide an example of this type. Both substances yield 1-phenylcyclohexene.[21] The trans isomer cannot do this by trans elimination since the only suitably located trans hydrogen atom is the one that would be lost

to give 3-phenylcyclohexene. It has been shown that 3-phenylcyclohexene does not isomerize rapidly enough under the reaction conditions to account for its absence in the reaction products.[22] Conclusive evidence that a direct elimination to form 1-phenylcyclohexene must be involved was provided by a study of the reaction using trans-2-phenylcyclohexyltrimethylammonium hydroxide bearing deuterium atoms on carbon atoms 3 and 6. The 1-phenylcyclohexene formed in 91% yield contained no detectable amount of the 3-phenyl isomer and had the same deuterium content as the quaternary base from which it was prepared.[23] The difference between the direction of elimination in this compound and that in the structurally similar menthylamine has been attributed to the effect of the phenyl group in increasing the acidity of the β hydrogen atom. It is also true that trans elimination in trans-2-phenylcyclohexylamine would require both the phenyl and trimethylamino groups to assume axial

[21] Arnold and Richardson, J. Am. Chem. Soc., **76**, 3649 (1954).
[22] Weinstock and Bordwell, J. Am. Chem. Soc., **77**, 6706 (1955).
[23] A. C. Cope, G. A. Berchtold, and D. L. Ross (in press, 1960).

positions, and this should be an important factor in raising the energy of the E2 transition state so that an alternative mechanism is favored. The isomeric *cis*-2-phenylcyclohexylamine may react by an E2 mechanism

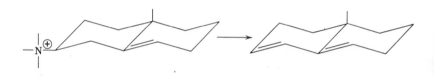

forming 1-phenylcyclohexene. The observation mentioned earlier,[18] that introduction of a double bond into the 5 position of a steroid nucleus enabled elimination to proceed using the otherwise unreactive 3β amino group suggests that allylic hydrogen atoms may be sufficiently acidic to enter into the two-step mechanism when the concerted process is not possible. If these examples are correctly interpreted, the intermediate carbanion mechanism may be expected to apply to compounds containing

allylic or benzylic β hydrogen atoms, but probably only when the *trans* elimination process is unfavorable. The mechanism in such cases is best described as non-stereospecific in that no particular geometry is required of the reactant. The reaction proceeds to give the more stable olefin, which, in the alicyclic compounds described immediately above, is *cis* and conjugated.

However, Hofmann elimination reactions that cannot proceed by a *trans* elimination mechanism are known in which the β hydrogen atoms are activated only by the positive nitrogen center. For these cases, it is possible to suggest the β carbanion mechanism, but an alternative is available.

It has been shown[20,24] that exchange of hydrogen for deuterium can occur in the α positions of quaternary ammonium bases. Such an exchange must involve ylides (α carbanions) as short-lived intermediates. It has also been shown[25,25a,26] that ylides are intermediates in elimination reactions

[24] Doering and Hoffmann, *J. Am. Chem. Soc.*, **77**, 521 (1955).
[25] Wittig and Polster, *Ann.*, **599**, 13 (1956).
[25a] Grob, Kny, and Gagneux, *Helv. Chim. Acta*, **40**, 130 (1937).
[26] Cope, Ciganek, and Le Bel, *J. Am. Chem. Soc.*, **81**, 2799 (1959).

forming olefins, presumably by a cyclic *cis* mechanism similar to the one proposed for the decomposition of tertiary amine oxides (p. 362). Consequently, ylides could be intermediates in the Hofmann elimination reaction.

It has been reported[27] that decomposition of β-tritioethyltrimethylammonium hydroxide at *ca.* 150° in the presence of excess superheated steam (introduced to minimize the introduction of tritium by exchange at the α positions) led to formation of trimethylamine containing 7.8% of the tritium that had been present in the quaternary base. It was concluded that the tritium was introduced into the trimethylamine by an intramolecular ylide elimination mechanism and not by exchange in the methyl groups of the quaternary ammonium hydroxide.

Similar tracer experiments with β deuterium labeling have led to results that are not in agreement with this conclusion.[27a] In the decomposition of 1-cyclohexylmethyl-1-d-trimethylammonium hydroxide at 90–110° and of β,β,β-trideuterioethylammonium hydroxide at *ca.* 115°, the trimethylamine formed initially contained no deuterium. As the decomposition progressed, the trimethylamine produced was found to contain increasing amounts of deuterium, paralleling exchange in the methyl groups of the quaternary hydroxide with the DOH formed by β elimination,

When β,β,β-trideuterioethyltrimethylammonium hydroxide was decomposed to the extent of 70% at 150–160° in the presence of a large excess of superheated steam, the trimethylamine formed contained less than 0.3% of monodeuteriotrimethylamine. These results appear to rule out a significant role for the ylide reaction path for the Hofmann elimination reaction of these two quaternary bases, and by inference for Hofmann eliminations in other simple compounds. With structures in which *trans* elimination cannot occur, the ylide mechanism may become important.[26]

Another possible reaction path leading to elimination is a two-step process in which the carbon-nitrogen bond breaks, first forming a carbonium ion and an amine (E1 mechanism). Base is not required for these

$$RN^{\oplus}(CH_3)_3 \rightarrow R^{\oplus} + N(CH_3)_3$$

$$R^{\oplus} \rightarrow \text{olefin} + H^{\oplus}$$

processes, and the quaternary iodides themselves undergo elimination. Pavinemethine,[28] N-methylemetinètetrahydromethine mono- and dimethiodides,[29] and the model compound IV[30] react in this way. In these

[27] Weygand, Daniel, and Simon, *Chem. Ber.*, **91**, 1691 (1958).
[27a] A. C. Cope, N. A. Le Bel, P. T. Moore, and W. R. Moore, to be published.
[28] Battersby and Binks, *J. Chem. Soc.*, **1955**, 2888.
[29] Battersby and Openshaw, *J. Chem. Soc.*, **1949**, S59.
[30] Norcross and Openshaw, *J. Chem. Soc.*, **1949**, 1174.

cases the carbonium ion postulated is benzylic and stabilized by a methoxyl group in the para position. Reaction with the solvent to form an alcohol

$$\underset{\text{IV}}{\underset{CH_3O}{\overset{CH_3O}{\diagdown}}\diagup\text{---}\underset{N(CH_3)_3 \oplus I^\ominus}{\overset{CHCH(CH_3)_2}{|}}} \xrightarrow[\text{ketone, }100°]{\text{Diethyl}} \underset{CH_3O}{\overset{CH_3O}{\diagdown}}\diagup\text{---}CH{=}C(CH_3)_2$$

or ether is an important side reaction in this process unless a non-hydroxylic solvent such as a ketone is used. The decomposition of the methiodides in the absence of base does not occur when the nitrogen atom is heterocyclic, as in emetine itself, in many other alkaloids containing the tetrahydroisoquinoline nucleus, and in such model compounds as V.[31]

$$\underset{\text{V}}{\underset{CH_3O}{\overset{CH_3O}{\diagdown}}\diagup\diagdown\diagup\underset{\underset{C_{15}H_{31}\text{-}n}{|}}{N(CH_3)_2 \oplus I^\ominus}}$$

The molecular rearrangements typical of carbonium ion reactions usually are not observed in Hofmann eliminations even with systems of the neopentyl type.[32] However, neobornyltrimethylammonium iodide in the presence of base in aqueous ethylene glycol yields camphene as the major product plus some tricyclene and bornylene.[33] Dry distillation of bornyl- or neobornyl-ammonium hydroxide produces bornylene without rearrangement.[33]

One reaction that is not readily accommodated by any of the preceding mechanisms is the formation of 1-methylcyclopentene during the decomposition of cyclopentylmethyltrimethylammonium hydroxide.[34] The proportion of 1-methylcyclopentene in the olefin mixture formed was as great as 29%. In some way, migration of a hydrogen atom to the α carbon atom has occurred, and experiments with cyclopentylmethylamine labeled with deuterium at the β position have shown that this atom is not the one which shifts.[35]

$$\text{cyclopentyl-}CH_2\overset{\oplus}{N}(CH_3)_3\overset{\ominus}{O}H \longrightarrow \text{cyclopentylidene}{=}CH_2 + \text{1-methylcyclopentene-}CH_3$$

---

[31] Pailer and Bilek, *Monatsh.*, **79**, 135 (1948).
[32] Stevens and Richmond, *J. Am. Chem. Soc.*, **63**, 3132 (1941).
[33] McKenna and Slinger, *J. Chem. Soc.*, **1958**, 2759.
[34] Cope, Bumgardner, and Schweizer, *J. Am. Chem. Soc.*, **79**, 4729 (1957).
[35] N. A. Le Bel, unpublished results.

DIRECTION OF ELIMINATION

Predictions of the olefins which will be formed from unsymmetrical quaternary bases can be based upon the many studies of decompositions with compounds of the type $RR'N^{\oplus}(CH_3)_2OH^{\ominus}$ or $R_2R'_2N^{\oplus}OH^{\ominus}$ in which the ratios of olefins derived from R and R' have been compared.[36,37] Similar information can be obtained from studies of the decomposition of compounds of the type $RCH_2CHN^{\oplus}(CH_3)_3OH^{\ominus}$ or from comparison of

$$\underset{\underset{CH_2R'}{|}}{}$$

the ratio of elimination to displacement in a series of quaternary hydroxides such as $RCH_2CH_2N^{\oplus}(CH_3)_3OH^{\ominus}$ and $R'CH_2CH_2N^{\oplus}(CH_3)_3OH^{\ominus}$, etc.[11,38,39] The goal in most of this research has been to contribute to an understanding of the reaction mechanism rather than to prepare olefins. The results have been summarized in the various expressions of the Hofmann rule for elimination reactions of "onium" compounds. However, no simple expression of this rule will apply to a very wide range of amines, and discussion of the rule will be deferred until the results of eliminations with different types of amines have been presented.

For many years the only evidence on which to base a discussion of the Hofmann elimination reaction was knowledge of the general reaction conditions and the direction of elimination. Largely because of the reaction conditions, the mechanism was assumed to be of the E2 type, yet the olefin formed from a quaternary base is very often not the one that would be produced by an E2 elimination of the corresponding halide. In providing explanations for the course that elimination will take in a given

$CH_3CH_2CHBrCH_3 + NaOC_2H_5 \rightarrow$ 2-butene, 81% + 1-butene, 19%   (ref. 40)

$$\underset{\underset{N(CH_3)_3OH}{|\oplus \quad \ominus}}{CH_3CH_2CHCH_3} \xrightarrow{97\%} \text{2-butene, 5.4\%} + \text{1-butene, 94.6\%} \quad \text{(ref. 36)}$$

case, three general factors are considered to be of importance, although there is some area of disagreement about the weighting of these factors. They are: the extent to which the olefin being formed may be stabilized by conjugation or hyperconjugation; the acidity of the β hydrogen atom that is to be eliminated; and the influence of steric interactions of the various groups in the rather rigid transition state assumed for the concerted elimination. The operation of the steric factor in particular is

---

[36] Cope, LeBel, Lee, and Moore, *J. Am. Chem. Soc.*, **79**, 4720 (1957).
[37] Smith and Frank, *J. Am. Chem. Soc.*, **74**, 509 (1952).
[38] Ingold and Voss, *J. Chem. Soc.*, **1928**, 3125.
[39] von Braun, *Ann.*, **382**, 1 (1911).
[40] Dhar, Hughes, and Ingold, *J. Chem. Soc.*, **1948**, 2058.

quite different in aliphatic, alicyclic, and heterocyclic amines and, for simplicity in this respect, these types will be given separate consideration.

Aliphatic Amines

In the study of quaternary ammonium hydroxides containing various primary alkyl groups, Hofmann[1, 2] observed that the ethyl group is the most readily eliminated (as ethylene). There is no exception to this generalization, which is one expression of the Hofmann rule, when it is restricted to primary alkyl groups. With methods such as gas chromatography[36] and mass spectrometry[37] it has been possible to obtain quite precise analyses of the olefin mixtures prepared in this way. In Table I

TABLE I

RELATIVE EASE OF ELIMINATION OF ALKYL GROUPS AS OLEFIN[36]

Alkyl Group	Not Corrected for Number of β Hydrogen Atoms	Corrected for Number of β Hydrogen Atoms
Ethyl	(100)	(100)
Isopropyl	143	72
t-Butyl	1280	427
n-Propyl	2.45	3.7
n-Butyl	1.6	2.4
n-Decyl	1.65	2.5
Isoamyl	0.8	1.2
β-t-Butylethyl[37]	0.16	0.24
Isobutyl	0.9	2.7
2-Phenethyl	2.6×10^6	3.9×10^6

values are given which express the relative ease of elimination of a given group as an olefin versus the ethyl group in terms of parts of olefin from "R" per 100 parts of ethylene. In the third column, correction has been made for the number of hydrogen atoms on the β carbon atom; i.e., three for ethyl, two for other n-alkyl groups, six for the isopropyl group and so on. A striking difference among simple alkyl groups is observed when the first three examples in Table I, in which the β hydrogen atoms are located on methyl groups, are compared with the others. Differences among other alkyl groups are slight; in particular it is interesting to note that the difference between the n-butyl and isobutyl groups is almost entirely a question of the number of available β hydrogen atoms. From the figures 1.6 and 0.9 given for these groups it would be predicted that the olefin mixture produced by pyrolysis of n-butylisobutyldimethylammonium hydroxide would contain 64% 1-butene and 36% isobutylene, which is exactly the composition found.[36] Branching at the γ carbon atom seems to have a greater effect than branching at the β position, to judge by the results of the decomposition of compounds containing isoamyl (β-isopropylethyl) and 3,3-dimethylbutyl (β-t-butylethyl) groups.[37]

These results illustrate the degree of validity of the Hofmann rule for elimination as applied to alkyl groups. The ease of elimination of isopropyl and t-butyl groups can be accommodated to the rule if it is stated that in elimination reactions of ammonium bases, β hydrogen atoms are lost most readily from a methyl group. To explain why the introduction of an alkyl group at the β position causes removal of a β hydrogen atom to become slower, an inductive effect was assumed to decrease its acidity.[41] However, the values above show that the introduction of a second alkyl group at the β position (compare n-propyl and isobutyl) has little additional effect on the rate of elimination but that an alkyl group which is branched at the γ carbon atom shows considerably decreased ease of elimination. In a study designed to test the susceptibility of the Hofmann reaction to inductive effects, a series of quaternary bases of the type $R_2CHCH_2N(CH_3)_3OH$, where $R = C_2H_5$, n-C_3H_7, i-C_3H_7, and t-C_4H_9, was pyrolyzed to give the following yields of the corresponding olefins: 77% ($R = C_2H_5$), 73% ($R = n$-C_3H_7), 67% ($R = i$-C_3H_7), and 81% ($R = t$-C_4H_9). The lowering of yield as R increases in branching from ethyl to isopropyl appears to be too small to be attributable to inductive effects. The high yield when R is t-butyl may be explained as the result of reaction by *cis* elimination. An examination of molecular models indicated that normal *trans* elimination is prohibited by interaction between the t-butyl groups and the trimethylammonium group.[42]

The dependence on size of the group rather than the number of groups is suggestive of a steric rather than an inductive influence on the reaction.[43, 44] The way in which the steric factor might operate is indicated in the following representations of transition states which involve the elimination of ethylene (VI) as compared with the elimination of $RCH{=}CH_2$ (VII) from $RCH_2CH_2N^\oplus(C_2H_5)(CH_3)_2OH^\ominus$. In formula VII the R group has one skew interaction with the quaternary ammonium group, and the decrease in ease of elimination as R changes in the sequence hydrogen, methyl, ethyl, isopropyl, t-butyl (i.e., with the ethyl, n-propyl, n-butyl, isoamyl, 3,3-dimethylbutyl groups attached to the nitrogen atom) is readily understood. Actually, formulas VI and VII are representations of specific conformations of the ground states. In the transition states the bonds to the hydrogen and nitrogen atoms are being broken

[41] Ingold, *Structure and Mechanism in Organic Chemistry*, Cornell University Press, Ithaca, New York, 1953, pp. 427 *et seq*.
[42] A. C. Cope and D. L. Ross, to be published.
[43] Schramm, *Science*, **112**, 367 (1950).
[44] Brown and Moritani, *J. Am. Chem. Soc.*, **78**, 2203 (1956).

$$
\begin{array}{cc}
\text{VI} & \text{VII}
\end{array}
$$

Newman projections: VI has H,H,H front; CH₃, CH₃, CH₂CH₂R on N⁺ rear. VII has H,R,H front; CH₃, CH₃, C₂H₅ on N⁺ rear.

and should be somewhat lengthened while the remaining groups should be somewhat flattened toward the planar arrangement that they will assume in the olefin. These modifications do not affect the nature of the argument, although the fact that the bond between the carbon atoms α and β to the nitrogen atom has some double bond character means that R could have a stabilizing effect on the transition state if it could conjugate with this developing unsaturation. When the substituent on the β carbon atom is a phenyl group, the steric factor is unimportant relative to the acidity of the β hydrogen atom and the elimination of styrene is so much more rapid than ethylene formation that it is usually reported as the only olefin produced.[37] Other groups such as the carbonyl group and the vinyl group which also can enter into conjugation with the new double bond greatly enhance the rate of elimination.[45] Such compounds must be considered as outside the scope of the Hofmann rule.

By a rather easy extension the Hofmann rule may be applied to predict which isomer is to be expected in the greater amount when the elimination reaction involves a group branched at the α carbon atom so that the double bond might be formed in either branch. The *sec*-butyl group affords a simple example of this type in which the choice involves removal of a β hydrogen atom from a methyl or a methylene group. This example is

$$\underset{\underset{N(CH_3)_3OH}{\oplus \quad \ominus}}{CH_3CH_2CHCH_3} \xrightarrow{97\%} \begin{array}{c} CH_3CH_2CH{=}CH_2,\ 95\% \\ + \\ \textit{cis-} \text{ and } \textit{trans-}\text{2-butene, } 5\% \end{array} \qquad \text{(ref. 36)}$$

similar to one in which ethyl and *n*-propyl groups are attached to the same nitrogen atom and, in accord with the preference shown previously, the less highly substituted olefin is formed in the greater amount. Here the choice between rotational forms (and presumably also between transition states) leading to elimination from the methyl and the ethyl branches (VIII and IX, respectively) is in favor of the former because the most bulky group [N⊕(CH₃)₃] would encounter less hindrance in VIII. As with the

---

[45] Wieland, Koschara, Dane, Renz, Schwarze, and Linde, *Ann.*, **540**, 103 (1939).

compounds discussed previously, a phenyl group on the β carbon atom directs elimination toward the conjugated olefin even in competition with a methyl group.

$$C_6H_5CH_2\underset{\overset{|}{\oplus N(CH_3)_3}}{C}HCH_3 \rightarrow C_6H_5CH=CHCH_3$$

$$C_6H_5CH_2\underset{\overset{|}{\oplus N(CH_3)_3}}{C}HCH_2OH \rightarrow C_6H_5CH=CHCH_2OH$$

Relatively little evidence is available concerning the stereochemistry of the olefin produced by the Hofmann elimination when *cis* and *trans* isomers may be formed. In the decomposition of 3-pentyltrimethylammonium hydroxide the 2-pentene obtained is a mixture containing 55.5% *cis* and 44.5% *trans* isomer.[36] sec-Butyltrimethylammonium hydroxide forms 5.4% of 2-butene of which 59% is *cis* and 41% is *trans*.[36] It

$$CH_3CH_2\underset{\overset{|}{\oplus N(CH_3)_3 \overset{\ominus}{O}H}}{C}HCH_2CH_3 \xrightarrow{96\%} \underset{(cis \text{ and } trans)}{CH_3CH=CHCH_2CH_3}$$

appears that in aliphatic cases there is produced a mixture considerably richer in the *cis* isomer than the equilibrium ratio of *cis* to *trans*. However, the quaternary hydroxide prepared from 1,2-diphenylethylamine forms *trans*-stilbene,[46] while quaternary bases of 1-phenyl-2-propylamine[47] and 1-phenyl-1-propylamine[48] give 1-phenylpropene which is largely the *trans* isomer, and ring-substituted derivatives of phenylalanine give derivatives of *trans*-cinnamic acid.[49, 50] These results suggest that when a phenyl group is present the more stable *trans* isomer is formed preferentially.

[46] Thomson and Stevens, *J. Chem. Soc.*, **1932**, 1932.
[47] Doering and Meislich, *J. Am. Chem. Soc.*, **74**, 2099 (1952).
[48] E. R. Trumbull and G. L. Willette, unpublished results.
[49] Körner and Menozzi, *Gazz. chim. ital.*, **11**, 549 (1881).
[50] Johnson and Kohmann, *J. Am. Chem. Soc.*, **37**, 1863 (1915).

Alicyclic Amines

As contrasted with aliphatic amines, the most important factor in the elimination reaction of alicyclic amines, at least those having rings of six carbon atoms or less, is the availability of a *trans* β hydrogen atom. This factor has been discussed as evidence for the *trans* nature of the elimination process. When there are *trans* β hydrogen atoms available on both sides of the amino group, as with neomenthylamine[16, 17] (X) and neoisomenthylamine[17] (XI), the tendency seems to be for elimination to produce

the more highly substituted 3-menthene by loss of the tertiary hydrogen atom. The ratio of 3-menthene to 2-menthene from neomenthylamine is about 9 : 1, showing a greater preference for tertiary over secondary hydrogen than is found in the aliphatic series. However, the greater reactivity of the methyl hydrogen atoms is still demonstrated by the results shown in Table II with a series of 1-methylcycloalkylamines. With the

TABLE II

n	Total Olefin Yield, %	Relative Amounts of Olefins Formed, %	
5	71	91	9
6	85	98.6	1.4
7	84	78.2	21.8
8	82	63.5	36.5 *cis*, 0.0 *trans*
9	83	48.0	51.0 *cis*, 1.0 *trans*
10	92	66.4	31.4 *cis*, 2.2 *trans*

exception of the nine-membered ring compound, the principal products are the less stable[51, 52] exomethylene compounds.[34, 52a] The very low

---

[51] Turner and Garner, *J. Am. Chem. Soc.*, **79**, 253 (1957).

[52] Cope, Ambros, Ciganek, Howell, and Jacura, *J. Am. Chem. Soc.*, **81**, 3153 (1959); **82**, 1750, (1960).

[52a] Cope, Ciganek, Howell, and Schweizer, *J. Am. Chem. Soc.*, **82**, (in press, 1960).

proportion of 1-methylcyclohexene ($n = 6$) may be accounted for by the fact that the orientation required for *trans* elimination within the ring would place the bulky trimethylammonium group in the axial position. The suggestion that cyclopentene derivatives are formed more readily than cyclohexene compounds is supported by a study of the decomposition of cyclopentylcyclohexyldimethylammonium hydroxide, which gave mostly cyclopentene[53] (95% of the product corresponded to the compounds formulated in the equation).

$$\text{cyclohexyl-N}^{\oplus}(CH_3)_2\text{-cyclopentyl} \longrightarrow \text{cyclohexyl-N}(CH_3)_2 + \text{cyclopentene}$$

When a phenyl group is located on the β carbon atom, elimination to give the conjugated olefin is preferred and, as indicated in the discussion of the mechanism of the reaction, there is some reason to believe that this is so even when the hydrogen atom to be removed is *cis* to the amino group.

The problem of explaining the stereochemistry of the olefin produced in these reactions is a difficult one. In alicyclic compounds with seven-membered or smaller rings only the *cis* form of the olefin is known, so the question does not arise. Both the *cis* and *trans* forms of cycloöctene,[9, 54] cyclononene,[55, 56] and cyclodecene[56] are known, and the Hofmann elimination reaction leads to a mixture in which the *trans* isomer predominates in each case, Table III. However, in all these compounds the *cis* isomer is the more stable,[57, 58] and it will be of interest to find an explanation for the

TABLE III

$$\text{CH}_2\text{---CHN}^{\oplus}(CH_3)_3 \text{O}^{\ominus}\text{H} \longrightarrow \text{CH}=\!\!=\!\!\text{CH}$$
$$\text{---}(CH_2)_{n-2}\text{---} \qquad\qquad \text{---}(CH_2)_{n-2}\text{---}$$

n	Olefin Yield, %	*trans*, %	*cis*, %	References
8	89	60	40	54
9	83	100[a]	—	55, 56
10	90	98	2	56, 58

[a] Based on infrared analysis. The product may contain a small amount of the *cis* isomer not detected by that method.

[53] Jewers and McKenna, *J. Chem. Soc.*, **1958**, 2209.
[54] Ziegler and Wilms, *Ann.*, **567**, 1 (1950).
[55] Blomquist, Liu, and Bohrer, *J. Am. Chem. Soc.*, **74**, 3643 (1952).
[56] Cope, McLean, and Nelson, *J. Am. Chem. Soc.*, **77**, 1628 (1955).
[57] Cope, Moore, and Moore, *J. Am. Chem. Soc.*, **81**, 3153 (1959).
[58] Cope, Moore, and Moore, *J. Am. Chem. Soc.*, **82**, 1744 (1960).

formation of the less stable *trans* form when a path is available that would yield the more stable *cis* isomer.

Even when there is a double bond already in the ring and the system is presumably less flexible, the tendency of the Hofmann elimination to yield the *trans* product is observed. Thus the decomposition of *cis*-cycloöcten-3-yltrimethylammonium hydroxide gives 15% of *cis-trans*-1,3-cycloöctadiene and 41% of *cis-cis*-1,3-cycloöctadiene;[59] the ratio of *trans* to *cis* changes from 3 : 2 in cycloöctylamine to 0.73 : 2 in cycloöctenylamine. With *cis*-cyclodecen-3-yltrimethylammonium hydroxide, *cis-trans*-1,3-cyclodecadiene was reported to be the only diene formed,[60] the new double bond apparently being introduced in the *trans* configuration exclusively, as is essentially the case with cyclodecylamine. Both of these *cis-trans* dienes are much more reactive than the *cis-cis* isomers and are sterically strained.

Heterocyclic Amines

Most of the useful applications of the Hofmann elimination reaction have been with alkaloids containing the nitrogen atom in a ring, usually five- or six-membered. In this work the structure of the alkaloid has been the primary concern and the structures of intermediates between the alkaloid and the final nitrogen-free product usually have not been investigated in detail. If the elimination reaction forms a mixture of olefins, the mixture may be subjected to a second Hofmann elimination reaction, or the isomers may be converted to a single compound by hydrogenation. Thus these reactions often do not provide information about the direction of elimination. Fewer model compounds have been studied in the heterocyclic series than in those previously treated. Such data as are available are explained by the assumptions of *trans* elimination,[61] preference for the formation of a conjugated olefin when possible, and preferential loss of hydrogen from a methyl group in competition with other alkyl groups.

There seems to be no record of the Hofmann elimination reaction as applied to a derivative of ethylene imine. Decompositions of some highly substituted compounds containing four-membered heterocyclic rings have been studied. 1,1,2-Trimethyl-4-isobutyltrimethyleneimonium hydroxide[62] (XII) is reported to yield an olefin whose structure was not established, and 1,1,2,2,4-pentamethyltrimethyleneimonium hydroxide (XIII) also undergoes ring opening to give a product for which two structures

[59] Cope and Bumgardner, *J. Am. Chem. Soc.*, **78**, 2812 (1956).
[60] Blomquist and Goldstein, *J. Am. Chem. Soc.*, **77**, 998 (1955).
[61] McKenna, *Chem. & Ind.* (*London*), **1954**, 406.
[62] Kohn and Giaconi, *Monatsh.*, **28**, 461 (1907).

have been suggested.[63, 64] Either of these isomers would be expected to produce 4-methyl-1,3-pentadiene (the observed product) in a second step, as indeed would other isomers. The observation that the N-ethyl-N-methyl derivative of XIII undergoes ring opening rather than elimination

$$\underset{\text{XII}}{\begin{array}{c}\text{CH}_3\\|\text{CH}_3\\\text{H}_3\text{C}|\overset{\oplus}{\text{N}}\text{--CH}_3\\|\\\text{CH}_2\text{CH}(\text{CH}_3)_2\end{array}}$$

$$\underset{\text{XIII}}{\begin{array}{c}\text{CH}_3\text{CH}_3\\||\overset{\oplus}{\text{}}\\\text{H}_3\text{C}\text{N--CH}_3\\|\\\text{CH}_3\end{array}} \to \begin{array}{c}\text{CH}_3\\|\\\text{CH}_3\text{--C--N}(\text{CH}_3)_2\\|\\\text{CH}_2\text{CH=CH}_2\\\text{or}\\\text{CH}_3\\|\\\text{CH}_3\text{--C}\text{N}(\text{CH}_3)_2\\\||\\\text{CH--CHCH}_3\end{array} \to (\text{CH}_3)_2\text{C=CHCH=CH}_2$$

of ethylene might be explained as a manifestation of ring strain or of the fact that one of the positions is rather similar to a *t*-butyl group. If a hydrogen atom is removed from the ring, a strictly *trans* orientation of the hydrogen and nitrogen atoms is not possible but, if the hydrogen atom comes from one of the methyl groups, this geometry could be attained. Trimethyleneimonium compounds without substituents in the 2 or 4 position do not appear to have been subjected to the conditions of the Hofmann elimination reaction.

Examples of the Hofmann elimination reaction with compounds containing five-membered heterocyclic rings are more numerous. By analogy with cyclopentane, the pyrrolidine ring should have a slightly puckered conformation in which a β hydrogen atom is coplanar with the nitrogen atom.[61] Pyrrolidinium compounds undergo the elimination reaction without difficulty. Decomposition of the 2-bromomethyl compound is of interest because of the long-standing question of the nature of

---

[63] Kohn and Morgenstern, *Monatsh.*, **28**, 479 (1907).
[64] Kohn and Morgenstern, *Monatsh.*, **28**, 529 (1907).

the final product, pirylene.[65, 66] The decomposition of the quaternary salt is accompanied by loss of hydrogen bromide, and an acetylenic amine is formed.[67] A second elimination yields methylvinylacetylene (pirylene).[68]

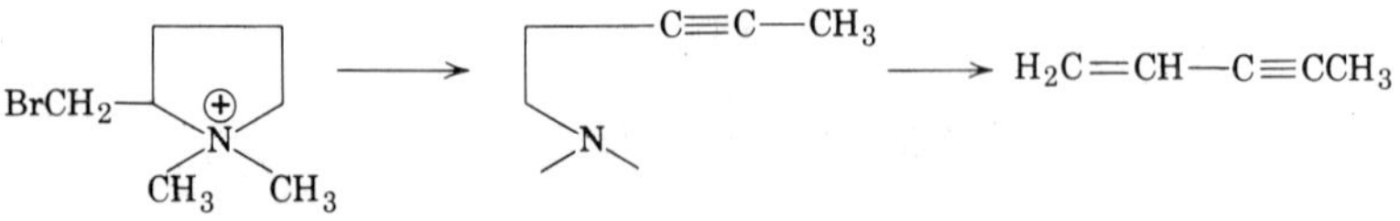

$$BrCH_2\text{-pyrrolidine}(CH_3)_2^+ \longrightarrow (CH_3)_2N\text{-}CH_2CH_2CH_2\text{-}C\equiv C\text{-}CH_3 \longrightarrow H_2C=CH\text{-}C\equiv CCH_3$$

Some measure of the relative reactivity of five- and six-membered rings is provided by the spiro compounds XIV and XV. In direct competition the pyrrolidinium and piperidinium rings appear about equally reactive, giving XVI and XVII in equal amounts.[53]

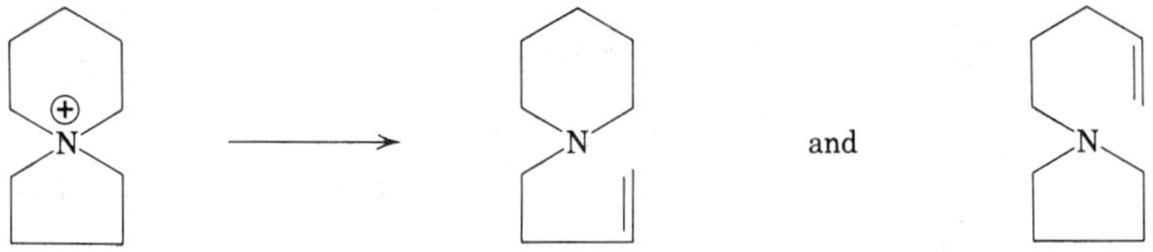

XIV ⟶ XVI and XVII

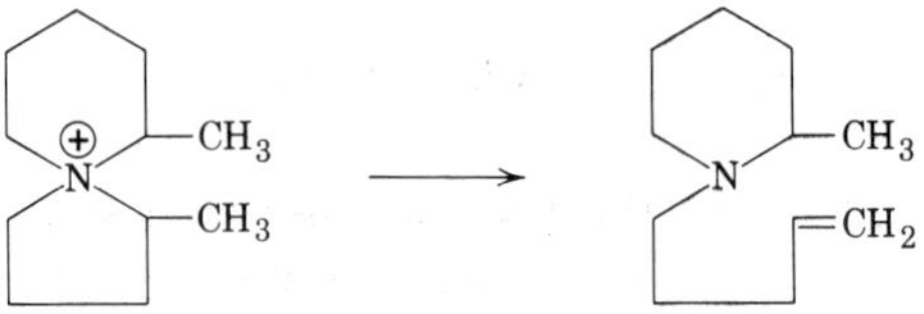

XV ⟶

When an α methyl group is available, elimination occurs with loss of a hydrogen atom on the methyl group of the five-membered ring. Attack at the methyl group might be expected, but the marked preference for the one attached to the pyrrolidinium ring is surprising.[53]

Elimination reactions in the octahydroindole series afford some interesting examples. *cis*-Octahydroindole is cleaved between the six-membered ring and the nitrogen atom, but the position of the double bond was not determined because the product was identified by reduction to N,N-dimethyl-β-cyclohexylethylamine.[69] With the 2-methyl compound,

[65] Ladenburg, *Ann.*, **247**, 1 (1888).
[66] von Braun and Teuffert, *Ber.*, **61**, 1902 (1928).
[67] E. R. Buchman, private communication.
[68] Sargent, Buchman, and Farquhar, *J. Am. Chem. Soc.*, **64**, 2692 (1942).
[69] King, Bovey, Mason, and Whitehead, *J. Chem. Soc.*, **1953**, 250.

however, cleavage occurs within the five-membered ring, presumably by attack at the methyl group, although again the position of the double

bond was not established.[70] The stereochemistry of *cis*-octahydroindole should be similar to that of *cis*-hydrindane, and the nitrogen atom can be

located on an axial bond of the cyclohexane ring where it is *trans* to neighboring axial hydrogen atoms. However, in *trans*-octahydroindole the nitrogen atom is probably in the equatorial position and no hydrogen atom in the cyclohexane ring is coplanar with it. One of the hydrogen atoms on the heterocyclic ring is removed, and the product is *trans*-N,N-dimethyl-2-vinylcyclohexylamine.[71]

2,3-Dihydroindole and hexahydrocarbazole react normally with cleavage of the five-membered ring to give ortho-substituted derivatives of dimethylaniline.[72]

[70] Fujise, *Sci. Papers Inst. Phys. Chem. Research (Tokyo)*, **8,** 185 (1927) *Chem. Zentr.*, **99, II,** 993 (1928).
[71] Booth and King, *J. Chem. Soc.*, **1958,** 2688.
[72] Booth, King, and Parrick, *J. Chem. Soc.*, **1958,** 2302.

Piperidinium compounds should exist mainly in the chair form analogous to cyclohexane, and in this situation equatorial hydrogen atoms at the β position are coplanar with the bond between the α carbon atom

and the nitrogen atom.[61] The ring is opened smoothly by the Hofmann procedure, although if the process is continued to the diene an allylic shift occurs and 1,3-pentadiene (piperylene) is the product.[4] When α-methylpiperidine is subjected to Hofmann exhaustive methylation, the first elimination is toward the methyl group and in the second step isomerization does not occur, so that 1,5-hexadiene (biallyl) is obtained.[73]

Some indication of the ease of opening of the piperidine ring in relation to elimination of simple alkyl groups is provided by the observation that N-ethyl-N-methylpiperidinium hydroxide yields 71% of ethylene and 18% of open-chain amine while the N-propyl, N-butyl, N-hexyl, and

---

[73] Merling, *Ann.*, **264**, 310 (1891).

N-octyl compounds give about the statistical ratio of 2 : 1 for ring opening versus loss of the alkyl group.[74]

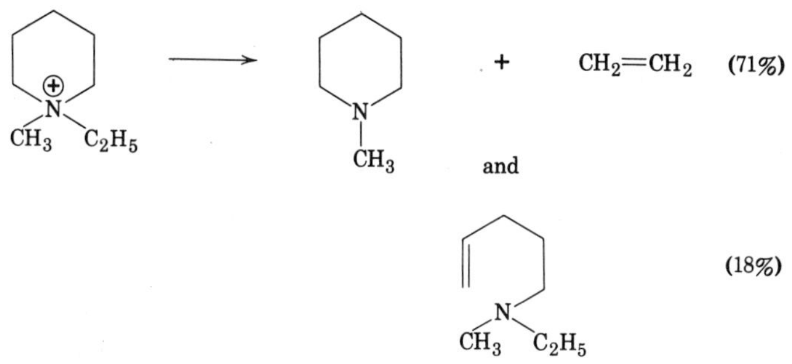

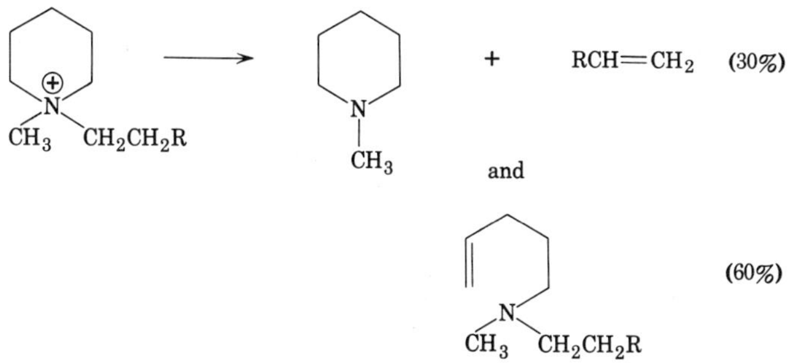

Two cases in which the Hofmann elimination fails are reported with the piperidine derivatives lobelan (XVIII)[75, 76] and lobelanidine (XIX).[75]

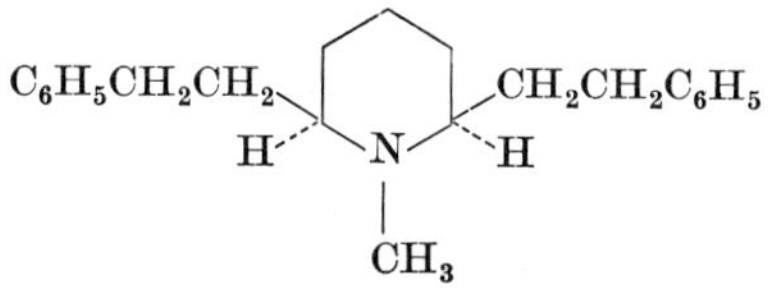

XVIII

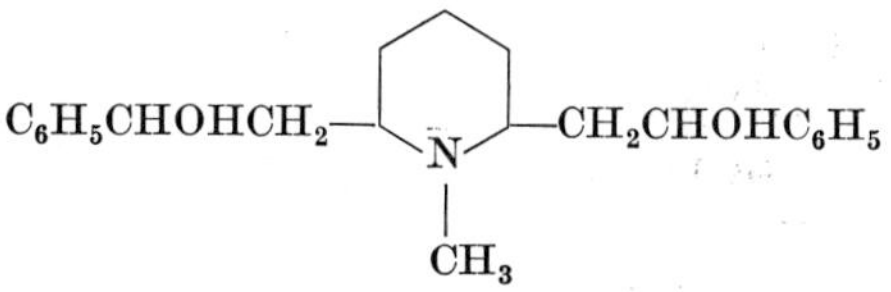

XIX

[74] von Braun and Buchman, *Ber.*, **64,** 2610 (1931).
[75] Wieland, Schöpf, and Hermsen, *Ann.*, **444,** 40 (1925).
[76] Schöpf and Boettcher, *Ann.*, **448,** 1 (1926).

Even the diketone corresponding to lobelanidine in which the hydrogen atoms β to the nitrogen atom are especially acidic does not give a good yield in the first step, although once the ring is opened the final elimination of the amino group is very easy.[77] When there is a double bond in the piperidine nucleus, as with lobinine (XX), ring opening is extremely

$$C_6H_5COCH_2-\overset{\underset{CH_3}{|}}{N}-CH_2CHOHCH_2CH_3$$

XX

facile.[45] The poor results obtained in the Hofmann elimination reaction of α,α'-disubstituted piperidine compounds has not been explained.

The tetrahydroisoquinoline ring is opened especially easily by the Hofmann procedure,[66] presumably because it is of the phenethyl type. This structural unit occurs commonly in alkaloids, and many examples of its activity in the Hofmann reaction are available. One especially interesting case is afforded by certain alkaloids of the protoberberine type XXI

XXI

which have a methyl group at one benzylic position. When the amine is converted to the N-methyl quaternary compound, two products are obtained and, for simplicity, these can be considered to arise by introduction of the N-methyl group on one side or the other of the plane of the molecule, creating a new asymmetric center at the nitrogen atom. In one of the diastereoisomers thus formed, the methyl group of the amino group and the one at the benzylic position are *cis* and, in the other, they are *trans* (XXII and XXIII). In the *cis* form the hydrogen and nitrogen atoms are suitably positioned for elimination and reaction occurs to form a dibenzazacyclodecene.[78] In the *trans* form the corresponding hydrogen atom is not in the correct orientation, so elimination occurs with the other β hydrogen atom forming a vinyl group.[78] Apparently a hydrogen atom

[77] Wieland and Dragendorff, *Ann.*, **473**, 83 (1929).
[78] Bersch, *Arch. Pharm.*, **283**, 36 (1950).

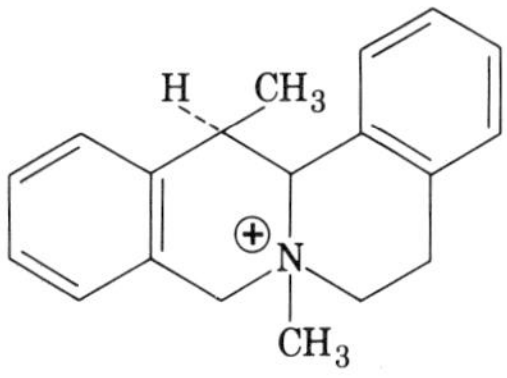

OLEFINS FROM AMINES

XXII (cis) XXIII (trans)

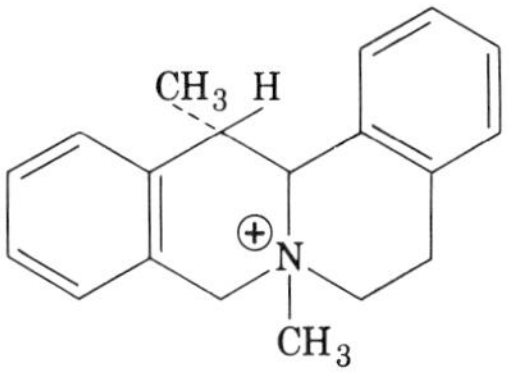

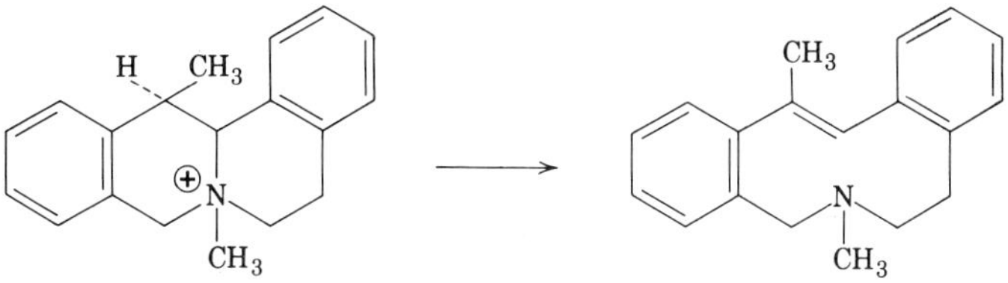

is eliminated from the tertiary rather than the secondary position when the stereochemistry of the amine allows a choice.

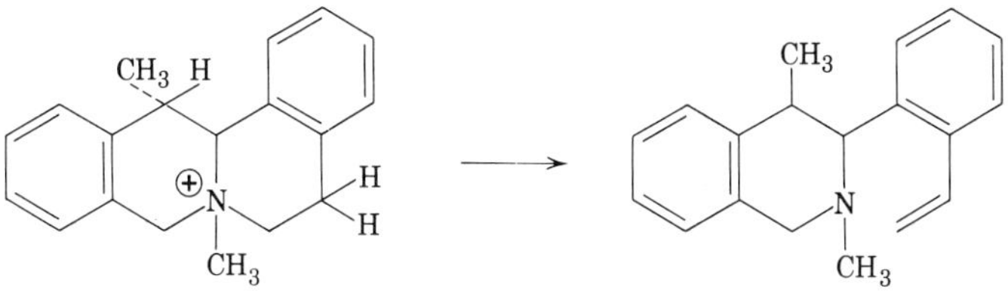

The following reactions may be considered illustrations of the principle that elimination will proceed in such a way as to yield a conjugated olefin when the stereochemistry is suitable.

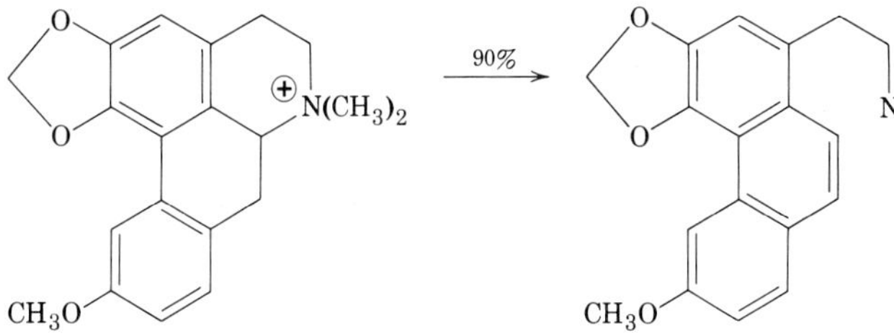

90%

(ref. 79)

[79] Schlittler, *Helv. Chim. Acta*, **15**, 394 (1932).

In marked contrast to tetrahydroisoquinolines, tetrahydroquinolinium compounds do not undergo elimination even when an α methyl group is available.[82, 83] Instead, the principal reaction is the attack of hydroxide ion on the N-methyl groups to form methanol. This is a common side reaction in the Hofmann procedure. It occurs to some extent with most compounds, but here it becomes the sole reaction. It seems unlikely that the effect is steric since both *cis*- and *trans*-decahydroquinoline react to

(position of the double bond uncertain)

---

[80] Schöpf, Schmidt, and Braun, *Ber.*, **64**, 683 (1931).
[81] Witkop, *J. Am. Chem. Soc.*, **71**, 2559 (1949).
[82] Feer and Koenigs, *Ber.*, **18**, 2388 (1885).
[83] Moller, *Ann.*, **242**, 313 (1887).

OLEFINS FROM AMINES 347

give ring opening by cleavage between the cyclohexyl ring and the nitrogen atom.[70, 84]

A number of bicyclic compounds with nitrogen as the bridging atom have been opened successfully by the Hofmann method. Tropidine,[85] granatanine,[86] and pavine[28] may be mentioned as examples of this type.

Tropidine methohydroxide → (90%)

N-Methylgranatanine methohydroxide → (~50%)

Pavine methiodide → N(CH$_3$)$_2$

The example of pavine is especially interesting because the second step, which should yield a derivative of dibenzcyclooctatetraene, does not proceed normally but results in replacement of the amine function by a hydroxyl group. Yet the dihydro derivative reacts normally to form a dibenzcyclooctatriene,[28] and a model system without the four methoxyl groups gives dibenzcyclooctatetraene in "satisfactory" yield.[87]

[84] Fujise, *Sci. Papers Inst. Phys. Chem. Research (Tokyo)*, **9**, 91 (1928) [*Chem. Zentr.*, **99, II,** 2359 (1928)].
[85] Merling, *Ber.*, **24,** 3108 (1891).
[86] Willstätter and Veraguth, *Ber.*, **40,** 957 (1907).
[87] Wittig, *Angew. Chem.*, **63,** 15 (1951).

CH₃O-[structure with N(CH₃)₃⁺]-OCH₃ → CH₃O-[structure with OH]-OCH₃

Many compounds which have the nitrogen atom at a bridgehead have been degraded by the Hofmann procedure. Quinuclidine[88] and 1-azabicyclo[2.2.1]heptane[89] do not afford olefins in good yield; the main products are the recovered amines. Alkaloids containing pyrrolizidine, quinolizidine, and other fused ring systems with a nitrogen atom at the ring juncture have been degraded successfully. Several examples are to be found in Table XVIII.

The Hofmann Rule

As a means of summarizing the previous information, the extent to which different types of ammonium compounds adhere to a general rule for elimination will be considered. A simple expression of the Hofmann rule will be used, as follows: "In elimination reactions of ammonium compounds the β hydrogen atom is removed most readily if it is located on a CH_3 group, next from RCH_2, and least readily from R_2CH."

With simple alkyl groups this rule holds, although the difference between RCH_2 and R_2CH is not striking and is largely a matter of the number of β hydrogen atoms. If R is phenyl, vinyl, carbonyl, or a similar group, the rule does not hold.

With alicyclic compounds containing an external methyl group in the appropriate position, the rule seems to hold. Within the ring, the necessity of having the amino group and hydrogen atom *trans* to each other is most important. Given *trans* hydrogen atoms in both β positions, the hydrogen atom is eliminated from the R_2CH groups; thus the rule is not followed. Whenever possible, a conjugated olefin will be formed.

Comparable generalizations may be made for heterocyclic compounds.

The Hofmann Rule as expressed here applies only to alkyl groups without unsaturated functions attached directly to the β carbon atom. Compounds containing bulky, highly branched alkyl groups may not react according to the prediction of the rule.

Application of the Hofmann rule depends on the assumption, which is usually valid, that the ratio of olefins formed in the elimination is determined by the relative rates of the competing reactions which lead to the different olefins and that, once formed, they do not equilibrate. Since

---

[88] Lukeš, Štrouf, and Ferles, *Collection Czechoslov. Chem. Communs.*, **22**, 1173 (1957).
[89] Lukeš, Štrouf, and Ferles, *Collection Czechoslov. Chem. Communs.*, **24**, 212 (1959).

the ratio of styrene to ethylene, for example, obtained in the Hofmann elimination reaction of ethyl phenethyl quaternary bases is very large, the rate of formation of styrene is much greater than the rate of formation of ethylene. It would be expected that decomposition of a salt containing a phenethyl group would occur at a lower temperature than the decomposition of a compound containing only alkyl groups, and that in general the ease with which elimination reactions occur will be dependent on the substituents in the ammonium compound. Indeed, quaternary salts bearing only alkyl substituents usually decompose slowly if at all in boiling aqueous solution, but reactions of phenethyl compounds and derivatives of tetrahydroisoquinoline occur readily at steam bath temperatures. Quaternary hydroxides derived from β amino ketones are still more reactive and decompose rapidly in solution at room temperature or lower. In some instances, therefore, the conditions necessary to bring about elimination serve as evidence concerning the structure of the quaternary compound.

REACTION WITH DIAMINES

The Hofmann elimination reaction has not been used widely for the synthesis of simple olefins, although cyclopropene,[8] cyclobutene,[90] *trans*-cyclooctene,[9] and a few other alicyclic olefins are best prepared in this way. In addition, some polyenes are most easily prepared from diamines by way of the quaternary hydroxides. For example, 1,12-diaminododecane gave 1,11-dodecadiene in 65% yield,[91] and similar dienes have been prepared in fair yield by this method. The interesting derivative of dimethylenecyclobutene XXIV was prepared from a diamine,[92, 93] and a

[90] Roberts and Sauer, *J. Am. Chem. Soc.*, **71**, 3925 (1949).
[91] von Braun and Anton, *Ber.*, **64**, 2865 (1931).
[92] Blomquist and Meinwald, *J. Am. Chem. Soc.*, **79**, 5317 (1957).
[93] Blomquist and Meinwald, *J. Am. Chem. Soc.*, **81**, 667 (1959).

number of alkaloids, e.g., of the bisbenzylisoquinoline type such as dauricine (XXV) are degraded at both functions simultaneously in good yield.[94] If the amino groups are sufficiently close together in the molecule, a conjugated olefin is usually produced. Thus 1,5-pentanediamine gives 1,3-pentadiene, not 1,4-pentadiene.[95]

SIDE REACTIONS: ALKYLATIONS BY QUATERNARY COMPOUNDS

Alcohol Formation

The most common process that competes with elimination when a quaternary ammonium compound reacts with hydroxide ion is a displacement reaction at the α carbon atom. Unlike the exchange reaction of α hydrogen atoms, which does not interfere with elimination, attack at the α carbon atom by hydroxide ion forms an alcohol and a tertiary amine, which are usually stable products under the reaction conditions. This side reaction may be important. In a few cases (tetrahydroquinoline, pavinemethine) the formation of an alcohol and a tertiary amine is the only reaction reported.

Attack at the α carbon atom by hydroxide ion is apparently a bimolecular displacement reaction with most compounds, although this is not the only possible mechanism.[10, 96] A unimolecular reaction which does

Trimethylpiperitylammonium hydroxide (ref. 97)

Tetrahydroharmine methiodide (ref. 98)

[94] Kondo, Narita, and Uyeo, *Ber.*, **68**, 519 (1935).
[95] von Braun, *Ann.*, **386**, 273 (1911).
[96] Ingold and Patel, *J. Chem. Soc.*, **1933**, 67.
[97] Read and Storey, *J. Chem. Soc.*, **1930**, 2770.
[98] Perkin and Robinson, *J. Chem. Soc.*, **115**, 933 (1919).

OLEFINS FROM AMINES

[Reaction scheme: Eserethole methiodide with ethoxy-substituted indoline bearing N-CH$_3$ and N$^+$(CH$_3$)$_2$ I$^-$ groups, converted by base to ethoxy-indoline with CH$_2$CH$_2$N(CH$_3$)$_2$ and OH substituents] (ref. 99)

Eserethole methiodide

$$C_6H_5CH_2\overset{\oplus}{N}(CH_3)_3\overset{\ominus}{OH} \longrightarrow C_6H_5CH_2OH + CH_3OH \quad (\text{ref. 100})$$
$$(65\%) \qquad (35\%)$$

[Reaction of dihydroisoindolium derivative XXVI with N$^+$(CH$_3$)$_2$ OH$^-$ giving N-methyl dihydroisoindole + CH$_3$OH]

XXVI

not require hydroxide ion has been demonstrated to occur with certain benzylamines having methoxyl substituents in the ring.[30] This is an exceptional case in which the carbonium ion would be especially well stabilized, but in most instances a nucleophile is required. The following examples illustrate this type of reaction with hydroxide and methoxide ions. It is interesting that the benzyl group does not have this high reactivity when it is part of a heterocyclic ring; the dihydroisoindolium derivative XXVI reacts mainly at the methyl groups.[101]

There is no way to avoid completely the side reaction which forms an alcohol, because the rate of this displacement and the rate of the elimination vary with hydroxide concentration in the same way. If anions less basic than hydroxide or alkoxide, such as acetate, phenoxide or carbonate, are used, the displacement reaction becomes more important.[11] For this reason solutions of quaternary hydroxides should be protected from carbon dioxide and should always be concentrated under reduced pressure rather than in an open vessel.[102] If no benzyl or allyl groups are attached to the nitrogen atom, most of the attack on carbon will occur at the methyl groups to regenerate the original tertiary amine. Thus the starting

$$RN(CH_3)_3^{\oplus}OH^{\ominus} \rightarrow RN(CH_3)_2 + CH_3OH$$

material is not lost, and it may be remethylated and the degradation

[99] Stedman and Barger, *J. Chem. Soc.*, **127**, 247 (1925).
[100] Hughes and Ingold, *J. Chem. Soc.*, **1933**, 69.
[101] Fränkel, *Ber.*, **33**, 2808 (1900).
[102] von Braun, Teuffert, and Weissbach, *Ann.*, **472**, 121 (1929).

repeated. Since attack at the methyl group does not affect the bond between the alkyl group and the nitrogen atom, the regenerated amine is not changed in stereochemical configuration.

Ethers and Epoxides

In addition to the alkylation of hydroxide ions by the quaternary compounds to form an alcohol, other hydroxyl groups may be alkylated to produce ethers. This reaction is the predominant one when β amino alcohols are subjected to the Hofmann elimination procedure and leads to the formation of epoxides. Examples of this reaction are collected in

$$\begin{array}{c} \text{OH} \\ | \\ \text{R}_2\text{C}-\text{CH}-\text{R} \\ | \\ \overset{\oplus}{\text{N}}(\text{CH}_3)_3\overset{\ominus}{\text{OH}} \end{array} \rightarrow \text{R}_2\text{C}\overset{\displaystyle\text{O}}{\overset{\displaystyle\diagup\ \diagdown}{}}\text{CHR} + (\text{CH}_3)_3\text{N} + \text{H}_2\text{O}$$

Table XI, p. 389. As would be expected from the general nature of the reaction, trimethylamine is displaced with inversion at the carbon atom to which it was attached. Thus the quaternary hydroxide prepared from ephedrine gives *trans-β*-methylstyrene oxide and the quaternary hydroxide

from pseudoephedrine yields the *cis* oxide.[103] Also, the *erythro* and *threo* forms of 1,2-diphenylethanolamine yield *trans-* and *cis*-stilbene oxides respectively.[104] The stereochemistry of the molecule may preclude the formation of an oxide by this process as in the case of *cis*-2-dimethylamino-cyclohexanol. When the methohydroxide of this compound is heated, the main products are recovered amino alcohol and its methyl ether; no cyclohexene oxide is obtained. The methyl ether may be produced by intramolecular alkylation.[105] With cyclic β amino alcohols containing twelve-, thirteen- and sixteen-membered rings in which the substituents can assume a *trans* conformation, the *cis* amino alcohol yields the *trans*

[103] Witkop and Foltz, *J. Am. Chem. Soc.*, **79**, 197 (1957).
[104] Rabe and Hallensleben, *Ber.*, **43**, 884 (1910).
[105] A. C. Cope, E. J. Ciganek, and J. Lazar, to be published.

oxide and the *trans* amino alcohol the *cis* oxide.[106] Compounds with the

cyclohexane with OH (up) and N(CH$_3$)$_3$OH$^\ominus$ (⊕) (cis) → cyclohexane with OH and N(CH$_3$)$_2$ + cyclohexane with OCH$_3$ and N(CH$_3$)$_2$

hydroxyl group farther removed from the nitrogen atom may also give oxygen-containing heterocycles. Thus the quaternary hydroxide from isomethadol (XXVII) gives a derivative of tetrahydrofuran in good yield.[107]

$$(C_6H_5)_2C\text{—}CHOHC_2H_5$$
$$|$$
$$CH\text{—}CH_2N(CH_3)_3{}^\oplus OH^\ominus \rightarrow$$
$$|$$
$$CH_3$$
$$\text{XXVII}$$

$$(C_6H_5)_2C\text{——}CH\text{—}C_2H_5$$
$$|\qquad\qquad\quad\diagdown$$
$$\qquad\qquad\qquad O$$
$$|\qquad\qquad\quad\diagup$$
$$CH\text{—}CH_2$$
$$|$$
$$CH_3$$

Compounds containing phenolic and enolic hydroxyl groups also are alkylated internally to give cyclic products if the hydroxyl and amino groups are in suitable proximity. The following examples illustrate this reaction.

<chemical structure: Tetrahydrothebainonemethine with N(CH$_3$)$_3{}^\oplus$ I$^\ominus$, CH$_3$O, OH, O → KOH → Thebenone with CH$_3$O, O, O> (ref. 108)

<chemical structure: $(C_6H_5)_2C\text{—}C(=O)\text{—}CH_2CH_3$ with CH$_2$CH$_2$N(CH$_3$)$_3{}^\oplus$OH$^\ominus$ → $(C_6H_5)_2C\text{—}C(=CHCH_3)\text{—}O$ with CH$_2$—CH$_2$> (ref. 109)

In order to avoid their alkylation by the quaternary base, phenolic hydroxyl groups are commonly converted to methyl or ethyl ethers before

[106] Svoboda and Sichee, *Collection Czechoslov. Chem. Communs.*, **23**, 1540 (1958).
[107] Easton and Fish, *J. Am. Chem. Soc.*, **77**, 2547 (1955).
[108] Rapoport and Lavigne, *J. Am. Chem. Soc.*, **75**, 5329 (1953).
[109] Easton, Nelson, Fish, and Craig, *J. Am. Chem. Soc.*, **75**, 3751 (1953).

application of the Hofmann elimination reaction. When the stereochemistry is not favorable or when elimination is facilitated by structural factors, the alkylation reaction is not important.

$$\underset{\underset{N(CH_3)_3^{\oplus}OH^{\ominus}}{|}}{C_6H_5CH_2CHCH_2OH} \rightarrow C_6H_5CH=CHCH_2OH \qquad \text{(ref. 110)}$$

(ref. 111)

The alcohols that are often formed as by-products in the Hofmann procedure may themselves be alkylated by the unreacted quaternary compound to produce ethers. Small amounts of such products have been observed in several instances and may have been overlooked in others.

(ref. 53)

Groups other than the oxygen-containing ones described above might be alkylated by quaternary ions, but compounds with structures suitable for testing such reactions have not been studied. There are a few examples in which the products are most easily explained by assuming alkylation of carbon by the quaternary nitrogen.

$$\underset{\underset{NHCOCH_3}{|}}{(CH_3)_3\overset{\oplus}{N}CH_2CH_2C(CO_2C_2H_5)_2} \rightarrow \underset{CH_2}{\overset{CH_2}{\diagdown}}\underset{NHCOCH_3}{\overset{CO_2C_2H_5}{\diagup}} \qquad \text{(ref. 112)}$$
$$OH^{\ominus}$$

$$\underset{\underset{CH_2C_6H_5}{|}}{(CH_3)_3\overset{\oplus}{N}CH_2CH_2C(CO_2C_2H_5)_2} \rightarrow \underset{CH_2}{\overset{CH_2}{\diagdown}}\underset{CH_2C_6H_5}{\overset{CO_2C_2H_5}{\diagup}} \qquad \text{(refs. 113–115)}$$
$$OC_2H_5^{\ominus}$$

[110] Karrer and Horlacher, *Helv. Chim. Acta*, **5**, 571 (1922).
[111] Stork, Wagle, and Mukharji, *J. Am. Chem. Soc.*, **75**, 3197 (1953).
[112] Rinderknecht and Niemann, *J. Am. Chem. Soc.*, **73**, 4259 (1951).
[113] Ingold and Rogers, *J. Chem. Soc.*, **1935**, 722.
[114] Weinstock, *J. Org. Chem.*, **21**, 540 (1956).
[115] Rogers, *J. Org. Chem.*, **22**, 350 (1957).

An unusual alkylation on nitrogen is reported with the alkaloid gelsemine and its dihydro and octahydro derivatives.[116, 117]

Gelsemine methohydroxide[118] → N(a)-Methylgelsemine

A few β amino alcohols have been observed to undergo a cleavage reaction instead of elimination or epoxide formation. This reaction is illustrated with quinine with the formulation suggested by Turner and Woodward.[119] Narcotine undergoes an analogous reaction.[120]

ISOMERIZATION OF OLEFINS FORMED

The Hofmann elimination reaction often leads to the formation of an olefin which is not the most stable isomer. For instance, at temperatures below 200° almost any terminal olefin is less stable than an isomeric non-terminal olefin. However, the olefins from the Hofmann elimination are obtained free of isomerized products except when there is the possibility of an allylic shift of a proton that would move the double bond into conjugation with another unsaturated system. Several examples of this

[116] Habgood, Marion, and Schwarz, *Helv. Chim. Acta*, **35**, 638 (1952).
[117] Prelog, Patrick, and Witkop, *Helv. Chim. Acta*, **35**, 640 (1952).
[118] Lovell, Pepinsky, and Wilson, *Tetrahedron Letters*, No. 4, p. 1, 1959.
[119] Turner and Woodward, in Manske and Holmes, *The Alkaloids*, Vol. III, Academic Press, New York, 1953, pp. 9. 10.
[120] Stevens, Creighton, Gordon, and MacNicol, *J. Chem. Soc.*, **1928**, 3193.

type have been mentioned, for instance, formation of piperylene and pirylene. The reaction of the methylenecyclobutane derivative XXVIII provides another instance of such an isomerization.[121]

$$\underset{\text{XXVIII}}{H_2C=\square-CH_2\overset{\oplus}{N}(CH_3)_3OH^{\ominus}} \xrightarrow{50\%} H_2C=\square-CH_3$$

In the case of 3-phenylpropylammonium salts which yield *trans*-1-phenylpropene, initial reaction to form 3-phenylpropene followed by isomerization has been assumed, and the isomerization of 3-phenylpropene has been shown to occur rapidly.[114] The decomposition of *trans*-2-phenylcyclohexyltrimethylammonium hydroxide to 1-phenylcyclohexene was assumed to involve a similar rearrangement, but it is now clear that this reaction proceeds instead by *cis* elimination.[22, 23]

MOLECULAR REARRANGEMENTS

Usually the Hofmann elimination procedure does not cause a change in the carbon skeleton of the molecule. In particular, carbonium-type rearrangements of quaternary ammonium hydroxides are not found even with structures such as XXIX;[32] however, see p. 330.[33] With

$$(CH_3)_3CCHCH_3 \rightarrow (CH_3)_3CCH=CH_2$$
$$\underset{\text{XXIX}}{|\ N(CH_3)_3 \oplus OH^{\ominus}}$$

N-benzyl derivatives of phenacylamines, the Stevens rearrangement is observed.[120, 122] A similar rearrangement has been observed with the spiro quaternary compound XXX[87] and with similar compounds.[123]

$$C_6H_5COCH_2\overset{\oplus}{N}(CH_3)_2 OH^{\ominus} \rightarrow C_6H_5COCH\underset{|}{N}(CH_3)_2$$
$$\underset{CH_2C_6H_5}{|} \qquad \underset{CH_2C_6H_5}{|}$$

[121] Caserio, Parker, Piccolini, and Roberts, *J. Am. Chem. Soc.*, **80**, 5507 (1958).
[122] Stevens, *J. Chem. Soc.*, **1930**, 2107.
[123] Wittig, Koenig, and Clauss, *Ann.*, **593**, 127 (1955).

In all these cases the normal elimination reaction could not occur for structural reasons.

ANALOGOUS "ONIUM" COMPOUNDS

Although quaternary ammonium compounds are the only ones which have been used in degradative and synthetic work, sulfonium hydroxides have been studied carefully and have been found to react in a manner similar to the ammonium analogs.[124] Phosphonium hydroxides usually decompose in a different way to form a hydrocarbon and a phosphine oxide.[125] Ammonium compounds rarely decompose in this way, the

$$R_4\overset{\oplus}{P}\overset{\ominus}{OH} \rightarrow RH + R_3P\rightarrow O$$

only reported instance being that of the nitrobenzylammonium compounds which apparently give some nitrotoluene.[126] Sulfones also

$$NO_2C_6H_4CH_2\overset{\oplus}{N}(CH_3)_3\overset{\ominus}{NO_3} \rightarrow NO_2C_6H_4CH_3$$

undergo an elimination reaction in the presence of base, although decomposition to give a paraffin has been observed as well.[124, 127]

$$C_2H_5SO_2R + KOH \rightarrow CH_2{=}CH_2 + RSO_2K + H_2O$$

EXPERIMENTAL CONSIDERATIONS

The Hofmann elimination reaction has usually been conducted by heating and concentrating an aqueous solution of the quaternary hydroxide until decomposition occurs. The base necessary for the reaction is often the quaternary hydroxide itself, and, depending on how much water is removed by distillation before the decomposition takes place, the reaction may proceed in aqueous solution or without a solvent. Variations of this procedure have been investigated and will be described below; none of them in general has proved more useful than concentrating aqueous solutions of the quaternary hydroxides under reduced pressure and raising the temperature until elimination occurs.

NATURE OF THE BASE

In the preparation of olefins from quaternary ammonium salts, hydroxide ion usually is the basic anion of choice. Instead of preparing the

---

[124] Ingold, Jessop, Kuriyan, and Mandour, *J. Chem. Soc.*, **1933**, 533.
[125] Fenton and Ingold, *J. Chem. Soc.*, **1929**, 2342.
[126] Ing and Robinson, *J. Chem. Soc.*, **1926**, 1655.
[127] Fenton and Ingold, *J. Chem. Soc.*, **1928**, 3127.

quaternary hydroxide, an alternative way of providing the base is to add excess potassium hydroxide to a solution of a quaternary chloride or iodide directly and pyrolyze this mixture.[128-131] This method has most often been applied to substances that undergo reaction easily, but no study has been made that would indicate whether better yields are to be expected from this method or from pyrolysis of the quaternary hydroxide itself.

The concentration of base can be controlled either by regulating the concentration of the quaternary hydroxide or by adding excess base to the solution. Since kinetic investigations[132] have shown that the rate of reaction is proportional to the concentration of hydroxide ion, this would seem to be one way of controlling the course of the reaction. Unfortunately, the most common side reaction, substitution by hydroxide ion to form an alcohol, is usually affected in the same way so that the yield of olefin is not improved by this method. The results in Table IV, obtained

TABLE IV

Decomposition of $n\text{-}C_{10}H_{21}\overset{\oplus}{N}(CH_3)_3\overset{\ominus}{O}H$ at $200°$ and 26 Atmospheres for 10 Hours

Conc. of $R\overset{\oplus}{N}(CH_3)_3\overset{\ominus}{O}H$	Decene, %	CH_3OH, %	Ratio of Elimination to Displacement
2%	8	14	0.57 : 1
6%	23	42	0.55 : 1
16%	29	49	0.59 : 1
Syrup, distilled	62	30	2.1 : 1

by conducting the reaction for a fixed length of time but at different concentrations, illustrate both the increase in rate and the fixed ratio of elimination to substitution.[102] However, in very concentrated solution this ratio is no longer constant.

When the effect of excess base was tested by adding four equivalents of potassium hydroxide to a syrup of the quaternary hydroxide, the results as shown in Table V indicated that excess base may favor the elimination reaction.[102]

Other basic anions have been tested with quaternary salts, including alkoxides, phenoxides, and carbonates.[11, 133] Again, two courses of reaction are possible, one leading to elimination by attack at the β hydrogen

[128] Manske, J. Am. Chem. Soc., **72**, 55 (1950).
[129] Woodward and Doering, J. Am. Chem. Soc., **67**, 860 (1945).
[130] Willstätter, Ber., **29**, 393 (1896).
[131] Freund and Becker, Ber., **36**, 1521 (1903).
[132] Hughes and Ingold, J. Chem. Soc., **1933**, 523.
[133] Ingold and Patel, J. Chem. Soc., **1933**, 68.

TABLE V

Decomposition of $\overset{\oplus}{R}N(CH_3)_3\overset{\ominus}{O}H$

Compound	Olefin, %	CH_3OH, %	Ratio
n-$C_4H_9\overset{\oplus}{N}(CH_3)_3\overset{\ominus}{O}H$	77	12	6.4 : 1
Same + 4KOH	81	12	6.7 : 1
n-$C_{10}H_{21}\overset{\oplus}{N}(CH_3)_3\overset{\ominus}{O}H$	62	30	2.1 : 1
Same + 4KOH	79	13	6.1 : 1

atom and the other leading to substitution at the α carbon atom. The relative importance of these paths is determined by the relative reactivity of the anion with a β hydrogen atom and an α carbon atom. Anions such

$$-\underset{|}{\overset{|}{C}}-\underset{H}{\overset{|}{C}}-\overset{\oplus}{N}\diagdown + B^\ominus \rightarrow BH + \diagup C=C\diagdown + N\diagup$$

$$-\underset{|}{\overset{|}{C}}-\overset{\oplus}{N}\diagdown + B^\ominus \rightarrow B-\underset{|}{\overset{|}{C}}- + N-$$

as phenoxide, acetate, carbonate, and halide preferentially attack carbon rather than hydrogen and give much less olefin than does hydroxide ion (Table VI).[11]

TABLE VI

Effect of the Anion on the Decomposition of n-$C_3H_7\overset{\oplus}{N}(CH_3)_3X^\ominus$

$X^\ominus$	Propylene, %	CH_3X, %
$OH^\ominus$	81	19
$CO_3^\ominus$	26	
$C_6H_5O^\ominus$	15	65
$I^\ominus$	13	
$Cl^\ominus$	10	
$CH_3CO_2^\ominus$	Trace	

The alkoxide ions cannot be compared with hydroxide ion in aqueous solution, but in two instances neither the methoxide nor the ethoxide derivative prepared in the corresponding alcohol led to higher yields of olefins than the hydroxide prepared in water (Table VII).[133]

An important result of these studies of the effect of various anions has been the recognition that carbon dioxide absorbed from the atmosphere seriously reduces the yield of olefin.[11, 102] The results of experiments in

TABLE VII
Effect of Alkoxide Ions on the Decomposition of $\overset{\oplus}{R}N(CH_3)_3X^{\ominus}$

Compound	$X = OH^{\ominus}$	$X = OCH_3^{\ominus}$	$X = OC_2H_5^{\ominus}$
$C_2H_5\overset{\oplus}{N}(CH_3)_3$	Ethylene, 94%	90%	88%
$i\text{-}C_4H_9\overset{\oplus}{N}(CH_3)_3$	Isobutylene, 63%	57%	55%

which the quaternary hydroxide solution was concentrated under reduced pressure as compared with concentration on a steam bath in air emphasize this point (Table VIII).[102]

TABLE VIII
Decomposition of $\overset{\oplus}{R}N(CH_3)_3OH^{\ominus}$

R	Under Reduced Pressure		In Air	
	Olefin, %	Alcohol, %	Olefin, %	Alcohol, %
$n\text{-}C_4H_9$	77	10	23	50
$n\text{-}C_{10}H_{21}$	62	30	25	72
(piperidyl)	82	Small	65	ca. 20

Some of the low yields reported in the early literature may be accounted for by consideration of this factor. Strangely enough, in a few special cases, especially with strychnine, decomposition of the carbonate gives better yields than any other method, although in no case is the yield good.[134, 135]

As stated previously, most Hofmann reactions have been conducted in aqueous solution or with the residue obtained when the water is distilled from these solutions. However, a few solvents have been employed to advantage in special instances, most of these being hydroxylic solvents such as glycerol, ethylene glycol, cyclohexanol, and amyl alcohol. Unfortunately, hydroxide ion can react with such solvents to form an alkoxide ion plus water and, while the position of equilibrium may be such that the alkoxide is not present in large amount at low temperature, when the water is removed by distillation this equilibrium will be displaced. Furthermore, the concentration of reactants will be lower in solution than in the syrupy quaternary ammonium hydroxide. Hence it is not clear which effect is responsible for the different results observed. A comparison of some decompositions of quaternary hydroxides alone and in

[134] Achmatowicz and Robinson, *J. Chem. Soc.*, **1934**, 581.
[135] Achmatowicz, Lewi, and Robinson, *J. Chem. Soc.*, **1935**, 1685.

glycerol solution indicates that in general this solvent lowers the yield of olefin (Table IX).[74, 102]

TABLE IX

DECOMPOSITION OF QUATERNARY BASES IN GLYCEROL

Quaternary Base	Free Hydroxide		Glycerol Solution	
	Olefin, %	Alcohol, %	Olefin, %	Alcohol, %
$n\text{-}C_4H_9\overset{\oplus}{N}(CH_3)_3OH^{\ominus}$	77	10	17	69
$n\text{-}C_{10}H_{21}\overset{\oplus}{N}(CH_3)_3OH^{\ominus}$	62	30	14	76
(N,N-dimethylpiperidinium hydroxide)	82	Small	32	49

In other cases the use of potassium hydroxide in ethylene glycol[136] or sodium cyclohexoxide in cyclohexanol[137] is reported to give better yields than pyrolysis of the quaternary hydroxide. Amyl and isoamyl alcohol also have been used[138, 139] but seem to offer little advantage.

Because of the effect of the ion-solvating power of the medium on bimolecular elimination and substitution reactions (ref. 41, p. 453), it would be expected that the ratio of olefin to alcohol would be increased by the use of non-aqueous solvents. This generalization might not be expected to extend to the very concentrated solutions employed in the usual conditions for the Hofmann elimination, and the results available do not constitute a fair test of this prediction. From what information is now at hand there seems to be little evidence to recommend the use of a solvent.

PYROLYSIS OF AMINE OXIDES

The oxides of tertiary amines decompose when heated to yield an olefin plus a derivative of hydroxylamine. Examples of this reaction are

$$R_2CHCR_2 \quad \rightarrow \quad R_2C{=}CR_2 + (CH_3)_2NOH$$
$$\;\;\;\;\;\;|$$
$$O{\leftarrow}N(CH_3)_2$$

reported in the early literature,[140, 141] but the utility of the reaction as a means for synthesizing olefins was not emphasized until 1949.[142] The

---

[136] Julian, Meyer, and Printy, *J. Am. Chem. Soc.*, **70**, 887 (1948).
[137] Mosettig and Meitzner, *J. Am. Chem. Soc.*, **56**, 2738 (1934).
[138] Cahn, *J. Chem. Soc.*, **1930**, 702.
[139] Ing, *J. Chem. Soc.*, **1931**, 2195.
[140] Wernick and Wolffenstein, *Ber.*, **31**, 1553 (1898).
[141] Mamlock and Wolffenstein, *Ber.*, **33**, 159 (1900).
[142] Cope, Foster, and Towle, *J. Am. Chem. Soc.*, **71**, 3929 (1949).

method is useful for preparing certain olefins and may also be used for the preparation of N,N-disubstituted derivatives of hydroxylamine.

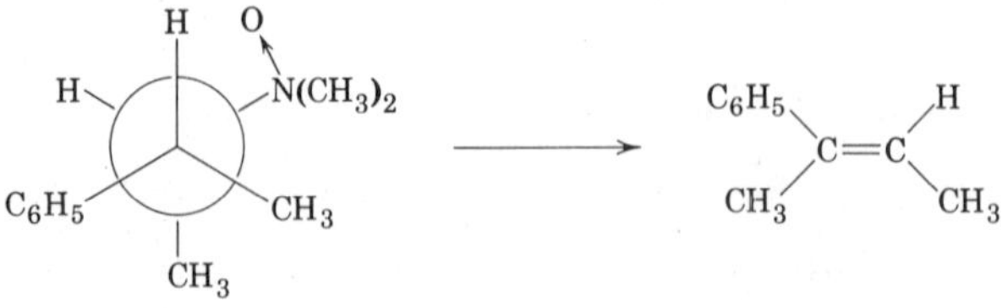

(threo) → (cis)

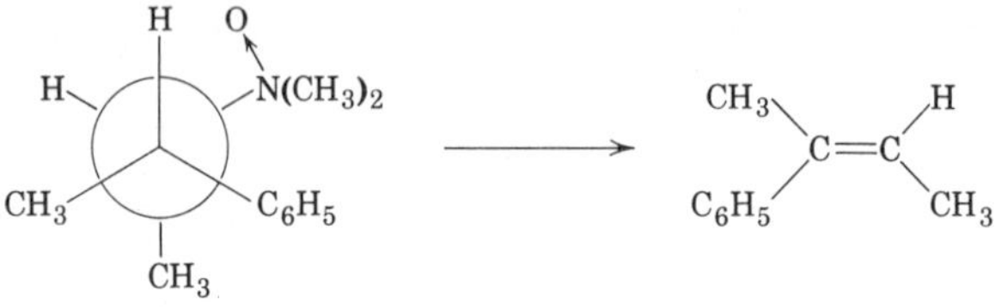

(erythro) → (trans)

Mechanism

There is good evidence that the pyrolysis of amine oxides involves *cis* elimination. The evidence has been obtained by the decomposition of *threo* and *erythro* derivatives of 2-amino-3-phenylbutane.[143] The *threo* isomer reacts to give predominantly the *cis* conjugated olefin, the ratio of *cis*- to *trans*-2-phenyl-2-butene being at least 400 to 1. With the *erythro* form the *trans* isomer is favored by a ratio of at least 20 to 1. The *threo* form, reacting through a transition state that involves less steric interaction than does the transition state for the *erythro* isomer, reacts more readily than the *erythro* form. There are several examples of pyrolysis of alicyclic amines oxides which show the *cis* nature of the elimination reaction. This evidence establishes an intramolecular mechanism involving a planar, five-membered cyclic transition state. The pyrolysis of amine oxides accordingly resembles the Chugaev reaction and the pyrolysis of esters.

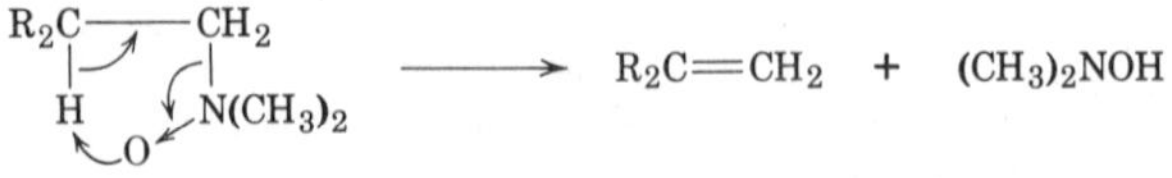

$$R_2C\text{---}CH_2 \longrightarrow R_2C=CH_2 + (CH_3)_2NOH$$

A few examples of a low-temperature decomposition of amine oxides have been described which may be base catalyzed. Salts of amine oxides

[143] Cram and McCarty, *J. Am. Chem. Soc.*, **76**, 5740 (1954).

derived from β-aminopropionic esters or nitriles undergo the reaction, which has been described as a reversal of the Michael addition, facilitated by the formal positive charge on nitrogen.[144]

$$R_2NCH_2CH_2CO_2C_2H_5 \xrightarrow{\text{Base}} R_2NOH + CH_2{=}CHCO_2C_2H_5$$
$$\downarrow \qquad\qquad\qquad\qquad \text{(not isolated)}$$
$$O$$

DIRECTION OF ELIMINATION

Acyclic Amines

With simple alkyl-substituted amine oxides the direction of elimination seems to be governed almost entirely by the number of hydrogen atoms at the various β positions. The marked preference for attack at a β methyl group in the Hofmann reaction finds no parallel in the amine oxide decomposition. Table X gives the ease of elimination of some alkyl groups relative to ethyl groups.[36]

TABLE X

RELATIVE EASE OF ELIMINATION OF ALKYL GROUP AS OLEFIN

Alkyl Group	Not Corrected for Number of β Hydrogen Atoms	Corrected for Number of β Hydrogen Atoms
Ethyl	100	100
Isopropyl	264	132
t-Butyl	606	202
n-Propyl	60	90
n-Butyl	80	120
Isoamyl	76	114
n-Decyl	88	132
Isobutyl	44	133
Phenethyl	7×10^3	1.0×10^4

Significant variations from the general value of 100 ± 30 are shown by the t-butyl group and the phenethyl group in which the relief of steric interactions and acidity of the β hydrogen atom, respectively, are factors that favor their elimination as olefins as compared with the ethyl group. The data were obtained by analysis of the olefin mixtures obtained by pyrolysis of compounds such as methylethylisopropylamine oxide and

$$\begin{array}{c} C_2H_5 \quad CH_3 \\ \diagdown \quad \diagup \\ N \\ \diagup \quad \diagdown \\ (CH_3)_2CH \quad O \end{array} \xrightarrow{90\%} \begin{array}{l} CH_2{=}CH_2 \quad 27.5\% \\ \text{and} \\ CH_3CH{=}CH_2 \quad 72.5\% \end{array}$$

[144] Rogers, J. Chem. Soc., **1955**, 769.

can be used to predict the ratio of olefins which would be formed in such a reaction. They may be extended to other cases with some sacrifice of accuracy. For example, with the use of the values of 100 and 60 for the ethyl and n-propyl groups respectively, the ratio of isomers predicted from the decomposition of dimethyl-*sec*-butylamine oxide is 62.5% of butene-1 and 37.5% of butene-2. The actual amounts of isomers produced in this decomposition are 67.3% of butene-1 and 32.7% of *cis*- and *trans*-butene-2.[36]

$$CH_3CH_2\underset{\underset{O}{\underset{\downarrow}{N(CH_3)_2}}}{C}HCH_3 \xrightarrow{91\%} CH_3CH{=}CHCH_3 + CH_3CH_2CH{=}CH_2$$

Use of the values for phenethyl and ethyl, and their application to the decomposition of 2-amino-3-phenylbutane, leads to the prediction that 97% of 2-phenyl-2-butene and 3% of 3-phenyl-1-butene will be formed, whereas the actual results are 92–93% and 7–8%, respectively.[143] For many purposes such predictions would be sufficiently accurate.

$$C_6H_5\underset{CH_3}{C}H{-}\underset{\underset{O}{\underset{\downarrow}{N(CH_3)_2}}}{C}HCH_3 \rightarrow C_6H_5\underset{CH_3}{C}{=}CHCH_3 + C_6H_5CH\underset{CH_3}{C}H{=}CH_2$$

Addition of unsymmetrical secondary amines (RR'NH) to α,β-unsaturated carbonyl compounds, followed by conversion of the product to an amine oxide and decomposition, provides a method for preparing unsymmetrical dialkylhydroxylamines (RR'NOH).[144]

$$\underset{R'}{\overset{R}{\diagdown}}\underset{\underset{O}{\underset{\downarrow}{}}}{N}CH_2CH_2COC_6H_5 \rightarrow \underset{R'}{\overset{R}{\diagdown}}NOH + CH_2{=}CHCOC_6H_5$$
(not isolated)

$$\underset{R'}{\overset{R}{\diagdown}}\underset{\underset{O}{\underset{\downarrow}{}}}{N}CH_2CH_2CO_2R \rightarrow \underset{R'}{\overset{R}{\diagdown}}NOH + CH_2{=}CHCO_2R$$
(not isolated)

In general, with acyclic amines which could undergo elimination forming either a *cis* or a *trans* olefin, the more stable *trans* form is obtained. Thus N,N-dimethyl-3-pentylamine oxide gives 86% of 2-pentene which consists of 29.2% of *cis*- and 70.8% of *trans*-2-pentene. Pyrolysis of N,N-dimethyl-2-butylamine oxide forms 91% of a mixture of 1-butene (67.3%) and 2-butene (33.7%). The 2-butene contains 35.8% of the *cis* isomer and 64.2% of the *trans* isomer.[36] Presumably the more stable *trans* olefins are formed because the steric factors which operate to influence the relative stabilities of the olefins also operate in the transition states leading to these olefins.

Alicyclic Amines

With alicyclic amines the pyrolysis has been shown to follow the pattern of *cis* elimination in the case of menthyl and neomenthyl compounds and with *cis*- and *trans*-2-phenylcyclohexylamine.[16, 145] Neomenthylamine

[145] Cope and Bumgardner, *J. Am. Chem. Soc.*, **79**, 960 (1957).

has only the *cis* hydrogen atom at the 2 position available and only 2-menthene is formed, whereas menthylamine has *cis* hydrogen atoms at the 2 and 4 positions and both menthenes are isolated. The preference for 2-menthene in the latter instance has been explained in terms of the eclipsing of the isopropyl group in the 4 position with the hydrogen atom in the 3 position that is required in the cyclic transition state if elimination takes this path.[16]

Pyrolysis of *trans*-2-phenylcyclohexyldimethylamine oxide gives 85% of 1-phenylcyclohexene and 15% of 3-phenylcyclohexene, showing less preference for elimination toward phenyl than is observed in an acyclic case. With the *cis* amine oxide, an olefin mixture containing 98% of 3-phenylcyclohexene and 2% of 1-phenylcyclohexene was obtained.[145] The small amount of 1-phenylcyclohexene may have been formed from a small amount of *trans* amine in the starting material; it is not formed by isomerization since 3-phenylcyclohexene does not isomerize under the

reaction conditions. Cycloheptyl- and cyclooctyl-dimethylamine oxide yield *cis*-cycloheptene and *cis*-cyclooctene,[9] respectively, and *cis*-cycloöcten-3-yldimethylamine oxide yields *cis-cis*-1,3-cyclooctadiene.[60] However, cyclononyl- and cyclodecyl-dimethylamine oxides form the *trans* olefins almost exclusively.[56] The thermal decompositions of cyclodecyl acetate and xanthate also form principally *trans*-cyclodecene.[146]

When an exocyclic branch in which the double bond may be formed is present, product stability parallels the direction of elimination, except in the cyclohexyl compounds. The examples below show the results with such amines.[34] Preference for the formation of the endocyclic double

---

[146] Blomquist and Goldstein, *J. Am. Chem. Soc.*, **77**, 1001 (1955).

bond in the cyclopentyl and cycloheptyl systems may simply be a reflection in the transition state of the greater stability of endocyclic olefins.

[Cyclopentyl compound with CH₃ and N(CH₃)₂→O groups] —77%→ [cyclopentylidene=CH₂] (2.5%) + [1-methylcyclopentene] (97.5%)

[Cyclohexyl compound with CH₃ and N(CH₃)₂→O groups] —84%→ [methylenecyclohexane] (97.2%) + [1-methylcyclohexene] (2.8%)

[Cycloheptyl compound with CH₃ and N(CH₃)₂→O groups] —84%→ [methylenecycloheptane] (15.2%) + [1-methylcycloheptene] (84.8%)

With the cyclohexyl derivative, however, elimination to form an endocyclic olefin through a planar five-membered transition state would require the ring to bend toward a more nearly planar, cyclohexene-like structure. This would introduce eclipsed interactions between the groups at the

[Chair cyclohexane with N→O(CH₃)₂ substituent and CH₃, showing hydrogens] ⟶ [transition state with O···H and N(CH₃)₂ arrangement]

1, 2, 3, and 6 positions which are not present in cyclohexene. Elimination toward the methyl group will not change the geometry of the cyclohexane ring if the double bond character of the transition state is not great. This effect may be unimportant with the cyclopentyl compound because the ring is already nearly planar and there would be little additional interaction introduced by endocyclic elimination. Because the geometries of the cycloheptyl and cycloheptenyl systems are less well known than those of the smaller rings, these arguments cannot be extended with certainty to the seven-membered ring at present.

Heterocyclic Amines

Pyrolysis of N-methylpiperidine oxide does not result in ring opening. However, the seven- and eight-membered cyclic amines do undergo ring opening in 53% and 79% yield, respectively.[147] Presumably, with azacycloalkanes containing larger rings, the ring system would also be sufficiently flexible to permit the formation of the cyclic transition state and elimination with ring opening should occur. N-Methyl-α-pipecoline oxide, which contains a six-membered ring, reacts to give a mixture of the unsaturated hydroxylamine and the saturated bicyclic compound XXXI.[147] Only the *trans* isomer forms these products; the *cis* isomer does not undergo the elimination reaction. N-Methyl- and N-ethyl-tetrahydroquinoline oxide are reported to yield tetrahydroquinoline plus formaldehyde and acetaldehyde, respectively.[148]

Side Reactions

One of the most attractive features of the synthesis of olefins by pyrolysis of amine oxides is the stability of the product under the reaction conditions. Migration of the double bond into conjugation with other unsaturated systems in the molecule is not observed in the first two examples given below.[145]

$$CH_2=CHCH_2CH_2CH_2N(CH_3)_2 \xrightarrow{61\%} CH_2=CHCH_2CH=CH_2$$
$$\downarrow O$$

$$C_6H_5CH_2CH_2CH_2N(CH_3)_2 \xrightarrow{91\%} C_6H_5CH_2CH=CH_2$$
$$\downarrow O$$

However, the dimethylenecyclobutane formed by pyrolysis of the amine oxide XXXII contains a small amount of the conjugated isomer,[121] and in a similar series of cyclobutane derivatives (XXXIII) having phenyl

---

[147] Cope and Le Bel. *J. Am. Chem. Soc.*, **82** (in press, 1960).
[148] Dodonov, *J. Gen. Chem. U.S.S.R.*, **14**, 960 (1944) [*C.A.*, **39**, 4612 (1945)].

substituents the olefin mixture produced contains equal parts of the isomers XXXIV and XXXV.[149]

[Reaction of cyclobutane-CH$_2$N(CH$_3$)$_2$ oxide (XXXII) giving 68% yield → methylenecyclobutane (=CH$_2$, principally) + methylcyclobutene (CH$_3$)]

[Reaction of 1,1-diphenyl-CH$_2$N(CH$_3$)$_2$ oxide (XXXIII) → diphenyl compounds XXXIV (=CH$_2$) + XXXV (=CH$_2$ with CH$_3$)]

If an allyl or a benzyl group is attached to the nitrogen atom of an amine oxide, these groups may rearrange from nitrogen to oxygen with the formation of O-substituted hydroxylamines. Apparently this

$$C_6H_5CH_2\diagdown \atop R \diagup N \diagup ^O \diagdown CH_3 \quad \rightarrow \quad R \diagdown \atop CH_3 \diagup N-OCH_2C_6H_5$$

process can compete favorably with elimination since allyldiethylamine oxide and benzyldiethylamine oxide as well as cycloöcten-3-yldimethylamine oxide give considerable amounts of the rearranged products.[142, 60]

[Cyclooctenyl-N(CH$_3$)$_2$ oxide → cyclooctadiene (50%) + cyclooctenyl-ON(CH$_3$)$_2$ (19%)]

$$CH_2{=}CHCH_2\overset{O}{\overset{\uparrow}{N}}(C_2H_5)_2 \;\rightarrow\; (C_2H_5)_2NOCH_2CH{=}CH_2 + CH_2{=}CH_2$$
(59%)

$$C_6H_5CH_2\overset{O}{\overset{\uparrow}{N}}(C_2H_5)_2 \;\rightarrow\; (C_2H_5)_2NOCH_2C_6H_5 + CH_2{=}CH_2$$
(31%)

[149] Blomquist and Meinwald, *Abstracts, A.C.S. Meeting*, April 1958, 77 N.

In the case of benzyldiethylamine oxide the normal product XXXVI expected from elimination of ethylene was isolated in 34% yield as well as products which may arise by alkylation of XXXVI by the amine oxide.[142] The conversion of dihydrothebainonedihydromethine oxide to

$$\text{C}_6\text{H}_5\text{CH}_2\text{N} \begin{array}{c} \text{OH} \\ \diagup \\ \diagdown \\ \text{C}_2\text{H}_5 \end{array} \quad + \quad \text{C}_6\text{H}_5\text{CH}_2\text{N} \begin{array}{c} \text{O} \\ \diagup \\ \diagdown \\ (\text{C}_2\text{H}_5)_2 \end{array}$$

XXXVI

$$\rightarrow \text{C}_6\text{H}_5\text{CH}_2\text{N} \begin{array}{c} \text{OCH}_2\text{C}_6\text{H}_5 \\ \diagup \\ \diagdown \\ \text{C}_2\text{H}_5 \end{array} \quad + \quad (\text{C}_2\text{H}_5)_2\text{NOH}$$

thebenone[150] illustrates the formation of a heterocycle by this alkylation process. The formal similarity between amine oxides and quaternary salts has been suggested earlier, and the use of the latter as alkylating agents is well known.

Dihydrothebainonedihydromethine oxide → Thebenone

Commonly a small amount of tertiary amine is recovered from the pyrolysis of the amine oxide.[42, 151]

An unexplained side reaction is involved in the pyrolysis of n-propylisoamylmethylamine oxide where the pentene fraction (55.9%) was found to contain 49.1% of 3-methyl-1-butene and two unexpected products, 11.2% of 2-methyl-2-butene and 1% of 2-methyl-1-butene.[36] Isoamylene was not isomerized under the reaction conditions, and the starting amine must have been pure since it reacted by the Hofmann elimination to give pure 3-methyl-1-butene.

[150] Bentley, Ball, and Ringe, *J. Chem. Soc.*, **1956**, 1963.
[151] Cope and Ciganek, *Org. Syntheses*, **39**, 40 (1959).

DECOMPOSITION OF AMINE PHOSPHATES

A third method of converting an amine to an olefin involves the distillation of the amine from crystalline phosphoric acid. This method was discovered and developed to some extent by Harries,[152, 153] but apparently it has found little use in other laboratories. Most of the amines Harries investigated were derivatives of cyclohexylamine related to various terpenes,[154, 155] and in several instances a diamine was used to prepare a diene in one step. The yields rarely exceeded 50%, and since the method apparently does not lend itself to the degradation of heterocyclic amines (which has been the main use of the Hofmann elimination reaction) it has received little attention. Formally, this method is similar to the dehydration of alcohols with phosphoric acid, but it is not possible at present to determine how closely this analogy applies. Primary amines may be used directly; apparently secondary and tertiary amines have not been investigated.

DECOMPOSITION OF ACYL DERIVATIVES OF AMINES

A few olefins have been obtained by heating N-acyl amines with phosphorus pentoxide in boiling xylene. This method apparently was discovered in the study of colchicine, and it is the method of choice in converting N-acetylcolchinol methyl ether to deaminocolchinol methyl ether.[156] Since the reaction seemed novel, it was investigated by Cook and applied to some simpler amines such as diphenylethylamine and

N-Acetylcolchinol methyl ether

---

[152] Harries, *Ber.*, **34**, 300 (1901).
[153] Harries and Johnson, *Ber.*, **38**, 1832 (1905).
[154] Harries and Antoni, *Ann.*, **328**, 88 (1903).
[155] Harries, *Ann.*, **328**, 322 (1903).
[156] Cook, Graham, Cotten, Lapsley, and Lawrence, *J. Chem. Soc.*, **1944**, 322.

cyclohexylamine.[157] In the latter case acetonitrile was isolated, and this is presumably the fate of the acyl group in other instances as well. The reaction is an extension to the N-alkyl amides of the dehydration of amides to nitriles. In this respect it is of interest that the reverse reaction,

$$\text{C}_6\text{H}_{11}\text{NHCOCH}_3 \rightarrow \text{C}_6\text{H}_{10} + \text{CH}_3\text{CN}$$

addition of an olefin to a nitrile, has been observed with a number of reactive olefins in the presence of sulfuric acid.[158] The N-alkyl amides obtained in this way were observed to undergo decomposition to an olefin on acid hydrolysis if the N-alkyl group was tertiary.

$$(\text{CH}_3)_3\text{CNHCOCH}_3 \xrightarrow{\text{H}_2\text{O, H}^\oplus} (\text{CH}_3)_2\text{C}=\text{CH}_2 + \text{CH}_3\text{CO}_2\text{H} + \text{NH}_3$$

From the results at hand it would seem that this type of decomposition depends strongly on the degree of branching of the N-alkyl group. N-Ethyl- and N-*n*-propyl-acetamide are reported to yield no olefin; N-cyclohexylacetamide gives cyclohexene when treated with phosphorus pentoxide in boiling xylene;[157] and N-tertiary alkyl acetamides form olefins when boiled with 15% hydrochloric acid.[158]

The use of phosphorus pentoxide in xylene for the degradation of amides involves reaction conditions identical with those often employed in the Bischler-Napieralski synthesis of dihydroisoquinolines.[159] With a properly constituted amine this type of reaction may be observed. For example, the acetyl derivative of 1,3-diphenyl-2-aminopropane (XXXVII) gives some of the dihydroisoquinoline (XXXVIII) as well as 1,3-diphenylpropene;[157] and the colchinol analog (XXXIX) undergoes ring closure exclusively.[157] (See formulas on p. 373.)

With the exception of the study by Cook and one application to a derivative of colchicine,[160] the preparation of olefins from N-acyl amines has not been studied in detail, and it is not possible to make any general statement concerning the scope or mechanism of the reaction.

The acetyl derivatives of amines have been pyrolyzed to olefins in the absence of phosphorus pentoxide by using temperatures of 500–600°.[161] Olefins were obtained in 14–67% conversion by one passage through the heated column. Better yields were obtained using an N-phenyl-N-alkyl derivative of acetamide than with compounds in which the aromatic

---

[157] Cook, Dickson, Ellis, and Loudon, *J. Chem. Soc.*, **1949**, 1074.
[158] Ritter and Minieri, *J. Am. Chem. Soc.*, **70**, 4045 (1948).
[159] Whaley and Govindachari, *Org. Reactions*, **6**, 75 (1951).
[160] Tarbell, Frank, and Fanta, *J. Am. Chem. Soc.*, **68**, 502 (1946).
[161] Bailey and Bird, *J. Org. Chem.*, **23**, 996 (1958).

OLEFINS FROM AMINES 373

[Structure XXXVII: benzyl group with CH₂-CHCH₂C₆H₅-NH-C(=O)-CH₃] → [Structure XXXVIII: dihydroisoquinoline with CH₂C₆H₅ and CH₃] + [styrene derivative: C₆H₅-CH=CHCH₂C₆H₅]

[Structure XXXIX: trimethoxy-tetrahydro compound with N-CH₂ group, C=O-CH₃, and OCH₃] → [cyclized product with N=CCH₃, trimethoxy and OCH₃ groups]

XXXIX

group was replaced by a methyl group or a hydrogen atom. In the cases reported, the direction of elimination was similar to that observed in the pyrolysis of esters.

REACTION OF QUATERNARY SALTS WITH ORGANOMETALLIC COMPOUNDS OR ALKALI METAL AMIDES

Olefins can be prepared from quaternary salts by treatment with phenyllithium in ether, potassium amide in liquid ammonia, or other strong bases.[25, 26, 162-164] These reactions involve an ylide intermediate and may yield a product which differs from that obtained by the usual Hofmann procedure. For example, the ratio of *trans-* to *cis-*cyclooctene is 5.7 : 1 when the mixture is prepared from cyclooctyltrimethylammonium bromide and potassium amide[163] but 1.5 : 1 when it is prepared from the quaternary hydroxide.[9] In a variant of this method the ylide is generated by treatment of a halomethyl quaternary derivative with phenyllithium. This process presumably involves halogen-metal interchange.[162]

$$R\overset{\oplus}{N}(CH_3)_2CH_2X + C_6H_5Li \rightarrow R\overset{\oplus}{N}(CH_3)_2CH_2Li + C_6H_5X$$

Cyclohexylmethyltrimethylammonium bromide containing deuterium at the β-position gave methylenecyclohexane free of deuterium, and

---
[162] Wittig and Polster, *Ann.*, **599**, 13 (1956).
[163] Wittig and Polster, *Ann.*, **612**, 102 (1958).
[164] Rabiant and Wittig, *Bull. soc. chim. France*, **1957**, 798.

trimethylamine which contained all the deuterium originally present,[165] thus confirming the postulated mechanism.

$$\text{(cyclohexyl)}-CH_2\overset{\oplus}{N}(CH_3)_3 Br^{\ominus} \xrightarrow{C_6H_5Li} \text{(cyclohexylidene)}=CH_2 + (CH_3)_2NCH_2D$$

(with D on cyclohexyl ring α-carbon)

However, the reaction of ethyltrimethylammonium bromide labeled with tritium at any of the positions in the ethyl group or in the methyl group showed extensive proton exchange among these positions when treated with phenyllithium.[25]

The composition of the products obtained from a quaternary salt may depend on whether potassium amide in liquid ammonia or phenyllithium in ether is used.[163]

Two cases are reported in which treatment of a quaternary halide with sodium amide in liquid ammonia forms cyclopropyl derivatives.[165a] In these instances the γ-hydrogen atom is benzylic. In other instances the

$$C_6H_5CH_2CH_2CH_2N(CH_3)_3Br \rightarrow H_5C_6CH\underset{CH_2}{\overset{CH_2}{<|}}$$

reaction of sodium amide in ammonia with quaternary bromides produced olefins.

COMPARISON OF METHODS

Of the four ways discussed for bringing about the conversion of an amine to an olefin it is obvious that the Hofmann exhaustive methylation procedure has been most extensively studied. As long as there is a β hydrogen atom in the quaternary base, the Hofmann method will almost always give some olefin, the important competing reaction being displacement to form an alcohol. The amine oxide method offers some advantages in experimental ease and usually does not cause isomerization of the olefin. However, the fact that it does not open the common nitrogen-containing rings is a limitation on its use as a tool in alkaloid investigations. In some instances the amine oxide method may lead to a geometrical isomer of the olefin different from that obtained from the quaternary hydroxide.

The pyrolyses of amines or their N-acyl derivatives in the presence of phosphoric acid have received so little attention that it is difficult to assess

---

[165] A. C. Cope and N. A. LeBel, unpublished results.
[165a] Bumgardner, *Chem. & Ind. (London)*, **1958**, 1555.

their utility. It is questionable whether heterocyclic amines would form olefins by these methods. As a method of preparing an olefin from a given primary amine, these reactions avoid the alkylation and subsequent procedures common to the Hofmann and amine oxide pyrolyses which may compensate for the somewhat lower yields obtained. (Olefin isomerization would be expected under the acidic reaction conditions employed.)

If the amine elimination reactions are considered as methods of degradation rather than syntheses, then the Hofmann reaction is the most useful, since it is most generally applicable. In this field there are two other methods which may accomplish the same sort of cleavage. The von Braun cyanogen bromide reaction[166] will open heterocyclic rings, but the relative reactivities of various groups differ from those observed in the exhaustive methylation procedure since attack at the α carbon atom rather than at the β hydrogen atom is involved. Methyl groups, for example, are readily removed and other substituents with no β hydrogen atoms may be cleaved. Reductive cleavage and especially the Emde reduction of quaternary salts to an amine and a hydrocarbon is the other general method.[167, 168] However, this process does not usually succeed unless the group to be removed is of the benzylic or allylic type. Lithium aluminum hydride may be used to reduce a quaternary salt to a tertiary amine and, with this reagent, alkyl groups may be removed from the nitrogen atom.[145, 167, 169, 170] In alkaloid degradations the Emde reduction may be used to remove the amino group from compounds of the tetrahydroisoquinoline type after a Hofmann step. Of course, this final cleavage cannot be accomplished by the Hofmann method. As the three methods

of degradation are complementary rather than competitive in most instances, it is meaningless to discuss their relative utility.

[166] Hageman, *Org. Reactions*, **7**, 198 (1953).
[167] Kenner and Murray, *J. Chem. Soc.*, **1950**, 406.
[168] Emde, *Helv. Chim. Acta*, **15**, 1330 (1932).
[169] Gaylord, *Reduction with Complex Metal Hydrides*, Interscience Publishers, New York, 1956, pp. 781–789.
[170] Cope, Ciganek, Fleckenstein, and Meisinger, *J. Am. Chem. Soc.*, **82** (in press, 1960).

EXPERIMENTAL CONDITIONS AND PROCEDURES

The fully alkylated amine required in the Hofmann and amine oxide procedures can be prepared in several ways. It is not our purpose here to include a comprehensive survey of methods of alkylation,* but to indicate the more commonly used techniques. In the application of the Hofmann reaction to alkaloids, methyl iodide has most often been used to prepare the tertiary amine and then the quaternary iodide in one reaction. For synthetic purposes, especially where a primary amine is to be degraded, there may be considerable advantage in using the formaldehyde-formic acid procedure[171] to prepare the tertiary amine. Other reagents that have been used to alkylate amines to obtain quaternary compounds for use in the Hofmann elimination reaction include dimethyl sulfate,[91] methyl p-toluenesulfonate,[172] ethyl chloroacetate,[173] and trimethyloxonium fluoborate[174].

To prepare the quaternary salt from a tertiary amine, the alkyl halides or sulfates are useful. Most commonly, methyl iodide has been used. Although there is no difficulty in preparing quaternary iodides with methyl

$$R_3N + R'X \rightarrow R_3R'NX$$

iodide, it might be pointed out that the general reaction (cf. refs. 37, 175, 176) does not always proceed easily. Ethyl acetate and methyl ethyl ketone have proved to be useful solvents in cases where equilibration of the quaternary halide with the various possible tertiary amines and alkyl halides is to be avoided. When dimethyl sulfate is used, only one methyl group is transferred to nitrogen per mole of sulfate, so that the salt formed is a quaternary methosulfate, $R_4\overset{\oplus}{N}(\overset{\ominus}{SO_4CH_3})$.

The quaternary hydroxide may be prepared from the iodide by using a base such as silver oxide that forms an insoluble iodide. This method suffers from the expense of the reagent and in some instances from the oxidizing power of silver salts in basic solution, but it is still most generally used. Thallous hydroxide may be used to obviate the oxidation effect, if not the cost of the silver salt.[75, 81, 177] If the quaternary methosulfate is used, it may be hydrolyzed to the sulfate and then converted to the hydroxide with barium hydroxide.[91] Perhaps the most promising method

---

* For such a survey see J. Goerdeler in Houben-Weyl, *Methoden der organischen Chemie* 4th ed., Vol. XI, part 2, Georg Thieme, Stuttgart, 1958.

[171] Moore, *Organic Reactions*, **5**, 301 (1947).
[172] Reynolds and Kenyon, *J. Am. Chem. Soc.*, **72**, 1597 (1950).
[173] Read and Hendry, *Ber.*, **71**, 2544 (1938).
[174] Meerwein, Battenberg, Told, Pfeil, and Willfang, *J. prakt. Chem.*, **154**, 83 (1940).
[175] Hey and Ingold, *J. Chem. Soc.*, **1933**, 66.
[176] Hughes, *J. Chem. Soc.*, **1933**, 75.
[177] von Bruchhausen, Oberembt, and Feldhaus, *Ann.*, **507**, 144 (1933).

of effecting the exchange of hydroxide ion for halide ion with a sensitive compound is the use of a basic ion exchange resin.[178, 36] The solutions obtained in this way are more dilute than those formed by other methods, and the apparatus takes longer to assemble, but this procedure seems to avoid most of the objectionable features of the precipitation methods.

Once the quaternary hydroxide has been prepared, the clear aqueous solution is decomposed directly. Depending on the ease with which the elimination reaction occurs, this may be accomplished by warming on a steam bath or by distillation at higher temperatures. The most recent practice seems to be to remove most of the water under reduced pressure with gentle heating. If decomposition does not occur during this process, the residual syrup or solid is heated in an oil bath under reduced pressure until it does decompose. This should rarely require a temperature as high as 200°. With some difficult decompositions very low pressures have been used to advantage,[178-180] but in general pressures readily attained with an oil pump or water aspirator have proved satisfactory. The importance of excluding carbon dioxide has been pointed out, and the early practice of concentrating the basic solution by allowing it to evaporate in an open vessel should not be employed.

In many instances the quaternary salt has not been converted to the hydroxide, but instead has been treated directly with excess base and then pyrolyzed. Usually, 10-20% aqueous sodium or potassium hydroxide has been used and the solution heated on a steam bath until decomposition seems to be complete. Not all compounds will decompose under such mild conditions, but, from a consideration of the compounds with which this method is useful, it appears that when more drastic conditions have been needed the previously described technique of preparing the quaternary hydroxide has been employed. However, decompositions of quaternary iodides by direct treatment with excess base have been carried out at temperatures up to 250°,[7] and the method may be quite generally applicable. With amines of high molecular weight the quaternary iodide may have a very low solubility, and it may be useful to prepare the quaternary chloride instead in order to obtain its solution in the basic reaction medium. This can be accomplished by digesting the iodide with freshly precipitated silver chloride.[181-183]

Isolation of Products. Because of the great differences in physical properties of the olefins formed in the Hofmann degradation it is not

[178] Weinstock and Boekelheide, *J. Am. Chem. Soc.*, **75,** 2546 (1953).
[179] Small and Lutz, *J. Am. Chem. Soc.*, **56,** 1738 (1934).
[180] Späth and Tharrer, *Ber.*, **66,** 904 (1933).
[181] Gadamer and Sawai, *Arch. pharm.*, **264,** 401 (1926).
[182] von Bruchhausen and Stippler, *Arch. Pharm.*, **265,** 152 (1927).
[183] Ghose, Krishna and Schlittler, *Helv. Chim. Acta*, **17,** 919 (1934).

possible to describe a method of isolation that will apply to all cases. Decomposition of the water-soluble quaternary base gives rise to olefins and amines that are usually less soluble and which may distil, steam distil, or remain as a residue, depending on the conditions of the pyrolysis. Usually some of the quaternary base will undergo displacement to regenerate the original tertiary amine, which will then be present as a contaminant. If the olefinic product is non-basic an easy separation is possible, but if nitrogen is retained in this portion of the molecule, as is the case when the original amine was heterocyclic, the problem of separating these amines may result. Faced with this situation, many investigators have simply remethylated the crude product and repeated the degradation until a nitrogen-free product was obtained. If it is necessary to separate the mixture of tertiary amines, this usually is achieved by taking advantage of a difference in solubility of the amines themselves or of one of their salts.

It is frequently possible for the degradation to yield a mixture of isomeric unsaturated amines. Furthermore, allylic rearrangement of the double bond may give rise to still more isomers. If several steps are to be carried out consecutively, the mixtures obtained add to the experimental difficulties. In such a case, the problem is simplified by hydrogenating the product after each step until the amino group is removed. Of course, less information about the structure of the original amine is obtained by this procedure, but the number of steps required to remove the amino group may still be used to determine its situation in the original compound.

In the investigation of alkaloids it is of interest to know when trimethylamine has been evolved during a pyrolysis. Usually the odor or a test with moistened litmus paper is sufficient indication of the liberation of an amine. When the decomposition is carried out under reduced pressure, the amine may be trapped in a receiver cooled in solid carbon dioxide or liquid nitrogen or in a trap containing acid.[8, 37] Occasionally dimethylamine is eliminated in a decomposition and, if the amines are collected in a trap containing hydrochloric acid, the melting point of the hydrochloride serves to distinguish between trimethylamine and dimethylamine. When different tertiary amines may be formed, the mixture may be trapped and separated by the methods described in ref. 184 or analyzed by gas chromatography.[53]

Preparation of Amine Oxides. Tertiary amines may be converted to the corresponding oxides by the use of 35% aqueous hydrogen peroxide in water or methanol solution at room temperature. Since the oxidation of the amine at room temperature may be a slow process, it is convenient to follow the conversion by spot tests with phenolphthalein; the amine

---

[184] Schryver and Lees, *J. Chem. Soc.*, **79**, 563 (1901).

oxides are not sufficiently basic to give a color test with this reagent.[9] The excess peroxide must be completely destroyed before pyrolysis to avoid the danger of explosion during concentration; this destruction is accomplished by the addition of platinum black[9] or of catalase.[121] The decomposition of the excess peroxide can be followed by periodic tests with lead sulfide paper, which is whitened immediately by hydrogen peroxide in low concentrations but not by solutions of amine oxides.[9] Amines such as tri-n-propylamine and those with larger alkyl groups are converted to the oxides with hydrogen peroxide very slowly, and stronger reagents such as 40% peroxyacetic acid[185] or monoperoxyphthalic acid[144] are used for their oxidation.

The solution of amine oxide is concentrated under reduced pressure to a syrup which is then pyrolyzed by heating in an oil bath. The isolation procedure is essentially the same as would be used after the Hofmann decomposition. In a few cases, in which the amino group is attached to a tertiary carbon atom or the β carbon atom is highly branched, the elimination may occur spontaneously during oxidation of the amine.[186]

Phosphoric Acid Deamination. The examples of this reaction which have been found (see Table XII, p. 391) are almost entirely those reported by Harries. The experimental procedures were not described in detail, and the reaction is largely unexplored. In many of the cases investigated by Harries a dihydrobenzene derivative was isolated and, perhaps for this reason, the decompositions were carried out in a carbon dioxide atmosphere.

In cases in which the N-substituted acetamide was heated with phosphorus pentoxide it was necessary first to prepare the acyl derivative of the amine. The usual methods of acylating amines with acid chlorides, anhydrides, etc., will not be reviewed here.

It is also possible to prepare the desired amides by treating an alcohol with acetonitrile, benzonitrile, or other nitriles under acidic conditions.[158] However, if the starting material to be converted to an olefin is an alcohol, probably one of the usual dehydration procedures would be more suitable.

To bring about decomposition, the amide is heated with an excess of phosphorus pentoxide in boiling xylene. The number of examples of the procedure is so small that variations in this technique are untested.

Cycloheptyltrimethylammonium Iodide. Alkylation with Methyl Iodide.[187] A solution of 66 g. of cycloheptylamine hydrochloride in 400 ml. of methanol is prepared in a large round-bottomed flask fitted with an efficient reflux condenser and two dropping funnels. The solution

---

[185] Cope and Lee, *J. Am. Chem. Soc.*, **79**, 964 (1957).
[186] A. C. Cope, F. M. Acton, and R. A. Pike, unpublished work.
[187] Willstätter, *Ann.*, **317**, 204 (1901).

is cooled in ice water until the theoretical quantities of reactants have been added in a manner to be described, and then for one additional hour. One hundred grams of a solution of potassium hydroxide (25% by weight) in methanol is added through one funnel and 126 g. of a 50% solution of methyl iodide in methanol through the other. When the reaction mixture becomes neutral or acid to litmus, the same quantities of base and methyl iodide are added. This procedure is repeated until 300 g. of potassium hydroxide solution and 378 g. of methyl iodide solution have been added. After the mixture has warmed to room temperature, an additional 100 g. of methyl iodide is added and 140–150 g. of potassium hydroxide solution is added slowly in small portions until the reaction mixture is neutral.

The methanol is removed by distillation from a steam bath, and the methiodide is precipitated by the addition of concentrated sodium hydroxide solution. The product is collected by filtration and washed with a mixture of water, methanol, and acetone. The dried product weighs 119 g. (95%). It may be purified by extraction with chloroform or acetone in a Soxhlet apparatus, or it may be recrystallized from acetone (a large quantity of solvent is required because of the low solubility of the iodide in boiling acetone).

n-Propyltrimethylammonium Iodide.[37] **Alkylation of Trimethylamine.** Thirty milliliters of a 25% solution of trimethylamine in absolute methanol is added to 17.2 g. of n-propyl iodide in a glass-stoppered 125-ml. Erlenmeyer flask. The mixture is cooled in ice for one hour and allowed to stand at room temperature overnight. The solution is then warmed on a steam bath until the trimethylamine is driven off (odor); then 65 ml. of ethyl acetate is added, and the mixture is heated to boiling. On cooling, large needles separate and are collected by filtration, washed with cold ethyl acetate, and dried. The yield of n-propyltrimethylammonium iodide melting at 192.0–192.5° is 22 g. (96%).

Di-n-butyldiisoamylammonium Iodide.[37] **Alkylation of a Hindered Amine.** A solution of 19.9 g. (0.1 mole) of isoamyldi-n-butylamine, 19.8 g. (0.1 mole) of isoamyl iodide, and 25 ml. of methyl ethyl ketone is heated under slow reflux for eighteen hours. The white crystals that separate when the solution is cooled are collected, washed with pure solvent, and dried. The yield of crude material melting at 117.0–119.5° is 25 g. Addition of 50 ml. of dry ether to the filtrate precipitates an additional 3 g. of product. The fractions are combined and recrystallized from ethyl acetate, yielding 25 g. (63%) of material melting at 120.0–120.5°.

Preparation of Silver Oxide.[188] A solution of one part by weight of silver nitrate in 10 parts of water is heated to 85° on a steam bath and

[188] Helferich and Klein, *Ann.*, **450**, 219 (1926).

treated with an equally warm solution of 0.23 part by weight of pure sodium hydroxide in 10 parts of water. The precipitated oxide is washed by decantation with 5 portions of hot water. This freshly precipitated oxide may be used as such. For pure, dry silver oxide, the precipitate is suspended in 5 parts of absolute ethanol, collected on a hardened filter paper, and washed several times with ethanol. The product is dried in air and then in a desiccator over phosphorus pentoxide.

Di-n-butyldiisoamylammonium Hydroxide.[37] **Use of Silver Oxide.** A solution of 6 g. (0.015 mole) of di-n-butyldiisoamylammonium iodide in 40 ml. of water and 5 ml. of methanol is shaken for one hour with thoroughly washed silver oxide prepared as described above from 5.1 g. (0.03 mole) of silver nitrate. The mixture is filtered as rapidly as possible with suction, and the filtrate is standardized acidimetrically.

Decomposition of Di-n-Butyldiisoamylammonium Hydroxide.[37] A 100-ml. pear-shaped flask, fitted with a capillary nitrogen inlet tube, containing 52 ml. (0.0111 mole) of the quaternary hydroxide solution prepared as described above is connected by large-diameter tubing to a condenser set for distillation. The condenser leads to a train of two 125-ml. gas-washing bottles containing 20 ml. of $3N$ hydrochloric acid, a drying tube, a trap cooled in liquid nitrogen, and finally to a mercury bubbler. The system is swept with nitrogen for thirty minutes, and then the flask is immersed in an oil bath at 85° and the temperature raised to 175°. At the latter temperature most of the water will have distilled into the first wash bottle. When the temperature is raised to 200°, vigorous decomposition sets in as evidenced by frothing in the flask, the appearance of oil in the condenser, and a rapid increase in the flow of gas through the wash bottles. Decomposition is complete in twenty minutes. The system is swept with nitrogen and the trap is closed and weighed. The olefin weighs 0.631 g. (94%) and consists of 67% butylene and 33% isoamylene as shown by mass spectral analysis.

1-Hexene. Methylation with Dimethyl Sulfate and Decomposition of the Sulfate.[91] One mole of n-hexylamine is suspended in 9 moles of a 25% solution of sodium hydroxide in water and shaken for a short time with 4 moles of dimethyl sulfate, which is added in small portions with cooling. The quaternary salt appears as a thick oil floating on the solution and is separated in a separatory funnel. The oil may be crystallized by solution in chloroform and precipitation with ether; however, the crude product may be used directly for decomposition.

The oily quaternary salt is dissolved in 1.5 moles of 20% sulfuric acid solution and heated for one and one-half to two hours under reflux. The solution is cooled and treated with a slight excess of barium hydroxide solution, and the precipitate of barium sulfate is removed by filtration.

The filtrate is concentrated under reduced pressure at 50°, 4 moles of a 50% solution of potassium hydroxide is added, and the solution is distilled. The distillate is placed in a separatory funnel and the aqueous layer removed. The oily mixture of olefin and amine is washed with dilute sulfuric acid, and the olefin is collected by distillation after being washed and dried. The entire fraction (60%) boils at 66° and is pure 1-hexene. The amine recovered from the acid washing amounts to 20% of the starting material.

In general, 1 mole of dimethyl sulfate and 2 moles of base per mole of dimethyl sulfate are required for each methyl group to be introduced. In addition, an excess of dimethyl sulfate is usually employed; the procedure above uses a one molar excess of alkylating agent and a one molar excess of base over that required by the 2 : 1 ratio.

des-N-Methylaphylline.[189] **Decomposition of a Quaternary Hydroxide under Reduced Pressure.** Ten grams of aphylline methiodide is dissolved in water and treated with the freshly precipitated silver oxide prepared from 5 g. of silver nitrate. The mixture is allowed to stand for twenty-four hours, and the precipitate is removed by filtration and washed with hot water. The combined filtrates are concentrated on a water bath at 6–15 mm. The "des" base separates from solution as white needles during this process. The mixture is heated on a water bath for one hour to complete the Hofmann elimination reaction. The material in the flask is taken up in ether, dried over potassium carbonate, and the ether is removed by distillation, leaving 5.5 g. (82%) of an oil that solidifies on cooling. The "des" base is purified by recrystallization from petroleum ether and is obtained as colorless needles, m.p. 113–115°.

Dihydro-des-N-dimethylcytisine.[190] **Decomposition Followed by Hydrogenation.** Seventeen grams of methylcytisine methiodide is dissolved in water and digested with excess silver oxide. The precipitate is collected by filtration, washed with hot water, and the combined filtrate and washings are concentrated under reduced pressure. The solution of quaternary base is transferred to a hydrogenation flask, palladium on charcoal catalyst is added, and the mixture is further concentrated under reduced pressure to the consistency of a syrup. The flask is then immersed in water at 80–90° for ten minutes to complete the decomposition, and the reaction mixture is diluted with cold water and hydrogenated at once.

When uptake of hydrogen has ceased (500 ml.), the catalyst is removed by filtration, washed well, and the solution is extracted with four portions of chloroform. The aqueous portion is concentrated, heated, and hydrogenated once again (uptake 150 ml. of hydrogen) in the manner described.

[189] Orechoff and Menshikoff, *Ber.*, **65,** 234 (1932).
[190] Späth and Galinovsky, *Ber.*, **65,** 1526 (1932).

The catalyst is again removed, and the solution is extracted with chloroform. The combined extracts are distilled, finally at 1 μ pressure. Dihydro-des-N-dimethylcytisine (5.5 g.) collects as a viscous oil at an air bath temperature of 150–160° at 1 μ pressure. From the aqueous portion of the extracts 5.1 g. of undecomposed starting material is recovered so that the yield of product (based on material not recovered) is 72%.

Decomposition of Cyclopropyltrimethylammonium Hydroxide. High Temperature Decomposition.[8] A pyrolysis tube is made by sealing one end of a piece of 30-mm. Pyrex tubing 12 cm. in length. The open end is constricted to hold a small two-hole stopper containing a gas inlet tube and a short-stemmed dropping funnel. A condenser made from 8-mm. Pyrex tubing is sealed to the side of the pyrolysis tube 8 cm. from the bottom, and the closed end of the pyrolysis tube is lined with a layer of 20% platinized asbestos 3 mm. thick. The condenser is attached to a 100-ml. receiver, in series with which are a 100-ml. spiral gas washing bottle containing $3N$ hydrochloric acid and a gasometer containing a saturated solution of sodium chloride. After concentrating a solution of the quaternary hydroxide [prepared from 22.7 g. (0.1 mole) of cyclopropyltrimethylammonium iodide] under reduced pressure at 40° in a nitrogen-filled apparatus, the pyrolysis tube is swept with carbon dioxide and heated to 320–330°. The concentrated solution of the quaternary hydroxide is dropped into the pyrolysis tube under a positive pressure of 30 cm. of water over a period of ten to twelve minutes. The gas collected amounts to 1.6–1.8 l., which can be converted to 8.0–9.5 g. of cyclopropene dibromide, b.p. 57–58°/50 mm., m.p. −1 to +1°, n_D^{20} 1.5360, d_4^{25} 2.0838. Some dimethylcyclopropylamine may be recovered from the hydrochloric acid wash bottle. Bromination of the gas also forms 1.5–2.0 g. of a tetrabromide, indicating the presence of some methylacetylene in the pyrolysis product.

1-Benzoyl-7-propionylheptatriene.[45] **Decomposition of a β mino Ketone.** An ethereal solution of lobinanine is treated with an excess of methyl iodide and allowed to stand for two days. The solvent is decanted from the precipitated methiodide, which is then washed with

$$C_6H_5\overset{O}{\overset{\|}{C}}CH_2\underset{\underset{CH_3}{|}}{N}CH_2\overset{O}{\overset{\|}{C}}CH_2CH_3$$

Lobinanine

$$\rightarrow C_6H_5\overset{O}{\overset{\|}{C}}CH_2CH=CHCH=CHCH=CHCOCH_2CH_3 + (CH_3)_2NH$$

ether. The methiodide is suspended in water and shaken with ether and aqueous sodium bicarbonate. Dimethylamine is evolved, and the ether layer becomes intensely yellow in color. The layers are separated, and the ether layer is washed with 0.1N hydrochloric acid, water, and dried over calcium chloride. The ether is removed by distillation, leaving a yellow-brown crystalline residue which is recrystallized from ligroin as dark-yellow crystals, m.p. 81–82°.

N-Uramidohomomeroquinene.[129] **Decomposition of a Quaternary Iodide with Excess Base.** N-Acetyl-10-trimethylammoniumdihydrohomomeroquinene ethyl ester iodide (1.45 g.) is taken up in an

equal quantity of water and heated in a platinum or nickel crucible with vigorous stirring with 2.5 ml. of a solution of 5 g. of sodium hydroxide in 4 ml. of water. Vigorous evolution of trimethylamine commences at 140°. The temperature is gradually raised to 165–180° while stirring is continued and water is added from time to time to replace that lost by evaporation. When the evolution of amine has ceased (one-half to one hour), the mixture is allowed to cool and the excess base is removed with a pipette from the upper layer of product, which is a light-tan solid or semisolid material. The latter is taken up in 3 ml. of water, neutralized to litmus with concentrated hydrochloric acid, and decolorized with Norit.

The carbon is removed by filtration and the filtrate treated with 0.35 g. of potassium cyanate in a small quantity of water. The solution is heated on a steam bath for thirty minutes, then acidified with concentrated hydrochloric acid to Congo Red while hot. N-Uramidohomomeroquinene (0.30 g., 38%) crystallizes from the solution when cooled as small shining prisms, m.p. 163–164° dec.

Preparation and Decomposition of Cyclohexylphenethyldimethylammonium Hydroxide.[75] **Use of Thallous Hydroxide.** A solution of 2.5 g. of thallous sulfate in 25 ml. of water is treated with 0.85 g. of barium hydroxide. This cloudy solution is used directly to prepare the free base from a solution of 3 g. of cyclohexylphenethyldimethylammonium iodide in 30 ml. of ethanol. The precipitated thallous iodide makes the precipitate of barium sulfate more easily removed by filtration. The solution is protected from atmospheric carbon dioxide during filtration. The clear filtrate is allowed to drop into a distilling flask heated at 120° in an oil bath, whereupon it decomposes at once. The products, styrene and cyclohexyldimethylamine, distil with the water and are collected in a receiver containing hydrochloric acid. The styrene is extracted from this mixture with ether and converted to the dibromide, giving 1.05 g. (64%) of this derivative, m.p. 72°.

trans-1,2-Octalin.[191] **Use of Silver Sulfate and Barium Hydroxide.** Twenty-five grams of *trans*-α-decalyltrimethylammonium iodide is dissolved in water and treated with 13 g. of silver sulfate. The precipitated silver iodide and undissolved silver sulfate are removed by filtration. The silver remaining in solution is precipitated with hydrogen sulfide, and the excess hydrogen sulfide is expelled with a stream of carbon dioxide. Concentrated barium hydroxide is added dropwise until no further precipitation of barium sulfate and carbonate is observed. Finally the solution is filtered again and the quaternary base is concentrated and decomposed by heating in a water bath at 3–4 mm. pressure. A yield of 4.1 g. (40%) of *trans*-1,2-octalin is obtained, b.p. 185°, $d_4^{15.6}$ 0.8970, $n_{He}^{15.6}$ 1.48722.

des-N-Methyldihydro-β-erythroidinol. Use of an Ion Exchange Resin.[178] A solution of 1.29 g. of dihydro-β-erythroidinol and 3 ml. of

Dihydro-β-erythroidinol

methyl iodide in 15 ml. of methanol is allowed to stand overnight and is then boiled under reflux for one hour. After removal of the solvent under reduced pressure, the residue is taken up in 15 ml. of water and passed through an 8-mm. tube packed to a height of 30 cm. with Amberlite IRA-400 (basic form). The column is eluted with 15 ml. of water, and the combined eluates are concentrated under reduced pressure. Distillation of the residue in a molecular still at 0·03 mm. (pot temperature 130–150°) gives a viscous oil which is taken up in methanol and treated with hexane. This causes separation of 0.95 g. (78%) of a white solid, m.p. 93–97°. Recrystallization of this material from hexane gives white crystals, m.p. 96–98°.

[191] Huckel and Naab, *Ann.*, **502**, 136 (1933).

Cularinemethine.[128] **Decomposition in Aqueous Solution with Added Base.** A suspension of 5 g. of cularine in 5 ml. of methanol is treated at room temperature with 4 g. of methyl iodide. The alkaloid dissolves readily and the methiodide then slowly separates in colorless crystals which melt at 205° after recrystallization from hot methanol.

Cularine

The methiodide is dissolved in water, any remaining organic solvent is removed by boiling, and a turbidity is removed by filtration. The solution (ca. 75 ml.) is then heated for twenty-four hours on a steam bath with 10 g. of potassium hydroxide. The oil that separates is extracted with ether, and the ether is removed, leaving a residue that weighs 5.2 g. when dried under reduced pressure. The residue does not crystallize, but the picrate crystallizes readily from methanol in pale-yellow needles melting sharply at 167°.

Methylenecyclohexane and N,N-Dimethylhydroxylamine Hydrochloride.[151] This *Organic Syntheses* procedure illustrates the standard method used for the preparation and pyrolysis of amine oxides. Methylenecyclohexane is obtained in 79–88% yield and N,N-dimethylhydroxylamine hydrochloride in 78–90% yield from 0.35 mole of N,N-dimethylcyclohexylmethylamine.

N,N-Dimethylcycloöctylamine Oxide.[9] A solution of 5.0 g. (0.032 mole) of N,N-dimethylcycloöctylamine in 10 ml. of methanol is cooled in an ice bath, and 10.0 g. (0.094 mole) of 35% hydrogen peroxide is added slowly (thirty minutes). The solution is allowed to come to room temperature and stand for twenty-six hours, at which time it gives a negative spot test for the amine with phenolphthalein. The excess hydrogen peroxide is decomposed by stirring the solution with 0.25 g. of platinum black for five hours, at which time a drop of the solution fails to whiten lead sulfide paper (negative hydrogen peroxide test). The platinum black is separated and the filtrate is concentrated at 10–12 mm. with a bath temperature of 30–40°, leaving the amine oxide as a colorless, viscous syrup.

Hindered amines or amines of high molecular weight are not converted to amine oxides by this procedure and should be oxidized with a peroxy acid (see p. 379).

cis-Cyclooctene.[9] The N,N-dimethylcyclooctylamine oxide described above is heated in a nitrogen atmosphere at 10 mm. in a 100-ml. round-bottomed flask connected through a short Vigreux column to two traps in series, the first cooled with solid carbon dioxide (Dry Ice) and the second with liquid nitrogen. The flask is placed in an oil bath and the temperature is raised 1–2° per minute; decomposition of the amine oxide begins at 100° and is complete at 120° after twenty-five minutes, at which time practically no material remains in the flask. The distillate is acidified with dilute hydrochloric acid, and the aqueous layer is frozen by cooling with solid carbon dioxide. The layer of *cis*-cyclooctene is removed with a pipette and distilled through a semimicro column. The yield is 3.22 g. (90%), b.p. 65° (59 mm.), n_D^{25} 1.4684.

After removal of the *cis*-cyclooctene, the aqueous hydrochloric acid solution is concentrated under reduced pressure, and the residual N,N-dimethylhydroxylamine hydrochloride is dried by adding absolute ethanol and removing it under reduced pressure. After further drying in a vacuum desiccator over potassium hydroxide the N,N-dimethylhydroxylamine hydrochloride weighs 2.91 g. (95%), and melts at 100–103° (sealed capillary). The melting point is raised to 104.5–106° (sealed capillary) by two crystallizations from ethanol-ether.

TABULAR SURVEY

The following tables list examples of epoxides prepared from β amino alcohols (Table XI), and olefins prepared by the pyrolysis of amines in the presence of phosphoric acid or phosphorus pentoxide (Table XII), by the pyrolysis of acetyl derivatives of amines (Table XIII), by the pyrolysis of amine oxides (Tables XIV and XV), and by the Hofmann elimination reaction (Tables XVI, XVII, and XVIII). The literature through 1957 has been searched for examples of these reactions and many more recent references are included. In each table amines are listed in order of increasing carbon content of the amine considered to be the parent compound; within a given carbon content the amines are listed in the order primary, secondary, tertiary; and within these divisions in the order aliphatic, alicyclic, heterocyclic, and polyfunctional. Thus *n*-hexylamine, 2-methylpiperidine, and triethylamine are all located (in the above order) under C_6 in Table XV, with the understanding that the compound actually degraded was the exhaustively methylated quaternary derivative. The carbon content of the free amine, not its acetyl derivative, is listed in Table XIII, and in Table XIV the carbon content of the unmethylated amine is listed with the understanding that in each case a tertiary amine oxide was pyrolyzed. If the precursor of the amine oxide is a tertiary amine that does not contain a methyl group, the amine is listed separately in Table XV with other similarly constituted amines, because the product sought in the pyrolysis of such an amine oxide is usually the dialkylhydroxylamine rather than the olefin. In the tabulation the yield of the dialkylhydroxylamine is given in these instances.

The examples of the Hofmann elimination reaction are divided into two categories, alkaloids and non-alkaloids. Unfortunately, because of the problem of locating examples there are undoubtedly many instances of the application of this reaction which are not listed. In the tables of non-alkaloidal amines (Tables XVI and XVII) the amines are tabulated as indicated above. For the alkaloid section (Table XVIII) the Manske and Holmes treatise, *The Alkaloids*,[192] has been used as a guide for nomenclature and structure except (a) for morphine and its derivatives, where the conventions of Bentley's monograph, *The Chemistry of the Morphine Alkaloids*,[193] were used, and (b) where more recent information was available. Closely related alkaloids are tabulated together under a group name which indicates the basic structure such as quinolizidine alkaloids, or which names one member of the group, such as the morphine alkaloids. Within the table of degradations the group names are in alphabetical order and the individual alkaloids are in the same order within each group. When feasible, a general structural formula is given for the whole group. It is to be understood that substituents in the alkaloids such as methoxyl groups are present in the degradation products unless otherwise specified.

A list of alkaloids in alphabetical order is provided in Table XIX which indicates group under which a given alkaloid is listed. In addition there is given the page in Table XVIII on which each group of alkaloids first appears. Those alkaloids whose names clearly indicate their relationships to alkaloids listed in Table XIX are not included in that table. Thus acetocodeine, bromocodeine, and dihydrocodeine are not listed in Table XIX as their relationship to codeine, which is listed, is obvious.

[192] Manske and Holmes, *The Alkaloids*, Academic Press, New York, 1953.
[193] Bentley, *The Chemistry of the Morphine Alkaloids*, Oxford University Press, New York, 1954.

TABLE XI
EPOXIDES FROM β AMINO ALCOHOLS

Amino Alcohol	Epoxide	Yield, %	References
trans-2-Aminocyclopentanol	Cyclopentene oxide	50	194
trans-2-Aminocyclohexanol	Cyclohexene oxide	74	194, 195
cis-2-Aminocyclohexanol	No oxide	—	105
2-Amino-1-heptanol	1-Heptene oxide	80	196
β-Cyclohexyl-β-aminoethanol	Cyclohexylethylene oxide	70	197
1-Phenyl-2-amino-1-propanol	1-Phenyl-2-methylethylene oxide	40	198
4-Phenyl-2-amino-1-butanol	4-Phenyl-1-butene oxide	70	197
4-Phenoxy-2-amino-1-butanol	4-Phenoxy-1-butene oxide	Poor	197
erythro-1,2-Diphenylethanolamine	trans-Stilbene oxide	68	104
threo-1,2-Diphenylethanolamine	cis-Stilbene oxide	62	104
2-Amino-1-hexadecanol	1-Hexadecene oxide	—	197
cis-2-Aminocyclodecanol	No oxide	—	106
trans-2-Aminocyclodecanol	cis-Cyclodecene oxide	85	106
cis-2-Aminocyclododecanol	trans-Cyclododecene oxide	48	106
trans-2-Aminocyclododecanol	cis-Cyclododecene oxide	91	106
cis-2-Aminocyclotridecanol	trans-Cyclotridecene oxide	95	106
trans-2-Aminocyclotridecanol	cis-Cyclotridecene oxide	84	106
cis-2-Aminocyclohexadecanol	trans-Cyclohexadecene oxide	90	106
trans-2-Aminocyclohexadecanol	cis-Cyclohexadecene oxide	84	106

CHCH₂CH₃, conhydrine (structure with piperidine-N–H and OH)

CHCH₂CH₃ (structure with N(CH₃) and epoxide O) — 90 — 199–201

Note: References 194 to 391 are on pp. 489–493.

390   ORGANIC REACTIONS

TABLE XI—Continued
Epoxides from β Amino Alcohols

Amino Alcohol	Epoxide	Yield, %	References
Dihydropseudoconhydrinemethine [(CH$_3$)$_2$NCH$_2$CH(OH)(CH$_2$)$_5$CH$_3$]	CH$_2$—CH(CH$_2$)$_5$CH$_3$ (epoxide)	—	202
Ephedrine C$_6$H$_5$CH(OH)CH(NHCH$_3$)CH$_3$ (*erythro*)	C$_6$H$_5$CH—CHCH$_3$ (*trans*)	30	103, 203, 204
Pseudoephedrine (*threo*)	C$_6$H$_5$CH—CHCH$_3$ (*cis*)	25	103, 204
dl-α-1-p-Chlorophenyl-1,2-diphenyl-2-aminoethanol *erythro* form	*trans*-C$_6$H$_5$CH—C(C$_6$H$_5$)C$_6$H$_4$Cl-*p* (epoxide)	75	204a
threo form	*cis*-C$_6$H$_5$CH—C(C$_6$H$_5$)C$_6$H$_4$Cl-*p* (epoxide)	85	204a
Quinine	(quinine epoxide structure)	—	119, p. 11

Note: References 194 to 391 are on pp. 489–493.

TABLE XII
Pyrolysis of Amines with Phosphoric Acid or Phosphorus Pentoxide

No. of C Atoms	Amine	Olefin	Yield, %	References
C_4	Cyclobutylamine	Butadiene	23	205
C_5	Cyclobutylmethylamine	Cyclopentene	—	206
C_6	$(CH_3)_2C(NH_2)CH_2CH(NH_2)CH_3$	"Methylpentadiene"	60	152
	3-aminocyclohexylamine (1,3-diaminocyclohexane)	1,3-Cyclohexadiene	25	154
	1,4-diaminocyclohexane	1,4-Cyclohexadiene	25	154
C_7	1-methyl-1,3-diaminocyclohexane	1-methyl-1,3-cyclohexadiene	50	152, 154
C_8	1,1-dimethyl-3,5-diaminocyclohexane	1,1-dimethyl-2,4-cyclohexadiene	50	154

Note: References 194 to 391 are on pp. 489–493.

TABLE XII—*Continued*

PYROLYSIS OF AMINES WITH PHOSPHORIC ACID OR PHOSPHORUS PENTOXIDE

No. of C Atoms	Amine	Olefin	Yield, %	References
C_{10}	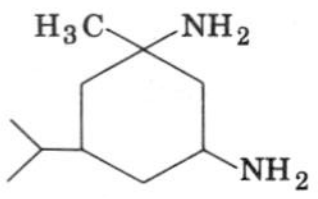	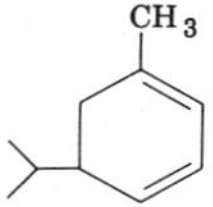 (?)	50	154
	Dihydroterpenylamine	Menthadiene	—	207
	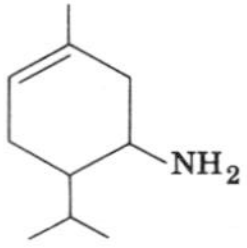	Menthadiene	73	153
	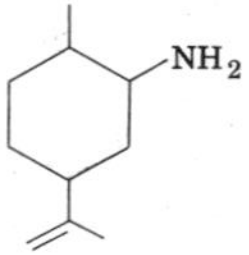	Menthadiene	30	155

Note: References 194 to 391 are on pp. 489–493.

TABLE XIII

OLEFINS FROM ACETYL DERIVATIVES OF AMINES

No. of C Atoms	Amine	Conditions	Product(s)	Yield, %	References
C_6	4-Methyl-2-pentylamine	590°	4-Methyl-1-pentene (largely), 4-methyl-2-pentene	14	161
	Cyclohexylamine	P_2O_5, xylene	Cyclohexene	36	157
	$C_2H_5NHCH_2CH_2OCOCH_3$	490°	Vinyl acetate	25	161
	Methyl-(4-methyl-2-pentyl)amine	570°	4-Methyl-1-pentene, 4-methyl-2-pentene (more than half)	27	161
C_7					
C_8	2,4,4-Trimethyl-2-pentylamine	510°	2,4,4-Trimethyl-1-pentene, 2,4,4-trimethyl-2-pentene (2:1)	35	161
C_{12}	Phenyl-(4-methyl-2-pentyl)amine	510°	4-Methyl-1-pentene, 4-methyl-2-pentene (1:1)	67	161
C_{14}	1,2-Diphenylethylamine	P_2O_5, xylene	trans-Stilbene	70	157
C_{15}	1,3-Diphenyl-1-propylamine	P_2O_5, xylene	1,3-Diphenylpropene	75	157
	1,3-Diphenyl-2-propylamine	P_2O_5, xylene	1,3-Diphenylpropene	10	157
	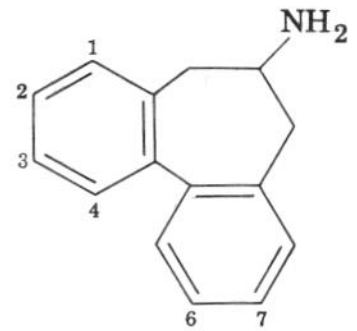 (structure with NH_2, positions 1,2,3,4,6,7)	P_2O_5, xylene	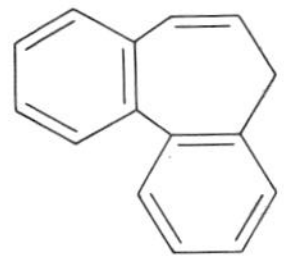 (fused tricyclic structure)	68	208
	Colchinol methyl ether (2,3,4,7-tetramethoxy derivative of structure above)	P_2O_5, xylene	Deaminocolchinol methyl ether	—	156
	Iodocolchinol methyl ether (2,3,4,7-tetramethoxy-6-iodo derivative of structure above)	P_2O_5, xylene	Deaminoiodocolchinol methyl ether	50–70	160
	1-Phenyl-3-p-methoxyphenyl-propylamine	P_2O_5, xylene	1-Phenyl-3-p-methoxyphenylpropene	Low	157

TABLE XIV
Pyrolysis of Amine Oxides

No. of C Atoms	Amine	Olefin(s) (Composition of Olefin Mixture)	Yield, %	References
C$_3$	n-C$_3$H$_7$NH$_2$	Propylene	—	144
C$_4$	(C$_2$H$_5$)$_2$NH	Ethylene	—	144
	Cyclobutylamine	Cyclobutene	50–60	90
	CH$_3$CH$_2$CH(NH$_2$)CH$_3$	1-Butene 67.3%, 2-butene (cis, 11.7%; trans, 21.0%)	91	36
C$_5$	CH$_3$CH$_2$CH(NH$_2$)CH$_2$CH$_3$	2-Pentene (cis, 29.2%; trans, 70.8%)	86	36
	CH$_3$CH$_2$NHCH$_2$CH$_2$CH$_3$	Ethylene (62.5%), propylene (37.5%)	82	36
	CH$_3$CH$_2$NHCH(CH$_3$)$_2$	Ethylene (27.5%), propylene (72.5%)	90	36
	CH$_2$=CH(CH$_2$)$_3$NH$_2$	1,4-Pentadiene	61	145
	piperidine (N–H)	No ring opening		147
C$_6$	(C$_2$H$_5$)$_2$CHCH$_2$NH$_2$	(C$_2$H$_5$)$_2$C=CH$_2$	80	42
	C$_2$H$_5$NHC$_4$H$_9$-n	Ethylene (55.5%), 1-butene (44.5%)	85	36
	C$_2$H$_5$NHC$_4$H$_9$-i	Ethylene (67.6%), isobutylene (32.4%)	85	36
	C$_2$H$_5$NHC$_4$H$_9$-t	Ethylene (14.2%), isobutylene (85.8%)	76	36
	(n-C$_3$H$_7$)$_2$NH	Propylene	—	144
	(methylenecyclobutane precursor: CH$_2$NH$_2$-cyclobutyl)	methylenecyclobutane (H$_2$C=cyclobutane)	69	121
	(cyclopentylmethylamine: cyclopentyl-CH$_2$NH$_2$)	methylenecyclopentane (CH$_2$=cyclopentane)	61	34

		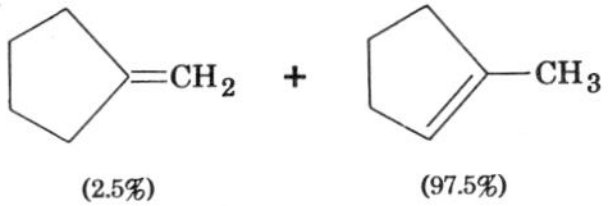	77	34
	Cyclohexylamine	Cyclohexene	83	9, 144
	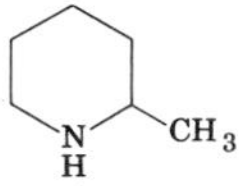	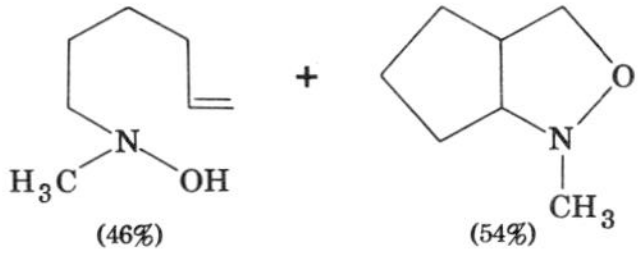	25.5	147
	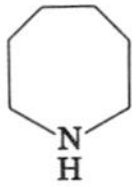	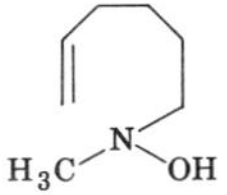	53	147
C_7	$(C_2H_5)_3CNH_2$	$(C_2H_5)_2C{=}CHCH_3$	57	41
	n-$C_3H_7NHC_4H_9$-n	Propylene (43.1%), 1-butene (56.9%)	62	36
	n-$C_3H_7NHC_4H_9$-i	Propylene (58.8%), isobutylene (41.2%)	73	36
	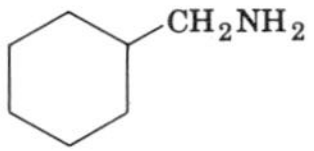	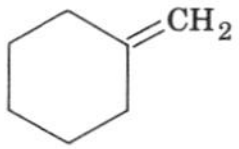	85	34
	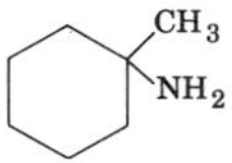	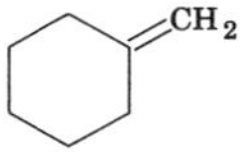	79–88	151, 34
	Cycloheptylamine	Cycloheptene	92	9
	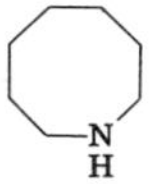	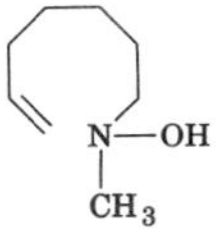	79	147

TABLE XIV—Continued

PYROLYSIS OF AMINE OXIDES

No. of C Atoms	Amine	Olefin(s) (Composition of Olefin Mixture)	Yield, %	References
C_7 (cont.)	endo — norbornyl-NH$_2$	Bicycloheptadiene	1.4	27
	exo — norbornyl-NH$_2$	Bicycloheptadiene	32	27
	endo — norbornyl-NH$_2$	Bicycloheptene	2.9	27
	exo — norbornyl-NH$_2$	Bicycloheptene	65	27
C_8	$(n\text{-}C_3H_7)_2CHCH_2NH_2$	$(n\text{-}C_3H_7)_2C=CH_2$	77	42
	$(i\text{-}C_3H_7)_2CHCH_2NH_2$	$(i\text{-}C_3H_7)_2C=CH_2$	80	42
	$n\text{-}C_4H_9NHC_4H_9\text{-}i$	1-Butene (64.8%), isobutylene (35.2%)	86	36
	$C_3H_7NHC_5H_{11}\text{-}i$	Propylene (38.7%), 3-methyl-1-butene (49.1%), 2-methyl-2-butene (11.2%), 2-methyl-1-butene (1.0%)	80	36
	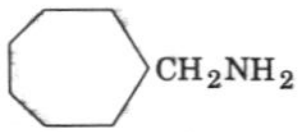 cycloheptyl-CH$_2$NH$_2$	Methylenecycloheptane	82	34
	1-methyl-1-aminocycloheptane	=CH$_2$ (15.2%) + methylcycloheptene—CH$_3$ (84.8%)	84	34

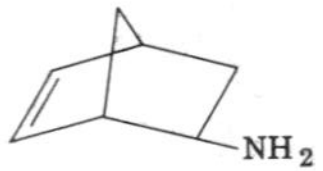

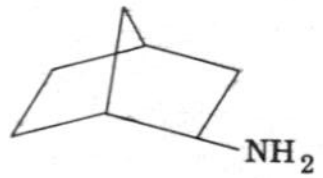

OLEFINS FROM AMINES

	Amine	Olefin	Yield (%)	Ref.
	Cyclooctylamine (cis-NH₂ cyclooctene)	cis-Cyclooctene	90	9
		cis-cis-1,3-Cyclooctadiene	48	59
	(cis-NH₂ cyclooctene)	cis-cis-1,5-Cyclooctadiene (91%), cis,cis-1,4-cyclooctadiene (6%), unidentified products (3%)	84	209
	$C_6H_5CH(NH_2)CH_3$	Styrene	70	142
	(bicyclic NH₂)	Bicyclooctadiene	67	26
C_9	$C_6H_5(CH_2)_3NH_2$	$C_6H_5CH_2CH=CH_2$	91	145
	$C_6H_5CH(NH_2)CH_2CH_3$	trans-$C_6H_5CH=CHCH_3$	60	48
	$C_6H_5C(NH_2)(CH_3)_2$	$C_6H_5C(CH_3)=CH_2$	78	142
	Cyclononylamine	trans-Cyclononene	90	56
	Cyclooctylmethylamine	Methylenecyclooctane	79	52a
	1-Methylcyclooctylamine	Methylenecyclooctane (1.4%), cis-1-methylcyclooctene (98.6%)	84	52a
C_{10}	$(t\text{-}C_4H_9)_2CHCH_2NH_2$	$(t\text{-}C_4H_9)_2C=CH_2$	73	42
	n-Decylamine	1-Decene	80	210
	Cyclononylmethylamine	Methylenecyclononane	80	52a
	$C_6H_5CH_2CH_2NHC_2H_5$	Ethylene (1.5%), styrene (98.5%)	85	36
	$C_6H_5C(CH_3)_2CH_2NH_2$	No olefin	—	142

Note: References 194 to 391 are on pp. 489–493.

TABLE XIV—Continued
PYROLYSIS OF AMINE OXIDES

No. of C Atoms	Amine	Olefin(s) (Composition of Olefin Mixture)	Yield, %	References
C_{10} (cont.)	$C_6H_5CHCH(NH_2)CH_3$ $\mid$ CH_3	$C_6H_5C{=}CHCH_3$ $\mid$ CH_3 $C_6H_5CHCH{=}CH_2$ $\mid$ CH_3		143
		cis trans		
	threo	93–94% 0.1–0.2% 7%		
	erythro	2–4% 89–90% 7–8%		
	Menthylamine	2-Menthene (65%), 3-menthene (35%)	85	16
	Neomenthylamine	2-Menthene (100%)	77	16
	Cyclodecylamine	trans-Cyclodecene (98%), cis-cyclodecene (2%)	90	56, 58
	1-Methylcyclononylamine	Methylenecyclononane (6%), 1-methylcyclononene (cis, 82%; trans, 12%)	72	52a
	Bornylamine	Bornylene and tricyclene	—	33
	Neobornylamine	Bornylene and camphene	—	33
C_{11}	Cyclodecylmethylamine	Methylenecyclodecane	74	52a
	1-Methylcyclodecylamine	Methylenecyclodecane (2.5%), 1-methylcyclodecene (cis, 64%; trans, 34%)	86	52a
C_{12}	[cyclohexane with NH_2 and C_6H_5 substituents]	[cyclohexene with C_6H_5 substituent]		
	cis	(2%) (98%)	72	145
	trans	(85%) (15%)	96	145
C_{13}	$n\text{-}C_3H_7NHC_{10}H_{21}\text{-}n$	Propylene (40.4%), 1-decene (59.6%)	55	36

C$_{15}$ and C$_{16}$	[structure: CH$_3$O-...-OH(OCH$_3$), NH$_2$]	α-Tetrahydrocodeimethine	[structure: CH$_3$O-...-OH(OCH$_3$)]	13-Vinyl morphenol derivative	45	150
		α-Tetrahydrocodeimethine methyl ether		13-Vinyl derivative	62	150
		Dihydrocodeimethine		13-Vinyl derivative	23	150
		α-Codeimethine		13-Vinyl derivative	18	150
		β-Codeimethine		13-Vinyl derivative	24	150
		α-Codeimethine methyl ether		13-Vinyl derivative	46	150
		β-Codeimethine methyl ether		13-Vinyl derivative	35	150
		Dihydrothebaine methine		13-Vinyl derivative	45	150
		Dihydrothebaine dihydromethine		13-Vinyl derivative	60	150
		Dihydro-14-hydroxycodeinone methine		13-Vinyl derivative	20	150
		Metathebainone methyl ether methine		13-Vinyl derivative	10	150
	[structure: CH$_3$O-...-OH, NH$_2$] Dihydrothebainone dihydromethine		[structure: CH$_3$O-...=O] Thebenone		42	150
C$_{18}$	[structure: C$_6$H$_5$, C$_6$H$_5$, CH$_2$NH$_2$, CH$_2$NH$_2$]		[structure: C$_6$H$_5$, C$_6$H$_5$, CH$_3$=CH$_2$] + [structure: C$_6$H$_5$, C$_6$H$_5$, =CH$_2$, =CH$_2$]		—	149

TABLE XV
Pyrolysis of Oxides of Tertiary Amines without N-Methyl Groups

Amine	Product(s)	Yield, %	References
$(C_2H_5)_3N$	$(C_2H_5)_2NOH$	69, 67	185, 144
N-Ethylpiperidine	(piperidine-N-OH)	42	140, 144
$(n\text{-}C_3H_7)_3N$	$(C_3H_7)_2NOH$	84	141, 185, 144
$(C_2H_5)_2NCH_2CH_2CO_2C_2H_5$	$(C_2H_5)_2NOH$	—	144
$O(CH_2CH_2)_2NCH_2CH_2CO_2C_2H_5$	(morpholine-NOH)	—	144
$(CH_2)_5NCH_2CH_2CO_2C_2H_5$	(piperidine-NOH)	—	144
$(n\text{-}C_3H_7)_2NCH_2CH_2CO_2C_2H_5$	$(n\text{-}C_3H_7)_2NOH$	—	144
$C_6H_5CH_2N(C_2H_5)_2$	$C_6H_5CH_2N(C_2H_5)OH$	34	142
$(n\text{-}C_4H_9)_3N$	$(n\text{-}C_4H_9)_2NOH$	79	185
$(n\text{-}C_4H_9)_2NCH_2CH_2CO_2C_2H_5$	$(n\text{-}C_4H_9)_2NOH$	—	144
$(n\text{-}C_5H_{11})_3N$	$(n\text{-}C_5H_{11})_2NOH$	75	185
	1-Pentene	65	
$(i\text{-}C_5H_{11})_3N$	$(i\text{-}C_5H_{11})_2NOH$	42	185
	Isoamylene	53	
$(n\text{-}C_6H_{13})_3N$	$(n\text{-}C_6H_{13})_2NOH$	70	185
	1-Hexene	64	
$(n\text{-}C_7H_{15})_3N$	$(n\text{-}C_7H_{15})_2NOH$	71	185
	1-Heptene	76	

TABLE XVI
DECOMPOSITION OF QUATERNARY AMMONIUM COMPOUNDS

No. of C Atoms	Amine	Derivative	Conditions	Elimination Product(s) (Composition of Mixture)	Yield, %	References
C_2	Ethylamine	OH	Distil	Ethylene	95–100	37, 211, 38
C_3	n-Propylamine	OH	200°	Propylene	83	37
			Distil	Propylene	High	211
			Distil	Propylene	84	38, 11
	Isopropylamine	OH	Distil	Propylene	"Mostly"	211
	Cyclopropylamine	OH	325°, Pd	Cyclopropene	45	8
	Allylamine	OH	320°	Methylacetylene (88%) +allene (12%) +oxygen-containing products	34	212
	1,3-Diaminopropane	OH	Distil	Allyldimethylamine*	—	95
	β-Alanine	Betaine	140°, aq. base	CH_2=$CHCO_2H$	—	213
C_4	n-Butylamine	OH	Distil	1-Butene	79	38, 11, 214
			200°	1-Butene	80	37, 102, 39
	Isobutylamine	OH	Distil	Isobutylene	63	11
			Distil		None	211
	sec-Butylamine	OH	150°/ 20 mm.	1-Butene (95%) + 2-butenes (5%)	97	36
	t-Butylamine	OH	Distil	1-Butene	65	11
	Diethylamine	OH	Distil	Isobutylene	High	11
	Cyclobutylamine	OH	Distil	Ethylene	High	39
			140°/ 50 mm.	Cyclobutene	73	90
			Distil	Cyclobutene	30	215, 205

Note: References 194 to 391 are on pp. 489–493.
* No allene was isolated.

TABLE XVI—Continued

DECOMPOSITION OF QUATERNARY AMMONIUM COMPOUNDS

No. of C Atoms	Amine	Derivative	Conditions	Elimination Product(s) (Composition of Mixture)	Yield, %	References
C_4 (cont.)	Pyrrolidine	Iodide + KOH	Distil	4-Dimethylamino-1-butene	—	5
	1-Amino-3-butene	Iodide, KOH	Distil	1,3-Butadiene	—	5
	1,2-Diaminobutane	Di OH	250°	Ethylacetylene	45	214
				Methylallene	55	
	1,3-Diaminobutane	Di OH	160°	Butadiene	—	216
	2,3-Diaminobutane	Di OH	250°	Butadiene, mixture of ethylacetylene and methylallene	45	214
	1,4-Diaminobutane	Di OH	Distil	1,3-Butadiene	—	217, 218
	1,3-Diamino-1-butene	tert-Methylated amine	160°/250 mm.	$CH_3CH=C=CN(CH_3)_2$	—	219
	1,4-Diamino-2-butene	Di OH	100–120°/13 mm.	Vinylacetylene	29	220
	trans-1,2-Diaminocyclobutane	Di OH	350°/0.1 mm.	No olefin, cyclobutanone + other products	—	221
	1,3-Diaminocyclobutane	Di OH	120–200°/20 mm.	Butadiene	—	222
	Piperazine	Chloride, OH	Distil	Acetylene, tetramethylethylenediamine, dimethylethanolamine	—	223
C_5	n-Amylamine	OH	Distil	1-Pentene	77	38, 39
	Isoamylamine	OH	200°	3-Methyl-1-butene	78	37, 38, 211
	2-Pentylamine	Iodide, KOC_2H_5	130°	1-Pentene (98%), 2-pentene (2%)	67	224
		OH	Distil	Pentene	—	11

OLEFINS FROM AMINES

Amine	Group	Conditions	Product	Yield (%)	Reference
3-Pentylamine	OH	85–150°/20 mm.	2-Pentene (cis, 55%; trans, 45%)	96	36
t-Amylamine	OH	Distil	Pentene	—	211
	Iodide, KOC₂H₅	130°	2-Methyl-1-butene (93%), 2-methyl-2-butene (7%)	84	44
	Iodide, 2,6-lutidine	Reflux	2-Methyl-1-butene (93%), 2-methyl-2-butene (7%)	79	44
Ethyl-n-propylamine	OH	85–150°/15 mm.	Ethylene (97.6%), propylene (2.4%)	94	36, 11
Ethylisopropylamine	OH	85–150°/15 mm.	Ethylene (41.2%), propylene (58.8%)	88	36
Cyclobutylcarbinylamine	OH	Distil	Methylenecyclobutane	Poor	206
Cyclopropylmethylcarbinylamine	OH	Distil	Vinylcyclopropane	68	225
Piperidine	OH	Distil	5-Dimethylamino-1-pentene	80	102, 91, 4, 65
1-Amino-4-pentene	OH	Distil	1,3-Pentadiene (piperylene)	—	4
1-Amino-2,4-pentadiene	OH	—	CH₃C≡CCH=CH₂ (pirylene)	—	226
⌷—CH₂—NH₂ (cyclobutyl)	OH	310°	⌷=CH₂	13	227
⌷(H₂N)—CH₂ (cyclopentyl)	OH	160°/40 mm.	⌷=CH₂	68	228
3-Methylpyrrolidine	Iodide, KOH	Distil	4-Dimethylamino-2-(or 3)-methyl-1-butene	—	229
4-Amino-2-(or 3)-methyl-1-butene	Iodide, KOH	Distil	Isoprene	—	230, 229

Note: References 194 to 391 are on pp. 489–493.

TABLE XVI—Continued

DECOMPOSITION OF QUATERNARY AMMONIUM COMPOUNDS

No. of C Atoms	Amine	Derivative	Conditions	Elimination Product(s) (Composition of Mixture)	Yield, %	References
C_5 (cont.)	2-Methylenepyrrolidine	Iodide, KOH	Distil	Mixture of bases	70	68, 66, 7
	Second step	Iodide, KOH	Distil	CH_3C=CCH=CH_2 (pirylene)	59	68, 66, 7
	1,5-Diaminopentane	Di OH	Distil	Piperylene	Good	95
	Second step	Mono OH	Distil	CH_2=$CH(CH_2)_3N(CH_3)_2$	ca. 40	39
C_6	n-Hexylamine	OH	Distil	1-Hexene	76	38, 39, 91
	3,3-Dimethylbutylamine	OH	200°	t-Butylethylene	20	37
	2,2-Dimethyl-3-aminobutane	OH	30°/1 μ	t-Butylethylene	"Only"	32
			160°	t-Butylethylene	48	32
	2-Ethylbutylamine	OH	Distil	$(C_2H_5)_2C$=CH_2	43	11
		OH	100°/ 10 mm.	$(C_2H_5)_2C$=CH_2	77	42
	5-Amino-1-hexene	OH	Distil	Biallyl and an isomer	—	73
	5-Methoxypentylamine	OH	Distil	Methoxypentene	ca. 30	39
	6-Amino-1-hexene	OH	160°	Biallyl and an isomer	80	73
	Cyclohexylamine	OH	105–120°/ 11 mm.	Cyclohexene	62	9
	![cyclopentylmethylamine structure] CH_2NH_2 (cyclopentyl)	OH	To 160°, vac.	=CH_2 (94%) —CH_3 (6%)	50	34

OLEFINS FROM AMINES

Amine		Conditions	Olefin	Yield (%)	Ref.
1-methyl-1-aminocyclopentane (CH₃, NH₂ on cyclopentane)	OH	To 160°, vac.	methylenecyclopentane =CH₂ (91%); 1-methylcyclopentene —CH₃ (9%)	71	34
Cyclobutylmethylamine (CH₂NH₂ on cyclobutane)	OH	160°/50 mm.	methylenecyclobutane (H₂C=, CH₃)	50	121
Cyclohexenylamine	OH	75°/115 mm.	1,3-Cyclohexadiene	Almost quant.	231
Cyclopentenylmethylamine (—CH₂NH₂ on cyclopentene)	OH	150°/45 mm.	methylenecyclopentene (=CH₂)	58	232
Ethyl-n-butylamine	OH	85–150°/20 mm.	Ethylene (98.4%), 1-butene (1.6%)	98	36
Ethylisobutylamine	OH	85–150°/20 mm.	Ethylene (99.1%), isobutylene (0.9%)	94	36
Ethyl-t-butylamine	OH	85–150°/20 mm.	Ethylene (7.2%), isobutylene (92.8%)	88	36
2-Methylpiperidine	OH	140°	6-Dimethylamino-1-hexene	—	73
(CH₃)₂CH-NH-CH(CH₃)₂ type	OH	Distil	Unsaturated base, C₈H₁₇N	—	233

Note: References 194 to 391 are on pp. 489–493.

TABLE XVI—Continued
DECOMPOSITION OF QUATERNARY AMMONIUM COMPOUNDS

No. of C Atoms	Amine	Derivative	Conditions	Elimination Product(s) (Composition of Mixture)	Yield, %	References
C_6 (cont.)	(tropane-like bicyclic amine with O)	OH	100°, vac.	(tetrahydrofuran with =CH$_2$ and CH$_2$CH$_2$N(CH$_3$)$_2$ side chain)	60	234
	Triethylamine	OH	Distil	Ethylene	High	211, 91
	(piperidine with CO$_2$H)	Betaine	Distil	Amine and CO_2	—	213
	N,N'-Diethylethylene diamine	Di OH	Boil	Ethylene	48	11
	(cyclohexene with two NH$_2$ groups)	OH	120–160°	Benzene	80–85	231
	(bicyclic amine)	OH	350°	1-Methyl-4-methylene-piperidine	46	89

	Amine		Conditions	Olefin	Yield, %	Refs.
C₇	n-Heptylamine	OH	Distil	1-Heptene	74	38, 39
	Cycloheptylamine	OH	Distil	Cycloheptene	87	9, 187
	3-Aminocycloheptene	OH	Distil	Cycloheptadiene	80	187
		Bromide, KOH	Distil	Cycloheptadiene	85–90	235
	cyclohexyl-CH₂NH₂	OH	To 160°, vac.	Methylenecyclohexane	69	34
	1-methyl-1-aminocyclohexane	OH	To 160°, vac.	Methylenecyclohexane (99%), 1-methylcyclohexene (1%)	85	34
	endo-Norbornylamine	OH	110–125°, vac.	Bicyclo[2.2.1]heptene	3.5	27
	exo-Norbornylamine	OH	90–110°, vac.	Bicyclo[2.2.1]heptene	77	27
	endo	OH	110–120°, vac.		3.1	27
	exo	OH	110–125, vac.	Bicyclo[2.2.1]heptadiene	58	27
	(bicyclic NH₂)	OH	—		—	236

Note: References 194 to 391 are on pp. 489–493.

TABLE XVI—Continued

DECOMPOSITION OF QUATERNARY AMMONIUM COMPOUNDS

No. of C Atoms	Amine	Derivative	Conditions	Elimination Product(s) (Composition of Mixture)	Yield, %	References
C₇ (cont.)	n-Butyl-n-propylamine	OH	85–105°/ 20 mm.	Propylene (59.8%), 1-butene (40.2%)	94	36
	n-Propylisobutylamine	OH	85–150°/ 20 mm.	Propylene (72.9%), isobutylene (27.1%)	92	36
	N-Ethylpiperidine	OH + KOH	Distil	Ethylene and ![structure: N-ethyl tetrahydropyridine-like ring with CH₃, C₂H₅]	71 18	74, 4
	Quinuclidine ![quinuclidine structure with CH₃]	OH	350°	1-Methyl-4-vinylpiperidine	Low	88
		OH	340°	1-Methyl-4-vinylpiperidine	50	237
	C₂H₅OCH₂N(C₂H₅)₂	OH	Room temp., vac.	Ethylene	82	238
			135°	Ethylene	10	238
	1,7-Diaminoheptane	Di OH	Distil	Heptadiene	ca. 5	95
	1,4-Diaminocyclohept-2-ene	Di OH	Distil	Cycloheptatriene	—	187
	![structure: cyclobutane with CH₃, CH₃, CH₂NH₂, H₂N substituents]	Di OH	160°/40 mm.	![structure: cyclobutane with H₃C, CH₃, =CH₂, CH₂]	52	228, 232

Amine	Reagent	Conditions	Product	Yield %	Reference
endo-5-Aminobicyclo-[2.2.1]hept-2-ene	Iodide + KOH	Distil	No olefin	—	239
C$_8$ n-Octylamine	OH	Distil	No olefin	—	239
	OH	Distil	1-Octene	75	38, 39
2-n-Propylpentylamine	OH	Distil	as-Di-n-propylethylene	31	11
2-Amino-2,4,4-trimethylpentane	OH + NaOH	100°	2,4,4-Trimethyl-1-pentene	70	44
	Iodide + pyridine	100°	2,4,4-Trimethyl-1-pentene (88%), 2,4,4-trimethyl-2-pentene (12%)	80	44
	Iodide + 2-picoline	100°	95% Δ^1- and 5% Δ^2-olefin	86	44
	Iodide + 2,6-lutidine	100°	99% Δ^1- and 1% Δ^2-olefin	96	44
(n-C$_3$H$_7$)$_2$CHCH$_2$NH$_2$	OH	100°/10 mm.	(n-C$_3$H$_7$)$_2$C=CH$_2$	73	42
(i-C$_3$H$_7$)$_2$CHCH$_2$NH$_2$	OH	100°/10 mm.	(i-C$_3$H$_7$)$_2$C=CH$_2$	67	42
Phenethylamine	OH	Distil	Styrene	"Completely"	39
	OH	100°	Styrene	"Completely"	132
β-(p-Nitrophenyl)ethylamine	Iodide + H$_2$O	100°	p-Nitrostyrene	High	132
Cycloöctylamine	OH	150°, vac.	trans-Cycloöctene	62	54
		120°/11 mm.	Cycloöctene (cis, 40%; trans, 60%)	89	9

Note: References 194 to 391 are on pp. 489–493.

TABLE XVI—Continued
Decomposition of Quaternary Ammonium Compounds

No. of C Atoms	Amine	Derivative	Conditions	Elimination Product(s) (Composition of Mixture)	Yield, %	References
C_8 (cont.)	cis-3-Aminocycloöctene	OH	70–185°/ 28–10 mm.	1,3-Cycloöctadiene cis-trans cis-cis	— 15 41	59
	cis-4-Cycloöctenylamine	OH	70–185°/ 3 mm.	cis,cis-1,3-Cycloöctadiene (10%), cis,trans-1,5-cycloöctadiene (90%)	64	209
	(cycloheptyl-CH₂NH₂)	OH OH	Distil, vac. To 160°, vac.	1,5-Cycloöctadiene Methylenecycloheptane	77 74	54 34
	(1-methyl-1-amino-cycloheptyl)	OH	To 160°, vac.	Methylcycloheptene Methylenecycloheptane (78%), methylcycloheptene (22%)	0·2 84	34
	(bicyclic-CH₂NH₂)	OH	120–140°, vac.	(bicyclic methylene compound)	61	240
	2-Aminobicyclo[2.2.2]octane 5-Aminobicyclo[2.2.2]oct-2-ene	OH, KOH OH	Distil 150–160°, vac.	Bicyclo[2.2.2]octene Bicyclo[2.2.2]octadiene	50 40	241 26
	n-Butylisobutylamine	OH	85–150°/ 20 mm.	1-Butene (64%), isobutylene (36%)	92	36
	n-Propylisoamylamine	OH	85–150°/ 20 mm.	Propylene (75%), isoamylene (25%)	95	36

Starting material	Reagent	Conditions	Product	Yield (%)	Ref.
cis-Octahydroindole	OH	100°, vac.	N,N-Dimethyl-2-cyclohexenyl-ethylamine	—	242
trans-Octahydroindole	OH	125–180°, vac.	trans-N,N-Dimethyl-2-vinyl-cyclohexylamine	65	71, 69
(cis) [bicyclic structure]	OH	120–130°, vac.	[cyclopentane with CH=CH₂ and CH₂N(CH₃)₂]	—	243
2,3-Dihydroindole	OH	80–110°, vac.	o-Vinyldimethylaniline	71	72
i-H₉C₄–NH–CH₃ [azetidine]	OH, KOH	Distil	$C_{10}H_{21}N$	—	62
Second step	OH, KOH	Distil	C_8H_{14}	35	62
N-Propylpiperidine	OH, KOH	Distil	Propylene and [N-methyl tetrahydropyridine with C₃H₇]	54	74, 244
CH₃–N(C₂H₅)–H₃C–CH₃	OH	Distil	$(CH_3)_2C=CH-CHCH_3-N(CH_3)(C_2H_5)$	77	63
Second step	OH	Distil	$(CH_3)_2C=CH-CH=CH_2$	—	63

Note: References 194 to 391 are on pp. 489–493.

TABLE XVI—Continued
DECOMPOSITION OF QUATERNARY AMMONIUM COMPOUNDS

No. of C Atoms	Amine	Derivative	Conditions	Elimination Product(s) (Composition of Mixture)	Yield, %	References
C₈ (cont.)	1-Methylpyrrolizidine (heliotridane)	OH	100°/20 mm.	![structure] pyrrolidine with CH₃, CH₂CH=CH₂, N-CH₃	60	245
	2-n-Propyl-3-methylpyrrolidine	—	—	CH₃, C₃H₇, N(CH₃)₂ with vinyl	—	246
	sec-Butylpyrrolidine	—	—	CH(CH₃)–, N(CH₃)₂ with vinyl	—	246
	2-Methylpyrrolizidine	OH	100°/20 mm.	H₃C-pyrrolidine, CH₂CH=CH₂, N-CH₃	66	247
	Diaminocycloöctadiene	Di OH	30–45°/0.2 mm.	Cycloöctatetraene	10–20	248–250
C₉	3-Phenylpropylamine	OH	Distil	1-Phenyl-1-propene,	28	40, 251
		OH	75–120°/0.5 mm.	1-Phenylpropene (*trans*, 94%; *cis*, 5%), 3-phenylpropene, 1%		145
	3-Phenoxypropylamine	OH	Distil	3-Phenoxy-1-propene	90	39

Amine	Reagent	Conditions	Olefin	Yield	Ref.
1-Phenyl-2-propylamine	Iodide + NaOH	80°	1-Phenyl-1-propene	—	47
3-Phenyl-2-amino-1-propanol	OH	Distil, vac.	3-Phenylallyl alcohol (cinnamyl alcohol)	—	110
3-Nitro-4-hydroxyphenyl-alanine	Iodide + NaOH	Boil	3-Nitro-4-hydroxycinnamic acid	78	50
HO—⬡(CH₃)—CH₂CH₂NH₂	OH	100°/2 mm.	HO—⬡(CH₃)—CH=CH₂	—	112
⬡(CH₃)—CH₂CH₂NH₂	OH	150°	⬡(CH₃)—CH=CH₂	High	111
⬡(C₂H₅)—CH₂NH₂ (cis or trans)	OH	120°, vac.	⬡(C₂H₅)=CH₂	Low	252
Cyclononylamine	OH	150°, vac.	trans-Cyclononene	69	55
				83	56
1-Methylcycloöctylamine	OH	80–90°, vac.	Methylenecycloöctane (64%), cis-1-methylcycloöctene (36%)	82	52a
Cycloöctylmethylamine	OH	95–110°, vac.	Methylenecycloöctane (99%), 1-methylcycloöctene (0.5%)	83	52a
n-Butylisoamylamine	OH	200°	Butylene (66%), isoamylene (34%)	83	37
cis-2-Methyloctahydroindole	OH	Distil	cis-2-n-Propyl-N,N-dimethyl-cyclohexylamine (after H₂)	—	84

Note: References 194 to 391 are on pp. 489–493.

TABLE XVI—Continued

DECOMPOSITION OF QUATERNARY AMMONIUM COMPOUNDS

No. of C Atoms	Amine	Derivative	Conditions	Elimination Product(s) (Composition of Mixture)	Yield, %	References
C_9 (cont.)	2,3-Dihydro-2-methylindole	OH	70°, vac.	o-Propenyldimethylaniline	77	72
	cis-Decahydroquinoline	OH	Distil	cis-2-n-Propyl-N,N-dimethyl-cyclohexylamine (after H_2)	—	70
	trans-Decahydroquinoline	OH	Distil	trans-2-n-Propyl-N,N-dimethyl-cyclohexylamine (after H_2)	—	70
	trans-2-Propylcyclohexylamine	OH	Distil	⟨cyclohexene with $C_3H_7\text{-}n$⟩	—	70
	Tetrahydroisoquinoline	OH	Distil	N,N-Dimethyl-2-vinyl-benzylamine	High	66
	cis-Decahydroisoquinoline	OH	120°, vac.	⟨cyclohexane with $CH{=}CH_2$ and $CH_2N(CH_3)_2$⟩ (cis)	81	252
	trans-Decahydroisoquinoline	OH	120°, vac.	⟨cyclohexane with $CH{=}CH_2$ and $CH_2N(CH_3)_2$⟩ (trans)	81	252
	Tetrahydroquinoline	OH	150°	No olefin	31	82
	N-Butylpiperidine	OH, KOH	Distil	Butylene and ⟨N with CH_3, $C_4H_9\text{-}n$⟩	59	74, 244

OLEFINS FROM AMINES

	OH	Distil, vac.	[structure: bicyclic with N–CH$_3$]	70	253
3-Ethylquinuclidine	OH	Distil	Unsaturated amine (position of unsaturation not determined)	Low	254
[structure: bicyclic amine]	Di OH	120–140°, vac.	[structure: norbornane with two =CH$_2$ groups]	57	240
CH$_2$NH$_2$ / CH$_2$NH$_2$ (trans)					
C$_{10}$ n-Amyldiethylamine	OH	Distil	Ethylene	—	1
C$_6$H$_5$N(CH$_3$)CH$_2$CH$_2$NH$_2$	OH	Distil	Vinylmethylaniline	—	255
n-Decylamine	OH	Distil	1-Decene	75	210, 102
4-Phenoxybutylamine	OH	Distil	4-Phenoxy-1-butene (?)	43	40
3,7-Dimethyloctylamine	OH, KOH	Distil	3,7-Dimethyl-1-octene	65	256
3,7-Dimethylocta-2,6-dienylamine	OH	Distil	Myrcene	—	219
(t-C$_4$H$_9$)$_2$CHCH$_2$NH$_2$	OH	100°/10 mm.	(t-C$_4$H$_9$)$_2$C=CH$_2$	81	42
3,7-Dimethylocta-6-enylamine	OH	Distil	β-Linalolene	—	219
1-Benzylallylamine	OSO$_2$OCH$_3$	Boil with NaOH	Phenylbutadiene	—	46
Cyclodecylamine	OH	Heat, vac.	Cyclodecene (cis, 2%; trans, 98%)	90, 64	56, 58, 257

Note: References 194 to 391 are on pp. 489–493.

TABLE XVI—Continued

DECOMPOSITION OF QUATERNARY AMMONIUM COMPOUNDS

No. of C Atoms	Amine	Derivative	Conditions	Elimination Product(s) (Composition of Mixture)	Yield, %	References
C_{10} (cont.)	3-Amino-cis-cyclodecene	OH	110°, vac.	cis-trans-1,3-Cyclodecadiene	20	60
	6-Hydroxycyclodecylamine	OH	—	6-Hydroxycyclodecene (cis, 60%; trans, 40%)	90	257a
	Cyclononylmethylamine	OH	140–180°, vac.	Methylenecyclononane (96%), 1-methylcyclononene (4%)	71	52a
	1-Methylcyclononylamine	OH	85–100°, vac.	Methylenecyclononane (48%), 1-methylcyclononene (cis-, 51%; trans, 1%)	83	52a
	Menthylamine	OH	130–140°	87% Δ^2- and 14% Δ^3-Menthene	80, 30	16, 17, 173
	Isomenthylamine	OH	145–200°, vac.	Δ^2-Menthene	58	17
	Neomenthylamine	OH	130–140°	8% Δ^2- and 92% Δ^3-Menthene	94, 86	16, 17, 173
	Neoisomenthylamine	OH	130–140°	Δ^3-Menthene "mainly"	30	17
	Carvomenthylamine	OH	166°/20 mm.	Δ^2-Menthene	3	258
	Piperitylamine	OH	Steam distil	Neopiperitol, α-Phellandrene	27 15	97
	Piperitylamine	Iodide	150–200°/30 mm.	α-Phellandrene, α-terpinene	68	97
	2-Aminotetralin	OH	50°/12 mm.	1,2-Dihydronaphthalene	60	259
	trans-α-Decalylamine	OH	100°/3–4 mm.	trans-$\Delta^{1,2}$-Octalin	40	191, 260, 261

Amine	Group	Conditions	Product	Yield	References
Bornylamine	OH	Pyrolyze	Bornylene	—	260, 261, 33
Neobornylamine	OH	180°	Bornylene and tricyclene (little)	—	33
Pinocamphylamine	OH	150°/0.02 mm.	α- and δ-Pinene	—	262, 263
α-Aminocamphene	OH	Distil	Camphinene	42	264
3-(p-Methoxyphenyl)-2-amino-1-propanol	OH	Vac. distil	3-(p-Methoxyphenyl)allyl alcohol	—	110
Phenethylethylamine	OH	85–150°/20 mm.	Styrene, 0.004% ethylene	93	36, 11
2-Methyltetrahydroquinoline	OH	Distil	No olefin, recovered amine	—	83
![structure: tetrahydroquinoline NH]	OH	100°, vac.	N,N-Dimethyl-1-naphthylamine	55	265
1,10-Diaminodecane	OH	Distil	1,9-Decadiene	6	95
![structure: bicyclic trans-diamine with two CH2NH2 groups] (trans)	Di OH	120–140° vac.	![structure: bicyclic diene with two =CH2 groups]	56	240
N-Amylpiperidine	OH	Distil	Amylene and CH3–N(C5H11-n)–(chain)	—	266
p-Methoxyphenylalanine	Iodide + KOH	100°	p-Methoxycinnamic acid	—	49

Note: References 194 to 391 are on pp. 489–493.

TABLE XVI—Continued

DECOMPOSITION OF QUATERNARY AMMONIUM COMPOUNDS

No. of C Atoms	Amine	Derivative	Conditions	Elimination Product(s) (Composition of Mixture)	Yield, %	References
C_{11}	1-Amino-5-phenylpentane	OH	Distil	5-Phenyl-1-pentene (?)	20	39
	5-Phenoxyamylamine	OH	Distil	5-Phenoxy-1-pentene	33	39
	1-(3,4-Dimethoxyphenyl)-propylamine	Iodide	100°, aq. soln.	$(CH_3O)_2C_6H_3CH=CHCH_3$, $(CH_3O)_2C_6H_3CHOHCH_2CH_3$	—	30
	1-(p-Methoxyphenyl)iso-butylamine	Free amine	Heat in formic acid	1-(p-Methoxyphenyl)-isobutylene (?)	—	30
	Cyclodecylmethylamine	OH	110–130°, vac.	Methylenecyclodecane (98%), 1-methylcyclodecene (2%)	74	52a
	1-Methylcyclodecylamine	OH	Vac.	Methylenecyclodecane (66%), 1-methylcyclodecene (cis, 31%; trans, 2%)	92	52a
	Isoamyl-(3,3-dimethylbutyl)-amine	OH	Distil, vac.	Isoamylene (91%), t-butylethylene (9%)	68	37
	Cyclopentylcyclo-hexylamine	OH	140°	Cyclopentene (95%), cyclohexene (5%)	—	53
		OH	35–110°, vac.	![product structure with N(CH3)2]	80	72
	N-Hexylpiperidine	OH	Distil, KOH	Hexene and ![structure with n-C6H13, CH3]	28 59	74

Compound			Product	"Entirely"	Ref.
N-Ethyltetrahydroquinoline	OH	Distil	Ethylene	—	244
N-Ethyl-N-propylaniline	OH	Distil, KOH	Ethylene (mostly), propylene	—	267
N-Cyclohexylpiperidine	OH	200°	(structure: cyclohexene with N–CH₃, C₆H₁₁)	80	268
(trans) [bicyclic diamine with two CH₂NH₂ groups]	Di OH	120–140°, vac.	(bicyclic bis-methylene structure)	61	240
C₁₂					
n-C₈H₁₇CH(C₂H₅)CH₂NH₂	OH, KOH	Distil	n-C₈H₁₇C(C₂H₅)=CH₂	45	256
5-Benzamido-1-pentylamine	OH	Distil	C₆H₅CONH(CH₂)₃CH=CH₂	40	39
trans-2-Phenylcyclohexylamine	OH	Distil	1-Phenylcyclohexene	73	21
trans-2-Phenyl-3,3,6,6-d₄-cyclohexylamine	OH	90°/1 mm.	1-Phenyl-3,3,6,6-d₄-cyclohexene	91	23
cis-2-Phenylcyclohexylamine	OH	Distil	1-Phenylcyclohexene	—	21
(carbazole-like fused structure with NH) (cis)	OH	85–90°, vac.	(phenyl-cyclohexene with N(CH₃)₂)	78	72
N-Benzylpiperidine	OH	Distil	C₆H₅CH₂–N(CH₃) structure	—	266
Second step	OH	Distil	1,4-Pentadiene	—	266

Note: References 194 to 391 are on pp. 489–493.

420   ORGANIC REACTIONS

TABLE XVI—Continued
DECOMPOSITION OF QUATERNARY AMMONIUM COMPOUNDS

No. of C Atoms	Amine	Derivative	Conditions	Elimination Product(s) (Composition of Mixture)	Yield, %	References
C_{12} (cont.)	N-Phenethylpyrrolidine	OH	Heat	Styrene	60	269
	N-Propyltetrahydoquinoline	OH	Distil	Propylene	25	244
	1,12-Diaminododecane	OH	Distil	1,11-Dodecadiene	65	91
	2-Phenyl-3-aminobicyclo-[2.2.1]heptane	OH	Distil	Neutral material	—	270
C_{13}	n-Decylpropylamine	OH	85–150°/20 mm.	Propene (59.7%), 1-decene (40.3%)	93	36
	(cyclopentylmethyl-NHCH₂-cyclohexyl)	OH	140°	Methylenecyclohexane, methylenecyclopentane, ca. 2:1 ratio	—	53
	(cis-hexahydrocarbazole)	OH	115–120°, vac.	(cyclohexene with N(CH₃)₂ substituent)	—	72
	(dimethoxy-tetrahydroisoquinoline with CH₃ groups)	OSO₂OCH₃	KOH, steam bath	(dimethoxybenzene with CH=CHCH₃ and CH(CH₃)N(CH₃)₃ substituents) (or isomer)	—	271
	Second step	OSO₂OCH₃	KOH, steam bath	(dimethoxybenzene with CH=CHCH₃ and CH=CH₂ substituents)	—	271

Starting material		Method	Product		Reference
N-Amyl-N-ethylaniline	OH	Distil, KOH	Ethylene	88	2
N-Octylpiperidine	OH	Distil, KOH	Octene (32%) and (68%) $\text{C}_8\text{H}_{17}\text{-N(CH}_3\text{)-CH}_2\text{CH}_2\text{...}$	—	74
C_{14} 1,2-Diphenylethylamine	OSO$_2$OCH$_3$, NaOH, boil		trans-Stilbene	—	46
(cyclohexenyl ketone with CH$_2$NH$_2$)	OH	Spont. decomp.	(cyclohexenyl ketone with vinyl)	—	272
(piperidine with CH$_2$CH$_2$CO$_2$C$_2$H$_5$, CHNH$_2$CH$_3$, COCH$_3$)	Iodide	60% KOH, 140–180°	Product isolated as (piperidine with CH$_2$CH$_2$CO$_2$H, CH=CH$_2$, CONH$_2$)	—	129
CH$_3$CHCH=CHN(piperidine)	Free amine	140°, vac.	CH$_3$CH=C=CHN(piperidine)	—	219

Note: References 194 to 391 are on pp. 489–493.

TABLE XVI—Continued

DECOMPOSITION OF QUATERNARY AMMONIUM COMPOUNDS

No. of C Atoms	Amine	Derivative	Conditions	Elimination Product(s) (Composition of Mixture)	Yield, %	References
C_{15}	$C_6H_5CH(CH_3)CH(NH_2)C_6H_5$					
	erythro	OC_2H_5	C_2H_5OH, reflux	*cis*-1,2-Diphenylpropene	85	15
	threo			*trans*-1,2-Diphenylpropene	90	15
	Either isomer	OC_4H_9-*t*	*t*-C_4H_9OH, 30°	*trans*-1,2-Diphenylpropene (mainly)	98	15
	$C_6H_5\overset{O}{\overset{\|}{C}}CHCH_2C_6H_4X$ $\quad\quad\quad NH_2$ X = *m*-Br, *p*-Br, *p*-NO_2, *p*-OCH_3	OSO_2OCH_3	KOH, boil	$C_6H_5\overset{O}{\overset{\|}{C}}CH=CHC_6H_4X$ X = *m*-Br, *p*-Br, *p*-NO_2, *p*-OCH_3	—	122, 273
	$C_6H_5CH_2CHCHC_6H_4X$ $\quad\quad\quad\quad NH_2$ X = *m*- or *p*-Cl = *m*-CH_3 = *p*-CH_3	OH	KOH, distil in vac.	$C_6H_5CH=CHCH_2C_6H_4X$ (I), $C_6H_5CH_2CH=CHC_6H_4X$ (II) X = *m*-Cl; 23% I, 77% II = *p*-Cl; 25% I, 75% II = *m*-CH_3; 27% I, 73% II = *p*-CH_3; 66% I, 34% II	90 93 95	274 274 274
C_{16}	*n*-Cetylamine	OH	Distil	Hexadecene	—	39

Starting material	Group	Conditions	Product	Yield	Ref.
$C_6H_5CH_2C(CO_2C_2H_5)_2$-$CH_2CH_2NH_2$	OC_2H_5	150–210°	$C_6H_5CH_2CCO_2C_2H_5$ ‖ $CHCH_3$ or isomer	—	113, 114
$C_6H_5COC(CH_3)CH_2C_6H_5$ | NH_2	OSO_2OCH_3	NaOH, reflux	$C_6H_5COC(CH_3)=CHC_6H_5$	—	46
N,N-bis-Phenethylamine	OH	Boil, aq. soln.	Styrene	—	11
N-β-(p-Nitrophenyl)-ethyl-N-phenethylamine	OH	Water bath	p-Nitrostyrene	—	11
(1-phenyl-7-methoxy-1,2,3,4-tetrahydroisoquinoline, C_6H_5, NH, CH_3O)	Iodide	KOH, steam bath	(styrylbenzene with C_6H_5, $CH_2N(CH_3)_2$, CH_3O)	78	275
(methylenedioxy-1-phenyl-tetrahydroisoquinoline, C_6H_5, NH, OCH_2O)	Iodide	KOH, steam bath	(methylenedioxy styryl with C_6H_5, $CH_2N(CH_3)_2$)	55	275
(dibenzazecine, N)	OH	—	(ring-opened olefin with N–CH_3)	Good	87
(dibenzobicyclic NH)	OH	—	(dibenzocycloheptene with $N(CH_3)_2$)	Good	87

Note: References 194 to 391 are on pp. 489–493.

TABLE XVI—Continued

DECOMPOSITION OF QUATERNARY AMMONIUM COMPOUNDS

No. of C Atoms	Amine	Derivative	Conditions	Elimination Product(s) (Composition of Mixture)	Yield, %	References
C_{16} (cont.)	[dibenzosuberene-NH$_2$ structure]	OH		[dibenzocycloheptene structure]	Satisfactory	87
	[dihydroacridine N–H structure]	OH	Heat	[naphthyl-phenyl-N(CH$_3$)$_2$ structure]	85	265
C_{17}	$C_6H_5COCH_2NH(CH_2)_3C_6H_5$ Amine, $C_{17}H_{35}N$, from naphthenic acid	Bromide OH	NaNH$_2$ Distil	$C_6H_5CH=CHCH_3$ Olefin	Small 63	276 91
C_{18}	[cyclobutane with H$_5$C$_6$, H$_5$C$_6$, CH$_2$NH$_2$, CH$_2$NH$_2$]	OH	100–140°/ 0.5 mm.	[cyclobutene with H$_5$C$_6$, H$_5$C$_6$, CH$_2$, CH$_3$]	—	149
	[cyclobutane with H$_5$C$_6$, H$_5$C$_6$, CH$_2$NH$_2$, CH$_2$NH$_2$]	OH	120–140°/ 0.5 mm.	[cyclobutadiene with H$_5$C$_6$, H$_5$C$_6$, CH$_2$, CH$_2$]	—	92
C_{19}	$C_6H_5(CH_2)_3CHCH_2NH_2$ $(CH_2)_2C_6H_5$	OH, KOH	Distil, vac.	$C_6H_5(CH_2)_3$ $\quad\quad$C=CH$_2$ $C_6H_5(CH_2)_2$	70	256

	277	100 46	278	28	28
	90	— —	91	86	86

C₂₀	1,1,2-Triphenylethylamine	OSO₂OCH₃ Heat, NaOH	Triphenylethylene
	N-Phenyl-1,2-diphenyl- ethylamine	OH 150–180° OSO₂OCH₃ CH₃OH, NaOH reflux	trans-Stilbene
		OH, KOH Reflux	
	Pavine	OH, KOH Reflux	
		OH, KOH Reflux	

Note: References 194 to 381 are on pp. 489–493.

TABLE XVI—Continued

DECOMPOSITION OF QUATERNARY AMMONIUM COMPOUNDS

No. of C Atoms	Amine	Derivative	Conditions	Elimination Product(s) (Composition of Mixture)	Yield, %	References
C_{20} (cont.)	[structure]	OSO_2OCH_3	KOH, 120°	[structure]	—	279
	[structure] (second step)	OSO_2OCH_3	KOH, steam bath	[structure]	—	279
C_{21}	$C_6H_5COCH(NH_2)CH(C_6H_5)_2$	OSO_2OCH_3	Alkali, reflux	$C_6H_5COCH=C(C_6H_5)_2$	—	122
	$3(\alpha), 12(\alpha)$-Dihydroxy-20-aminopregnane (partly acetylated)	Iodide	50% NaOH, 180°	Isolated as $3(\alpha),12(\alpha)$-diacetoxy-Δ^{20}-pregnene	35	280
	$3(\alpha)$-Aminoallopregnane	OH	200°/5 μ	Allopregnene, Δ^2- or Δ^3-	30	18
	$3(\beta)$-Aminoallopregnane	OH	200°/5 μ	Allopregnene, Δ^2- or Δ^3-	4	18
	3-Acetoxy-20-amino-Δ^5-pregnene	Iodide	KOH, ethylene glycol	3-Hydroxy-$\Delta^5, {}^{20}$-pregnadiene	65	136

OLEFINS FROM AMINES

	Amine	Group	Conditions	Product	Yield %	Ref.
	$C_6H_5CH_2CH_2NH(CH_2)_5$-$NHCH_2CH_2C_6H_5$	OH	Heat	Styrene	60	269
C_{23}	$3(\alpha),12(\alpha)$-Dihydroxy-23-amino-norcholane (partly acetylated)	Iodide	50% KOH, 200°	Isolated as $3(\alpha),12(\alpha)$-diacetoxy-Δ^{22}-norcholene	14	280
C_{24}	$3(\alpha)$-Hydroxy-12-aminocholanic acid	Iodide	60% KOH, 160°	$3(\alpha)$-Hydroxy-Δ^{11}-cholanic acid (isolated as the methyl ester)	35	280
	o-$C_6H_4(CH_2NHCH_2CH_2C_6H_5)_2$	OH	Heat	Styrene	65	269
C_{26}	[tetrahydroquinoline structure with CH_3O, CH_3O, NH, $(CH_2)_{14}CH_3$]	OH	100°	[styrene derivative with CH_3O, CH_3O, CH_3, $-N-CH_3$, CH, $C_{15}H_{31}$-n]	—	31
	[structure with CH_3O, CH_3O, C_2H_5, CH-$C_{15}H_{31}$, NH_2]	Iodide	100°	[structure with CH_3O, CH_3O, C_2H_5, $CH=CHC_{14}H_{29}$]	—	31
C_{27}	$3(\alpha)$-Aminocholestane	OH	170°/0.5 mm.	Δ^2- and Δ^3-Cholestene	50	18
	$3(\beta)$-Aminocholestane	OH	170°/0.5 mm.	Neutral product, not investigated	ca. 3	18
	$3(\beta)$-Amino-Δ^5-cholestene	OH	180°/0.1 mm.	$\Delta^{3,5}$-Cholestadiene	—	18
	6-(α)-Aminocholestane	OH	175–195°/0.02 mm.	5- and 6-Cholestene	Very low	19
	6-(β)-Aminocholestane	OH	Room temp., vac.	5-Cholestene	65	19

Note: References 194 to 381 are on pp. 489–493.

428   ORGANIC REACTIONS

TABLE XVI—Continued
Decomposition of Quaternary Ammonium Compounds

No. of C Atoms	Amine	Derivative	Conditions	Elimination Product(s) (Composition of Mixture)	Yield, %	References
C_{28}	(structure: bis-tetrahydroisoquinoline with OCH$_3$ groups, linked by (CH$_2$)$_4$)	OH	100°	[vinyl-dimethoxyphenyl-CH(N(CH$_3$)$_2$)-CH$_2$CH$_2$CH$_2$-]$_2$	—	31
	[CH$_3$O-C$_6$H$_3$(OCH$_3$)-CH(C$_2$H$_5$)-CHNH$_2$-CH$_2$CH$_2$CH$_2$-]$_2$	Di-iodide	100°	[CH$_3$O-C$_6$H$_3$(OCH$_3$)-CH(C$_2$H$_5$)-CH=CH-CH$_2$CH$_2$-]$_2$	—	31
	(dibenzazepine-type structure)	OH	180–190°	(N-methyl dibenzazocine/cyclophane structure)	82	280a

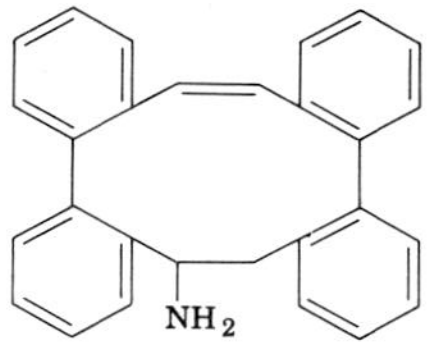

280a

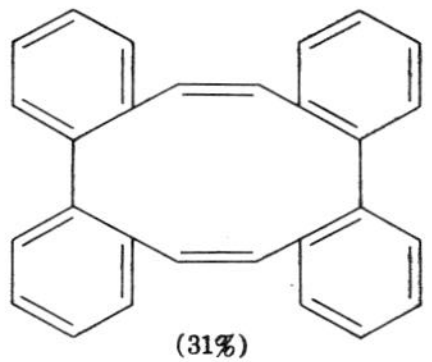

(31%)

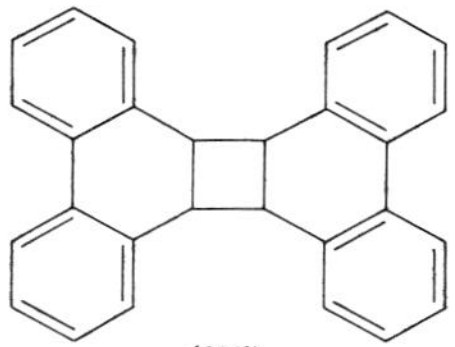

(29%) Phenanthrene (19%)

OH

NH₂

Note: References 194 to 391 are on pp. 489–493.

TABLE XVII
Quaternary Compounds that Contain No N-Methyl Groups

No. of C Atoms	Ammonium Ion	Derivative	Conditions	Elimination Product(s) (Composition of Mixture)	Yield, %	References
C$_8$	Tetraethyl	OH	Distil	Ethylene	—	1
C$_9$	(piperidine-ethyl)	OH	140°	(50%) and (50%)	—	53, 281
C$_{10}$	Diethyldi-n-propyl	OH	Distil	Ethylene (96%), propylene (4%)	99	37
	(morpholine-vinyl)	OH	170°, vac.		57	234
C$_{11}$	n-Amyltriethyl	OH	Distil	Ethylene	—	1
	Isoamyltriethyl	OH	Distil	Ethylene	—	39
	(piperidine with CH₃ and vinyl)	OH	140°		—	53

OLEFINS FROM AMINES

	OH	75–160°, vac.	[structure: bicyclic amine with O] and [structure: piperidine with CH₂ and furan-CH₂]	71	234
C₁₂ Phenyltriethyl	OH	Heat	Ethylene	—	2
C₁₃ n-Butyltri-n-propyl	OH	Distil	Propylene (83%), butylene (17%)	—	37
[structure: tetrahydroisoquinoline fused with pyrrolidine]	OH	Distil, vac.	[structure: styrene with CH₂N-pyrrolidine ortho]	—	281
C₁₄ Di-n-propyldi-n-butyl	OH	Distil	Propylene (63%), butylene (37%)	95	37
Phenethyltriethyl	OH	Distil	Styrene	79	37
[structure: tetrahydroisoquinoline fused with piperidine]	OH	Distil	[structure: styrene with CH₂N-piperidine ortho]	—	281
[structure: dipiperidinyl-methylene-amine]	Chloride	KOH, heat	[structure: dicyclohexyl-NCH₂CH₂N] and acetylene, β-hydroxyethylpiperidine	—	223
C₁₅ n-Propyltri-n-butyl	OH	Distil	Propylene (36%), butylene (64%)	98	37

Note: References 194 to 381 are on pp. 489–493.

TABLE XVII—Continued
Quaternary Compounds That Contain No N-Methyl Groups

No. of C Atoms	Ammonium Ion	Derivative	Conditions	Elimination Product(s) (Composition of Mixture)	Yield, %	References
C_{16}	Di-n-propyldiisoamyl	OH	Distil	Propylene (96%), isoamylene (4%)	94	37
C_{17}	(isoindoline structure)	OH	220°/20 mm.	(vinyl-benzyl-N-CH₂ structure)	60	281
C_{18}	Di-n-butyldiisoamyl	OH	Distil	Butylene (67%), isoamylene (33%)	94	37
	$C_6H_5COCH_2N(C_2H_5)_2C_6H_5$	Bromide	KOH, heat	Ethanol	—	276
C_{19}	(dinitro-diphenyl pyridinium structure)	Bromide	250°/0.01 mm.	2,4′-Dinitrodiphenylacetylene	55	282
C_{20}	Tetra-n-amyl	OH	Distil	Amylene	—	2
C_{21}	$(C_6H_5)_2CCH$ (pyridinium cyclopropane structure)	Iodide	NaOH, reflux	$(C_6H_5)_2C$ (cyclopropane structure)	—	283

Note: References 194 to 391 are on pp. 489–493.

TABLE XVIII
HOFMANN ELIMINATION REACTIONS WITH ALKALOIDS

Name	Derivative	Conditions	Product	Yield, %	References
Aporphine			Methine*		
Actinodaphnine, 3,4-dimethoxy-5,6-methylenedioxyaporphine —methine	Iodide	Aq. KOH, boil	Methine	—	183
Anolobine, —methine	Chloride	Base, heat	Vinylphenanthrene	—	183
2-hydroxy-5,6-methylenedioxyaporphine	Iodide, O-Methyl	Aq. base, 100°	Methine	—	284
—methine	Iodide	Aq. base	Vinylphenanthrene	—	284
Anonaine,	Iodide	Aq. base, heat	Methine	91	285
5,6-methylenedioxyaporphine —methine	Iodide	CH$_3$OH, base, heat	Vinylphenanthrene	76	285

Note: References 194 to 391 are on pp. 489–493.

* The methine nomenclature is explained on p. 321.

434   ORGANIC REACTIONS

TABLE XVIII—*Continued*

HOFMANN ELIMINATION REACTION WITH ALKALOIDS

Name	Derivative	Conditions	Product	Yield, %	References
Aporphine (Continued)					
Boldine, 2,6-dihydroxy-3,5-dimethoxyaporphine ——methine	O,O-Di-ethyl, OH	100°, vac.	Methine	—	180
Crebanine, 1,2-dimethoxy-5,6-methylenedioxy aporphine'	OH	100°, vac.	Vinylphenanthrene	—	180
	—	—	1,2-Dimethoxy-5,6-methylene-dioxyphenanthrene, after oxidation and decarboxylation	—	286
Dicentrine, 2,3-dimethoxy-5,6-methylenedioxyaporphine	Iodide	Aq. base, 100°	Methine	—	287
Glaucine, 2,3,5,6-tetramethoxy-aporphine	Iodide	Heat, base	Methine	—	288, 289
——methine	OH	Distil, base	Vinylphenanthrene	83	288, 289

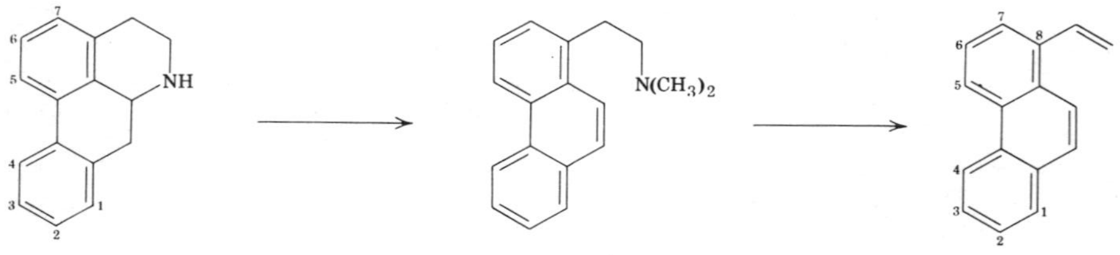

* Methine
8-Vinylphenanthrene

Compound	Reactant	Conditions	Product	Yield	References
Isothebaine, 3,5-dimethoxy-4-hydroxy-aporphine	O-Methyl, OSO$_2$OCH$_3$	Aq. base, heat	Methine	85	290, 291
——methine	OSO$_2$OCH$_3$	Base, CH$_3$OH, heat	Vinylphenanthrene	80	290, 291
Laureline, 3-methoxy-5,6-methylenedioxy-aporphine	Iodide	Aq. KOH, boil	Methine	—	292, 79
——methine	Chloride	Aq. KOH, boil	Vinylphenanthrene	89	292, 79
Laurotetanine, 2-hydroxy-3,5,6-trimethoxy-aporphine	N-Ethyl-, O-ethyl-, iodide	Aq. NaOH, heat	Methine	—	293
——methine	OH	Heat	Methine	88	294
	OH	Distil, vac.	Methine	—	295
	Chloride	Aq. NaOH	Vinylphenanthrene	—	293
	OH	Heat	Vinylphenanthrene	60	294
	OH	100°/10 mm.	Vinylphenanthrene	70, overall	295
	Iodide, O-methyl	Aq. KOH, heat	Methine	—	292
Pukateine, 4-hydroxy-5,6-methylenedioxyaporphine ——methine	Chloride	Aq. KOH, heat	Vinylphenanthrene	—	292
Tuduranine, 3-hydroxy-5,6-methylenedioxyaporphine	O-Methyl iodide	Aq. NaOH, heat	Methine	—	296
	O-Ethyl iodide	Aq. NaOH, heat	Methine	—	296
——methine	O-Methyl iodide	Aq. NaOH, heat	Vinylphenanthrene	—	296
	O-Ethyl iodide	Aq. NaOH, heat	Vinylphenanthrene	—	296

Note: References 194 to 391 are on pp. 489–493.

* The methine nomenclature is explained on p. 321.

TABLE XVIII—Continued
Hofmann Elimination Reaction with Alkaloids

Name	Derivative	Conditions	Product	Yield, %	References
Benzylisoquinoline					
Armepavine, 1,2,3,4-tetrahydro-2-methyl-6,7-dimethoxy-4′-hydroxybenzylisoquinoline ——methine	Iodide, O-Methyl	CH₃OH, base, heat	Methine	Quant.	297
	Iodide	CH₃OH, base, heat	Vinylstilbene	95	297
Coclaurine, 1,2,3,4-tetrahydro-6-methoxy-7,4′-dihydroxy-benzylisoquinoline	OSO₂OCH₃, O,O-dimethyl	Aq. base, 120–130°	Methine	—	298

——methine	OSO_2OCH_3	Aq. base, 80°	Vinylstilbene	—	298
Laudenine, 1,2,3,4-tetrahydro-2-methyl-6,7,4'-trimethoxy-3'-hydroxybenzylisoquinoline	O-ethyl, OSO_2OCH_3	Aq. base, heat	Methine	>90	299
Bisbenzylisoquinoline-A					
			Methine	Small	94
Dauricine	O-Methyl OSO_2OCH_3	Base	Methine	Small	94

Note: References 194 to 391 are on pp. 489–493.

438 ORGANIC REACTIONS

TABLE XVIII—Continued
Hofmann Elimination Reaction with Alkaloids

Name	Derivative	Conditions	Product	Yield, %	References
Bisbenzylisoquinoline-B			Methine → Ozonized methine → des-aza Product / des-aza Aldehyde	—	300
Cepharanthine $R_1 = R_4 = R_5 = R_6 = CH_3$ $R_2 = R_3 = -CH_2-$	Iodide	Aq. base, heat	α-Methine, 30 parts β-Methine, 1 part		

OLEFINS FROM AMINES

					Ref.
α-Methine, ozonized	Iodide	Aq. base,	des-aza Aldehyde	—	300
Daphnandrine $R_1 = R_3 = R_4 = CH_3$ $R_2 = H$ R_5 or $R_6 = CH_3$ R_6 or $R_5 = H$ ——methine	O-Methyl iodide	Aq. base, heat	Methine	—	301
	Iodide	Aq. base, heat	des-aza Product	—	301
Daphnoline (trilobamine) $R_1 = R_2 = CH_3$ $R_2 = R_4 = H$ R_5 or $R_6 = CH_3$ R_6 or $R_5 = H$	O,O-Diethyl iodide	Aq. base, heat	Methine	—	301, 302
Hydroepistephanine $R_1 = R_2 = R_3 = R_4 = R_5 = CH_3$ $R_6 = H$ ——methine	—	—	Methine	—	303
Oxyacanthine $R_1 = R_2 = R_3 = R_5 = R_6 = CH_3$ $R_4 = H$	O-Methyl OSO_2OCH_3	Aq. base, heat	des-aza Product	—	303
	O-Methyl OSO_2OCH_3	Aq. base, heat	Methine	—	304
——methine	O-Methyl OSO_2OCH_3	Aq. base, heat	des-aza Product	—	304
Repandine $R_1 = R_2 = R_3 = R_5 = R_6 = CH_3$ $R_4 = H$	O-Methyl OH	Aq. base, heat	Methine	—	305

Note: References 194 to 391 are on pp. 489–493.

TABLE XVIII—*Continued*

HOFMANN ELIMINATION REACTION WITH ALKALOIDS

Name	Derivative	Conditions	Product	Yield, %	References

Bisbenzylisoquinoline-B'

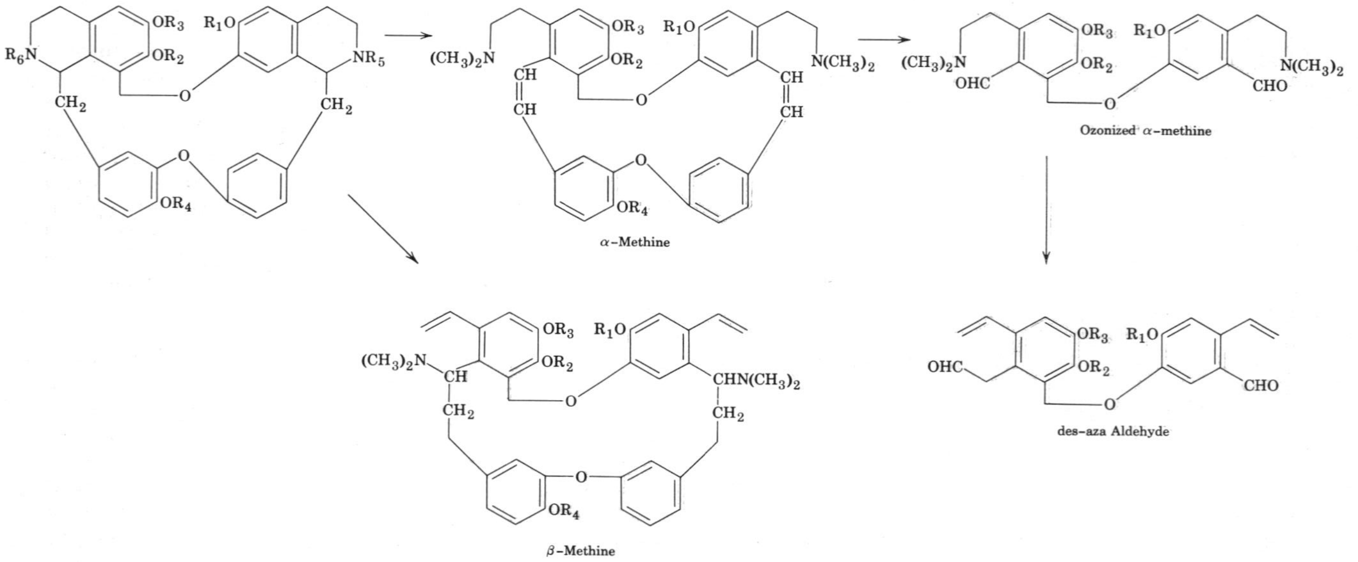

Compound				Product	Ref.
Berbamine $R_1 = R_2 = R_3 = R_5 = R_6 = CH_3$, $R_4 = H$	O-Methyl OH	Aq. base, heat	—	α-Methine	177
Pheanthine (1-isotetrandrine) $R_1 = R_2 = R_3 = R_4 = R_5 = R_6 = CH_3$	—	—	—	α- and β-Methine	306
Tetrandrine $R_1 = R_2 = R_3 = R_4 = R_5 = R_6 = CH_3$	OH	Aq. base, heat	—	α-Methine and β-methine Mixture of α- and β-methines	307 177
Ozonized-α-methine	—	Aq. base, heat	—	des-aza Aldehyde	307

Note: References 194 to 391 are on pp. 489–493.

TABLE XVIII—*Continued*

HOFMANN ELIMINATION REACTION WITH ALKALOIDS

Name	Derivative	Conditions	Product	Yield, %	References
Bisbenzylisoquinoline-C					
Bebeerine (chondodendrine) $R_1 = R_3 = CH_3$ $R_2 = R_4 = H$	O-Ethyl chloride	Aq. base, boil	des-aza Product	—	308
	O,O-Dimethyl chloride	Aq. base, boil	des-aza Product	—	309
	O,O-Dimethyl chloride	Aq. base, boil	des-aza Product	—	310
Chondrofoline R_1 or $R_3 = CH_3$ R_3 or $R_1 = H$ $R_2 = R_4 = CH_3$	O-Methyl chloride	Aq. base, boil	Methines	23 (as methiodide)	311

Compound	Reagent		Product	Ref.
Tubocurarine chloride $R_1 = R_3 = CH_3$ $R_2 = R_4 = H$ N,N-dimethyl	O,O-Diethyl chloride	Aq. base, boil	des-aza Product	17, 312
———methine mixture	O,O-Dimethyl chloride	Aq. base, boil	Mixture of 4 methines	309
	O,O-Dimethyl chloride	Aq. base, boil	des-aza Product	309
Bisbenzylisoquinoline-D				
Isochondodendrine $R_1 = R_3 = CH_3$ $R_2 = R_4 = H$	O,O-Dimethyl chloride	Aq. base, heat	Methine (opt. inact.)	313

Note: References 194 to 391 are on pp. 489–493.

444 ORGANIC REACTIONS

TABLE XVIII—Continued
Hofmann Elimination Reaction with Alkaloids

Name	Derivative	Conditions	Product	Yield, %	References
Bisbenzylisoquinoline-D (Continued)					
Oxidized methine	Chloride	Aq. base, heat	Methine	—	313
Neoprotocuridine $R_1 = R_3 = H$ $R_2 = R_4 = CH_3$	O,O-Dimethyl chloride	Aq. base, boil	Methine (opt. inact.)	—	314

OLEFINS FROM AMINES

Bisbenzylisoquinoline-E

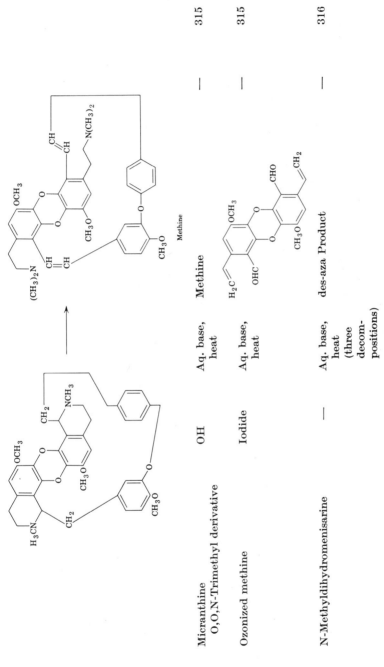

Micranthine O,O,N-Trimethyl derivative	OH	Aq. base, heat	Methine	—	315
Ozonized methine	Iodide	Aq. base, heat	Methine	—	315
N-Methyldihydromenisarine	—	Aq. base, heat (three decompositions)	des-aza Product	—	316

Note: References 194 to 391 are on pp. 489–493.

TABLE XVIII—Continued
HOFMANN ELIMINATION REACTION WITH ALKALOIDS

Name	Derivative	Conditions	Product	Yield, %	References
Bisbenzylisoquinoline-F		Methine → Ozonized methine → des-aza Aldehyde			
Trilobine	OSO₂OCH₃	Aq. base, heat	Methine (opt. inact.)	ca. 80	317
	Di-iodide	Aq. base, heat	des-aza Aldehyde	—	317
Ozonized methine	OSO₂OCH₃	Aq. base, heat	Methine (opt. act.)	—	317
Isotrilobine	OSO₂OCH₃	Base	Methine	—	318

OLEFINS FROM AMINES

Compound	Salt/Group	Conditions	Product	Yield	Ref.
Ozonized methine *Cephaeline*	Iodide	Aq. base, heat	des-aza Aldehyde	—	317
Cephaeline R = H	O-Ethyl OH	Heat, 90°/ 12 mm.	Methine	—	319
Emetine R = CH₃	OH	Heat, vac.	Methine	93	29
Tetrahydromethine	Iodide	NaOH or H₂O	des-N-(a)-Emetinetetrahydromethine methiodide	47	29

Note: References 194 to 391 are on pp. 489–493.

448 ORGANIC REACTIONS

TABLE XVIII—Continued
HOFMANN ELIMINATION REACTION WITH ALKALOIDS

Name	Derivative	Conditions	Product	Yield, %	References
Cephaeline (Continued)					
Emetine R = CH$_3$	N-Acetyl OH	Heat	Methine	—	320
			Acetyl des-N-methylemetine		
N-Acetyl dihydro-des-N-methylemetine	OH	Aq. base, heat/12 mm.	N-Acetyldihydro-des-N,N-dimethylemetine	Good	320

Compound		Conditions	Product	Yield	Ref.
N-Acetyldihydro-des-N,N-dimethylemetine	OH	Aq. base, heat/12 mm.	N-Acetyldihydro-des-aza-emetine	—	320
Hexahydro-des-aza-emetine	OH	Aq. base, heat/12 mm.	des-N-Methylhexahydro-des-aza-emetine	Poor	320
Dihydro-des-N-methyl-hexahydro-des-aza-emetine	OH	Heat	Octahydro-des-diaza-emetine	—	320

Note: References 194 to 391 are on pp. 489–493.

TABLE XVIII—Continued
HOFMANN ELIMINATION REACTION WITH ALKALOIDS

Name	Derivative	Conditions	Product	Yield, %	References
Colchinol					
Colchinol methyl ether	OH	100–260°/ 15 mm.	Deaminocolchinol methyl ether	—	321
Coniine			Coniine methine		
Conhydrine, 7-hydroxyconiine methine, 6,7-Epoxyconiine methine	OH	100°, vac.	6,7-Epoxyconiine methine	—	199–201
Pseudoconhydrine, 3-hydroxyconiine	OH	100°, vac.	5,6-Dihydroxyoctene, 5,6-epoxyoctene	Low	199–201
	OH	Base, 100°, vac.	3-Hydroxyconiine methine	32	202
Dihydromethine, 6,7-Dihydro-3-hydroxy-coniine methine	OH	Base, heat, vac.	1,2-Epoxyoctane	—	202
Coniine	OH	Heat	Coniine methine	—	322
Coniine methine	OH	Heat	Octadiene	—	322

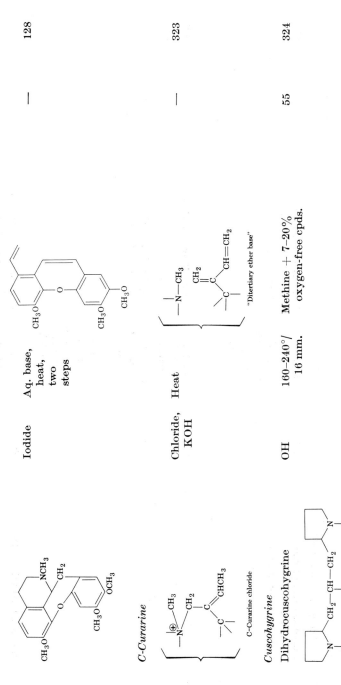

Cularine	Iodide	Aq. base, heat, two steps	—	128
C-Curarine	Chloride, KOH	Heat	—	323
Cuscohygrine				
Dihydrocuscohygrine	OH	160–240°/16 mm.	Methine + 7–20% oxygen-free cpds.	324, 55

Note: References 194 to 391 are on pp. 489–493.

TABLE XVIII—Continued

HOFMANN ELIMINATION REACTION WITH ALKALOIDS

Name	Derivative	Conditions	Product	Yield, %	References
Cuscohygrine (Continued)					
Dihydrocuscohygrine-methine (hydrogenated first)	OH	65–150°/17 mm. repeated until all N removed	After hydrogenation: undecan-6-ol and undecane	—	324
Cytisine	Iodide	Amyl alcohol, reflux	des-N-Dimethylcytisine	—	139
	OH	Evaporated, heat at 90°/5–10 mm. with Pd–C hydrogenate immediately	Dihydro-des-N-dimethylcytisine	72	190
des-N-Dimethylcytisine	OH	Amyl alcohol, reflux	$C_{22}H_{22}N_2O_2$ (bimolecular), des-aza-cytisine	—	139

Dihydro-des-N-dimethylcytisine	OH		Dihydrohemicytisylene	70	190
Tetrahydrodesoxycytisine	N-Acetyl OH	120°		—	325
		Distil at 140°/0.01 mm. (3 degradations)			
Tetrahydrodesoxycytisine	OH	Distil	des-N-Dimethyltetrahydrodesoxycytisine	90	190
Dihydro-des-N-dimethyltetrahydrodesoxycytisine	OH	Distil at 100° (3 degradations followed by hydrogenation)	$C_{13}H_{29}N$	—	190
Delphinine					
Delphinine	Iodide	Distil from aq. base	Methine base	—	326

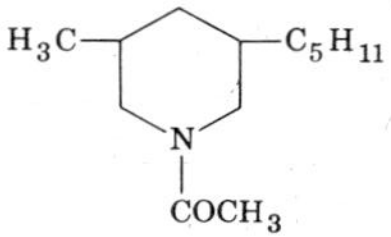

Structure shown: piperidine ring with H_3C and C_5H_{11} substituents and N–$COCH_3$.

Note: References 194 to 391 are on pp. 489–493.

TABLE XVIII—Continued
HOFMANN ELIMINATION REACTION WITH ALKALOIDS

Name	Derivative	Conditions	Product	Yield, %	References
Dioscorine	OH	Distil in vac.	Methine	—	327
———methine	OH	Distil in vac.		—	327
Ergothionene	Betaine	Aq. base, boil		75	328, 329
Erythroidine	OH	160–220°/ 0.03 mm.	des-N-Methyldihydro-α-erythroidinol	89	330, 331

Dihydro-α-erythroidinol

Compound			Conditions	Yield (%)	References
des-N-Methyldihydro-α-erythroidinol	OH	(structure: des-N-Dimethyldihydro-α-erythroidinol)	180–210°/0.01 mm.	90	330, 331
des-N-Dimethyltetrahydro-α-erythroidinol	OH	(structure: des-aza-Tetrahydro-α-erythroidinol)	80–100°/1 μ	87	330, 331
β-Erythroidinol	OH	(structure: des-N-Methyl-β-erythroidinol)	150–170°/0.05 mm.	70	332

Note: References 194 to 391 are on pp. 489–493.

TABLE XVIII—Continued
Hofmann Elimination Reaction with Alkaloids

Name	Derivative	Conditions	Product	Yield, %	References
Erythroidine (Continued)					
Dihydro-β-erythroidinol	OH	130–150°/ 0.03 mm.	des-N-Methylhydro-β-erythroidinol	78	178
des-N-Methyldihydro-β-erythroidinol	OH	160–170°/ 0.03 mm.	des-N,N-Dimethyldihydro-β-erythroidinol	90	178
des-N,N-Dimethyldihydro-β-erythroidinol	OH	150–160°/ 0.001 mm.	des-aza-Dihydro-β-erythroidinol	84	178

des-N-Methyldesoxydihydro-β-erythroidinol	OH	110–120°/0.003 mm.	des-N,N-Dimethyldesoxydihydro-β-erythroidinol	84	178
des-N,N-Dimethyldesoxytetrahydro-β-erythroidinol	OH	120°/0.02 mm.	des-aza-Desoxytetrahydro-β-erythroidinol	69	178
Desoxyapo-β-erythroidinol	OH	Heat	des-N-Methyldesoxyapo-β-erythroidinol	—	333

Note: References 194 to 391 are on pp. 489–493.

TABLE XVIII—Continued
HOFMANN ELIMINATION REACTION WITH ALKALOIDS

Name	Derivative	Conditions	Product	Yield, %	References
Erythroidine (Continued)					
Apo-β-erythroidine	Iodide	Aq. base	des-N-Methylapo-β-erythroidinol	54	334, 178
des-N-Methyldihydroapo-β-erythroidinol	OH	Heat, 1.5 mm.	des-N,N-Dimethyldihydroapo-β-erythroidinol	55	334
Erysotrine					
Tetrahydroerysotrine	OH	120°, vac.		—	335

OLEFINS FROM AMINES

Starting material	Substituent	Conditions	Product	Ref.
Tetrahydroerythraline	OH	Heat, vac.	(structure with N–CH₃, CH=CH₂, methylenedioxy, CH₃O)	335
Apoerysopine	OSO_2OCH_3	Aq. base, heat	des-N-Methylapoerysotrine	336
des-Methylapoerysotrine	OSO_2OCH_3	Aq. base, heat	des-Dimethylapoerysotrine, $C_{20}H_{23}NO_2$	336

Note: References 194 to 391 are on pp. 489–493.

TABLE XVIII—Continued
HOFMANN ELIMINATION REACTION WITH ALKALOIDS

Name	Derivative	Conditions	Product	Yield, %	References
Gelsemine	Iodide	Aq. base, 240–250°, vac.	N(a)-Methylgelsemine	—	7, 116, 117
Dihydrogelsemine	Iodide	Aq. base, 240–250° vac.	N(a)-Methyldihydrogelsemine	—	7, 116, 117
Octahydrogelsemine	Iodide	Aq. base, 240–250° vac.	N(a)-Methyloctahydrogelsemine	—	7, 116
Gramine	Iodide	Methanol or aq. base		—	337

OLEFINS FROM AMINES

Granatanine

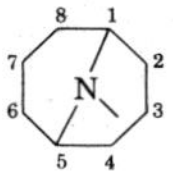 Compound	Group	Conditions	Product	Yield (%)	Refs.
N-Methylgranatanine	OH	Distil	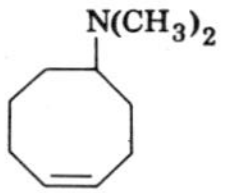 Δ⁴-des-Dimethylgranatanine	50	86, 338
Δ⁴-des-Dimethylgranatanine	OH	Distil	Mixture of cycloöctadienes	64	86, 339
Methylgranatoline (3-hydroxy-N-methylgranatanine)	OH	Distil	Mixture of products	—	338
Dimethylaminocycloöcta-2,4-diene	OH	Distil	Cycloöctatriene	—	339
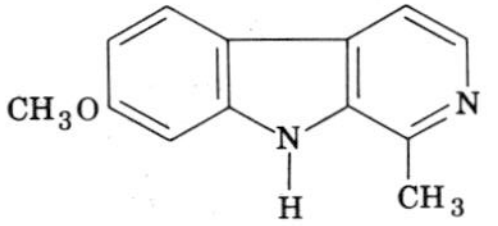 Isomethyltetrahydroharmine methohydroxide methyl ether	OSO₂OCH₃	Base, cold, then heat at 100°	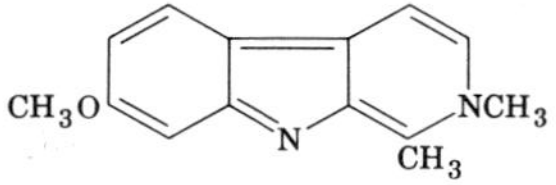 (structure shown)	—	340
(Harmine derivative)	OSO₂OCH₃	CH₃OH, base, boil	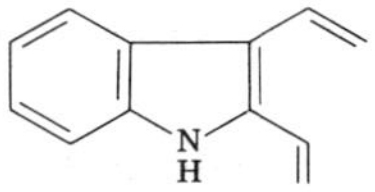 Polymer	—	98

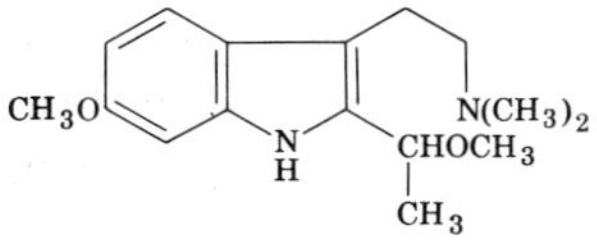

Note: References 194 to 391 are on pp. 489–493.

TABLE XVIII—Continued

HOFMANN ELIMINATION REACTION WITH ALKALOIDS

Name	Derivative	Conditions	Product	Yield, %	References
Hetisine $C_{20}H_{27}NO_3$	OH	160–200°/ 0.3 mm.	des-N-Methylhetisine	—	341
Dihydrohetisine, $C_{20}H_{29}NO_3$	OH	160–200°/ 0.3 mm.	Methine base	—	341
Hordenine ![HO-C6H4-CH2CH2-N(CH3)2]	OH	120–130°	![CH3O-C6H4-CH=CH2]	90	342
Hypaphorine ![indole-CH2CH(CO2⁻)N⁺(CH3)3]	Betaine	Aq. base, heat	Indole	—	343
Isoquinoline ![isoquinoline ring numbered 1-8, N at 2]					

(Several derivatives of tetrahydroisoquinoline are included that are converted to the corresponding phenethylamine methiodides prior to the Hofmann elimination.)

OLEFINS FROM AMINES

Starting material	Reagent	Product	Yield (%)	Ref.
Cotarnine / Cotarnine anil (The iodide is derived from the aldehyde form.)	Iodide, Base, heat	Cotarnone	—	131
Hydrastinine, 1-hydroxy-2-methyl-6,7-methylenedioxytetrahydroisoquinoline	Iodide, Aq. base, heat	Norcotarnone	—	131
	Iodide, Aq. base, heat	Hydrastal	—	344
Pellotine, 1,2-dimethyl-6,7-dimethoxy-8-hydroxy-1,2,3,4-tetrahydroisoquinoline (Emde degradation product)	Aq. base, reflux	1-Vinyl-2-ethyl-3-ethoxy-4,5-dimethoxybenzene	>90	345

Note: References 194 to 391 are on pp. 489–493.

464 ORGANIC REACTIONS

TABLE XVIII—*Continued*

HOFMANN ELIMINATION REACTION WITH ALKALOIDS

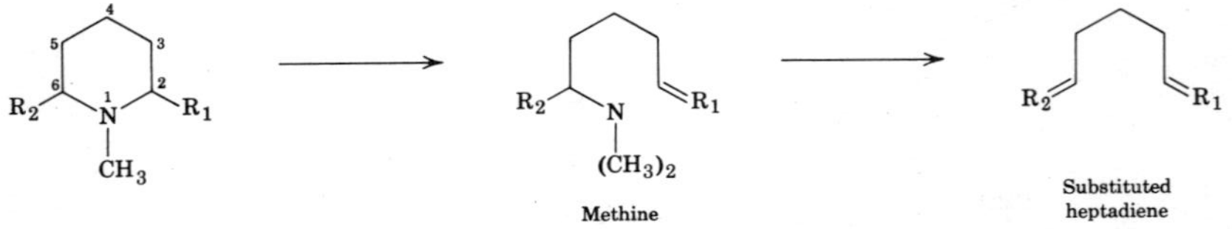

Name	Derivative	Conditions	Product	Yield, %	References
Laburnine 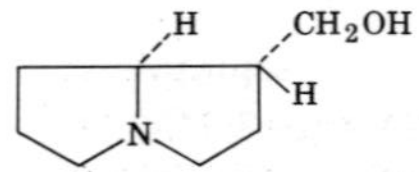			(3 degradations) N-Free product	—	346
Lobelia Alkaloids		OH	Methine → Substituted heptadiene	—	45
Lelobanine $R_1 = -CH_2CC_2H_5$ (=O) $R_2 = -CH_2CC_6H_5$ (=O)		Heat	$C_6H_5CH(CH_2)_7CHC_2H_5$ with OH groups (after hydrogenation)		
Lobelanidine $R_1 = R_2 = -CH_2CH-C_6H_5$ with OH		Heat, 270°	"N-Containing compound"	Poor	75

OLEFINS FROM AMINES

Lobelanine	OH	Heat	Methine	77
$R_1 = R_2 = -CH_2\overset{O}{\underset{\|}{C}}C_6H_5$				
——methine	OH	Cold	Substituted heptadiene	77
Isolobinanine, 3,4-dehydrolobanine	Iodide	Aq. bicarbonate, cold	1-Benzoyl-7-propionyl-heptatriene	45
Lobinine	Iodide	Aq. base	1-Benzoyl-7-propionyl-heptatriene	45
$R_1 = -CH_2CHCH_2CH_3$				
$\qquad\qquad\quad | $				
$\qquad\qquad\; OH$				
$R_2 = -CH_2COC_6H_5$ (3,4-dehydro)				
Lobinanine, 2,4-dehydrolobanine, lobinone	Iodide	Aq. bicarbonate, cold	$C_{17}H_{20}O_2$ (unsaturated diketone)	347
Lycorine	OH	100°, vac.	Methine	348, 349

Note: References 194 to 391 are on pp. 489–493.

TABLE XVIII—Continued

HOFMANN ELIMINATION REACTION WITH ALKALOIDS

Name	Derivative	Conditions	Product	Yield, %	References
Mavacurine E_2-Dihydromavacurine methochloride	Chloride	KOH, alcohol, heat	Methine	—	350
Morphine	α-Methine	Aq. base, heat	Morphenol or 13-Vinyl morphenol derivative	72	351
Codeine $R_1 = CH_3, R_2 = H$ —methine	O-Methyl iodide		Methyl codeinemethine		
OH		Aq. NaOH, reflux	No elimination	—	137

OLEFINS FROM AMINES

Acetocodeine (1-acetocodeine)	Iodide	Aq. base, heat	1-Acetomethine	70	352
——methine	OH	200°/0.1 mm.	1-Acetomethylmorphenol	Low	352
Bromocodeine (1-bromocodeine)	Iodide	Aq. base, heat	1-Bromomethine	87	353
——methine	OSO$_2$OCH$_3$	Aq. base, heat	1-Bromomethylmorphenol	—	353
Bromodesoxycodeine-C	OSO$_2$OCH$_3$	Aq. NaOH		—	353
Dihydrocodeine, R$_1$ = CH$_3$, R$_2$ = H, 7,8-dihydro					
——dihydromethine	Iodide	Aq. base, reflux	Dihydromethine	91	351
——tetrahydromethine	OH	140–190°/0.4 mm.	6-Hydroxy- and 6-methoxy-13-vinylhexahydromethylmorphenol	59	351
——tetrahydromethine	OH	140–190°/0.4 mm.	6-Hydroxy- and 6-methoxy-13-vinyloctahydromethylmorphenol	56	351
——tetrahydromethine	O-Methyl OH	140°/0.4 mm.	6-Methoxy-13-vinyloctahydromethylmorphenol	87	351

Note: References 194 to 391 are on pp. 489–493.

TABLE XVIII—Continued
HOFMANN ELIMINATION REACTION WITH ALKALOIDS

Name	Derivative	Conditions	Product	Yield, %	References
Morphine (Continued)			Morphenol or 13-Vinyl morphenol derivative		
Desoxycodeine-D	Iodide	Aq. base, heat	Methine	—	354
6-Methyldihydrocodeine	Iodide	Aq. base, heat	Methine	83	355

OLEFINS FROM AMINES

Compound	Group	Conditions	Product		References
Morphine $R_1 = R_2 = H$	OH	100°, vac.	6-Methyl-6-hydroxy-13-vinylhexahydromethylmorphenol ——methine	62	355
		Always degraded as methyl ether; see Codeine			
Isomorphine ——methine	Phenolic methyl ether iodide	Aq. base, heat	Methylisomorphimethine	—	184
Neopine	OH OSO$_2$OCH$_3$	160° Aq. base, heat	Methylmorphenol β-Codeinemethine	— —	184 356
Pseudocodeine	O-Methyl iodide	Aq. base, heat	Methine	—	357

Note: References 194 to 391 are on pp. 489–493.

TABLE XVIII—Continued
Hofmann Elimination Reaction with Alkaloids

Name	Derivative	Conditions	Product	Yield, %	References
Morphine (Continued)					
	α-Methine	Aq. base, heat	Dihydromethine (Morphenol or 13-Vinyl morphenol derivative)	—	358
	O-Methyl iodide				
	Iodide	NaOCH$_3$ CH$_3$OH, heat		—	359
6,7-Dihydropseudocodeine	OH	Reflux, isoamyl alcohol	Thebaol		
Thebaine			6-Methoxy-13-vinyl-tetrahydromethylmorphenol	30	138

OLEFINS FROM AMINES

Structure	Group	Conditions	Product	Yield (%)	Ref.
Dihydrodesoxycodeine-A	OH	140°/1 μ	des-N-Methyl base	79	179
(CH₃)₂N... Desoxytetrahydro-α-codeinemethine	Iodide	Aq. base, heat		—	360
	O-Methyl iodide	Base, amyl alcohol, reflux	Recovered tertiary amine	85	360
Dihydropseudocodeine-B	Iodide	Aq. base, heat	Methine base	—	361

Note: References 194 to 391 are on pp. 489–493.

TABLE XVIII—Continued
HOFMANN ELIMINATION REACTION WITH ALKALOIDS

Name	Derivative	Conditions	Product	Yield, %	References
Morphine (Continued)	Dihydropseudocodeine-C	Iodide	Aq. base, heat	α-Methine → Morphenol or 13-Vinyl morphenol derivative	—	361
		Iodide	Aq. base, heat	Methine	—	362

OLEFINS FROM AMINES

Methyldihydrothebaine

				Yield %	Ref.
δ-Isomer	Iodide	Aq. base, heat	Isomethine	85	363
O-Acetyl δ-isomer	OH	98°, vac.	Methine	12	363
O-Acetyl α-isomer	OH	98°, vac.	Isomethine	42	363
Methyldihydrothebaine methine	Iodide	Aq. base, heat		90	363
δ-Methyldihydrothebaine isomethine	OH	Aq. base, heat	Vinyldihydromethylthebaol	Low	363

Note: References 194 to 391 are on pp. 489–493.

TABLE XVIII—Continued
HOFMANN ELIMINATION REACTION WITH ALKALOIDS

Name	Derivative	Conditions	Product	Yield, %	References
Methyldihydrothebaine (Continued)			Methine / Isomethine	—	
Thebaizone (α)	OH	Heat	des-N-Methylthebaizone acid, $C_{19}H_{21}O_5N$	—	364
Dihydrodesoxythebaizone	OH	Heat	des-N-Methyldihydrodesoxy-thebaizone acid, $C_{19}H_{23}O_4N$	—	364

Compound	Group	Conditions	Product	Ref.	Ref.
Narcidonine	Iodide	NaOC₂H₅, heat	Narcidone	40	365
Physostigmine (eserine)	OH	Reflux, aq. base	Eseretole methine	—	366, 99
Dehydroeseretholemethine	OH	200°		—	99

Note: References 194 to 391 are on pp. 489–493.

TABLE XVIII—*Continued*

HOFMANN ELIMINATION REACTION WITH ALKALOIDS

Name	Derivative	Conditions	Product	Yield, %	References
Protoberberines					
Canadine, (tetrahydroberberine), 2,3-methylenedioxy-9,10-dimethoxyprotoberberine	N-Benzyl chloride	Alcoholic KOH	des-N-Benzyl base A	—	367
des-N-Benzyl base	OH Iodide	100°, vac. Alcoholic KOH	Mixture of des-N-methyl bases des-N-Benzyl-N-methyl base	— —	368 367
Corybulbine, 2,9,10-trimethoxy-3-hydroxy-13-methyl-protoberberine	O-Ethyl chloride	CH_3OH, KOH heat	des-N-Methyl base B	—	181

Corydaline, 2,3,9,10-tetramethoxy-13-methyl protoberberine	Chloride	Base, distil	des Base A from "meso" material; des base B from racemic material	—	78
	Chloride	CH_3OH, KOH, heat	des-N-Methyl base	—	182
Isocryptopine chloride, 2,3-dimethoxy-9,10-methylenedioxy-13,14-dehydro-N-methyl-protoberberine chloride	Chloride	CH_3OH, KOH, heat	des-N-Methyl base A	—	369
Dihydroisocryptopine chloride, 13,14-dihydro-isocryptopine -des-N-Methyl bases	Chloride	CH_3OH, KOH, heat	des-N-methyl bases A and B	—	369
	OSO_2OCH_3	CH_3OH, KOH, heat	des-N,N-Dimethyl base	—	369
Thalictricavine, 2,3-methylenedioxy-9,10-dimethoxy-13-methyl-protoberberine	Chloride	Base, distil	des-N-Methyl bases A or B depending on isomer of starting material used	—	78

Note: References 194 to 391 are on pp. 489–493.

TABLE XVIII—Continued
HOFMANN ELIMINATION REACTION WITH ALKALOIDS

Name	Derivative	Conditions	Product			Yield, %	References
Protopine			A	Methines	C		
Protopine, 2,3-methylenedioxy-9,10-methylenedioxy	OSO$_2$OCH$_3$	CH$_3$OH, KOH, heat		A and C methines		—	369
Cryptopine, 2,3-dimethoxy-9,10-methylenedioxy	OSO$_2$OCH$_3$	CH$_3$OH, KOH, heat		A and C methines		—	369
Anhydrodihydrocryptopine-A	OSO$_2$OCH$_3$	Base, CH$_3$OH, heat		des-N-Methylisoanhydrodihydrocryptopine		—	369

OLEFINS FROM AMINES

Compound	Substituent	Conditions	Product	Yield (%)	Ref.
Tetrahydroanhydrocryptopine-B	OSO₂OCH₃	Base, CH₃OH heat	des-N-Methylisoanhydro-dihydrocryptopine	—	369
	OH	Aq. base, heat	des-N-Methyltetrahydroanhydrocryptopine	—	370

Quinolizidine

Lupinine, 5-hydroxymethylquinolizidine	OH	165–170°/15 mm.	des-N-Methyllupinine	74	371
	OH	Heat, vac. 200°/20 mm.	des-N-Methyllupinine	90	372
des-N-Methyllupinine	OH		des-N-dimethyllupinine	—	371

Note: References 194 to 391 are on pp. 489–493.

480　ORGANIC REACTIONS

TABLE XVIII—Continued
HOFMANN ELIMINATION REACTION WITH ALKALOIDS

Name	Derivative	Conditions	Product	Yield, %	References
Quinolizidine (Continued)					
Dihydro-des-N-methyllupinine	OH	Distil, 180°/12 mm.	Dihydro-des-N,N-dimethyl-lupinine	83	372
des-N,N-Dimethyl-lupinine	OH	Distil, vac.	Unsaturated alcohol	—	371
Tetrahydro-des-N,N-dimethyllupinine	OH	120°/15 mm.	Unsaturated alcohol	46	372
5-Benzoylquinolizidine	Iodide	Aq. NaOH, heat	(structure: COC₆H₅, N-CH₃ ring)	Quant.	80
Scopoline (oscine)	O-Methyl OH	160°/13 mm.	$C_9H_{15}NO_2$, des-N-Methyl-scopolines (mixture of isomers)	—	373, 374

Sparteine

Anagyrine, 2-keto-3,4,5,6-dehydrosparteine	OH	Benzene, heat	Anagyrine methine	—	375
Dihydroanagyrine methine	OH	120°/10 mm.	Dihydroanagyrine bismethine	—	375
Tetrahydroanagyrine bismethine	OH	120°/10 mm.	Tetrahydroanagyrine tris-methine	—	375
Aphyllidine, 5,6-dehydro-10-ketosparteine	OH	Heat, vac.	des-N-Methylaphyllidine, $C_{16}H_{24}N_2O$	93	189
	Iodide	Base, CH_3OH, reflux	des-N-Methylaphyllidine	80	376
des-N-Methylaphyllidine	OH	Heat, vac.	des-N,N-Dimethylaphyllidine, $C_{17}H_{26}N_2O$	—	189
	Iodide	Base, CH_3OH, reflux	des-N,N-Dimethylaphyllidine	—	376
des-N,N-Dimethylaphyllidine	OH	250°/11 mm.	Hemiaphyllidylene, $C_{15}H_{19}NO$	—	189
	OH	CH_3OH, distil	Hemiaphyllidylene	—	376
Aphylline, 10-ketosparteine	Iodide, OH	Base, heat, vac.	des-N-Methylaphylline, $C_{16}H_{26}N_2O$	80	189

Note: References 194 to 391 are on pp. 489–493.

TABLE XVIII—*Continued*

HOFMANN ELIMINATION REACTION WITH ALKALOIDS

Name	Derivative	Conditions	Product	Yield, %	References
Sparteine (Continued)					
des-N-Methylaphylline	OH	Distil, vac.	des-N,N-Dimethylaphylline, $C_{17}H_{28}N_2O$	73	189
des-N,N-Dimethylaphylline	OH	Heat, vac.	Hemiaphylline, $C_{15}H_{21}NO$	—	189
Sparteine	OH	Heat, vac. (6 degradations)	Nitrogen-free product	—	377, 378
	OH	N_2, 40–50°, vac.	α- and β-des-N-Methylsparteine	(α) 45–55	379
Oxysparteine (isolupanine), 17-ketosparteine	OH	Heat	des-N-Methyloxysparteine, $C_{16}H_{26}ON_2$	98	380
des-N-Methyloxysparteine	OH	170°/0.05 mm.	des-N,N-Dimethyloxysparteine, $C_{17}H_{29}ON_2$	65	380
Dihydro-des-N-methyloxysparteine	OH	Heat	Dihydro-des-N-dimethyloxysparteine	—	380
Tetrahydro-des-N,N-dimethyloxysparteine	OH	150°	Tetrahydrohemioxyspartylene, $C_{15}H_{25}ON$	49	380

OLEFINS FROM AMINES

Strychnine

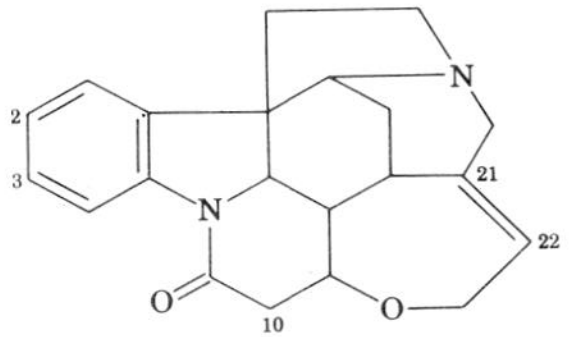

Dihydrostrychnidine-A, 21,22-dihydro-10-desoxystrychnine -des-Base A	Hydrogen carbonate	250°	—	des-Base D plus methyl-*chano*-dihydro-neostrychnidine	134
	Hydrogen carbonate, or chloride	Heat, NaOCH₃	—	Dimethyl-des-strychnidine-D plus dimethyl-des-neostrychnidine	381, 382
Dimethyl-des-strychnidine-D Dihydrobrucidine, 2,3-dimethoxy-21,22-dihydro-10-desoxystrychnine	Chloride Hydrogen carbonate	— Heat	— —	des-aza-Strychnidine-a and -b N-(b)-Methyl-des-dihydrobrucidine-a and -b	382 135

Tazettine

| 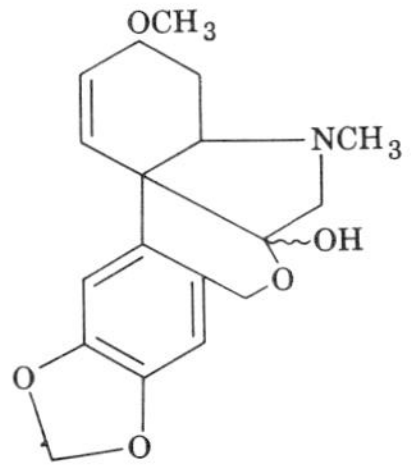 | OH | 100°, vac. | 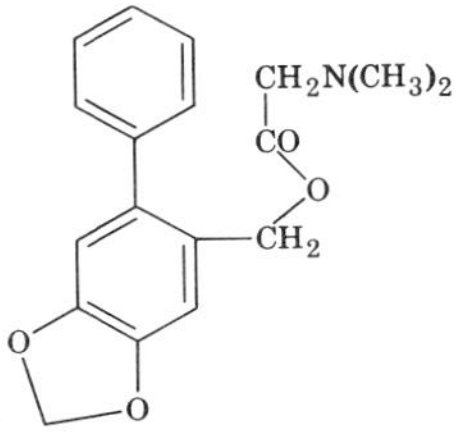 — | | 383 |

Note: References 194 to 391 are on pp. 489–493.

TABLE XVIII—Continued
HOFMANN ELIMINATION REACTION WITH ALKALOIDS

Name	Derivative	Conditions	Product	Yield, %	References
Tropane					
Ecgonine $R_1 = $ —OH $R_2 = $ —CO_2H	Ethyl ester iodide	Aq. base, heat	Cycloheptatriene carboxylic acid, $C_8H_8O_2$	—	384
Anhydroecgonine (ecgonidine) $R_1 = $ H $R_2 = CO_2H$	Ethyl ester iodide	Aq. base, heat	Cycloheptatrienecarboxylic acid, $C_8H_8O_2$	—	385
Dihydroanhydroecgonine	Ethyl ester iodide	Aq. base, heat	Cycloheptadienecarboxylic acid	—	386, 387
Hydroecgonidine $R_1 = $ H $R_2 = CO_2H$	Ethyl ester iodide	Aq. base	$(CH_3)_2N$—[ring]—CO_2H 2-Carboxy-5-dimethyl-aminocycloheptene	—	387
	Ethyl ester iodide	K_2CO_3, 75°	2-Carboxy-5-dimethyl-aminocycloheptene	60	386

OLEFINS FROM AMINES

2-Carbethoxy-5-dimethylamino-cycloheptene	Iodide	Base, heat	Cycloheptadienecarboxylic acid	85	386, 387
3,4-Dehydrotropidine	OH	Aq. base, distilled	Dimethylaminocycloheptadiene "methyl tropidine"	90	85
Hydrotropidine $R_1 = R_2 = H$	OH	100°	Dimethylaminocycloheptene "methyl hydrotropidine"	—	388
Methylhydrotropidine $R_1 = R_2 = H$	OH	100°	Cycloheptadiene	82	388
Methyltropidine	OH	Heat, aq. soln.	Cycloheptatriene	—	85
Tropinone $R_1 = O$, $R_2 = H$	OH iodide	Heat, base	Mixture of cycloheptadienones	80	389, 130
Tropinic acid	Dimethyl ester iodide	Aq. base	"des-N-Methyltropinic acid, dimethyl ester"	90	390
des-N-Methyltropinic acid dimethyl ester	Iodide	Aq. base	Unsaturated dicarboxylic acid	75	390
Yohimbine	OH	Distil, vac.	Methylyohimboic acid, $C_{21}H_{26}N_2O_3$	—	391

Note: References 194 to 391 are on pp. 489–493.

TABLE XVIII—Continued

HOFMANN ELIMINATION REACTION WITH ALKALOIDS

Name	Derivative	Conditions	Product	Yield, %	References
chano-Desoxyyohimbol	OH	Heat, vac.	(trans)	—	81
chano-Dihydrodesoxyyohimbol	Bicarbonate	180°/30 mm.		79	81

TABLE XIX

List of Alkaloids by Type

The parenthesized number following each entry in the second column indicates the page in Table XVIII on which each type of alkaloid first appears.

Alkaloid	Listed Under
Actinodaphnine	Aporphine (433)
Anagyrine	Sparteine (481)
Anhydrocryptopine	Protopine (478)
Anolobine	Aporphine (433)
Anonaine	Aporphine (433)
Aphyllidine	Sparteine (481)
Aphylline	Sparteine (481)
Apoerysopine	Erysotrine (458)
Armepavine	Benzylisoquinoline (436)
Bebeerine	Bisbenzylisoquinoline-C (442)
Berbamine	Bisbenzylisoquinoline-B' (440)
Boldine	Aporphine (433)
Brucidine	Strychnine (483)
Canadine	Protoberberine (476)
Cephaeline	Cephaeline (447)
Cepharanthine	Bisbenzylisoquinoline-B (438)
Chondodendrine, see Bebeerine	
Chondrofoline	Bisbenzylisoquinoline-C (442)
Coclaurine	Benzylisoquinoline (436)
Codeine	Morphine (466)
Colchinol	Colchinol (450)
Conhydrine	Coniine (450)
Coniine	Coniine (450)
Corybulbine	Protoberberine (476)
Corydaline	Protoberberine (476)
Cotarnine	Isoquinoline (462)
Crebanine	Aporphine (433)
Cryptopine	Protopine (478)
Cularine	Cularine (451)
C-Curarine	C-Curarine (451)
Cuscohygrine	Cuscohygrine (451)
Cytisine	Cytisine (452)
Daphnandrine	Bisbenzylisoquinoline-B (438)
Daphnoline	Bisbenzylisoquinoline-B (438)
Dauricine	Bisbenzylisoquinoline-A (437)
Delphinine	Delphinine (453)
Dicentrine	Aporphine (433)
Dioscorine	Dioscorine (454)
Ecgonidine	Tropane (484)
Ecgonine	Tropane (484)
Emetine	Cephaeline (447)
Epistephanine	Bisbenzylisoquinoline-B (438)

TABLE XIX—Continued

LIST OF ALKALOIDS BY TYPE

Alkaloid	Listed Under
Ergothionene	Ergothionene (454)
Erysotrine	Erysotrine (458)
α-Erythroidine	Erythroidine (454)
β-Erythroidine	Erythroidine (454)
Eserethole	Physostigmine (475)
Eserine, see Physostigmine	
Gelsemine	Gelsemine (460)
Glaucine	Aporphine (433)
Gramine	Gramine (460)
Granatanine	Granatanine (461)
Harmine	Harmine (461)
Hetisine	Hetisine (462)
Homotrilobine, see Isotrilobine	
Hordenine	Hordenine (462)
Hydrastinine	Isoquinoline (462)
Hypaphorine	Hypaphorine (462)
Isochondodendrine	Bisbenzylisoquinoline-D (443)
Isocryptopine chloride	Protoberberine (476)
Isolobinanine	Lobelia Alkaloids (464)
Isolupanine, see Oxysparteine	
Isomorphine	Morphine (466)
Isotetrandrine, see Pheanthine	
Isothebaine	Aporphine (433)
Isotrilobine	Bisbenzylisoquinoline-F (446)
Laburnine	Laburnine (464)
Laudenine	Benzylisoquinoline (436)
Laureline	Aporphine (433)
Laurotetanine	Aporphine (433)
Lelobanine	Lobelia alkaloids (464)
Lobelanidine	Lobelia alkaloids (464)
Lobelanine	Lobelia alkaloids (464)
Lobinanine	Lobelia alkaloids (464)
Lobinine	Lobelia alkaloids (464)
Lobinone, see Lobinanine	
Lupinine	Quinolizidine (479)
Lycorine	Lycorine (465)
Mavacurine	Mavacurine (466)
Menisarine	Bisbenzylisoquinoline-E (445)
Micranthine	Bisbenzylisoquinoline-E (446)
Morphine	Morphine (466)
Narcidonine	Narcidonine (475)
Neopine	Morphine (466)
Neoprotocuridine	Bisbenzylisoquinoline-D (443)
Oscine, see Scopoline	
Oxyacanthine	Bisbenzylisoquinoline-B (438)

TABLE XIX—Continued
List of Alkaloids by Type

Alkaloid	Listed Under
Oxysparteine	Sparteine (481)
Pellotine	Isoquinoline (462)
Pheanthine	Bisbenzylisoquinoline-B' (440)
Physostigmine	Physostigmine (475)
Protopine	Protopine (478)
Pseudocodeine	Morphine (466)
Pseudoconhydrine	Coniine (450)
Pukateine	Aporphine (433)
Repandine	Bisbenzylisoquinoline-B (438)
Scopoline	Scopoline (480)
Sparteine	Sparteine (481)
Strychnine (dihydrostrychnidine-A)	Strychnine (483)
Tazettine	Tazettine (483)
Tetrahydroberberine, see Canadine	
Tetrahydroerythraline	Erysotrine (458)
Tetrandrine	Bisbenzylisoquinoline-B' (440)
Thalictricavine	Protoberberine (476)
Thebaine	Morphine (466)
Thebaizone (α)	Morphine (466)
Trilobamine, see Daphnoline	
Trilobine	Bisbenzylisoquinoline-F (446)
Tropidine	Tropane (484)
Tropinic acid	Tropane (484)
Tropinone	Tropane (484)
Tubocurarine chloride	Bisbenzylisoquinoline-C (442)
Tuduranine	Aporphine (433)
Yohimbine	Yohimbine (485)

REFERENCES FOR TABLES XI-XIX

[194] Bersch and Hübner, *Arch. Pharm.*, **289**, 673 (1956).
[195] Wilson and Read, *J. Chem. Soc.*, **1935**, 1269.
[196] von Braun and Schirmacher, *Ber.*, **56**, 1845 (1923).
[197] von Braun, *Ber.*, **56**, 2178 (1923).
[198] Rabe and Hallensleben, *Ber.*, **43**, 2622 (1910).
[199] Späth and Adler, *Monatsh.*, **63**, 127 (1933).
[200] Sicher and Lichy, *Chem. & Ind. (London)*, **1958**, 16.
[201] Hill, *J. Am. Chem. Soc.*, **80**, 1609 (1958).
[202] Späth, Kuffner, and Ensfellner, *Ber.*, **66**, 591 (1933).
[203] Schmidt, *Arch. Pharm.*, **253**, 52 (1915).
[204] Rabe, *Ber.*, **44**, 824 (1911).
[204a] Curtin, Harris, and Pollak, *J. Am. Chem. Soc.*, **73**, 3453 (1957).
[205] Willstätter and Bruce, *Ber.*, **40**, 3979 (1907).
[206] Demjanow and Dojarenko, *Ber.*, **55**, 2727 (1922).
[207] Harries and Morrell, *Ann.*, **410**, 70 (1915).
[208] Cook, Dickson, and London, *J. Chem. Soc.*, **1947**, 746.

[209] A. C. Cope and C. F. Howell, to be published.
[210] C. L. Bumgardner, Ph.D. Thesis, Massachusetts Institute of Technology, 1956.
[211] Collie and Schryver, *J. Chem. Soc.*, **57,** 767 (1890).
[212] Howton, *J. Org. Chem.*, **14,** 1 (1949).
[213] Willstätter, *Ber.*, **35,** 584 (1902).
[214] Hurd and Drake, *J. Am. Chem. Soc.*, **61,** 1943 (1939).
[215] Willstätter and Schmaedel, *Ber.*, **38,** 1992 (1905).
[216] Mannich and Margotte, *Ber.*, **68,** 273 (1935).
[217] Konowalowa and Magidson, *Arch. Pharm.*, **266,** 449 (1928).
[218] Willstätter and Heubner, *Ber.*, **40,** 3869 (1907).
[219] Mannich, Handke, and Roth, *Ber.*, **69,** 2112 (1936).
[220] Willstätter and Wirth, *Ber.*, **46,** 535 (1913).
[221] Buchman, Schlatter, and Reims, *J. Am. Chem. Soc.*, **64,** 2701 (1942).
[222] Avram, Nenitzescu, and Marica, *Chem. Ber.*, **90,** 1857 (1957).
[223] Knorr and Rothe, *Ber.*, **39,** 1420 (1906).
[224] Brown and Wheeler, *J. Am. Chem. Soc.*, **78,** 2199 (1956).
[225] Demjanow and Dojarenko, *Ber.*, **55,** 2718 (1922).
[226] Lukes and Pliml, *Collection Czechoslov. Chem. Communs.*, **21,** 625 (1956).
[227] Howton and Buchman, *J. Am. Chem. Soc.*, **78,** 4011 (1956).
[228] Applequist and Roberts, *J. Am. Chem. Soc.*, **78,** 4012 (1956).
[229] Euler, *J. prakt. Chem.*, [2] **57,** 131 (1898).
[230] Euler, *Ber.*, **30,** 1989 (1897).
[231] Willstätter and Hatt, *Ber.*, **45,** 1464 (1912).
[232] Nazarov and Kuznezov, *J. Gen. Chem.* (U.S.S.R.), **29,** 767 (1959).
[233] Kohn, *Ann.*, **351,** 134 (1907).
[234] Cope and Schweizer, *J. Am. Chem. Soc.*, **81,** 4577 (1959).
[235] Kohler, Tishler, Potter, and Thompson, *J. Am. Chem. Soc.*, **61,** 1057 (1939).
[236] Vogel, *Fortsch. chem. Forsch.*, **3,** 469 (1955).
[237] Lukes, Strouf, and Ferles, *Collection Czechoslov. Chem. Communs.*, **23,** 326 (1958).
[238] Stewart and Aston, *J. Am. Chem. Soc.*, **49,** 1718 (1927).
[239] Parham, Hunter, Hanson, and Lahr, *J. Am. Chem. Soc.*, **74,** 5646 (1952).
[240] Alder, Hartung, and Netz, *Chem. Ber.*, **90,** 1 (1957).
[241] Seka and Tramposch, *Ber.*, **75,** 1379 (1942).
[242] King, Barltrop, and Walley, *J. Chem. Soc.*, **1945,** 277.
[243] Ayerst and Schofield, *J. Chem. Soc.*, **1958,** 4097.
[244] von Braun, *Ber.*, **42,** 2532 (1909).
[245] Menshikoff, *Ber.*, **68,** 1555 (1935).
[246] Menshikoff, *Bull. acad. sci. U.R.S.S., Classe sci. math., Sér. chim.*, **5,** 1035 (1937) [*C.A.,* **32,** 2944 (1938); *Chem. Zentr.*, **110, I,** 2790 (1939)].
[247] Menshikoff, *Ber.*, **69,** 1802 (1936).
[248] Willstätter and Waser, *Ber.*, **44,** 3423 (1911).
[249] Willstätter and Heidelberger, *Ber.*, **46,** 517 (1913).
[250] Cope and Overberger, *J. Am. Chem. Soc.*, **70,** 1433 (1948).
[251] Senfter and Tafel, *Ber.*, **27,** 2309 (1894).
[252] King and Booth, *J. Chem. Soc.*, **1954,** 3798.
[253] Prelog and Seiwerth, *Ber.*, **72,** 1638 (1939).
[254] Koenigs and Bernhart, *Ber.*, **38,** 3049 (1905).
[255] von Braun and Kirschbaum, *Ber.*, **52,** 2261 (1919).
[256] von Braun and Teuffert, *Ber.*, **62,** 235 (1929).
[257] Blomquist, Burge, and Sucsy, *J. Am. Chem. Soc.*, **74,** 3636 (1952).
[257a] A. C. Cope and P. N. Jenkins, to be published.
[258] McNiven and Read, *J. Chem. Soc.*, **1952,** 159.
[259] Willstätter and King, *Ber.*, **46,** 527 (1913).
[260] Shriner and Sutherland, *J. Am. Chem. Soc.*, **60,** 1314 (1938).
[261] Ruzicka, *Helv. Chim. Acta*, **3,** 748 (1920).
[262] Ruzicka and Trebler, *Helv. Chim. Acta*, **3,** 756 (1920).

[263] Ruzicka and Pontalti, *Helv. Chim. Acta*, **7**, 489 (1924).
[264] Nametkin and Zabrodin, *Ber.*, **61**, 1491 (1928).
[265] Wittig and Behnisch, *Chem. Ber.*, **91**, 2358 (1958).
[266] Schotten, *Ber.*, **15**, 421 (1882).
[267] Claus and Hirzel, *Ber.*, **19**, 2785 (1886).
[268] Mannich and Davidsen, *Ber.*, **69**, 2106 (1936).
[269] von Braun and Cahn, *Ann.*, **436**, 262 (1924).
[270] Parham, Hunter, and Hanson, *J. Am. Chem. Soc.*, **73**, 5068 (1951).
[271] Bruckner, Kovacs, and Kovacs, *Ber.*, **77**, 610 (1944).
[272] Karrer and Karanth, *Helv. Chim. Acta*, **33**, 2202 (1950).
[273] Stevens, Sneeden, Stiller, and Thomson, *J. Chem. Soc.*, **1930**, 2119.
[274] von Braun and Hamann, *Ber.*, **65**, 1580 (1932).
[275] Reichert and Hoffmann, *Arch. Pharm.*, **274**, 153 (1936).
[276] Dunn and Stevens, *J. Chem. Soc.*, **1934**, 279.
[277] Gensler and Samour, *J. Am. Chem. Soc.*, **74**, 2959 (1952).
[278] Battersby and Binks, *J. Chem. Soc.*, **1955**, 2896.
[279] Robinson and Sugasawa, *J. Chem. Soc.*, **1932**, 789.
[280] MacPhillamy and Scholz, *J. Org. Chem.*, **14**, 643 (1949).
[280a] Wittig, Koenig, and Clauss, *Ann.*, **593**, 127 (1955).
[281] von Braun, *Ber.*, **49**, 2629 (1916).
[282] Kröhnke and Meyer-Delius, *Chem. Ber.*, **84**, 941 (1951).
[283] Lipp, Buchkremer, and Seeles, *Ann.*, **499**, 1 (1932).
[284] Manske, *Can. J. Research*, **16B**, 76 (1938).
[285] Barger and Weitnauer, *Helv. Chim. Acta*, **22**, 1036 (1939).
[286] Tomita and Shirai, *J. Pharm. Soc. Japan*, **63**, 532 (1943) [*C.A.*, **45**, 3401a (1951)].
[287] Manske, *Can. J. Research*, **8B**, 592 (1933).
[288] Warnat, *Ber.*, **58**, 2768 (1925).
[289] Warnat, *Ber.*, **59**, 85 (1926).
[290] Klee, *Arch. Pharm.*, **252**, 211 (1914).
[291] Schlittler and Müller, *Helv. Chim. Acta*, **31**, 1119 (1948).
[292] Barger and Girardet, *Helv. Chim. Acta*, **14**, 481 (1931).
[293] Barger, Eisenbrand, Eisenbrand, and Schlittler, *Ber.*, **66**, 450 (1933).
[294] Späth and Tharrer, *Ber.*, **66**, 583 (1933).
[295] Späth and Suominen, *Ber.*, **66**, 1344 (1933).
[296] Goto, *Ann.*, **521**, 175 (1936).
[297] Marion, Lemay, and Portelance, *J. Org. Chem.*, **15**, 216 (1950).
[298] Kondo and Kondo, *J. prakt. Chem.*, **126**, 24 (1930).
[299] Schöpf and Thierfelder, *Ann.*, **537**, 142 (1939).
[300] Kondo and Keimatsu, *Ber.*, **71**, 2553 (1938).
[301] Bick, Ewen, and Todd, *J. Chem. Soc.*, **1949**, 2767.
[302] Kondo and Tomita, *J. Pharm. Soc. Japan*, **55**, 646 (1935) [*C.A.*, **33**, 627^4 (1939)].
[303] Kondo and Tanaka, *J. Pharm. Soc. Japan*, **63**, 273 (1943) [*C.A.*, **45**, 3400b (1951)].
[304] von Bruchhausen and Schultze, *Arch. Pharm.*, **267**, 567 (1929).
[305] Bick and Todd, *J. Chem. Soc.*, **1948**, 2170.
[306] Santos, *Ber.*, **65**, 472 (1932).
[307] Kondo and Yano, *Ann.*, **497**, 90 (1932).
[308] King, *J. Chem. Soc.*, **1939**, 1157.
[309] King, *J. Chem. Soc.*, **1935**, 1381.
[310] King, *J. Chem. Soc.*, **1936**, 1276.
[311] King, *J. Chem. Soc.*, **1940**, 737.
[312] King, *J. Chem. Soc.*, **1948**, 265.
[313] Faltis and Dieterich, *Ber.*, **67**, 231 (1934).
[314] King, *J. Chem. Soc.*, **1937**, 1472.
[315] Bick and Todd, *J. Chem. Soc.*, **1950**, 1606.
[316] Kondo and Tomita, *J. Pharm. Soc. Japan*, **55**, 637 (1935) [*C.A.*, **33**, 626^4 (1939)].
[317] Kondo and Tomita, *Ann.*, **497**, 104 (1932).

[318] Tomita and Tani, *J. Pharm. Soc. Japan*, **62**, 468 (1942) [*C.A.*, **45**, 4728e (1951)].
[319] Pailer and Porschinski, *Monatsh.*, **80**, 101 (1949).
[320] Pailer, *Monatsh.*, **79**, 127 (1948).
[321] Windaus, *Ann.*, **439**, 59 (1924).
[322] Hofmann, *Ber.*, **14**, 705 (1881).
[323] Philipsborn, Schmid, and Karrer, *Helv. Chim. Acta*, **38**, 1067 (1955).
[324] Hess and Bappert, *Ann.*, **441**, 137 (1925).
[325] Späth and Galinovsky, *Ber.*, **66**, 1338 (1933).
[326] Schneider, *Arch. Pharm.*, **283**, 86 (1950).
[327] Gorter, *Rec. trav. chim.*, **30**, 161 (1911).
[328] Heath, Lawson, and Rimington, *J. Chem. Soc.*, **1951**, 2215.
[329] Barger and Ewins, *J. Chem. Soc.*, **99**, 2336 (1911).
[330] Godfrey, Tarbell, and Boekelheide, *J. Am. Chem. Soc.*, **77**, 3342 (1955).
[331] Boekelheide and Morrison, *J. Am. Chem. Soc.*, **80**, 3905 (1958).
[332] Boekelheide, Weinstock, Grundon, Sauvage, and Agnello, *J. Am. Chem. Soc.*, **75**, 2550 (1953).
[333] Grundon, Sauvage, and Boekelheide, *J. Am. Chem. Soc.*, **75**, 2541 (1953).
[334] Grundon and Boekelheide, *J. Am. Chem. Soc.*, **75**, 2537 (1953).
[335] Kenner, Khorana, and Prelog, *Helv. Chim. Acta*, **34**, 1969 (1951).
[336] Folkers, Koniuszy, and Shavel, *J. Am. Chem. Soc.*, **73**, 589 (1951).
[337] Madinaveitia, *J. Chem. Soc.*, **1937**, 1927.
[338] Willstätter and Veraguth, *Ber.*, **38**, 1984 (1905).
[339] Willstätter and Veraguth, *Ber.*, **38**, 1975 (1905).
[340] Perkin and Robinson, *J. Chem. Soc.*, **115**, 933 (1919).
[341] Jacobs and Heubner, *J. Biol. Chem.*, **170**, 189 (1947).
[342] Léger, *Compt. rend.*, **144**, 488 (1907).
[343] Van Romburgh, *Koninkl. Akad. Wentenschap. Amsterdam, Wisk. Natuurk, Afdel.*, **19**, 1250 [*Chem. Zentr.*, **82, I**, 1548 (1911)].
[344] Freund, *Ber.*, **22**, 2329 (1889).
[345] Späth and Boschan, *Monatsh.*, **63**, 141 (1933).
[346] Galinovsky, Goldberger, and Pohm, *Monatsh.*, **80**, 550 (1949).
[347] Wieland and Ishimasa, *Ann.*, **491**, 14 (1931).
[348] Kondo, Katsura, and Uyeo, *Ber.*, **71**, 1529 (1938).
[349] Humber, Kondo, Kotera, Takagi, Takeda, Taylor, Thomas, Tsuda, Tsukamoto, Uyeo, Yajima, and Yanaihara, *J. Chem. Soc.*, **1954**, 4622.
[350] Bickel, Schmid, and Karrer, *Helv. Chim. Acta*, **38**, 649 (1955).
[351] Rapoport, *J. Org. Chem.*, **13**, 714 (1948).
[352] Small and Mallonee, *J. Org. Chem.*, **12**, 558 (1947).
[353] Small and Turnbull, *J. Am. Chem. Soc.*, **59**, 1541 (1937).
[354] Small and Mallonee, *J. Org. Chem.*, **5**, 350 (1940).
[355] Small and Rapoport, *J. Org. Chem.*, **12**, 284 (1947).
[356] Van Duin, Robinson, and Smith, *J. Chem. Soc.*, **1926**, 903.
[357] Small and Lutz, *J. Am. Chem. Soc.*, **57**, 361 (1935).
[358] Lutz and Small, *J. Am. Chem. Soc.*, **54**, 4715 (1932).
[359] Knorr, *Ber.*, **37**, 3499 (1904).
[360] Cahn, *J. Chem. Soc.*, **1926**, 2562.
[361] Lutz and Small, *J. Am. Chem. Soc.*, **56**, 1741 (1934).
[362] Lutz and Small, *J. Am. Chem. Soc.*, **56**, 2466 (1934).
[363] Small and Fry, *J. Org. Chem.*, **3**, 509 (1939).
[364] Wieland and Small, *Ann.*, **467**, 17 (1928).
[365] Freund and Oppenheim, *Ber.*, **42**, 1084 (1909).
[366] Polonovski and Polonovski, *Bull. soc. chim. France*, [4] **23**, 335 (1918).
[367] McDavid, Perkin, and Robinson, *J. Chem. Soc.*, **101**, 1218 (1912).
[368] Pyman, *J. Chem. Soc.*, **103**, 817 (1913).
[369] Perkin, *J. Chem. Soc.*, **109**, 815 (1916).
[370] von Bruchhausen and Bersch, *Ber.*, **63**, 2520 (1930).

[371] Willstätter and Fourneau, *Ber.*, **35,** 1910 (1902).
[372] Karrer, Canal, Zohner, and Widmer, *Helv. Chim. Acta*, **11,** 1062 (1928).
[373] Hess, *Ber.*, **52,** 1947 (1919).
[374] Hess and Wahl, *Ber.*, **55,** 1979 (1922).
[375] Ing, *J. Chem. Soc.*, **1933,** 504.
[376] Orechoff and Norkina, *Ber.*, **67,** 1974 (1934).
[377] Schirm and Bisendorf, *Arch. Pharm.*, **280,** 64 (1942).
[378] Karrer, Shibata, Wettstein, and Jacobowicz, *Helv. Chim. Acta*, **13,** 1292 (1930).
[379] Schöpf and Braun, *Ann.*, **465,** 132 (1928).
[380] Späth and Galinovsky, *Ber.*, **71,** 1282 (1938.)
[381] Achmatowicz and Dybowski, *J. Chem. Soc.*, **1938,** 1483.
[382] Achmatowicz and Dybowski, *J. Chem. Soc.*, **1938,** 1488.
[383] Ikeda, Taylor, Tsuda, and Uyeo, *Chem. & Ind.* (*London*), **1956,** 411; Ikeda, Taylor, Tsuda, Uyeo, and Yajima, *J. Chem. Soc.*, **1956,** 4749.
[384] Einhorn and Friedlaender, *Ber.*, **26,** 1482 (1893).
[385] Einhorn and Tahara, *Ber.*, **26,** 324 (1893).
[386] Willstätter, *Ber.*, **30,** 702 (1897).
[387] Willstätter, *Ber.*, **31,** 2498 (1898).
[388] Willstätter, *Ber.*, **30,** 721 (1897).
[389] Meinwald, Emerman, Yang, and Büchi, *J. Am. Chem. Soc.*, **77,** 4401 (1955).
[390] Willstätter, *Ber.*, **28,** 3271 (1895).
[391] Spiegel and Corell, *Ber.*, **49,** 1086 (1916).

AUTHOR INDEX, VOLUMES 1–11

Adams, Joe T., 8
Adkins, Homer, 8
Angyal, S. J., 8

Bachmann, W. E., 1, 2
Baer, Donald R., 11
Behr, Lyell C., 6
Bergmann, Ernst D., 10
Berliner, Ernst, 5
Blatt, A. H., 1
Blicke, F. F., 1
Brewster, James H., 7
Brown, Weldon G., 6
Bruson, Herman Alexander, 5
Buck, Johannes S., 4
Butz, Lewis W., 5

Carmack, Marvin, 3
Carter, H. E., 3
Cason, James, 4
Cope, Arthur C., 9, 11
Corey, Elias J., 9
Crounse, Nathan N., 5

Daub, Guido S., 6
DeTar, DeLos F., 9
Djerassi, Carl, 6
Donaruma, L. Guy, 11
Drake, Nathan L., 1
DuBois, Adrien S., 5

Eliel, Ernst L., 7
Emerson, William S., 4
England, D. C., 6

Fieser, Louis F., 1
Folkers, Karl, 6
Fuson, Reynold C., 1

Geissman, T. A., 2
Gensler, Walter J., 6
Gilman, Henry, 6, 8
Ginsburg, David, 10

Govindichari, Tuticorin R., 6
Gutsche, C. David, 8

Hageman, Howard A., 7
Hamilton, Cliff S., 2
Hamlin, K. E., 9
Hanford, W. E., 3
Hartung, Walter H., 7
Hassall, C. H., 9
Hauser, Charles R., 1, 8
Heldt, Walter Z., 11
Henne, Albert L., 2
Hoffman, Roger A., 2
Holmes, H. L., 4, 9
House, Herbert O., 9
Hudson, Boyd E., Jr., 1

Ide, Walter S., 4
Ingersoll, A. W., 2

Jackson, Ernest L., 2
Jacobs, Thomas L., 5
Johnson, John R., 1
Johnson, William S., 2, 6
Jones, Reuben G., 6

Kende, Andrew S., 11
Kloetzel, Milton C., 4
Kornblum, Nathan, 2
Kosolapoff, Gennady M., 6
Kulka, Marshall, 7

Lane, John F., 3
Leffler, Marlin T., 1

McElvain, S. M., 4
McKeever, C. H., 1
Magerlein, Barney J., 5
Manske, Richard H. F., 7
Martin, Elmore L., 1
Moore, Maurice L., 5
Morgan, Jack F., 2
Morton, John W., Jr., 8

Mosettig, Erich, 4, 8
Mozingo, Ralph, 4

Newman, Melvin S., 5

Pappo, Raphael, 10
Parmerter, Stanley M., 10
Phadke, Ragini, 7
Phillips, Robert R., 10
Price, Charles C., 3

Rabjohn, Norman, 5
Roe, Arthur, 5
Rondestvedt, Christian S., Jr., 11
Rytina, Anton W., 5

Sauer, John C., 3
Sethna, Suresh, 7
Sheehan, John C., 9
Shirley, David A., 8
Shriner, Ralph L., 1
Simonoff, Robert, 7
Smith, Lee Irvin, 1
Smith, Peter A. S., 3, 11

Spielman, M. A., 3
Spoerri, Paul E., 5
Struve, W. S., 1
Suter, C. M., 3
Swamer, Frederic W., 8
Swern, Daniel, 7

Tarbell, D. Stanley, 2
Todd, David, 4
Touster, Oscar, 7
Truce, William E., 9
Trumbull, Elmer R., 11

Wallis, Everett S., 3
Weston, Arthur W., 3, 9
Whaley, Wilson M., 6
Wilds, A. L., 2
Wiley, Richard H., 6
Wilson, C. V., 9
Wolf, Donald E., 6
Wolff, Hans, 3
Wood, John L., 3

Zaugg, Harold E., 8

CHAPTER INDEX, VOLUMES 1-11

Acetoacetic ester condensation and related reactions, 1
Acetylenes, 5
Acylation of ketones to β-diketones or β-keto aldehydes, 8
Acyloins, 4
Aliphatic fluorine compounds, 2
Alkylation of aromatic compounds by the Friedel-Crafts method, 3
Alkylation of esters and nitriles, 9
Amination of heterocyclic bases by alkali amides, 1
Arndt-Eistert synthesis, 1
Aromatic arsonic and arsinic acids, 2
Aromatic fluorine compounds, 5
Arylation of unsaturated compounds by diazonium salts, 11
Azlactones, 3

Baeyer-Villiger oxidation of aldehydes and ketones, 9
Beckmann rearrangement, 11
Benzoins, 4
Biaryls, 2
Bischler-Napieralski synthesis of 3,4-dihydroisoquinolines, 6
Bucherer reaction, 1

Cannizzaro reaction, 2
Carbon-carbon alkylation with amines and ammonium salts, 7
Catalytic hydrogenation of esters to alcohols, 8
Chloromethylation of aromatic compounds, 1
Claisen rearrangement, 2
Cleavage of non-enolizable ketones with sodium amide, 9
Clemmensen reduction, 1
Coupling of diazonium salts with aliphatic carbon atoms, 10
Curtius reaction, 3
Cyanoethylation, 5

Cyclic ketones by intramolecular acylation, 2

Darzens glycidic ester condensation, 5
Demjanov and Tiffendeau-Demjanov ring expansions, 11
Diels-Alder reaction: ethylenic and acetylenic dienophiles, 4
Diels-Alder reaction with cyclenones, 5
Diels-Alder reaction with maleic anhydride, 4
Direct sulfonation of aromatic hydrocarbons and their halogen derivatives, 3

Elbs reaction, 1
Epoxidation of ethylenic compounds with organic peracids, 7

Favorskii rearrangement of haloketones, 11
Friedel-Crafts reaction with aliphatic dibasic acid anhydrides, 5
Fries reaction, 1

Gattermann-Koch reaction, 5
Gattermann synthesis of aldehydes, 9

Halogen-metal interconversion reaction with organolithium compounds, 6
Hoesch synthesis, 5
Hofmann reaction, 3
Hydrogenolysis of benzyl groups, 7
Hydroxylation of ethylenic compounds with organic peracids, 7

Jacobsen reaction, 1
Japp-Klingemann reaction, 10

β-Lactams, 9
β-Lactones, 8
Leuckart reaction, 5

Mannich reaction, 1
Metalation with organolithium compounds, 8
Michael reaction, 10

Nitrosation of aliphatic carbon atoms, 7

Olefins from amines, 11
Oppenauer oxidation, 6

Pechmann reaction, 7
Periodic acid oxidation, 2
Perkin reaction and related reactions, 1
Pictet-Spengler synthesis of tetrahydroisoquinolines, 6
Pomeranz-Fritsch synthesis of isoquinolines, 6
Preparation of amines by reductive alkylation, 4
Preparation of benzoquinones by oxidation, 4
Preparation of ketenes and ketene dimers, 3
Preparation of phosphonic and phosphinic acids, 6
Preparation of thiazoles, 6
Preparation of thiophenes and tetrahydrothiophenes, 6
Pschorr synthesis and related ring closure reactions, 9

Reaction of diazomethane and its derivatives with aldehydes and ketones, 8
Reaction of halogens with silver salts of carboxylic acids, 9
Reduction with aluminum alkoxides, 2
Reduction with lithium aluminum hydride, 6
Reformatsky reaction, 1
Replacement of aromatic primary amino groups by hydrogen, 2
Resolution of alcohols, 2
Rosenmund reduction, 4

Schmidt reaction, 3
Selenium dioxide oxidation, 5
Skraup synthesis of quinolines, 7
Sommelet reaction, 8
Stobbe condensation, 6
Substitution and addition reactions of thiocyanogen, 3
Synthesis of aldehydes from carboxylic acids, 8
Synthesis of ketones from acid chlorides and organometallic compounds of magnesium, zinc, and cadmium, 8

Von Braun cyanogen bromine reaction, 7

Willgerodt reaction, 3
Wolff-Kishner reduction, 4

SUBJECT INDEX, VOLUME 11

Since the tables of contents of the individual chapters provide a quite complete index, only those items not readily found on the contents pages are indexed here. Numbers in **boldface** type refer to experimental procedures.

Acetanilide, **58**
N-Acylamines, conversion to olefins, 371–373, 393
Amides, conversion to olefins, 371–373, 393
Amine oxides, preparation, 378–379
  pyrolysis, 361–370, 394–399
Amine phosphates, conversion to olefins, 371, 379, 391–392
Aminomethylcycloalkanes, preparation, 172–173
Aminomethylcycloalkanols, preparation, 173–175
Arylation of unsaturated compounds by diazonium salts, 189–260
1-Arylbutadienes, preparation, 212
3-Arylcoumarins, preparation, 212
2-Arylquinones, preparation, 213

Beckmann rearrangement, 1–156
  effect of *ortho* substituents, 10, 18, 22
  formation of amidines, 19, 24
  formation of aromatic compounds, 28–30
  formation of benzimidazoles, 25
  formation of benzoxazoles, 20, 24–25
  formation of *peri*-benzoylene-9-morphanthridones, 25
  formation of 1-carboxy-2-methylanthraquinonecarboxylic acid, 26
  formation of furazans, 37
  formation of isoquinolines, 16, 43
  formation of 1,2,4-oxadiazoles, 37, 48
  formation of phenanthridones, 22
  formation of a dihydropyridine, 16
  formation of tetrazoles, 50
  imine intermediates, 6, 13
  second-order, 35, 38
1-Benzoyl-7-propionylheptatriene, **383**

1,4-Bis-(2'-chloro-2'-cyanoethyl)benzene, **221**
20-Bromo-17(20)-pregnen-3β-ol-21-oic acid, **290**

ε-Caprolactam, **58**
trans-p-Chlorocinnamic acid, **219**
2-*o*-Chlorophenylbenzoquinone, **221**
α-*p*-Chlorophenyl-N-isopropylmaleimide, **220**
Copper, catalyst in Beckmann rearrangement, 15, 18
Cularinemethine, **386**
Cuprous halides, catalysts in Beckmann rearrangement, 43
Cycloheptanol, **178**
Cycloheptanone, **178**
Cycloheptyltrimethylammonium iodide, **379**
Cyclohexanecarboxylic acid, **290**
Cyclohexanone oxime esters, 12
Cyclohexylphenethyldimethylammonium hydroxide, **384**
  Hofmann elimination, **385**
Cycloöctanone, **179**
cis-Cycloöctene, **387**
Cyclopropene, **383**

Deamination of diazonium salts, 210
Decahydroheptalene, **178**
Decarboxylation, during Meerwein arylation, 200, 206
Demjanov ring expansion, 157–188
"des," definition, 321–322
Diazo resins, 211
Di-*n*-butyldiisoamylammonium hydroxide, **381**
  Hofmann elimination, **381**
Di-*n*-butyldiisoamylammonium iodide, **380**

Dihydro-des-N-methylcytisine, **382**
N,N-Dimethylcycloöctylamine oxide, **386**
 pyrolysis, **387**
N,N-Dimethylhydroxylamine hydrochloride, **386**
N,N-Disubstituted hydroxylamines, preparation from amino oxides, 362

Epoxyethers, formation from α-haloketones, 263, 283–284
Ethyl α,α-dibenzylacetoacetate oxime, 15
Ethyl 3,3-diphenylpropionate, **289**
(+)-3-Ethylheptan-2-one oxime, 5
(+)-2-Ethylhexanoic acid amide, 5
Ethyl trimethylacetate, **289**
Exhaustive methylation, definition, 320

Favorskiĭ rearrangement, 261–317
Furans, arylation by diazonium salts, 201

Heptanamide, **58**
1-Hexene, **381**
Hofmann elimination reaction, 317–493
 deleterious effect of carbon dioxide, 359–360, 377
 failures, 343–344, 346
 with 1,2-diphenylpropylamines, 324
Hofmann rule, 331, 332–335, 348–349
Homodihydrocarbostyril, **57**
3-Hydroxydecahydroheptalene, **178**
α-Hydroxyketones from α-haloketones, 283–284, 287

Isoxazolines from oximes, 15, 19

Japanese acid earth, catalyst for Beckmann rearrangement, 18

Meerwein arylation reaction, 189–260
Methine nomenclature, 321
2-Methoxy-4'-phenylstilbene, **220**
des-N-Methylaphylline, **382**
Methyl cyclopentanecarboxylate, **289**
des-N-Methyldihydro-β-erythroidinol, **385**

Methylenecyclohexene, **386**
Methyl 3-methyl-2-butenoate, **290**

Neber rearrangement, 45
Nitriles from oximes, 16, 19–20, 28, 33, 40–41, 43
trans-p-Nitrocinnamonitrile, **220**
Nitroles, rearrangement, 48
Nitrones, rearrangement, 47
p-Nitrophenylbutadiene, **219**
3-p-Nitrophenylcoumarin, **219**
p-Nitrophenylmaleic anhydride, **221**

Octahydrophenazine from cyclohexanone oxime, 14, 30
trans-1,2-Octalin, **385**
Olefins, from amines, 317–493
 from ammonium salts and organometallic compounds or alkali metal amides, 373–374
Oximation, 51, 55
Oxime anhydrides, 8
Oximes, cleavage to nitriles, 16, 19–20, 28, 33, 40–41, 43
 isomerization, 53–54
 stereochemical designation, 3

Phenanthridone, **57**
Phosphonium hydroxides, thermal cleavage, 357
Phosphorus pentasulfide, catalyst in Beckmann rearrangement, 21
Picryl ethers of oximes, 7, 13
Pirylene, 340
Pivalanilide, **58**
n-Propylisoamylmethylamine oxide, pyrolysis, 370
n-Propyltrimethylammonium iodide, **380**

Ring contraction of cyclic ketones, 271

Sandmeyer reaction, 210
Silver oxide, **380**
Stevens rearrangement, 356
Sulfones, Hofmann-type elimination, 357

Thiophenes, arylation by diazonium salts, 201

Tiffeneau-Demjanov ring expansion, 157–188

N-Uramidohomomeroquinene, **384**

δ-Valerolactam, **57**

Wallach degradation, 280

Ylides, in decomposition of quaternary ammonium salts, 373–374
  in Hofmann elimination, 328–329